EPIGENETICS

EPIGENETICS

*Linking Genotype and Phenotype
in Development and Evolution*

Edited by

Benedikt Hallgrímsson
and Brian K. Hall

UNIVERSITY OF CALIFORNIA PRESS
Berkeley   Los Angeles   London

University of California Press, one of the most distinguished university presses in the United States, enriches lives around the world by advancing scholarship in the humanities, social sciences, and natural sciences. Its activities are supported by the UC Press Foundation and by philanthropic contributions from individuals and institutions. For more information, visit www.ucpress.edu.

University of California Press
Berkeley and Los Angeles, California

University of California Press, Ltd.
London, England

Library of Congress Cataloging-in-Publication Data

Hallgrímsson, Benedikt.
    Epigenetics : linking genotype and phenotype in development and evolution / Edited by Benedikt Hallgrímsson and Brian K. Hall.
        p.   cm.
    Includes bibliographical references and index.
    ISBN 978-0-520-26709-1 (cloth : alk. paper)
    1. Phenotype.   2. Epigenesis.   3. Genetic regulation.
4. Developmental genetics.   5. Evolutionary genetics.   I. Hall, Brian Keith, 1941-   II. Title.

QH438.5.H35   2010
576.5'3—dc22   2010030921

Manufactured in the United States

19 18 17 16 15 14 13 12 11

10 9 8 7 6 5 4 3 2 1

The paper used in this publication meets the minimum requirements of ANSI/NISO Z39.48-1992 (R 1997) (*Permanence of Paper*). ∞

Jacket illustrations, clockwise from top: coronal sections through the internasal suture of pigs, by A.K. Burn, modified; computer simulation of the reaction–diffusion wave, by Shigeru Kondo and Rihito Asai, reprinted by permission from Macmillan Publishers Ltd; beach sand under a rivulet of water, by Ellen W. Larsen; *Oliva porphyria*, © Scott Camazine; *Pomacanthus semicirculatus*, by Shigeru Kondo and Rihito Asai, reprinted by permission from Macmillan Publishers Ltd.

CONTENTS

CONTRIBUTORS

MARTHA ALONZO
Duke University
martha.alonzo@duke.edu

BROOKE AUTUMN ARMFIELD
Northeastern Ohio Universities
   College of Medicine
bgarner@neoucom.edu

JOEL ATALLAH
University of Toronto
joel.atallah@utoronto.ca

ALAN S. BEEDLE
The University of Auckland
a.beedle@auckland.ac.nz

TATJANA BUKLIJAS
The University Auckland
a.beedle@auckland.ac.nz

LISA NOELLE COOPER
Northeastern Ohio Universities
   College of Medicine
l.noelle.cooper@gmail.com

SUSAN A. FOSTER
Clark University
sfoster@clarku.edu

TAMARA A. FRANZ-ODENDAAL
Mount Saint Vincent University
Tamara.Franz-Odendaal@msvu.ca

PETER D. GLUCKMAN
The University of Auckland
pd.gluckman@auckland.ac.nz

ROOT GORELICK
Carleton University
Root_Gorelick@carleton.ca

JAMES GRIESEMER
University of California, Davis
jrgriesemer@ucdavis.edu

CHRISTOPH GRUNAU
Universitéde Perpignan
christoph.grunau@univ-perp.fr

BRIAN K. HALL
Dalhousie University
Bkh@dal.ca

BENEDIKT HALLGRÍMSSON
University of Calgary
bhallgri@ucalgary.ca

THOMAS F. HANSEN
University of Oslo
thomas.hansen@bio.uio.no

MARK A. HANSON
University of Southampton
m.hanson@soton.ac.uk

SUSAN W. HERRING
University of Washington
herring@u.washington.edu

BORIS KABLAR
Dalhousie University
bkablar@is.dal.ca

MARGARET L. KIRBY
Duke University
mlkirby@duke.edu

CHRIS KOVACH
Florida State University
ckovach@neuro.fsu.edu

ELLEN W. LARSEN
University of Toronto
ellie.larsen@utoronto.ca

MANFRED LAUBICHLER
Arizona State University
Manfred.Laubichler@asu.edu

DANIEL E. LIEBERMAN
Harvard University
dlieberman@oeb.harvard.edu

VETT LLOYD
Mt. Allison University
vlloyd@mta.ca

FELICIA M. LOW
The University of Auckland
f.low@auckland.ac.nz

RACHEL MASSICOTTE
Université de Montréal
Rachel.Massicotte@umontreal.ca

PIERRE MATTAR
University of Calgary
Pierre.Mattar@ircm.qc.ca

LORI A. McEACHERN
Dalhousie University
lamceach@dal.ca

ALESSANDRO MINELLI
University of Padova
alessandro.minelli@unipd.it

CHRISTOPHER J. NEUFELD
University of Alberta
rich.palmer@ualberta.ca

LENNART OLSSON
Friedrich-Schiller-Universität Jena
Lennart.Olsson@uni-jena.de

A. RICHARD PALMER
University of Alberta
chris.neufeld@ualberta.ca

CHRISTOPHER J. PERCIVAL
Pennsylvania State University
cjp216@psu.edu

JOAN T. RICHTSMEIER
Pennsylvania State University
jta10@psu.edu

CAROL SCHUURMANS
University of Calgary
cschuurm@ucalgary.ca

KATHLEEN K. SMITH
Duke University
kksmith@duke.edu

DONALD L. SWIDERSKI
University of Michigan
dlswider@umich.edu

J. G. M. THEWISSEN
Northeastern Ohio Universities
  College of Medicine
thewisse@neoucom.edu

MATTHEW A. WUND
Clark University
mwund@clarku.edu

MIRIAM LEAH ZELDITCH
University of Michigan
zelditch@umich.edu

1

Introduction

Benedikt Hallgrímsson and Brian K. Hall

Epigenetics is the study of emergent properties in the origin of the phenotype in development and in modification of phenotypes in evolution. Features, characters, and developmental mechanisms and processes are epigenetic if they can be understood only in terms of interactions that arise above the level of the gene as a sequence of DNA. Methylation and imprinting of gene sequences are examples of epigenetics at the level of the structure and function of the gene, sometimes referred to as the "phenotype" of the gene, a concept that blurs (appropriately, we believe) the distinction between genotype and phenotype established over 100 years ago. Inductive interactions between two cell populations that lead to the formation of a third are examples of epigenetic phenomena at the cell population level.

Aspects of bone morphology are epigenetic in origin because they arise as the result of interactions between muscle activity and bone. The result of those interactions is not predictable from the intrinsic development of either tissue but can be understood only at the level of their interaction. Interactions between individuals of the same or different species also are

epigenetic, as seen in the adjustment of the number of soldiers in an ant colony in response to depletion of soldier numbers and to hormonal cues or in the interactions between predator and prey species in plankton, where diffusible chemicals from the predator elicit the formation of features in the prey not seen in the absence of the predator.

Epigenetics falls broadly within the area of systems biology premised on the concept that system-level phenomena are essential as explanatory factors in biology. Concepts such as epigenetics are required not because biological systems are fundamentally indeterminate but because their complexity prohibits the construction of an exhaustive deterministic framework. Epigenetic explanations arise whenever we create theoretical constructs to make sense of the complex relationships between genetic and phenotypic variation and evolution. In these senses, canalization, developmental stability, and integration are epigenetic phenomena. Developmental pathways, networks, and processes also are epigenetic concepts. Explanations of development and evolution that focus on properties of processes or pathways, such as the

Atchley and Hall (1991) model for mandibular development or Salazar-Ciudad and Jernvall's (2005) interactome for tooth development, are epigenetic explanations. In both of these cases, the relevant parts of the explanations for phenotypic variation are at the level of the interactions among gene products, among cell populations, and among the processes generated that link the two levels.

Biological systems may or may not be completely deterministic above the level of subatomic particles. However, they are functionally indeterminate in the sense that their complexity prevents us in most cases from creating frameworks that predict phenotypic outcomes directly from DNA sequences or sequence variation. Epigenetics is the study of the construction of frameworks that allow us to bridge the gulf between genotype and phenotype and thus integrate the vast amount of information that is being generated about the roles of specific genes in developmental systems. Thus, epigenetics is more than the study of emergent properties in developmental systems. It defines an area of study and a level of processes that creates a coherent understanding of the emergent relationship between genotype and phenotype.

The term *epigenetics* has increased in use in the molecular, evolutionary, and developmental literature in recent years. Its varied use in evolutionary and developmental biology has made it meaningless to many. Our definition is very much in the spirit of Waddington's (1957) concept of the "epigenotype." We follow Hall (1990) and Salazar-Ciudad and Jernvall (2005) in defining epigenetics as the study of the epigenotype, which is the study of the properties of the pathways and processes that link the genotype and phenotype.

Molecular genetics, developmental biology, and, to a lesser degree, evolutionary developmental biology have deemphasized the study of epigenetics. This is due to technical advances that have greatly increased the pace at which gene-focused research compiles information about gene expression and uncovers gene function but also because of the prevailing notion that all explanations lie within the genome. We argue that while the pace of information accrual in developmental genetics has increased enormously, our understanding of the developmental basis for phenotypic variation has not. Integration of the concepts of epigenetics into research in mainstream developmental biology is essential if we are to answer fundamental questions about how development produces phenotypic traits such as skull shape and limb length and how it produces variation in those traits. Otherwise, we will continue to document correlations between variation in gene function and expression with phenotypic variation without a deeper understanding of how phenotypic variation is generated.

This volume explores the dimensions of epigenetics in different systems and contexts to show the central place and explanatory power of the concept for evolutionary and developmental biology. It seeks to rationalize and limit the meaning of the term *epigenetics* and its application so as to maintain and enhance its explanatory power and utility. The book is organized around themes and levels of interactions and processes in the biological hierarchy in which epigenetics acts.

The first part lays out the conceptual foundations for an epigenetic view of development and the implications of that view for evolutionary biology. This part consists of two chapters. In the first, the coeditors of this volume pull together themes from the various chapters to present a coherent conceptualization of epigenetics that is true to the original intent of Waddington but also to the usage of the term in modern biology. The argument made here is that Waddington's concept of epigenetics presaged a systems biology view of development and evolution and that much is lost to theoretical biology by throwing out this broader concept of epigenetics in favor of a narrower one focused only on chromatin modification. The third chapter, by James Griesemer, deals more specifically with how the discovery of epigenetic inheritance and epigenetic mechanisms has been incorporated into our understanding of the evolutionary process. Griesemer argues for a relative significance approach in weighing the role of these novel mechanisms against those based on genetic inheritance only.

The second part presents several different levels at which epigenetic concepts are applied to development and evolution. The first chapter in this part, by McEachern and Lloyd, deals with genomic imprinting, which is one of the key concepts in the modern study of epigenetics. The chapter explains this phenomenon; discusses its importance in plants, insects and mammals; and discusses its evolutionary significance. McEachern and Lloyd make the case that molecular epigenetic processes are amazingly conserved across metazoan phyla. The next chapter, by Christoph Grunau, provides an overview of methylation mapping in humans. In both chapters, the focus is on the phenotype of the gene and epigenetics is used in the sense in which it is used in molecular biology, referring to processes that influence heritable variation in gene expression without DNA sequence variation. The focus in both chapters is chromatin modification—in particular, methylation and acetylation.

Chapter 6, by Gorelick et al., presents what is, in some ways, a hybrid perspective between molecular biology and the older Waddingtonian view. Gorelick et al. define epigenetic mechanisms as those "molecular signals that are literally on top of DNA." Limiting the definition in this way, they retain a Waddingtonian view of the importance of mechanisms acting above the gene level and apply it to the evolution of asexual taxa.

The next chapter, by Larsen and Atallah, provides a contrasting and, we believe, complementary perspective on epigenetics, making the case for the need to consider emergent properties in development. Here, epigenetics is extended beyond the phenotype of the gene to include mechanisms and processes that can be understood only above the level of the gene. These include epigenetic inheritance, of course, but also emergent phenomena such as self-organization and variational properties of development (e.g.,

canalization, robustness). They also include the spatiotemporal context of gene expression and the interesting phenomenon of intergenomic interactions. Building on a similar perspective, the following chapter, by Allessandro Minelli, develops a theoretical framework for self-organization and inclusion of the physics of development into developmental and evolutionary theory. Minelli makes a case for defining null models in developmental biology and uses the phrase "developmental inertia" to describe the null conditions of developmental systems. Like Larsen and Atallah, this chapter draws extensively on well-worked examples and nicely illustrates the use of Waddington's epigenetic concept in a modern biological framework. These five chapters provide descriptions of very different conceptual approaches to epigenetics and examples and insight about epigenetic processes at very different levels of the epigenetic hierarchy. These chapters, which differ in perspective, are presented in the same part to make this contrast clear but also to draw connections between them.

The next part is a series of case studies in which epigenetic processes at various scales are discussed in different organ systems. The first chapter, by Kovach et al., takes a molecular perspective on epigenetics and deals with the role of chromatin modification in neural development. The second chapter, by Olsson, deals with the next level of the epigenetic hierarchy, the self-organization of tissues. The specific topic is pigment patterns in vertebrates, a system that illustrates beautifully the potential for self-organization in complex developmental systems. The remaining chapters deal with epigenetic processes at the organ level or higher. The third chapter, by Alonzo et al., discusses the role of epigenetic interactions in the cardiac neural crest and other tissues. Here, the concept is used in the Waddingtonian sense and the emphasis is on the emergent consequences of interactions between developmental components. The next chapter, by Franz-Odendaal, takes a similar perspective and applies it to the epigenetic processes that impinge on osteogenesis and chondrogenesis and the significance of these mechanisms for understanding the developmental basis for skeletal variation. The next chapter, by Sue Herring, continues the skeletal theme at later developmental stages and a higher phenomenological level and deals with the epigenetic role of mechanical and particularly muscle–bone interactions in skeletal development and the implications of these mechanisms for interpreting skeletal morphology in evolutionary contexts. The next chapter, by Cooper et al., focuses on epigenetic interactions in the developing limb, and in particular on the role of epithelial-mesenchymal interactions in the apical ectoderm. The theme of mechanical interactions produced by muscle activity is then continued in the final chapter, by Kablar, who deals with the role that these mechanisms play in lung, retinal, and early skeletal development.

The final part presents a series of chapters that apply epigenetic frameworks to the study of evolution and dysmorphology. The first chapter, by Dan Lieberman, makes the case for the evolutionary significance of epigenetic mechanisms as determinants of integrated and evolvable complex phenotypes. To make this point, Lieberman analyzes the vertebrate skull and builds on a body of work in which he and others have demonstrated the existence and importance of epigenetic mechanisms in determining the patterns of variation in craniofacial shape. The second chapter in this part develops a similar argument but in a somewhat different direction. Zelditch and Swiderski emphasize the distinction between epigenetic mechanisms and genetic pleiotropy as determinants of integration. Using the mammalian dentary as their example, they summarize what is known about the developmental basis for integration in that system. They argue that epigenetic pleiotropy needs to be recognized as a distinct phenomenon, somewhat akin to developmental plasticity. This argument echoes those made by both Waddington (1957) and Schmalhausen (1949), who argued independently for the epigenetic basis for phenotypic plasticity. This theme is continued in the next chapter, by Foster and Wund, who address phenotypic plasticity in the threespine stickleback, arguing for an evolutionary role for genetic assimilation ("Baldwin effect") in radiations of the threespine stickleback. For Waddington, genetic assimilation was one of the main ways in which epigenetic mechanisms were relevant to evolutionary change.

The discussion of genetic assimilation and phenotypic plasticity continues in a very different phylogenetic context in the chapter by Neufeld and Palmer, who discuss the interaction of developmental plasticity in morphology and behavior, expressed as learning in particular, and the relationship of this interaction to rates of morphological evolution. They make the novel argument that learning-enhanced morphological plasticity plays an underappreciated role in morphological evolution.

The next chapter, by Thomas Hansen, provides an interesting and challenging counterpoint to the previous four. He tackles the question of the adaptive significance of the epigenetic processes that underlie phenotypic variability from a quantitative genetic perspective. Hansen reiterates the often misunderstood point that selection does not always produce adaptation, and he examines critically the assumption that variational properties that result from such epigenetic interactions as integration, plasticity, canalization, and evolvability are adaptive and, thus, driven by selection for their adaptive value rather than indirectly by selection acting on other aspects of the phenotype. He shows that variational properties may commonly arise as an indirect and secondary result of selection acting directly on the traits themselves, a finding that has profound implications for our understanding of the role of variational properties in evolution.

Two chapters then discuss the role of epigenetic mechanisms in dysmorphology and disease. The first, by Percival and Richtsmeier, deals with craniofacial dysmorphology and involves mechanisms that, akin to those discussed by Lieberman and by Zelditch and Swiderski, involve interactions between developmental components that produce emergent properties. The second chapter in this pair and the penultimate chapter of the volume takes a more molecular approach to the study of human disease. In this chapter, Gluckman et al. provide an overview of the epigenetics of human disease which, while limiting the definition of epigenetic mechanisms to chromatin modification, also draws upon many of the observations that informed the early discussion of epigenetic mechanisms.

The motivation behind this volume is to pull together the divergent approaches through which the mechanisms that link the genotype and phenotype maps are explored. A thorough treatment of epigenetics, however defined, is well beyond the scope of a single volume such as this. What we have done, however, is to pull together scholars who are using the concept of epigenetics in a variety of contexts to show the common root of these approaches, mostly in the work of Waddington and Schmalhausen, and to show the connections that still exist between these approaches. It is likely that as the field of molecular epigenetics grows, those who use epigenetics in the context of chromatin modification will be increasingly unaware of its roots in the developmental and evolutionary biology of the 1940s and 1950s. It is also possible that systems biology will recreate much of the theoretical framework of higher-level epigenetics without reference to its conceptual roots. An important aim of this book is an appeal against both of those trends. It is clear from the work presented in this volume that in the minds of many scholars the concept of epigenetics in the broad sense envisioned by Waddington still has theoretical coherence in the face of the monumental complexity of modern biology. Indeed, only an epigenetic approach will allow us to unravel that complexity. We hope that the reader will agree.

REFERENCES

Atchley, W. R., and B. K. Hall. 1991. A model for development and evolution of complex morphological structures. *Biol Rev* 66:101–57.

Hall, B. K. 1990. Genetic and epigenetic control of vertebrate embryonic development. *Neth J Zool* 40 (1–2): 352–61.

Salazar-Ciudad, I., and J. Jernvall. 2005. Graduality and innovation in the evolution of complex phenotypes: Insights from development. *J Exp Zool B Mol Dev Evol* 304 (6): 619–31.

Schmalhausen, I. I. 1949. *Factors of Evolution*. Chicago: University of Chicago Press.

Waddington, C. H. 1957. *The Strategy of the Genes*. New York: MacMillan.

PART ONE

Historical and Philosophical Foundations

2

A Brief History of the Term and Concept *Epigenetics*

Brian K. Hall

CONTENTS

This chapter provides a brief evaluation of the history of epigenetics as a term and as a concept. Although the term was not coined until the 1940s, the concept that genes are influenced by factors beyond the genome (the "epi" in *epigenetics*) is much older and can be traced to late nineteenth-century discussions of whether the nucleus or the cytoplasm "controlled" development, to earlier nineteenth-century discussions of whether the sperm or egg provided the primary material for development and therefore for life, and even earlier to the eighteenth-century concepts of organismal structure as either preformed and arising by unfolding—the original use of the term *evolution*—or arising gradually by epigenesis.

EPIGENESIS OR PREFORMATION

Two parallel approaches to how organisms arise can be traced back to the 4th century BCE and the writings of the Greek philosopher Aristotle (384–322 BCE). Developed specifically to explain animal embryogenesis (development), the concepts have become known as *preformation*—the gradual unfolding through growth of features preformed in the egg or sperm—and *epigenesis*—the successive differentiation of features during development leading to increasing complexity and the formation of the adult form. Charles Bonnet (1720–1793) enshrined preformation in the eighteenth century with his *emboîtment*, or encapsulation, theory, in which all the members of all future generations were present in an early developmental stage: cotyledons within the seeds of plants, future generations of insects in the pupa. The ability of *Hydra* to regenerate the entire body or of newts to regenerate their tails was interpreted as the unfolding of preformed features (Farley, 1982; Dinsmore, 1991; Hall, 1998b).

The term *epigenesis* may have first been used by the German anatomist and embryologist Caspar Friedrich Wolff (1733–1794), whose dissections of chicken embryos revealed the progressive development of the tubular gut and led him to conclude that "When the formation of the intestine in this manner has been duly weighed, almost no doubt can remain, I believe, of the truth of epigenesis" (Wolff, 1768–1769, 460–61). Wolff also proposed a series of causal links between developing parts of the embryo: "each part is first of all an effect of the preceding part, and itself becomes the cause of the following part" (Wolff, 1764, 211). Preformation lost further support in the early 1800s with the publication of the investigations of chicken embryological development by Louis Sébastien Tredern and Christian Pander and the even more extensive studies by the comparative embryologist Karl von Baer (1792–1876), who demonstrated that vertebrate embryos differentiate from a foundation of fundamental germ layers (primary differentiation), followed by histological and then morphological differentiation (Hall, 1997; Horder, 2008).

NUCLEUS OR CYTOPLASM

The transition from preformation to epigenesis ruled out gradual unfolding as the basis for development but provided neither a proximate mechanism for how embryos develop nor an ultimate mechanism for how the same type of embryo appears generation after generation when individuals from a single species are bred. Part of the resolution is that some preformed features are passed on from generation to generation, including the cytoplasmic constituents of the egg, mitochondrial DNA, and the nuclear genetic constituents from the combined contributions of the male and female parents. Indeed, much of the latter part of the nineteenth century and the first decades of the twentieth were taken up with endeavors to determine whether cytoplasm or nucleus "controlled" embryonic development and so controlled life (Wilson,

1925; see Hall, 1983, 1998a, 1998b; and the papers in Laubichler and Maienschein, 2007, for evaluations).

The leading cell biologist of the early twentieth century, E. B. Wilson (1856–1939), equated preformation and epigenesis with nuclear or cytoplasmic control. On the basis of extensive and exhaustive analyses of cell lineages in invertebrate embryos, Wilson assigned determination to the nucleus and an initiating role to the cytoplasm:

> Fundamentally, however, we reach the conclusion that in respect to a great number of *characters heredity is effected by the transmission of a nuclear preformation which in the course of development finds expression in a process of cytoplasmic epigenesis.*

(Wilson, 1925, 1112; *emphasis in original*)

EPIGENETICS · *An Integrated Approach*

The definition of epigenetics that emerges is that epigenetics is the sum of the genetic and nongenetic factors acting upon cells to control selectively the gene expression that produces opment and evolution.

As a result of studies on the genetic basis of features during embryogenesis in fruit flies (*Drosophila*) and to emphasize the role of genes and genetic control in and over embryonic development, the British embryologist and geneticist Conrad Waddington (1900–1975) coined the term and invoked the concept of *epigenetics*. Waddington (1942, reprinted in 1975, 218) defined epigenetics as "causal interactions between genes and their products which bring the phenotype into being."

Genes are regulated by a multitude of mechanisms including products of other genes (transcription factors), other organisms (presence of a predator, population density), and environmental factors such as temperature or the uterine environment. Epigenetics includes all these levels and more. Waddington developed the idea of epigenetic control of gene regulation most clearly in his concept of the canalization of developmental pathways, which allows embryos to "buffer" changes in individual genes epigenetically (see Hall, 1993, 2008, for overviews).

Medawar and Medawar (1983) took a broadbrush approach in their definition when they stated the following:

> In the modern usage "epigenesis" stands for all the processes that go into implementation of the genetic instructions contained within the fertilized egg. "Genetics proposes; epigenetics disposes."
>
> (p. 114)

Maclean and Hall (1987), defined *epigenetics* as encompassing increasing hierarchical complexity and the influences of the environment on phenotypic expression through control of gene expression, the genotype as the starting point and the phenotype as the end point of epigenetic control.

Epigenetic or epigenetics does not mean nongenetic. It is not a Lamarckian form of inheritance, although it has been claimed to be in the past. Nor is it appropriate to speak of genetic versus epigenetic. As summarized by the mathematician and theoretical biologist René Thom,

> If you were to follow Aristotle's theory of causality (four types of causes: material, efficient, formal, final) you would say that from the point of view of material causality in embryology, every thing is genetic—as any protein is synthesised from reading a genomic molecular pattern. From the point of view of efficient causality, everything is also "epigenetic", as even the local triggering of a gene's activity requires—in general—an extra-genomal factor.
>
> (1989, 3)

Because epigenetic control includes both genetic and environmental factors, an important research agenda is to understand heritable and environmental components and how environmental signals can, o ver time and with selection, become heritable genetic factors (Hall et al., 2004). "Phenotypic variability results from intrinsic genetic effects, heritable epigenetic effects and non-genetic environmental effects, some of which act epigenetically" (Hall, 1998b, 323).

EPIGENETIC INHERITANCE

Transmission of epigenetic states of inheritance from generation to generation has been demonstrated in a number of systems and organisms. Some, such as cortical inheritance, have been known for decades. Others, such as patterns of DNA methylation, have come into prominence much more recently. All attest to the changing nature of the concept of epigenetics and to the critical importance of being aware of the multiple levels at which genes are regulated. Because these are discussed extensively in the remainder of the book, I introduce only three here, and then only briefly: cortical (cytoplasmic) inheritance, maternal cytoplasmic control, and patterns of methylation.

Cortical cytoplasm is the superficial cytoplasm of a cell, either a cell from a multicellular organism (plant, animal, or fungus) or a single-celled organism such as *Paramecium*. *Cortical inheritance* is the transmission of information via organelles within the cortical cytoplasm. All of the rows of cilia in paramecia are oriented in the same direction and beat in coordinated waves. Grafting a piece of cortical cytoplasm in reverse orientation reverses the direction of the wave in the graft, a reversal that is inherited when the paramecium divides (Beisson and Sonneborn, 1965; Preer, 2006).

Early embryonic development is not primarily controlled by the embryo's genome but by the products of maternal genes deposited into the egg during oogenesis. This early phase of development (which differs in duration in different taxa) is known as maternal cytoplasmic control and is an example of epigenetic control. At another level of epigenetic control, the products of maternal and zygotic genes interact with effects that persist into adult life (Reik et al., 1993).

Methylation is the addition of a methyl group to cytosine DNA residues. The nucleotide sequence is not changed by this process, but highly methylated DNA is less transcriptionally active than less methylated DNA; epigenetics today is increasingly applied to such heritable changes in gene expression that do not involve alterations in nucleotide sequence. Methylation is inherited as a stable state of gene expression that differs from cell type to cell type. In a fascinating and little understood process, DNA is demethylated in early embryos and patterns of methylation are reestablished as cell types differentiate (Reik et al., 1990; Holliday, 1994; Trasler et al., 1996; Jaenisch and Bird, 2003).

In evaluating a large body of evidence Jablonka and Lamb (1995) concluded that epigenetic inheritance systems are less stable, are more sensitive to the environment (and so more adaptive), are more directed, and have more predictable variation but more limited alternate states than genetic inheritance. The extent to which such epigenetic systems are heritable in the sense of being *replicators*—a hereditary unit from which copies are made (Szathmáry and Maynard Smith, 1995; Russo et al., 1996; Hall, 1998b)—is one of the many important and fundamental aspects of epigenetics addressed in this book and awaiting future research.

CONCLUSION

As a continuation of the concept that development unfolds and is not preformed (or ordained), epigenetics is the latest expression of epigenesis. The term *epigenetics* was coined in the 1940s by Waddington to reflect the discovery of the roles of genes in development and the (then) growing and hardening thesis that genes control development. As the past 70 years has made abundantly clear, genes do not control development. Genes themselves are controlled in many ways, some by modification of DNA sequences, some through regulation by the products of other genes and/or by context, and others by external and/or environmental factors. A sure sign that epigenetics is in the ascendancy is the appearance of articles in the popular press: a large spread in the *Toronto Globe & Mail* in 2009, three pages in *The Guardian Weekly* of April 2, 2010. A sure sign that epigenetics has a long way to go is the letter in a subsequent issue of *The Guardian Weekly* in which all evolutionary change is claimed to be based on mutations and epigenetics to have nothing at all to do with evolution. As the chapters in this book attest, epigenetics is real, epigenetic control mechanisms evolve, and epigenetics is central to understanding development, evolution, and many disease states and malformations.

REFERENCES

Beisson, J., and T. M. Sonneborn. 1965. Cytoplasmic inheritance of the organization of the cell cortex in *Paramecium aurelia*. *Proc Natl Acad Sci USA* 53:275–82.

Dinsmore, C. E., ed. 1991. *A History of Regeneration Research. Milestones in the Evolution of a Science.* Cambridge: Cambridge University Press.

Farley, F. 1982. *Gametes & Spores: Ideas about Sexual Reproduction. 1750–1914.* Baltimore: Johns Hopkins University Press.

Hall, B. K. 1983. Epigenetic control in development and evolution. In *Development and Evolution*. The Sixth Symposium of the British Society for Developmental Biology. ed. B. C. Goodwin, N. Holder, and C. C. Wylie, 353–379. Cambridge: Cambridge University Press.

Hall, B. K. 1993. Waddington's legacy in development and evolution. *Am Zool* 32:113–22.

Hall, B. K. 1997. Germ layers and the germ-layer theory revisited: Primary and secondary germ layers, neural crest as a fourth germ layer, homology, demise of the germ-layer theory. *Evol Biol* 30:121–86.

Hall, B. K. 1998a. Epigenetics: Regulation not replication. *J Evol Biol* 11:201–05.

Hall, B. K. 1998b. *Evolutionary Developmental Biology.* 2nd ed. Dordrecht, the Netherlands: Kluwer Academic Publishers.

Hall, B. K. 2008. Evo–devo concepts in the work of Conrad Waddington. *Biol Theory* 3:198–203.

Hall, B. K., R. Pearson, and G. B. Müller. 2004. *Environment, Development, and Evolution: Toward a Synthesis.* Cambridge, MA: MIT Press.

Holliday, R. 1994. Epigenetics: An overview. *Dev Genet* 15:453–57.

Horder, T. J. 2008. A history of evo–devo in Britain. *Ann Hist Phil Biol* 13:101–74.

Jablonka, E., and M. J. Lamb. 1995. *Epigenetic Inheritance and Evolution: The Lamarckian Dimension.* Oxford: Oxford University Press.

Jaenisch, R., and A. Bird. 2003. Epigenetic regulation of gene expression: How the genome integrates intrinsic and environmental signals. *Nat Genet* 33 (Suppl): 245–54.

Laubichler, M. D., and J. Maienschein, eds. 2007. *From Embryology to Evo–Devo: A History of Developmental Evolution.* Symposium of the Dibner Institute for the History of Science, MIT. Cambridge, MA: MIT Press.

Maclean, N., and B. K. Hall. 1987. *Cell Commitment and Differentiation.* Cambridge: Cambridge University Press.

Medawar, P. B., and J. S. Medawar. 1983. *Aristotle to Zoos. A Philosophical Dictionary of Biology.* Cambridge, MA: Harvard University Press.

Preer, J. R., Jr. 2006. Sonneborn and the cytoplasm. *Genetics* 172:1373–77.

Reik, W., S. K. Howlett, and M. A. Surani. 1990. Imprinting by DNA methylation: From transgenes to endogenous gene sequences. *Development* (Suppl): 99–106.

Reik, W., I. Romer, S. C Barton, et al. 1993. Adult phenotype in the mouse can be affected by epigenetic events in the early embryo. *Development* 119:933–42.

Russo, V. E. A., R. A. Martienssen, and A. D. Riggs, eds. 1996. *Epigenetic Mechanisms of Gene Regulation.* Monograph 22. Woodbury, NY: Cold Spring Harbor Laboratory Press.

Szathmáry, E., and J. Maynard Smith. 1995. The major evolutionary transition. *Nature* 374: 227–32.

Thom, R. 1989. An inventory of Waddingtonian concepts. In *Theoretical Biology*, ed. B. Goodwin and P. Saunders, 1–7. Edinburgh: Edinburgh University Press.

Trasler, J. M., D. G. Trasler, T. H. Bestor, et al. 1996. DNA methyltransferase in normal and $Dnmt^n/Dnmt^n$ mouse embryos. *Dev Dyn* 206: 239–47.

Waddington, C. H. 1942. The epigenotype. *Endeavour* 1:18–20.

Waddington, C. H. 1975. *The Evolution of an Evolutionist.* Ithaca, NY: Cornell University Press.

Wilson, E. B. 1925. *The Cell in Development and Heredity*, 3rd ed. New York: Macmillan.

Wolff, C. F. 1764. *Theorie vonder Generation in zwo Abhandlungen.* Friedrich Wilhelm Birusteil, Berlin.

Wolff, C. F. 1768–1769. De formatione intestinorum praecipue. *Novi Commentarii Academiae Scientiarum Imperialis Petropoliteae* 12:403–507, 13: 478–530.

3

Heuristic Reductionism and the Relative Significance of Epigenetic Inheritance in Evolution

James Griesemer

CONTENTS

The role of epigenetic inheritance in evolution is hotly contested. Some claim that recently discovered epigenetic mechanisms of gene regulation constitute a nongenetic inheritance system that underwrites a "Lamarckian dimension" of inheritance and therefore of evolution. Others judge epigenetic inheritance to be relatively insignificant in evolution, even in principle, due to disanalogies with the genetic system (unstable states, high mutation rates, non-Mendelian, Lamarckian). I argue for a role for relative significance arguments and reductionism in heuristic strategies for investment in epigenetics research. I argue that "biologists argue the way they do" (Beatty, 1997) because of differing goals and commitments of distinct research specialties or lines of work with overlapping domains. Possible mechanisms are key to the forward-looking, investment-oriented heuristic strategies that are the subject of current debates about relative significance.

PHILOSOPHY FOR SCIENCE

Prospects are limited for philosophical contributions at the front line of rapidly advancing fields of biology such as epigenetic inheritance. The motivations, goals, and methods of biologists are as different from those of philosophers as are the motivations, goals, and methods of fruit flies in choosing mates from those of biologists studying flies and mate choice. Because biologists, like philosophers but unlike fruit flies, sometimes find reflection valuable in their work, there are some ways in which philosophy might be of service to the advancement of biology.

Three possible contributions with relevance to the field of epigenetic inheritance are (1) offering new organizing descriptions of phenomena to help articulate the scientific research agenda, (2) describing theory structure in relevant specialties to clarify differences in explanatory strategy, and (3) clarifying heuristic research strategies, in particular the differing nature of reductionism in molecular epigenetics and in evolutionary dynamics. This chapter concerns the second and third points. Beyond offering new organizing descriptions of phenomena leading to proposals for new research programs that diverge from "business as usual," philosophers can make observations about conceptual differences among theoretical enterprises in arenas involving disparate specialties. Epigenetic inheritance is such an arena. Molecular and cellular biologists have claimed for 20 years that epigenetic phenomena have significant implications for evolution, not only as adaptations but also as inheritance systems that could fuel evolution at a level above the genetic level (e.g., Jablonka and Lamb, 1989, 1995; see Holliday, 2006). Ecologists and evolutionists working on phenotypic plasticity, evo–devo, and phenotypic evolution have begun to respond (e.g., Van Speybroeck et al., 2002; Pigliucci, 2007; Bossdorf et al., 2008; NESCent, 2009). Evolutionists sometimes support and sometimes doubt the implications claimed (Jablonka and Lamb, 1998; Maynard Smith, 1990; Maynard Smith and Szathmáry, 1995; Jablonka and Szathmáry, 1995; Pigliucci, 2007; cf. Walsh, 1996; Hall, 1998; Wolpert, 1998; Pál and Hurst, 2004). Evolutionary implications of epigenetic phenomena have been explored in most detail where connections to long-standing evolutionary problems are clearest, such as parent–offspring conflict in the case of parent-specific genomic imprinting; but the results have not always promoted further or broader exploration because they seem to indicate only very limited scope for adaptive evolution based on epigenetic effects (Haig and Westoby, 1989; cf. Hurst and McVean, 1998).[1]

The epigenetic inheritance arena is now becoming a meeting ground for scientists working on molecular mechanisms and quantitative evolutionary dynamics. Molecular, cellular, and developmental biologists; developmental geneticists; and evo–devo researchers are interested in a wide range of epigenetic phenomena— from bacterial immune responses to foreign DNA to chromatin remodeling, regulation of gene expression, imprinting and paramutation, and transmission of cortical structures, organelles, and metabolic steady states. Some of them are beginning to meet up with epidemiologists, cancer biologists, ecologists, population geneticists, quantitative geneticists, and evolutionists interested in disease propensities, heritable maternal and environmental effects, phenotypic plasticity, Baldwin effects, genetic assimilation,

---

1. However, since imprints and in many cases methylation marks are reestablished de novo in each generation rather than transmitted intact through multiple generations, some authors argue that these do not qualify as transgenerational epigenetic *inheritance* phenomena (Rakyan and Whitelaw, 2003). In this narrow interpretation of inheritance, only structures that persist unchanged through transmission count.

phenotypic evolution, and speciation (see, e.g., Gilbert and Epel, 2009).

Philosophers of science are interested in, and routinely observe, conceptual "mismatches" between fields attempting to interact within such arenas in their studies of conceptual variety among the sciences; formal modeling of the structure of scientific laws, models, theories, and explanations; and attempts to articulate and codify modes of scientific reasoning. They are sometimes in a position to do philosophy *for* science (Griesemer 2008), i.e., to make observations about the state of the (conceptual) art that may help scientists think about proposals for new kinds of research in multidisciplinary arenas.

Here, I offer two specific observations about theories and reductionism in the arena of epigenetic inheritance. (1) Concepts of inheritance and evolution have different significance and implications in mechanistic molecular sciences (MMS) and quantitative dynamical evolutionary sciences (QDES) because these sciences construct models and theories in very different ways.[2] Below, I characterize these different

ways to show why it has proved difficult to coordinate research investments in epigenetic inheritance from the molecular and evolutionary sides of the arena. (2) "Reductionism" comes in different styles, concordant with distinct ways of theory-making, that can pull in different directions in MMS and QDES. This divergence of explanatory strategies has been noted by biologists. Haig (2002, 67) contrasts molecular and adaptive explanations, pointing out that MMS tends to explain the nature of the phenotype by appeal to molecular mechanisms while QDES tends to explain the presence of genotypes (DNA sequences) in populations by appeal to natural selection operating in ancestral generations. These observations together suggest that the integration of molecular and evolutionary studies of epigenetic inheritance will not simply be a reductionistic molecular explanation of phenotypes from the underlying mechanisms of epigenetic heritability but a transformation with implications for theories on both sides. Understanding the conceptual tensions may be of use to biologists as they consider investing in specific new theoretical and empirical research programs aimed at articulating and evaluating the relative significance of epigenetic inheritance in evolutionary dynamics rather than only recognizing and claiming that transgenerational epigenetic inheritance mechanisms may be significant for evolution.

The agenda for research in the arena of epigenetic inheritance and evolution has begun to take shape recently in workshops designed to bring molecular and evolutionary biologists and even philosophers together (e.g., Van Speybroeck et al., 2002; Gilbert and Epel, 2009). A recent NESCent (2009) workshop included the following questions:

1. What is epigenetics?

2. What methodologies are available to investigate epigenetic variation and inheritance in model systems?

3. How can we assess the frequency of heritable epigenetic processes in natural populations?

---

2. This is not to say that molecular sciences cannot be quantitative-dynamical or that evolutionary sciences cannot be mechanistic. Biochemistry is both a quantitative and a mechanistic molecular science, although there are important distinctions to be made between nonquantitative molecular biology and quantitative biochemistry. Some philosophers argue that natural selection is a mechanism but that existing theories of mechanistic science are inadequate to account for it as such (see Skipper and Millstein, 2005). My contrast here is between the clockwork models of much of molecular biology (articulations of parts and sequences of changes in parts' configurations, see Kauffman 1971) and the recursion equation and changing quantities models of mathematical ecology, population and quantitative genetics. I also do not mean to suggest that dynamical models of changing phenotype or genotype (or environment) frequencies are all there is to evolutionary theory (mathematical or otherwise). Evolutionary theory includes, in addition to accounts of gene, genotype, and phenotype frequency change, accounts of adaptation, distribution, and abundance; speciation and cladistic diversification; and the evolution of form and novelty. I am here contrasting molecular epigenetics only with the quantitative dynamical theory of trait or gene frequency change. Other parts of evolutionary theory, e.g., phylogenetics, evo–devo, and evolutionary ecology, require different treatments from those the contrast here can illuminate.

4. How do we go from studying epigenetic variation to assessing its ecological relevance?

5. How can we separate genetic from epigenetic effects in natural populations?

6. How do we evaluate the relative importance of epigenetic effects for phenotypic evolution?

This agenda (see also Pál and Hurst 2004; Bossdorf et al., 2008) marks a departure from the majority of claims in the molecular epigenetics literature since the 1980s in that it focuses on specific questions for the evolutionary dynamics project and moves beyond the claim that epigenetic mechanisms which transmit variation in structures or information across cell or organism generations might or must have such implications.[3]

NEW DESCRIPTIONS, NEW PROPOSALS

I mention here the first way philosophers may contribute to the advancement of the study of epigenetic inheritance. In subsequent sections, I focus on issues of theory structure in molecular and evolutionary sciences and on their divergent strategies of reductionism. The first, and best, way philosophers can contribute to fast-moving empirical sciences in multidisciplinary arenas like epigenetic inheritance is to organize descriptions of empirical phenomena and theoretical accounts in the scientific literature in order to draw attention to them in new and different ways. Variety of conceptualizations in a field fuels thoughtful development of perspectives and results that are robust to the idealizations that are inevitably introduced for practical reasons into modeling and empirical studies (Van der Weele, 1999; Griesemer, 2002b; Wimsatt, 2007).

Jablonka and Lamb (e.g., 1989, 1995, 1998, 2005) have perhaps done the most to organize descriptions of epigenetic phenomena and articulate the implications for evolution in ways that call for new, specific programs of research that diverge from and challenge traditional thinking embodied in neo-Darwinism and the modern evolutionary synthesis. It is noteworthy that Jablonka, whose career started in cytology and genetics, had to develop expertise in evolutionary biology and, indeed, in the history and philosophy of science in order to press their case that the evolutionary implications of the full range of epigenetic inheritance mechanisms must be taken seriously and not only for cases like imprinting, where the connection to specific dynamical problems was easily made to fit preexisting theoretical puzzles (see Pál and Hurst, 2004). They called attention to the need to understand how epigenetic inheritance may play a "Lamarckian" role as a family of mechanisms through which inheritance systems are responsive to environments, such that variations originate nonrandomly with respect to fitness consequences. This is perhaps the most radical implication of epigenetic inheritance for evolutionary theory, which was founded on the notion that genetic variation originates at random with respect to fitness or the "direction" of adaptation (see Lamm and Jablonka, 2008, on additional relevant senses of randomness in neo-Darwinian theory that are violated by epigenetic inheritance). Early work on epigenetic inheritance, such as paramutation in plants (e.g., Brink 1956, 1960), identified the directed nature of such non-Mendelian mechanisms but focused on the somatic consequences of epimutations and, thus, their implications for development rather than for transgenerational inheritance and evolution.

Jablonka and Lamb (1995, 2005) also explored a variety of evolutionary implications for adaptation, speciation, and the coevolution of genetic and epigenetic inheritance systems, mainly in verbal models. A key project for the integration of molecular epigenetic inheritance into quantitative dynamical evolutionary theory

---

3. In biological science, as in the biological world, there are always exceptions. Jablonka and Lamb have, since 1989, argued for considering the kinds of questions posed by the NESCent workshop. See, e.g., Jablonka and Lamb (1995, chaps. 7–9).

will be to develop mathematical models for evolutionary response to selection involving epigenetic as well as genetic mechanisms for heritability (Maynard Smith, 1990; Jablonka et al., 1992, 1995; Lachmann and Jablonka, 1996).

Jablonka and Lamb's work has been important in the arena of epigenetic inheritance not so much for contributions to new empirical knowledge of specific molecular epigenetic mechanisms but as organizers of key descriptions of these mechanisms and as promoters of evolutionary modeling in a way that prepares the path for an "expanded evolutionary synthesis" (Pigliucci, 2007). Whether that synthesis is indeed an "expansion" or something more radical is a topic I will return to in the context of a discussion of reductionism in MMS and QDES. It is an important and unresolved question whether epigenetic inheritance systems can be integrated into existing quantitative dynamical evolutionary theory as merely more molecular mechanisms realizing the quantity of heritability (and others) or whether, because of Lamarckian or at least non-Mendelian properties, epigenetic inheritance mechanisms require the transformation of quantitative dynamical evolutionary theory into a new and different kind of theory.[4]

I frame the question of the evolutionary significance of epigenetic inheritance in terms of theory structure and styles of reductionism involving assessment of the potential costs, risks, and benefits of investment in novel research programs. This has several dimensions: (1) assessment of the potential significance of epigenetic inheritance for evolution relative to other inheritance mechanisms; (2) the different nature of theories and models in molecular and evolutionary biology; (3) the level of risk of investing in research into mechanisms (molecular biology) and quantities (evolutionary

dynamics) that may not exist, may not be competent even if they exist, and may not be responsible in nature even if they are competent in laboratory or simulation studies; (4) heuristic strategies of reductionism in molecular and evolutionary biology; and (5) the ways in which molecular and evolutionary biologists may misunderstand what is at stake for the other side in a program of research that seeks to integrate the two.

EPIGENETIC INHERITANCE MECHANISMS AND EVOLUTION

Some epigenetic mechanisms involve covalent chemical modification of DNA, posttranslational modification of chromatin proteins, or DNA-binding proteins. Because chromosomes are replicated and pass through cell division, epigenetic "marks" such as cytosine methylation, histone acetylation, and bound transcription factors might be transmitted as well. There is a correlation between degree of DNA methylation and transcription activity—more methylation, less transcription. Moreover, methyltransferase enzymes recognize the hemimethylated state of recently replicated DNA and complete the replication of methylation patterns at CpG or CNG pairs (Jablonka and Lamb, 1995, fig. 4.12; Turner, 2001, fig. 10.1). Thus, methylation patterns can track the semiconservative replication of DNA. Acetylation of nucleosomal histone protein tails relaxes highly compacted chromatin, which increases access of transcription factors that regulate gene expression. DNA-binding protein complexes like Polycomb and Trithorax in *Drosophila* regulate homeotic gene expression. Their epigenetic effects on the transcriptional potential of genes can be stably transmitted through many cell generations and through female meiosis (see Turner, 2001, 234–5). Since patterns of epigenetic marks can in principle vary at least quasi-independently of associated DNA sequences, these patterns, if transmitted, could constitute inheritance systems parallel to the genetic inheritance system.

---

4. Consideration of parallels to the transformation of evolutionary theory effected by models of niche construction (Odling-Smee et al. 2003) would be instructive but are beyond the scope of this chapter.

There are other kinds of epigenetic mechanisms that can pass through cell division besides those that interact directly with chromatin, e.g., metabolic steady-state systems which maintain the relative concentrations of metabolites in autocatalytic cycles such that variation can be transmitted to daughter cells (Novick and Weiner, 1957; discussed in Jablonka and Lamb, 1995; Pál and Hurst, 2004; see also Keller, 1995). There are also structural inheritance systems in which naturally varying patterns of cell organelles such as cortical structures in ciliates can be maintained through membrane growth and stably transmitted through cell division for many generations (e.g., Nanney, 1968). Moreover, there seem to be behavioral and symbolic inheritance systems which can propagate informational state variations through social learning or linguistic communication between organisms (see reviews by Jablonka and Lamb, 1995, 2005).[5]

The possibility of an evolutionary role in virtue of trans-cell or trans-organism transmission is a far cry from the full prospect of an inheritance system that rivals the genetic inheritance system in scope and capacity. Here, I examine the *possibility* of an evolutionary role for this kind of developmental phenomenon, *presuppositions* of strategies for deciding whether to invest research effort in epigenetic phenomena, and the role of possible mechanisms in differing *evaluations of relative significance* of epigenetic versus genetic inheritance and investment risk by molecular epigeneticists and evolutionists. My goal is not to offer a detailed historical reconstruction of epigenetics or a theory of inheritance that generalizes evolutionary theory but, rather, to capture a sense of the grounds for controversy in a very dynamic and rapidly changing empirical and theoretical landscape.

THE POSSIBILITY OF AN EVOLUTIONARY ROLE FOR EPIGENETIC MECHANISMS

Objections to a significant evolutionary role for epigenetic mechanisms frame considerations of investment in research and heuristic strategies of theory construction but, as I discuss below, *not in the same way* for molecular epigeneticists as for evolutionists. It is unknown whether any of the known epigenetic mechanisms supports a sufficient set of combinatorial possibilities to sustain effectively unlimited heredity (Maynard Smith and Szathmáry, 1999). Without it, there could be epigenetic inheritance of a sort but only modest prospects for adaptive evolution because in limited systems of heredity the global optimal phenotype is quickly found and then all remaining or new variation eliminated by purifying selection.

It has often been asserted that epigenetic marks may be transmitted through mitosis in the development of multicellular organisms but not through meiosis. Hence, due to the discontinuity of soma and the continuity of the germ line, according to Weismann's doctrine (Figure 3.1; see also Griesemer and Wimsatt 1989), epigenetic mechanisms may play a role in development but not in transgenerational inheritance. It has also been pointed out frequently that early germ–soma differentiation is characteristic of only some animals and not of most of life (e.g., Buss, 1987); but the relevant issue of relative significance comes down to which taxa are the "important" ones, and this obviously differs for those molecular biologists focused on human disease, or metazoan model organisms, and for those evolutionary biologists interested in the distribution of epigenetic mechanisms across the whole tree of life. Although one can parse this particular debate in terms of the kinds of generalizations biologists seek, whether

5. There are also contemporary views of epigenesis beyond the scope of this chapter distinct from what might be called "molecular epigenetics." Epigenesis is a much broader category than molecular epigenetics and deals with processes through which character variation among organisms can arise through developmental interactions at any level, though typically studied at the cell or tissue level, without being supported by genetic variation. Developmental interactions are then "assimilated" into heritable patterns, whether by Waddington's mechanism or something else (e.g., Newman and Müller, 2000; discussed in Griesemer, 2002a).

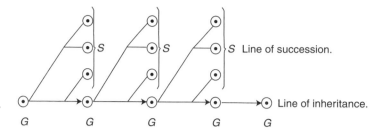

FIGURE 3.1 E. B. Wilson's diagram of Weismann's doctrine.

Line of succession.

Line of inheritance.

causal regularities within a given taxon or distributions among taxa or over space and time (Waters, 1998), I suggest it is fruitful to consider the issue with respect to forward-looking research strategies rather than (or in addition to) reflections on explanations and the nature of biological generalizations.

Even if epigenetic marks pass through meiosis, they are (or were recently believed to be) completely "reset" or "erased" in embryogenesis each generation in order to restore totipotency to germ-line cells, as is the case with chromosomal imprinting to inactivate maternal or paternal chromosomes. Reset epigenetic states must be reestablished de novo in the daughter cells or offspring organisms. Whitelaw and colleagues have shown (e.g., Rakyan et al., 2001) that in mammals (mice, humans) marks often are not reset after passage through meiosis, although some of their cases concern disease etiology (kinky tail) rather than adaptive phenotypes. Some taxa, such as *Drosophila* species, exhibit little or only localized activity of a given epigenetic mechanism while displaying a substantial amount of another: *Drosophila* species were thought not to methylate DNA at all (until about 1999, see Lyko, 2001) but are known to make extensive use of chromatin remodeling mechanisms for gene regulation that can pass through mitosis and (female) meiosis (Cavalli and Paro, 1998). Because the methylation system's tight coupling of epigenetic marks to DNA make it a strong candidate for an epigenetic inheritance system, its lack in some taxa that rely instead on epigenetic systems with seemingly less evolutionary potential is reason enough for skepticism of a general, significant

role in transgenerational inheritance. However, as empirical knowledge of epigenetic mechanisms rapidly advances, assessments of the relative significance of possible mechanisms of epigenetic inheritance seem to be shifting, so there could be a downside to a conservative investment strategy of assuming the sufficiency of genetic inheritance for evolution until there is overwhelming evidence of abundant transgenerational epigenetic inheritance.

Moreover, the de novo synthesis of epigenetic marks at high, variable rates compared to genetic sequence mutation rates suggests a level of instability which some regard as an unlikely basis for adaptive evolution. Epigenetic transmission can occur, but failure to reset marks is interpreted as an aberrant, pathological condition rather than a reliable and regular inheritance system. Another source of skepticism derives from the observation that epimutation may be limited to genes that happen to be close to retrotransposons because it evolved from an ancestral bacterial immune system for silencing foreign DNA (Bestor, 1990). Increasing evidence that epigenetic regulatory states in humans due to the nutritional environment can be stably transmitted for multiple generations suggests that, although there may be a link to pathologies like diabetes, there may be (positive) adaptive significance as well, e.g., for longevity (Kaati et al., 2002; Gallou-Kabani and Junien, 2005; Jablonka, 2004). An epigenetic mechanism in mammals that may have evolved in bacteria to silence viral retrotransposons hardly sounds like a plausible basis for a reliable mechanism for adaptive epigenetic inheritance in mammals until it is pointed out that as much as 42% of mammalian genomes are of retroviral origin (see Rakyan et al., 2001).

Finally, the enzymes that cause chromatin marking states are themselves gene-encoded, so there is the in-principle possibility that explanations of patterns of epigenetic states can be "reduced to" (i.e., explained in terms of mechanisms for determining) the states of genes, either in the present or in a previous generation, via maternally acting genetic effects. One generation's development is the next generation's evolutionary effect if genotype × environment interaction in one generation has transmissible fitness consequences in the offspring. The possibility to explain effects in offspring on the basis of causes operating in the parents is part and parcel of conservative evolutionary explanation, but it is also the root of a particular kind of parsimonious significance assessment: If a novel mechanism is not necessary to explain a phenomenon, then it is not significant. Maternal effects can in principle explain epigenetic effects as developmental consequences of gene action, so epigenetic inheritance need not be postulated. The problem is that in the biological world of evolved mechanisms, no particular mechanism is necessary and all are contingent (Beatty, 1995, 1997), so the apparent lack of lawlike necessity of biological generalizations undermines the conservative explanatory strategy when it appeals to "standard" theory as though it were standard because it had discovered a biological "law." In biology, need may be the mother of adaptive invention, but necessity is not the measure of significance.

RELATIVE SIGNIFICANCE ARGUMENTS AND THEORY STRUCTURE IN BIOLOGY

The foregoing description of the issues around epigenetic inheritance contains many modal qualifiers and conditional statements: Epigenetic marks *might* be transmitted; and *if* they are, they *might* serve as an inheritance system; and *if* they do, they *might* contribute to adaptive evolution. For at least some epigenetic phenom-

ena, there is little doubt that they exist and indeed are regulators of gene expression; but whether they are also competent to pass through mitosis or meiosis and are in fact responsible for any cases of adaptive evolution is in question. In this section, I introduce the notion of relative significance arguments and then consider the character of theory in evolutionary biology and molecular epigenetics. In subsequent sections, I suggest that relative significance arguments are shaped by the structure of theories and heuristic strategies for theory construction in disparate lines of scientific work. I also consider the role of risk strategies for research investment in light of theoretical differences between molecular epigenetics and quantitative dynamical evolutionary theory.

RELATIVE SIGNIFICANCE

John Beatty describes certain arguments in biology about the extent of applicability of a given model or theory within a domain as relative significance arguments, e.g., the extent to which gene regulation is correctly explained by Jacob and Monod's model of negative regulation (Beatty, 1995).[6] Jacob and Monod famously claimed that what is true for the colon bacillus is true for the elephant (quoted in Beatty, 1995). They imply that their model is correct for effectively all taxa in the domain of gene regulation and, therefore, that negative regulation is highly or exclusively significant. Put differently, the claim is that their theory of negative regulation is *sufficient* to explain all (or nearly all) gene-regulation phenomena in the domain. The truth of their claim of significance relative to other *possible* regulation mechanisms depends on empirical facts about what other mechanisms can and (regularly) do cause the same kind of phenomenon, but it also depends on assumptions about the domain in question: that the domain

---

6. Mitchell (1992, 2003) examines another kind: disputes about the relative contribution of different kinds of causes to a given effect, e.g., genes vs. environment or selection vs. drift in evolution.

includes all mechanisms or modes of gene regulation and all taxa that exhibit the phenomenon of gene regulation. If the domain were taken to include only *Escherichia coli* and elephants, their claim might be true; but it would be significant only for an odd, gerrymandered domain covering one bacterium and one metazoan clade with two extant species rather than the domain of all taxa that exhibit gene regulation.[7] Treating the domain as all (extant) life makes the truth of their claim much more empirically significant but false. Other mechanisms in addition to negative regulation are required to cover that domain, and Beatty's point is that no one generalization, hence no one theory, is likely to cover any important domain because evolution tends to diversify the properties of biological mechanisms.[8]

Claims that heritable traits can be due to epigenetic, rather than genetic, mechanisms assert that the supposed (near) universal significance of genetic inheritance mechanisms for the domain of inheritance phenomena is open to challenge. Claims that molecular epigenetic mechanisms are more widespread than previously thought and that there is mounting evidence of epigenetic inheritance in a wide variety of taxa for an increasing number of traits shifts the balance of relative significance among those inheritance mechanisms that exist, assuming

they are competent to cause transgenerational heritability, whether across cell divisions or mitotic or meiotic generations. Empirical studies of epigenetic inheritance further purport to show that epigenetic, rather than (or in addition to) genetic, mechanisms have actually been responsible for trait heritabilities in nature.

Beatty argues that biology tends to be theoretically pluralistic, with different theories invoked to account for different kinds of phenomena within a domain, because of a special feature of the biological realm. The properties of biological systems, including the deepest ones such as the behavior of the genetic system itself, are *evolved* properties; and therefore, any generalizations about them are evolutionarily *contingent*. They could have been otherwise, if evolution had run differently and, indeed, they could become different in the future. Evolution is a chancy process, and contingency runs deep in biology. Mendel's "laws" do not hold in the same way or to the same universal extent as the fundamental laws of physics. Because of this disanalogy, due to the deep contingency of biological generalizations, Beatty argues that there are no laws in biology.

Beatty's argument focuses on the relative explanatory significance of the plurality of correct theories needed to account for the variety of evolutionarily contingent phenomena comprising a domain. He did not address, however, an additional important feature of many relative significance arguments: They are not always directly about the need for multiple theories to correctly explain all the phenomena of a given kind in a domain (e.g., gene regulation or transgenerational inheritance). Rather, they are arguments about the potential for scientific progress due to research investments that could be made in order to explore possible alternative mechanisms to those currently deemed significant. Science is a risky business, and investing limited research time and money to find out whether a mechanism for an uncertain phenomenon exists, is competent to produce the phenomenon, or is in fact responsible for it

---

7. Waters (1998) argues that many of Beatty's cases really concern distribution generalizations rather than claims about law-like generalities governing a phenomenon. I believe that their disagreement cannot be settled without establishing the extent to which "conservative" explanations in biology appeal to the tacit assumption that biological generalities are intended to express law-like generalizations, regardless of which philosopher is correct about what the correct assumptions should be.

8. Of course, interest and importance are in the eye of the beholder. The domain *Homo sapiens* is important biologically for all sorts of reasons, but many philosophers of biology assume that species are individuals and, therefore, that generalizations cannot be about them per se. Phenomena exhibited by members of a given species are likely to be exhibited by members of other species as well due to genealogical relatedness; hence, (most) biological generalizations of any significance are unlikely to concern only members of a single species.

cannot be taken lightly. Some scientists are risk-averse and others risk-tolerant, so it would not be surprising to find different judgments about the value of exploring domain extensions that would add mechanisms. More important here, however, is the fact that relative significance claims in arenas where different specialties or lines of work meet, such as epigenetic inheritance in evolution, also depend on perceived risk. Specialties, like individual scientists, can differ in their risk assessments, which means that evaluations of relative significance, taken as claims about the potential benefits of research investment, can also differ.

Considerations of the disparate ways of theory-making in molecular and evolutionary biology (see Winther, 2006) lead to differing assessments of the implications of research into what a possible epigenetic mechanism can or might do, what it is competent to do, and what it is actually responsible for. These differences in turn affect how relative significance arguments between specialties transition from debates about research investment when phenomena are novel or poorly understood to debates about the extent of applicability of a given causal mechanism (relative to others) when the phenomena are known to exist but their distribution and extent of applicability are unclear to debates about the relative contributions of different causes to a given effect in combined accounts of a single theory for a domain when the phenomena are well established and mechanisms are known to be competent to produce them.

In sum, arguments about relative significance range across a continuum of cases from issues of investing in future research, evaluating which cases are correctly covered by models and theories of which mechanisms, and evaluating relative contributions of causal mechanisms to produce complex phenomena such as regulated genes or heritable traits in any given instance. Moreover, and centrally to the case of epigenetic inheritance, when specialties with different ways of theory-making engage in questions of relative significance, risk–benefit assessments can differ as well, leading to a situation where something that is patently and obviously novel and important from the perspective of specialty A can seem just as obviously unlikely to be significant and unworthy of serious investment from the perspective of specialty B. The question of the relative significance of epigenetic inheritance in evolution has been like this for roughly the last 20 years. Mechanistic studies of molecular epigenetic phenomena have claimed a potentially significant role in evolution for epigenetic mechanisms that can cause transgenerational epigenetic heritability across cell division, mitosis, and meiosis in a variety of taxa across the tree of life. However, few evolutionary theorists have invested in examining the implications, and of those who have, it has mainly been to express skepticism of the prospects for research investment.

Questions about the importance of epigenetic inheritance in evolution are at present more aptly characterized as questions about potential research investment risks and benefits than as questions of the extent of applicability of any given model, although what evidence there is about the extent of applicability of models of epigenetic inheritance has a direct bearing on arguments concerning the value of research investment in both molecular mechanistic and theoretical evolutionary research. Considerations of relative causal contributions of genetic and epigenetic mechanisms to trait heritability, to gene-regulatory networks that affect developmental dynamics, or to the list of nonrandom "biasing forces" (generalizing from natural selection) within a single, quantitative, dynamical evolutionary theory will surely be important to an eventual "extended evolutionary synthesis" (Pigliucci, 2007), especially one that articulates evidence about molecular epigenetic mechanisms with quantitative dynamics of evolutionary theory. However, the very different qualities of theories in mechanistic molecular biology and quantitative dynamical evolutionary biology result in different assessments of the risks and potential benefits of research investment. To the extent that those assessments guide research investment, they influence the prospects for answering questions about extent of applicability and causal contribution as a kind of "ascertainment bias."[9]

THEORIES AND REDUCTIONISMS

Theories are collections of models together with their robust consequences (Levins, 1968). Theories in molecular biology, insofar as anyone formulates *theories* in this field rather than individual models of particular mechanisms (see Bechtel, 2006) within a domain (*pace* Monod), typically have the form of lists of mechanisms from which to choose the elements of causal narrative explanations of phenomena. (Think of the Watson-Crick explanation of protein sequence structure in terms of a concatenation of sequentially operating mechanisms of DNA transcription and RNA translation.) The grail of highest significance in mechanistic biology (molecular or not) goes to models of universally distributed mechanisms that are so deeply entrenched that they are evolutionarily highly conserved (e.g., the genetic code). Even if they are evolutionarily contingent, once evolved they are very hard to change (on generative entrenchment in evolution, see Wimsatt, 2007). That is as close to necessity as biological mechanisms ever get, yet it is not the same sense of significance as in evolutionary biology.

Quantitative dynamical theories of evolution, building on Darwin's principles of heritable variation in fitness (Darwin, 1859; Lewontin, 1970), generally have the form of recursion equations specifying mathematical relations among quantities changing over time. Quantities are properties of causal capacities, represented by variables (and sometimes parameters) in models. When a causal capacity is realized by a mechanism, it can be represented by a variable taking on a (range of) value(s). The model

in turn represents the fulfillment of a function. The quantity *heritability* represents a capacity of a population in an environment for a trait correlation among relatives; and when it takes a value as a consequence of the operation of a genetic inheritance mechanism, it fulfills the function of transmitting trait values from parents to offspring, e.g., as represented in a response to selection equation. The "opportunity for selection" in quantitative evolutionary genetics (Arnold and Wade, 1984) is a quantity represented by a variable for the variance in relative fitness and takes a value as a consequence of the operation of natural selection so as to fulfill, in joint operation with inheritance, an evolutionary response to selection.

The disparate nature of theories in the arenas of mechanistic molecular biology and evolutionary dynamics leads to different strategies of reduction, which in turn fuels not only relative significance debates, as mechanisms and quantities are added to existing theories, but also contrasting senses of what kinds of research investments are judged to be conservative or transformative and therefore less or more risky. Research is *conservative* if it involves empirical work to support the specification of current theory (adding mechanisms to the list in the case of mechanistic molecular theories, filling in values for variables and parameters or gaining insights into detailed mathematical consequences through mathematical derivation or computer simulation in the case of quantitative dynamical evolutionary theories). Research is *transformative* if it forces change in what we already understand, e.g., in MMS adding mechanisms to a molecular theory that necessitate alteration of models of other mechanisms already on the list or forcing accounts of interactions among mechanisms that themselves constitute second-order mechanisms which must be added to the list. In QDES, transformative research involves adding quantities to dynamical theories that require changing the form of the equations rather than only increasing the number of terms, e.g., in recognition that fitness is frequency-dependent or that heritability

---

9. I use "bias" here as a term expressing a heuristic research strategy rather than something morally or epistemically flawed (see Wimsatt, 2007).

depends on epistatic interactions of genes mediated by epigenetic mechanisms that constitute a parallel system of inheritance.[10]

To see relative significance debates as concerning research investment and not only as questions of extent of applicability or causal contribution of a cause already modeled or empirically established, it is helpful to view reductionism itself as a heuristic research strategy for theory construction (Griesemer, 2002b) rather than, as philosophers usually do, as an account of explanation by derivation of less from more general theories or, as scientists usually do, as an account of higher-level phenomena explained in terms of lower-level mechanisms (see Wimsatt, 2007, on reductionism in science). However, different ways of theory-making lead to different concepts of heuristic reductionism and different assessments of whether research investment need only be conservative or will require transformative efforts. Asserting the potential significance of a novel epigenetic inheritance mechanism as a call for research investment reverses the usual logic of justification and discovery—*justification* (to a sponsor, in peer review, or as an appeal for research by specialists in another field) is what you do in order to be enabled to make the discovery that the mechanism exists, that it is competent to produce heritable effects, or that it is in fact responsible for adaptive evolution in empirical cases.[11] That order switches back to the usual one, with justification following discovery, after the discoveries are made and questions of relative significance turning to applicability across the domain or to causal contribution relative to other causes acting in any particular instance.[12]

DARWINIAN THEORY

The great nineteenth-century philosophical debate between John Herschel and William Whewell, on the proper conduct of scientific investigation into causes of natural phenomena, is instructive for current controversies over the nature and power of epigenetic causes to contribute to evolutionary change. Herschel's views followed Newton's rules of reasoning and fueled an "empiricist" philosophy of science in which proper causal explanation should appeal only to actual causes and requires demonstration that alleged causes (1) actually exist (by which Herschel meant they could be observed acting), (2) are competent to produce the effects observed, and (3) are actually responsible for the effects in the cases observed. Whewell worried that Herschel's approach would preclude the discovery of any genuinely new kind of cause, the discovery of any known kind of cause acting in degrees unwitnessed by scientific observers, or the discovery of any causes responsible for shaping our world not now acting. Whewell offered a methodology of "consilience," in which the jumping together of many different kinds of disparate facts could be used to go beyond the facts to suggest new kinds of causes, acting in degrees and ways not (yet) observed, to produce effects otherwise inexplicable.

Herschel's philosophy of science and Charles Lyell's uniformitarian *Principles of Geology* were

10. My terms *conservative* and *transformative* are similar in some ways to Kuhn's (1970) *normal science* and *revolutionary science*, respectively. However, in Kuhn's account, the practice of normal science leads to revolution in the face of anomalies that build to a crisis. My focus is on scientists' forward-looking assessment of the need for transformative research in order to account for a phenomenon modeled by a mechanism from another specialty very likely produced in that other specialty by conservative research in my sense. Transformative research need not be revolutionary either. It can easily be the case that all the mathematics necessary to change the form of an equation is known to the scientists involved or that the empirical tools and technologies are available to molecular biologists who want to study an interaction among known mechanisms. The issue is not "normal" vs. "revolutionary" but rather risk. Transformative research requires a change to "business as usual," so scientists must assess in advance whether the benefit is likely to be worth the risk.

11. I thank Chris di Teresi and Elihu Gerson for this astute observation.

12. Thanks to the audience at Pittsburgh's Center for Philosophy of Science for pressing me to relate the research investment sense of relative significance arguments with claims about objective relations between causal contributions to an effect (Mitchell, 2003) and claims about extent of applicability (Beatty, 1995, 1997).

a quantity of the theory—is competent to cause evolutionary change. If it is competent in this theoretical sense, then it is a potentially responsible cause in nature.

Thus, the discovery by molecular biologists of epigenetic inheritance mechanisms, as they have pointed out for 20 years, is significant for evolutionary theory just by the fact of the existence of such mechanisms. Whether such mechanisms are significant relative to genetic inheritance mechanisms *for evolution* depends on more than just whether they exist. It depends on whether epigenetic mechanisms can be shown to be competent to cause (adaptive) evolution. However, if that means competent to serve as an inheritance system, it may well require transforming evolutionary theory and not merely adding a new instantiating mechanism of heritability. In contrast, to be significant relative to other molecular mechanisms *for molecular biology*, an epigenetic mechanism only has to be shown to be competent to produce epigenetic effects in order to belong on the list of mechanisms. Thus, relatively conservative, low-risk theoretical consequences in molecular biology may coincide with relatively transformative, high-risk theoretical consequences in evolutionary biology.

Considering relative significance arguments as a question of research investment in this case brings out the asymmetrical impact on different specialties of the assessment that epigenetic inheritance is significant for evolution: Should a theorist invest time and effort in exploring the mathematical consequences of instantiating an evolutionary model with an epigenetic mechanism for heritability, given that the mechanism does not follow Mendelian rules and may exhibit high (and highly variable) epimutation rates and that epimutation may be directly induced by the environment and therefore not random with respect to fitness? Putting epigenetic mechanisms "on the list" of instantiators of heritability requires more than just noting it—it (probably) requires changing the mathematical representation of the quantity. It is also important to note here that if an epigenetic mechanism for heritability exists that is shown to be theoretically competent through mathematical modeling, then it becomes an empirical obligation to test for it. The reason is evolutionary contingency. If a mechanism exists but is demonstrably not responsible for a given effect (of a relevant type) or even competent to produce the effect in a particular instance, it could well be because evolution attenuated its effects through countervailing mechanisms. Retroviruses may not be competent to cause disease under normal cell conditions because the methylation system evolved to suppress them. Somatic mutations in animals may not be transmissible because development evolved to restrict germ-line access to only one or a few cell lines (Buss, 1987; Michod, 1999).

I have argued elsewhere (Griesemer and Wade, 1988; Griesemer, 2002a) that laboratory studies of group selection aimed to demonstrate that group selection can occur and is competent to cause evolutionary change. Whether it occurs in nature and is both competent and actually responsible for measurable evolutionary change in traits of interest is a problem of extrapolation or inference from the laboratory, together with empirical investigation guided by research protocols designed in accordance with an understanding of the capacities of the mechanism and ecological conditions. It has been argued since 1960 (Lewontin and Dunn, 1960) that ignoring population structure (which is a key condition for the operation of group selection) can lead to dynamically insufficient models (see Lewontin 1974) and empirically incorrect inferences about mechanisms operating in selection processes (Wade 1978). Similarly, it can be expected that if transgenerational epigenetic inheritance mechanisms operate but are not represented in the models, dynamically insufficient and empirically incorrect results may ensue if scientists infer from a lack of demonstrated responsibility that the mechanism does not exist or is not competent. Whether to invest in complicating the models or empirically investigating phenomena that may or may not occur, may or may not be competent, and may or may not be responsible is a major question for scientists actively engaged in research.

MOLECULAR EPIGENETICS

Molecular epigeneticists have rather different concerns about the potentiality of epigenetic inheritance mechanisms. Their theories, for one thing, are so unlike neo-Darwinian dynamical equations that molecular biology is often treated as theory-less, but a more accurate characterization is that molecular theories are kinetic rather than dynamic: They describe the structural transformation of inputs to outputs in molecular processes, but they do not consider the time rate of change or energetics of transformations. Thus, molecular explanations can often be temporal narratives without the need for quantitative mathematical forms to express change. This is, of course, very crude and not intended to offend biochemists or molecular dynamicists. Watson and Crick's model of the mechanism of DNA replication is a good exemplar. It describes the structure of the DNA molecule and outlines how the strands go through a sequence of steps, including some details of interactions with proteins, but in a schematic way. Time is considered only in the sense of ordering the steps in the process.

To expand the model of theory structure in molecular biology previously sketched, a molecular theory consists of a list of molecular mechanism descriptions or representations (e.g., in diagrams) that can be used singly or in combination to narrate causal processes producing outcomes (effects) which can be observed, measured, or otherwise taken as data in empirical studies. A more detailed theory might include more items on the list or a list of interactions between mechanisms, comprising a sublist of second- (or higher-) order mechanisms. Theories in molecular biology are open so long as biologists continue to search for mechanisms to go on the list. Pragmatic explanatory goals delimit what counts toward the coherence of a

given theory: Something is permitted on the list if it is useful, in combination, to narrate a causal process of interest.

CONSERVATIVE AND TRANSFORMATIVE RESEARCH

Research can be conservative or transformative, as mentioned above. In terms of the distinction between ways of theory-making that I attributed to molecular epigenetics and to evolutionary theory, conservative theory-making in molecular biology involves either specifying details of an existing mechanism or adding a mechanism to "the list" that comprises a theory. A step along the way toward molecular theory-making involves modeling. Modeling a molecular mechanism involves taking note that a phenomenon exists (usually as the result of experiments that establish conditions for reliably producing it in the lab, i.e., establishing a causal relationship between an outcome or effect and the set of experimental conditions taken as causal). Demonstration of the existence of a phenomenon, P, licenses incorporating its description into a model of a mechanism, M, a representation of the mechanism (typically in a diagram or narrative description). By virtue of P's incorporation into M, M is potentially competent to produce the outcome it is hypothesized to be "for." Whether P in M is causally competent and whether P is actually responsible for some part of M's behavior in any actual circumstances are separate questions requiring independent research.

An example from the molecular epigenetics literature is a pair of models for how the incomplete erasure and stochastic reestablishment of epigenetic marks in offspring could generate the variably expressive phenotypes observed in crosses of agouti- and yellow-colored mice (Rakyan et al. 2001). One model assumes that epigenetic marks are reestablished in the preimplantation embryo (rather than in the primordial germ cells of the parents). To justify this model of the erasure and reestablishment

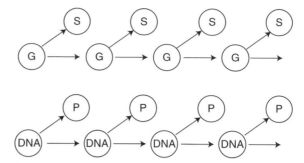

FIGURE 3.3 Maynard Smith's diagram showing the isomorphism of Crick's central dogma and Weismannism.

mechanism, the authors cite evidence (Mayer et al. 2000) that maternal and paternal genomes are demethylated at different times in zygotic DNA, "so parent-of-origin-specific erasure does exist" (Rakyan et al. 2001, 7). The existence of the phenomenon of parent-of-origin-specific erasure licenses using it in a model of a mechanism which could generate the incomplete erasure pattern, provided the mechanism is "inefficient."

Transformative theory-making in molecular biology occurs when additions or specifications involve claims of interaction with other mechanisms already on the list that necessitate revision or elimination of mechanisms previously understood.

In molecular epigenetics, working out the details of how a methyltransferase enzyme causes DNA methylation is specification work. Adding histone acetylation to the list of epigenetic mechanisms for gene regulation (via chromatin remodeling rather than DNA alteration) is "addition work," i.e., theory- as opposed to mechanism-specification work. Resolving the way in which histone acetylation interacts with DNA methylation and the possibility that only the two acting in concert are competent to regulate genes is potentially transformative in the sense that all those causal explanations of how DNA methylation silences genes and even what methylation is for must be rethought. We can see the transformative effect of discovering new kinds of interactions among mechanisms in that causal narratives, which string mechanisms together, have to be "restrung" if the interactions dictate different "assembly rules" for

narratives; e.g., in classical genetics, discovering that DNA is the carrier of genetic information dictates that inheritance narratives assemble mechanisms to follow the paths described by Crick's central dogma (DNA to RNA to protein; see Figure 3.3).

EXAMPLE OF CONSERVATIVE MODELING STRATEGIES IN GENETICS AND EPIGENETICS

Experiments in genetics revealing variable "expressivity" or "penetrance" of traits were puzzling to classical geneticists who expected Mendelian mechanisms to yield clean, mathematically explicable results. The fur color of mice that are genetically identical and raised in similar environments should be the same, but it is not always so. Studies through the 1970s and 1980s, using classical genetic crossing techniques, e.g., reciprocal crosses between heterozygotes and wild-type outbred mice, attributed the variability to "unlinked modifier genes" (see Rakyan et al. 2001, 5). This is a conservative strategy in the sense that the ways modifier genes work had been investigated empirically and mathematically (by population and evolutionary geneticists) so the explanation merely applied a known kind of interaction, already on the list and already accounted for in evolutionary dynamics, to new cases. Jablonka and Lamb (1995), in their argument for the significance of epigenetic inheritance in evolution, reviewed many cases of variable expression from classical genetics experiments and argued that the conservative strategy swept the epigenetic phenomena under the rug rather than faced up to the need to transform genetic theory.

Emma Whitelaw and colleagues conducted experiments on the same traits (agouti fur color and kinked tail shape) in mice using inbred strains so that it would be unlikely that modifier gene differences could explain the variations among offspring (Rakyan et al. 2001). They found correlations in fur color between parents and offspring (though not as strong as would be expected if the traits were determined by Mendelian genes). They also did transplant experiments to rule out metabolic differences in intrauterine environments of mothers with different fur color as the cause of the variable offspring traits. They found that transplanting fertilized eggs of yellow females into black females did not alter the proportion of fur colors among offspring compared to those gestated by yellow mothers (Rakyan et al. 2001, 5). Although environmental causes prior to transplantation could not be ruled out, they concluded the inheritance pattern "is likely to be due to an epigenetic modification" (Rakyan et al. 2001, 6). The work is conservative in the sense described above, in that it was aimed to show that additional mechanisms to transmission through the genetic system can be involved in trait heritability but no special interaction requiring modified understanding of genetic transmission resulted.

One other interesting conservative feature of the Rakyan et al. work is their denial that their models are Lamarckian. They suppose that epigenetic marks in the parents are erased either in the formation of germ cells or in the preimplantation zygote. Marks are then, according to the models, randomly reestablished at either gamete maturation or early embryogenesis. This occurs either because the erasure mechanism is not 100% effective at erasing all epigenetic marks (model 1 for "silent" traits) or, for genes that are unmarked (model 2 for "penetrant" traits) in the parent, the reestablishment mechanism is not 100% effective. This incomplete erasure or reestablishment explains the degree of heritability of the trait (Rakyan et al. 2001, 7). The randomness of reestablishment is interpreted to mean not that the environment directs the production of heritable variation but rather

that variation arises as a result of "random failure to completely erase marks at certain alleles during development" (Rakyan et al. 2001, 8). The speculation that the mechanism is stochastic is plausible, given that biomolecular mechanisms are generally less than 100% efficient; but it introduces a measure of conservatism into evolutionary considerations about what sort of investment in theoretical research would be needed to accommodate epigenetic inheritance mechanisms. If the molecular epigenetic mechanism generates random variation, as does the genetic system, then there would seem to be "no need" to consider radical, nonrandom, Lamarckian alternatives. Rakyan et al. do cite one source indicating that an environmental change might induce a shift in phenotype in the advantageous direction, so they do not close the door on Lamarckian epigenetic inheritance; but in molecular biology, citing a single work serves to leave the existence of the phenomenon in doubt, thus leaving the licensing of modeling efforts uncertain as well.

I focused on this article from 2001 because it seems to be a pivotal time in the history of molecular epigenetics. The basic phenomena of molecular epigenetics are discoveries of the late 1970s and 1980s; their articulation into a theory is more or less a product of the 1990s, especially with the development of bisulfite sequencing methods and a broadening survey of taxa for methylation (and other epigenetic) mechanisms by application of bioinformatics tools to the growing list of sequenced taxa. In the early 2000s, it appears that epigenetic mechanisms for gene regulation and especially for the maintenance of differentiated cell types in development became widely accepted in the molecular community (e.g., included in textbooks such as Turner, 2001). The potential evolutionary significance of epigenetic inheritance dawned on many in the molecular epigenetics community as the evidence for transmission through meiosis mounted, particularly in *Drosophila*, yeast, mice, and humans (Klar, 1998). In 2001, it made sense to be a bit conservative about evolutionary implications, particularly because the

one line of work can have a bearing on the evaluation of relative significance in the other.[14] For example, adding a mechanism for transmission of epigenetic marks through meiosis to the molecular theory can be conservative, risk-averse work, while it could provoke transformative research in evolutionary biology to specify how epigenetic inheritance changes heritability rather than merely instantiates it. Conversely, evaluations of evolutionary models showing that epigenetic inheritance is unlikely to be effective in adaptive evolution (Pál and Hurst, 2004) may have little impact on molecular investigations of transgenerational epigenetic inheritance since such mechanisms can still be deemed significant for evolution by molecular biologists on their standard of demonstration of transmission through meiosis or "maternal" effects.

What one line of work sees as risky another may not, so the actions of risk-averse parsimonists and risk-tolerant heuristic reductionists can induce patterns of research investment in other lines of work due to differing relative significance assessments. Reconciliation of this kind of "entangled" disagreement between lines of work may be central to scientific arenas such as evo–devo and epigenetic inheritance, but reconciliation is a process beyond the scope of this chapter.

It may be that the continued operation of "laws" of biology undermines the stability of the very properties we appeal to in biological explanation (Beatty, 1995). Nevertheless, relative significance debates even here depend on there being differences of investment in research strategies because of the ascertainment "biases" that affect claims about extent of applicability.

Jacob and Monod got away with their significance claim at a time when investments had not been made to discover alternative mechanisms. Such differences are likely to arise between any two research specializations that have overlapping domains of interest. I agree with Mitchell (2003) that the relative significance of causes, phenomena, and mechanisms is a pragmatic matter. As a matter of investment strategy, relative significance will almost certainly be specific to lines of work, specialties, and disciplines.

From an investment point of view, the question is, Should a molecular lab conduct empirical research on epigenetic effects, incorporate epigenetics results of other labs into its explanations, or include nongenetic mechanisms of inheritance in its theories; or should it stick to better-understood mechanisms of gene regulation and expression in development and rely on traditional models of transgenerational inheritance through the genetic inheritance system? Should evolutionary biologists add new variables or build new state spaces in their theoretical work and redesign experiments and field protocols to accommodate possible novel evolutionary forces due to epimutation, environment-inducible epigenetic variation, and nongenetic transmission; or should they stick to better-understood accounts and mechanisms of DNA sequence change, genetic variation, phenotypic variation, environmental variation, and Mendelian patterns of genetic transmission?

Consider two kinds of investment in new phenomena: (1) the exploration of new mechanisms and quantities from the perspective of established research programs and (2) the transformation of established research programs into substantially new ones in light of the discovery of new mechanisms. For molecular biologists, the key issue has been how epigenetic mechanisms can be added to established research on gene expression in regard to direct or indirect interactions with DNA, a nexus of genetic inheritance and development. Whether epigenetic mechanisms also constitute new, nongenetic inheritance systems might be a transformative question for established molecular research

---

14. The thought here is that evolutionists can see a particular epigenetic mechanism as potentially significant so long as there is no evidence that it cannot be transgenerationally inherited. Since potential significance calls for investment in evolutionary biology but not so much in molecular biology, lack of investment in molecular biology can drive investment decisions in theoretical evolutionary biology and shifts in investment patterns in molecular biology could lead to investment in evolutionary biology drying up.

programs, but only if it is understood to challenge the underlying assumption that inheritance is exclusively genetic and grounded in mechanisms of nucleic acid sequence variation, replication, transcription, and transmission. The idea that there are inheritance systems beyond the genes would thus place epigenetic mechanisms in the context of a new and quite different research program only if genetic inheritance mechanisms had to be revised in light of epigenetic discoveries.

These two kinds of investment, putting new epigenetic mechanisms in old theoretical bottles and fashioning new research programs in light of new mechanisms, result in quite a different risk trade-off for evolutionary biologists. The question of whether epigenetic mechanisms are nongenetic inheritance systems calls conservatively for putting the new mechanisms, as heritability instantiators, into the framework of Darwinian theory.

The crux of the difference with molecular biology over the significance of new epigenetic phenomena and how to put them in old theoretical bottles is how much and in what ways to invest in possibilities. Above, I characterized causal possibilities in terms of Darwin's argument for evolution by natural selection. For the evolutionist, if a mechanism exists, it is potentially significant because if it can be shown in theory to be competent, then it is important to test for its responsibility in nature. Thus, establishment of a mechanism in a molecular line of work in a risk-averse, conservative way can result in investment in risk-tolerant, transformative theoretical research in evolutionary theory prior to establishing the molecular competence of the mechanism to produce evolutionary effects. What is mere phenomenon in molecular biology is potentially transformative for evolutionary theory.

Turning to the second kind of investment, epigenetics describes a variety of regulatory states and effects—phenotypes. These have rather different properties from the sorts of states treated by classical Darwinian biology, ones which cannot be adequately formalized while keeping the "black box" of development closed. From the evolutionary perspective, this has meant treating the mapping from genotype to phenotype purely abstractly and mathematically rather than in terms of detailed empirical evidence of the operation of molecular mechanisms. If gene regulation by epigenetic networks operates essentially through interaction with environments (a Lamarckian dimension in which changing environments cause changes in organism "needs"), it may require transformation of the Darwinian program of evolutionary research in so far as epigenetic variations may be neither "random" with respect to fitness consequences nor stable on the time scale that sequence variations are.

For molecular biology, epigenetic mechanisms are just more items to be considered for the list of regulators of gene expression, even if they are nonrandomly influenced by environments. Eco–devo (Gilbert, 2001; Sultan, 2007) need not be eco–evo–devo. On the other hand, the idea of transgenerational epigenetic inheritance may challenge the core theory (or dogma) that inheritance is exclusively genetic. For evolutionary biology, epigenetic inheritance could be just one more kind of mechanism potentially on the lists of producers of variability and instantiators of heritability (if the heuristic reduction strategy worked), while the idea that epigenetic states are, by their nature, produced adaptively in response to changing environments challenges the core neo-Darwinian assumption that inheritable variation is random and stable.

Skepticism of the existence, competence, and responsibility of epigenetic inheritance from within QDES amounts to an assumption that epigenetic inheritance either does not occur or is, in principle, reducible to genetic theory; hence, there is no justification for research investment. For molecular epigeneticists, only mechanisms that are not only demonstrated to exist but shown to be empirically competent to produce effects are significant enough to put on the theory list. This difference in where potential significance gets a purchase—existence and theoretical competence (for QDES) versus empirical competence (for MMS)—is at the root of an asymmetry between molecular and evolutionary lines of work, making relative significance arguments between them complex.

Moreover, investment strategies for theory construction in the two lines of work depend on rather different conceptions of theory and roles for mechanisms in theory structure. Reductionism and parsimony lead scientists in many lines of work to explain the new in terms of the old—it is a conservative strategy. However, the concept of reductionism has too often been framed retrospectively, in terms of the logical, explanatory relations between theories and data or between more and less general or higher- and lower-level theories. Science, like business, is forward-looking and therefore much more concerned with investment strategies and conditions that facilitate or inhibit the work yet to be done. Reductionism in science is better viewed as a heuristic strategy than a logic of explanation (Griesemer 2000, 2002b), and it should not be confused with mechanistic research per se (see Brandon, 1996, chap. 11). Scientific controversies around issues of reductionism are therefore better viewed as strategic rather than reflective.

The challenge of heuristic reduction of inheritance phenomena for quantitative dynamical evolutionary theory is to construct a version of evolutionary theory that admits epigenetic inheritance as a mechanism for heritability which could reduce the classical, exclusively genetic inheritance evolutionary theory. It is heuristic in so far as the construction relies on features of the genetic inheritance system to search heuristically for those features of epigenetic phenomena that provide a possible mechanism for epigenetic heritability. If the theory is successful, a new wave of reductionist research with a new, more powerful theory of inheritance would result. But if it fails, we might better learn how to go about transforming evolutionary theory into something quite new and different.

CONCLUSION • The Irony of the Relative Significance of Epigenetic Inheritance in Evolution

The foregoing considerations of theory structure, investment strategy, and heuristic reduction suggest that relative significance arguments about the potential role of epigenetic inheritance in evolution are arguments across lines of work about whether to invest in research and in what type of research effort to invest. There is an irony in the case I have used to illustrate these ideas, but I think it makes my point in the end. At this moment in the history of research into epigenetic inheritance, evolutionary theorists are mostly skeptical or uninterested. Only a few have paid much attention to what is going on in molecular epigenetics. On the other hand, the discoveries of molecular epigeneticists like Holliday, Klar, Jablonka and Lamb, and Whitelaw and colleagues suggest potential evolutionary significance for transgenerational epigenetic inheritance.

However, I have been arguing for just the opposite, in that we would expect, in my account, that evolutionary theorists would argue for the potential theoretical significance of epigenetic inheritance, as a justification of investment of research effort, on the grounds that if the phenomena exist, we need to find out if in principle epigenetic inheritance could be competent to cause evolutionary change. Yet evolutionists resist. Molecular epigeneticists are the ones who are supposed to resist putting mechanisms on their theory lists unless empirical competence for the effects is already demonstrated, yet they have been pushing the idea that epigenetic mechanisms could be inheritance systems with each step toward recognizing transmission of epigenetic marks through meiosis in an ever wider array of taxa and circumstances. What gives?

One answer is that it is easy to be risk-tolerant with respect to the heuristic reduction of traditional genetics-based evolutionary theory by some sort of generalized inheritance theory if you are a molecular epigeneticist expecting only to be a consumer of evolutionary theory who does not have to sweep up the broken equations after it fails. To study the fragments of failed conservative attempts to articulate epigenetic mechanisms within the structure of existing theory or to pursue the transformative consequences of constructing alternatives to the standard model would take a major investment. Only risk-tolerant theorists looking for a transformative theory would find it worth the risk. (Or maybe only philosophers oblivious to risk would since their line of work protects them from abject failure.)

Likewise, it is easy to be risk-tolerant with respect to the heuristic reduction of traditional genetics-based molecular theory if you are an evolutionary biologist expecting only to be a consumer of molecular theory who does not have to sweep up the broken glass after a failed empirical program. To study the fragments of failed conservative attempts to specify molecular theory by adding mechanisms and interactions without upsetting current understanding of how known mechanisms work would take a major investment. Only a risk-tolerant molecular biologist looking for a transformative theory would find it worth the risk, and there are no philosophers willing to do the lab work; thus, it seems that molecular biologists are immune to the directives of evolutionary theorists who call for this or that possible, but yet to be demonstrated, mechanism to be investigated. That ease, of urging risk tolerance in the other research program and risk aversion in one's own, is, I believe, the basis for relative significance arguments about research investment. You know better not only how hard the specification work is but also what the consequences of heuristic reduction failures are like in your line of work.

Risk-averse, conservative epigenetics research aims to fill in details of molecular epigenetic mechanisms, exploring how different mechanisms may interact to produce the kinds of phenomena already familiar to the field: maintenance of differentiated cell types through cell division, silencing of foreign DNA, chromosomal imprinting, role in disease etiology. Risk-averse, transformative research could be called "accidental" or "serendipitous." Making a mistake in the lab and noticing it so as to see that an unexpected phenomenon results from the mistake, if it leads to addition of a mechanism that forces reconsideration of existing ones, can be transformative, although the aim of the research investment was to avoid taking on risk by doing things "as usual." The best "normal science" is conservative but risk-tolerant. This is the kind of science that Wimsatt (2007) describes in detail as following an elaborate set of heuristic research strategies. Heuristics are rules of thumb that can guide action but, unlike an algorithm or deductive procedure, do not guarantee a correct result or even a result at all. More importantly, Wimsatt (1987) emphasizes that in scientific research following heuristic strategies in the risk-tolerant sense means using "false models" as a means to "truer theories." That is, the researcher expects the hypothesis, model, theory, or experiment to fail in a literal sense and the main payoff is that heuristic failure tends to be systematic rather than random. The particular pattern of systematic failure is a fruitful failure—one that can guide research to a better understanding. It is the ethos of the engineer who works in exemplars and concrete structures rather than the physicist who works in universal laws of nature. However, the consideration of whether epigenetic inheritance is causally significant in evolution seems to demand risk-tolerant, transformative research. It seems to require consideration of new kinds of mechanisms which would disrupt the rules by which the dynamical theory of neo-Darwinian evolution was put together. Although the most sensible approach would be heuristic reductionism—seeking to use as much of the structure of current theory to guide construction of a theory that can accommodate epigenetic inheritance and reduce (in a traditional sense) genetic neo-Darwinism—both require such major investments that few evolutionists are ready to try.

Thus, despite the irony of molecular biologists arguing for evolutionary significance and evolutionists resisting, I think the main conclusion still stands. Because Darwinian evolutionary theory is framed in terms of a requirement for a capacity or disposition of heritability, it leaves open questions of which inheritance mechanisms realize heritability in nature and the possibility that different mechanisms are involved in different instances. Because molecular biology is concerned with mechanisms that frequently or generally operate in gene regulation and development (though not without interest in variation), it leaves open questions of which mechanisms are significant in evolution and in what ways. Evolutionary biology is concerned with both the distribution of traits (including the presence of mechanisms) among taxa and the dynamical role of mechanisms in evolutionary processes of change within and among lineages. Thus, in asking questions about reductionism and relative significance of epigenetic phenomena in relation to well-established theories of genetic inheritance, we have to ask, "Reduction for whose purpose(s)?" and "Significance relative to whose problem(s)?"

ACKNOWLEDGMENTS

Thanks to Elihu Gerson and Chris di Teresi for comments on the manuscript; Steve Tonsor, Sue Kalisz, and Rasmus Winther for helpful discussion; and the Center for Philosophy of Science, University of Pittsburgh, for providing a critical audience.

REFERENCES

Arnold, S., and M. Wade. 1984. On the measurement of natural and sexual selection: Theory. *Evolution* 38:709–19.

Beatty, J. 1995. The evolutionary contingency thesis. In *Concepts, Theories and Rationality in the Biological Sciences*, ed. G. Wolters and J. Lennox, 45–81. Pittsburgh: University of Pittsburgh Press.

Beatty, J. 1997. Why do biologists argue like they do? *Philos Sci* 64 (Proceedings): S432–43.

Bechtel, W. 2006. *Discovering Cell Mechanisms: The Creation of Modern Cell Biology*. New York: Cambridge University Press.

Bestor, T. 1990. DNA methylation: Evolution of a bacterial immune function into a regulator of gene expression and genome structure in higher eukaryotes. *Philos Trans R Soc Lond B* 326:179–87.

Bossdorf, O., C. Richards, and M. Pigliucci. 2008. Epigenetics for ecologists. *Ecol Lett* 11:106–15.

Brandon, R. 1996. *Concepts and Methods in Evolutionary Biology*. Cambridge: Cambridge University Press.

Brink, R. A. 1956. A genetic change associated with the R locus in Maize which is directed and potentially reversible. *Genetics* 41:872–89.

Brink, R. A. 1960. Paramutation and chromosome organization. *Q Rev Biol* 35 (2): 120–37.

Buss, L. W. 1987. *The Evolution of Individuality*. Princeton, NJ: Princeton University Press.

Cavalli, G., and R. Paro. 1998. The *Drosophila* Fab-7 chromosomal element conveys epigenetic inheritance during mitosis and meiosis. *Cell* 93:505–18.

Crick, F. 1958. On protein synthesis. *Symposia of the Society for Experimental Biology* 12:138–63.

Crick, F. 1970. Central dogma of molecular biology. *Nature* 277: 561–63.

Darwin, C. 1859. *On the Origin of Species*. Facsimile of the 1st edition, 1964. Cambridge, Mass.: Harvard University Press.

Darwin, C. 1871. *The Descent of Man*. London: John Murray.

Gallou-Kabani, C., and C. Junien. 2005. Nutritional epigenomics of metabolic syndrome: New perspective against the epidemic. *Diabetes* 54:1899–1906.

Gerson, E. 1976. On quality of life. *American Sociological Review* 41:793–806.

Gilbert, S. F. 2001. Ecological developmental biology: Developmental biology meets the real world. *Dev Biol* 233:1–12.

Gilbert, S. F., and D. Epel. 2009. *Ecological Developmental Biology: Integrating Epigenetics, Medicine, and Evolution*. Sunderland, MA: Sinauer Associates.

Griesemer, J. 2000. Reproduction and the reduction of genetics. In *The Concept of the Gene in Development and Evolution, Historical and Epistemological Perspectives*, ed. P. Beurton, R. Falk, and H.-J. Rheinberger, 240–85. New York: Cambridge University Press.

Griesemer, J. 2002a. What is "epi" about epigenetics? In *From Epigenesis to Epigenetics: The Genome in Context*, ed. G. Vandevijver, L. Vanspeybroeck, and D. Dewaele. *Ann N Y Acad Sci* 981:97–110.

Griesemer, J. 2002b. Limits of reproduction: A reductionistic research strategy in evolutionary biology. In *Promises and Limits of Reductionism in the Biomedical Sciences*, ed. M. H. V. Van Regenmortel and D. Hull, 211–31. Chichester, UK: John Wiley & Sons.

Griesemer, J. 2006. Genetics from an evolutionary process perspective. In *Genes in Development*, ed. Eva M. Neumann-Held and Christoph Rehmann-Sutter, 199–237. Durham, NC: Duke University Press.

Griesemer, J. 2007. Tracking organic processes: Representations and research styles in classical embryology and genetics. In *From Embryology to Evo-Devo: A History of Developmental Evolution*, ed. M. Laubichler and J. Maienschein, 375–433. Cambridge, MA: MIT Press.

Griesemer, J. 2008. Philosophy and tinkering. *Biol Philos* doi: 10.1007/s10539-008-9131-0.

Griesemer, J., and M. Wade. 1988. Laboratory models, causal explanation and group selection. *Biol Philos* 3:67–96.

Griesemer, J., and W.C. Wimsatt. 1989. Picturing Weismannism: A case study of conceptual evolution. In *What the Philosophy of Biology Is, Essays for David Hull*, ed. M. Ruse, 75–137. Dordrecht, the Netherlands: Kluwer Academic Publishers.

Haig, D. 2002. *Genomic Imprinting and Kinship*, chap. 6. Piscataway, NJ: Rutgers University Press.

Haig, D., and M. Westoby. 1989. Parent-specific gene expression and the triploid endosperm. *Am Nat* 134 (1): 147–55.

Hall, B.K. 1998. Epigenetics: Regulation not replication. *J Evol Biol* 11:201–205.

Hodge, M. 1977. The structure and strategy of Darwin's 'long argument'. *J Hist Sci* 10:237–46.

Holliday, R. 2006. Epigenetics: A historical overview. *Epigenetics* 1 (2): 76–80.

Hurst, L., and G. McVean. 1998. Do we understand the evolution of genomic imprinting? *Curr Opin Genet Dev* 8:701–708.

Jablonka, E. 2004. Epigenetic epidemiology. *Int J Epidemiol* 33:1–7.

Jablonka, E., and M. Lamb. 1989. The inheritance of acquired epigenetic variations. *J Theor Biol* 139 (1): 69–83.

Jablonka, E., and M. Lamb. 1995. *Epigenetic Inheritance and Evolution*. Oxford: Oxford University Press.

Jablonka, E., and M. Lamb. 1998. Epigenetic inheritance in evolution. *J Evol Biol* 11 (2): 159–83.

Jablonka, E., and M. Lamb. 2005. *Evolution in Four Dimensions: Genetic, Epigenetic, Behavioral, and Symbolic Variations in the History of Life*. Cambridge, MA: MIT Press.

Jablonka, E., and E. Szathmáry. 1995. The evolution of information storage and heredity. *Trends Ecol Evol* 10 (5): 206–11.

Jablonka, E., M. Lachmann, and M. Lamb. 1992. Evidence, mechanisms and models for the inheritance of acquired characters. *J Theor Biol* 158:245–68.

Jablonka, E., B. Oborny, I. Molnar, E. Kisdis, J. Hofbauer, and T. Czaran. 1995. The adaptive advantage of phenotypic memory in changing environments. *Philos Trans R Soc Biol Sci* 350 (1332): 133–41.

Kaati, G., L.O. Byorgen, and S. Edvinsson. 2002. Cardiovascular and diabetes mortality determined by nutrition during parents' and grandparents slow growth period. *Eur J Hum Genet* 10:682–8.

Kauffman, S.A. 1971. Articulation of parts explanations in biology and the rational search for them. *Boston Studies in the Philosophy of Science* 8: 257–72.

Keller, A.D. 1995. Fixation of epigenetic states in a population. *J Theor Biol* 176:211–19.

Klar, A. 1998. Propagating epigenetic states through meiosis: Where Mendel's gene is more than a DNA moiety. *Trends in Genet* 14:299–301.

Kuhn, T. 1970. *The Structure of Scientific Revolutions* (2nd ed). Chicago: University of Chicago Press.

Lachmann, M., and E. Jablonka. 1996. Inheritance of phenotypes: An adaptation to fluctuating environments. *J Theor Biol* 181:1–9.

Lamm, E., and E. Jablonka. 2008. The nurture of nature: Hereditary plasticity in evolution. *Philos Psychol* 21 (3): 305–19.

Levins, R. 1968. *Evolution in Changing Environments*. Princeton: Princeton University Press.

Lewontin, R. 1970. The units of selection. *Annual Review of Ecology and Systematics* 1:1–17.

Lewontin, R. 1974. *The Genetic Basis of Evolutionary Change*. New York: Columbia University Press.

Lewontin, R., and L. Dunn. 1960. The evolutionary dynamics of a polymorphism in the house mouse. *Genetics* 45:705–22.

Lyku, F. 2001. DNA methylation learns to fly. *Trends Genet* 17:169–72.

Mayer, W., A. Niveleau, J. Walter, R. Fundele, and T. Haaf. 2000. Demethylation of the zygotic paternal genome. *Nature* 403:501–502.

Maynard Smith, J. 1990. Models of a dual inheritance system. *J Theor Biol* 143:41–53.

Maynard Smith, J., and E. Szathmáry. 1995. *The Major Transitions in Evolution*. Oxford: W.H. Freeman Spektrum.

Maynard Smith, J., and E. Szathmáry. 1999. *The Origins of Life: From the Birth of Life to the Origin of Language*. Oxford: Oxford University Press.

Michod, R. 1999. *Darwinian Dynamics, Evolutionary Transitions in Fitness and Individuality*. Princeton: Princeton University Press.

Mitchell, S.D. 1992. On pluralism and competition in evolutionary explanations. *Am Zool* 32: 135–44.

Mitchell, S. D. 2003. *Biological Complexity and Integrative Pluralism*. New York: Cambridge University Press.

Nanney, D. L. 1968. Cortical patterns in cellular morphogenesis. *Science* 160 (3827): 496–502.

NESCent. 2009. http://www.nescent.org/cal/calendar_detail.php?id=282.

Newman, S. A., and G. B. Müller. 2000. Epigenetic mechanisms of character origination. *J Exp Zool* 288:304–17.

Novick, A., and M. Weiner. 1957. Enzyme induction as an all-or-none phenomenon. *Proc Natl Acad Sci USA* 43:553–66.

Odling-Smee, F. J., K. N. Laland, and M. W. Feldman. 2003. *Niche Construction: The Neglected Process in Evolution*. Princeton, NJ: Princeton University Press.

Pál, C., and L. Hurst. 2004. Epigenetic inheritance and evolutionary adaptation. In *Organelles, Genomes and Eukaryotic Phylogeny: An Evolutionary Synthesis in the Age of Genomics*, ed. R. P. Hirt and D. S. Horner, 353–70. London: Taylor & Francis.

Pigliucci, M. 2007. Do we need an extended evolutionary synthesis? *Evolution* 61 (12): 2743–9.

Rakyan, V., and E. Whitelaw. 2003. Transgenerational epigenetic inheritance. *Curr Biol* 13 (1): R6.

Rakyan, V., J. Preis, H. Morgan, and E. Whitelaw. 2001. The marks, mechanisms and memory of epigenetic states in mammals. *Biochem J* 356:1–10.

Ruse, M. 1971. *The Darwinian Revolution*. Cambridge: Cambridge University Press.

Skipper, R., and R. Millstein. 2005. Thinking about evolutionary mechanisms: Natural selection. *Stud Hist Philos Biol Biomed Sci* 36:327–47.

Star, S. and E. Gerson. 1986. The management and dynamics of anomalies in scientific work. *Sociological Quarterly* 28:147–69.

Sultan, S. E. 2007. Development in context: The timely emergence of eco–devo. *Trends Ecol Evol* 22:575–82.

Turner, B. M. 2001. *Chromatin and Gene Regulation: Molecular Mechanisms in Epigenetics*. Oxford: Blackwell.

van der Weele, C. 1999. *Images of Development: Environmental Causes in Ontogeny*. Albany: State University of New York Press.

Van Speybroeck, L., G. Van de Vijver, and D. De Waele, eds. 2002. From Epigenesis to Epigenetics: The Genome in Context. *Ann N Y Acad Sci* 981 (whole issue).

Wade, M. 1978. A critical review of the models of group selection. *Quarterly Review of Biology* 53: 101–14.

Wade, M. 1985. Soft selection, hard selection, kin selection and group selection. *Am Nat* 125:61–73.

Walsh, J. B. 1996. The emperor's new genes. *Evolution* 50 (5): 2115–18.

Waters, C. K. 1998. Causal regularities in the biological world of contingent distributions. *Biol Philos* 13:5–36.

Wimsatt, W. C. 1987. False models as means to truer theories. In *Neutral Models in Biology*, ed. M. Nitecki and A. Hoffman, 23–55. London: Oxford University Press.

Wimsatt, W. C. 2007. *Re-Engineering Philosophy for Limited Beings: Piecewise Approximations to Reality*. Cambridge, MA: Harvard University Press.

Winther, R. G. 2006. Parts and theories in compositional biology. *Biol Philos* 21:471–99.

Wolpert, L. 1998. Comments on "Epigenetic inheritance in evolution." *J Evol Biol* 11:239–40.

PART TWO

Approaches to Epigenetics

4

The Epigenetics of Genomic Imprinting

CORE EPIGENETIC PROCESSES ARE CONSERVED IN MAMMALS, INSECTS, AND PLANTS

Lori A. McEachern and Vett Lloyd

CONTENTS

WHAT IS GENOMIC IMPRINTING?

Genomic imprinting is an epigenetic process in which an allele is marked according to the sex of the parent transmitting it. These sex-specific marks may affect single genes, gene clusters, or entire chromosomes and result in maternal and paternal alleles or chromosomes that are epigenetically distinct from one another. This difference in epigenetic status can lead to differential transcriptional activity, chromosome loss, or chromosome inactivation. In an organism, allelic differences that result from genomic imprinting can be observed as the exclusive or preferential expression of a gene when it is inherited from one parent but not the other. Thus, in contrast to classic mechanisms of gene expression and regulation, in genomic imprinting it is the allele's parent-of-origin, and not the underlying DNA sequence, that determines its activity. In this chapter, we therefore use the term *epigenetic* in the limited but specific sense employed in molecular biology to mean processes that affect gene expression without changing DNA sequence.

Genomic imprinting has been most extensively studied in mammals (Morison et al., 2005; Wood and Oakey, 2006) but has been observed in a wide range of organisms, including plants (Alleman and Doctor, 2000; Scott and Spielman, 2006), insects (Khosla et al., 2006; Lloyd, 2000), *Caenorhabditis elegans* (Bean et al., 2004), and zebrafish (Martin and McGowan, 1995). Many similarities exist in the epigenetic mechanisms that underlie genomic imprinting in these species. These mechanisms include, but are not limited to, DNA methylation, histone modifications, changes in higher-order chromatin structure, noncoding RNA, and RNA interference (RNAi). Accumulating evidence suggests that these epigenetic mechanisms are frequently interrelated and mutually reinforcing.

DNA methylation is an epigenetic process in which methyl groups are added to nucleotides, often cytosines present in CpG dinucleotides, without affecting the underlying DNA sequence. When DNA methylation encompasses the promoter of a gene, it frequently results in transcriptional repression. However, methylation-sensitive enhancers, repressors, and protein-binding sequences are also common and important in mediating epigenetic gene expression.

In the cell, DNA is wrapped around *nucleosomes,* protein structures that consist of two copies of four different histone proteins (H2A, H2B, H3, and H4). Chemical modifications including methylation, acetylation, and phosphorylation of amino acids in the histone sequences can contribute to the formation of inactive or active chromatin structures (Figure 4.1A). Evidence suggests that DNA methylation and histone modifications are intimately linked and exhibit extensive epigenetic "crosstalk," with information flowing from DNA to histones and from histones to DNA. Given that DNA methylation can guide histone modifications and histone modifications can influence DNA methylation, it is likely that these processes function in a mutually reinforcing epigenetic loop that ensures maintenance of a repressive chromatin state (Fuks, 2005; Vaissiere et al., 2008).

Epigenetic gene regulation by noncoding RNAs can involve RNAi-mediated pathways, in which the noncoding RNAs are processed into small RNAs but can also be RNAi-independent. RNAi-mediated epigenetic gene regulation may occur at the posttranscriptional level, with the small RNAs guiding degradation of an mRNA transcript or inhibiting its translation; or it may occur at the transcriptional level, with the small RNAs mediating chromatin modifications that inhibit transcription. The exact function of many noncoding RNAs remains elusive, but a variety of evidence connects noncoding RNAs with other epigenetic mechanisms, including histone modifications, DNA methylation, and heterochromatin formation (Bernstein and Allis, 2005; Matzke and Birchler, 2005; Zaratiegui et al., 2007). Thus, it is likely that many noncoding RNA transcripts are important in mediating higher-order chromatin structure. For example, in mammalian X-chromosome inactivation, a 17 kb noncoding RNA called *Xist* is expressed from the X chromosome that will be inactivated and subsequently coats that chromosome. This initiates a variety of chromatin remodeling events, including histone modifications and the incorporation of a specialized histone variant, which ensure silencing of the inactive X (Bernstein and Allis, 2005).

Recent work in yeast has revealed further details of the mechanism by which RNAi can direct heterochromatin formation. This analysis demonstrated that transcripts from heterochromatic regions of the genome accumulate during

A

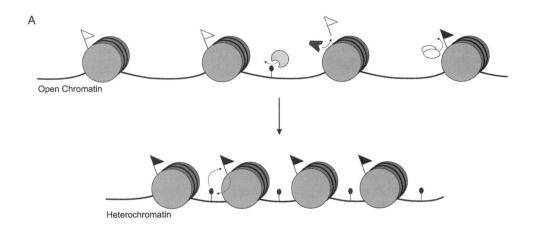

Open Chromatin

Heterochromatin

B

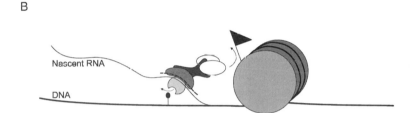

Nascent RNA

DNA

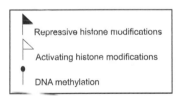

Repressive histone modifications

Activating histone modifications

DNA methylation

FIGURE 4.1 Genomic imprinting utilizes several interrelated and conserved epigenetic mechanisms. (A) Epigenetic gene regulation is often mediated by histone and DNA modifications that contribute to higher-order chromatin structure. In the cell, DNA is wrapped around nucleosomes (gray cylinders), each of which contains eight histone proteins. Histones can acquire activating or repressive modifications. Activating histone modifications lead to an open chromatin structure that promotes gene transcription. Removal of activating histone modifications and the addition of repressive histone modifications and DNA methylation lead to a condensed heterochromatic structure that hinders transcription. DNA methylation and histone modifications are often mutually reinforcing, with each modification influencing and contributing to the other. (B) Noncoding and small RNAs can regulate gene expression and contribute to the formation of a compacted heterochromatin structure by directing histone modifications and DNA methylation. Illustrated is a possible model for small RNA–mediated chromatin modifications. Small RNAs (dashed line) interact with a nascent transcript at the target gene locus according to sequence homology and tether chromatin-modifying enzymes to the locus. These may include DNA methyltransferases (light gray "Pacman") or histone-modifying enzymes (white ellipses). These RNA-directed DNA and histone modifications may then alter the chromatin structure of the locus.

S phase of the cell cycle and are processed into short interfering RNAs (siRNAs), which then recruit histone methylation that contributes to heterochromatin formation (Kloc et al., 2008).

RNAi-mediated heterochromatin formation has also been reported in plants and animals, and de novo DNA methylation in plants is RNA-directed (Lippman and Martienssen, 2004;

Matzke and Birchler, 2005; Zaratiegui et al., 2007). How small or noncoding RNAs direct DNA modifications is not yet fully understood, but recent evidence in yeast favors a model in which small RNA interaction with nascent RNA transcripts recruits a complex of chromatin- and DNA-binding and modifying proteins (Figure 4.1B) (Buhler et al., 2006; Irvine et al., 2006). The discovery that DNA methylation of a group of genes in *Arabidopsis* is directed by a small RNA that targets an exon–exon junction also supports an RNA–RNA interaction model (Bao et al., 2004). Alternatively, small or non-coding RNAs may direct these modifications via base-pairing interactions with genomic DNA (Grewal and Moazed, 2003; Mayer et al., 2006).

The observation of imprinting in such an extensive range of animal and plant species, combined with the utilization of many of the same mechanisms to establish and maintain imprinted expression, suggests that genomic imprinting is a widespread occurrence based on phylogenetically conserved core epigenetic processes that can be adapted to serve different functions in different species. The conservation of these epigenetic processes is further emphasized by transgenic experiments in which an epigenetic control region from one organism successfully functions in another. In this chapter we will examine examples of genomic imprinting from mammals, insects, and plants, with a focus on imprinting mechanisms and the conservation of core epigenetic processes.

THE EVOLUTION OF GENOMIC IMPRINTING

Genomic imprinting has been most extensively studied in mammals, and thus, the majority of hypotheses about the selective forces leading to imprinted gene expression are based on mammalian imprinted genes. The *parental conflict hypothesis* is the most thoroughly debated and is based on a reproductive mode involving multiple paternity within a litter of mammals (Moore and Haig, 1991). In this reproductive scenario, it is beneficial to the mother to distribute nutri-

ents evenly to her offspring as all share her genes, while it is in the father's genetic interests for only his offspring to receive maximal resources. This hypothesis predicts that paternal imprinting should enhance the expression of fetal growth promoters, while maternal imprinting should have the opposite effect. While the early identification of several imprinted genes involved in fetal growth resulted in great enthusiasm for this hypothesis, the discovery of imprinted genes with a variety of functions makes it increasingly unlikely that the parental conflict hypothesis can account for the imprinting of all genes.

The *ovarian time bomb hypothesis,* based on the supposition that genomic imprinting was selected to prevent ovarian trophoblastic disease (Varmuza and Mann, 1994) and the role of imprinting in preventing parthenogenesis, has also been debated. In mice, imprinting is a major barrier to parthenogenesis (Kono, 2006), and uniparental inheritance of chromosomal regions containing imprinted genes can lead to embryonic lethality and postnatal growth and developmental defects (Cattanach and Kirk, 1985). However, at least six imprinted genes have been identified in chromosomal regions that have no obvious phenotype when inherited uniparentally (Peters and Beechey, 2004), and additional imprinted genes have been shown to have only behavioral or cognitive effects (Davies et al., 2005; Plagge et al., 2005). Thus, it is likely that imprinting affects a wide range of genes in the mouse and other mammals, not only those involved in growth and development. Prevention of parthenogenesis or ovarian trophoblastic disease is therefore unlikely to be the selective force behind the imprinting of all mammalian genes.

Additional hypotheses relate to the benefit of imprinting in establishing appropriate gene dosages or functional haploidy (Holliday, 1990; Ohlsson et al., 2001; Okamura and Ito, 2006), enhancing the "evolvability" of a population (Beaudet and Jiang, 2002) and modifying expression of genes that have different optimal expression levels and selective pressures in males and females (Day and Bonduriansky,

2004; Iwasa and Pomiankowski, 1999). Overall, as each new imprinted gene is discovered and as imprinting is discovered in animals and plants with reproductive strategies differing from those in mammals, it seems increasingly unlikely that a single evolutionary hypothesis will explain the occurrence of all imprinted genes, even within a single species. It is more likely that a variety of selective forces have contributed to the evolution of imprinted gene expression.

GENOMIC IMPRINTING IN MAMMALS

Genomic imprinting has been observed in the eutherian (Khatib et al., 2007) and marsupial (Suzuki et al., 2005) lineages of mammals and is most frequently studied in mice and humans. Imprinting in humans has been of particular interest to the medical and research communities due to the association of imprinted genes and aberrant imprinting with a variety of diseases and cancers. In both mice and humans, many of the identified imprinted genes occur in clusters that contain shared regulatory regions and/or transcripts that control the imprinted expression of multiple genes in the cluster. Imprinted genes that reside within introns or have originated from retrotransposition events are also common (Morison et al., 2005). Many of the identified mammalian imprinted genes are not imprinted in all tissues at all times, indicating that tissue- and time-specific imprinting may be frequent and may complicate the identification of imprinted genes with unique patterns of expression.

At the molecular level, the protein-coding imprinted genes identified in mice and humans participate in a wide range of cellular processes, with no obvious function or theme in common (Morison et al., 2005; Peters and Beechey, 2004). In addition, approximately 30% of imprinted genes correspond to noncoding RNA transcripts, many of which are involved in regulating the imprinted expression of other genes. The imprinted genes influenced by the transcript may overlap, be located nearby or at a distance from the transcript, and may be imprinted in the same or opposite direction. Other imprinted RNAs encode small nucleolar RNAs (snoRNAs) and microRNAs, and still others have no known function.

While there is much overlap between the genes that are imprinted in mice and humans, there are also significant differences. Several genes are reported to be imprinted in mice but not in humans or vice versa, and at least two genes are reported to be oppositely imprinted in the two species (*COPG2* and *ZIM2*). Additional genes are imprinted in one species but lack an orthologue in the other (Morison et al., 2005). These differences may suggest that, for many genes, the loss or gain of imprinting during a species' evolution may occur somewhat easily and without drastic effects. Thus, while the epigenetic processes underlying imprinting may be conserved, there likely exist many species-specific differences in the genes that are affected.

IMPRINTING AT THE *H19/IGF2* LOCUS

H19 and *Insulin-like growth factor 2* (*Igf2*) are perhaps the most extensively studied and best-characterized imprinted genes. Located approximately 90 kb apart, these two genes are reciprocally imprinted, with the noncoding *H19* transcript expressed only from the maternal allele and *Igf2* expressed only from the paternal allele (Bartolomei et al., 1991; DeChiara et al., 1991). Imprinted expression of *H19* and *Igf2* is controlled by a shared imprint control region (ICR) located approximately 2 kb upstream of the *H19* transcription start site. Deletion of this ICR results in a loss of imprinting of both genes (Thorvaldsen et al., 1998). In addition to the ICR, expression of both *Igf2* and *H19* requires several tissue-specific enhancers spread over at least three regions 10–120 kb downstream of the *H19* gene (Ainscough et al., 2000a; Davies et al., 2002; Kaffer et al., 2000; Leighton et al., 1995).

DNA METHYLATION

The *H19/Igf2* ICR contains multiple binding sites for the enhancer-blocking, insulator protein CTCF, which can bind only when these sites are unmethylated (Bell and Felsenfeld, 2000; Hark et al., 2000). Methylation of the ICR is present in sperm but not ova (Tremblay et al., 1995), enabling CTCF to bind to the maternally inherited ICR but not the paternally inherited ICR. CTCF binding to the unmethylated maternal ICR prevents the downstream enhancers from activating *Igf2*, and instead the enhancers stimulate transcription of *H19*. Conversely, CTCF cannot bind to the methylated ICR on the paternal allele, and the downstream enhancers activate expression of *Igf2* on the paternally inherited chromosome (Bell and Felsenfeld, 2000; Hark et al., 2000).

Once imprinted expression is established, the ICR is required to maintain *Igf2* silencing on the maternal chromosome but not *H19* silencing on the paternally inherited chromosome (Srivastava et al., 2000). The *H19* promoter acquires methylation on the paternal allele during embryogenesis (Bartolomei et al., 1993; Tremblay et al., 1995, 1997), which is likely sufficient to maintain its silenced state. Mutation or deletion of the CTCF-binding sites does not affect differential methylation of the ICR in sperm and ova, and thus, the methylated paternal allele imprints appropriately in the absence of CTCF-binding sites. However, in the absence of CTCF-binding sites on the maternal allele, the *H19* promoter and gene region acquire methylation postimplantation, *H19* expression is reduced, and *Igf2* is expressed *biallelically* (transcribed from both parental alleles) (Engel et al., 2006; Szabo et al., 2004).

CONSERVED EPIGENETIC MECHANISMS

The epigenetic processes that cause *H19/Igf2* imprinting appear to be conserved between mammals and *Drosophila*, as demonstrated by transgenic experiments. A silencer element within the mouse *H19* ICR was originally discovered using transgenic *Drosophila* containing the mouse *H19* upstream region adjacent to *lacZ* and mini-*white* reporter genes (Lyko et al., 1997). Deletion constructs delineated the silencing element to a 1.2 kb region (Lyko et al., 1997), and subsequent experiments showed

that this 1.2 kb element also functions specifically in *H19* silencing, and not imprinting, at the endogenous mouse locus (Drewell et al., 2000). Targeted deletion of the silencer resulted in a loss of *H19* silencing following paternal transmission, while paternal *Igf2* expression, differential methylation, and expression of both genes following maternal transmission were unaffected. A similar 1.5 kb silencer element appears to exist at the 3' end of the human *H19* ICR (Arney et al., 2006). This region silenced a mini-*white* reporter gene in transgenic *Drosophila* and functioned as a silencer in transient transfection assays using a human embryonic kidney cell line (Arney et al., 2006), while additional regions from the human ICR did not.

Additional insight into the complexity of the *H19* ICR and imprinting mechanism stems from recent evidence demonstrating that the mouse ICR is biallelically transcribed and produces both sense and antisense RNA (Schoenfelder et al., 2007). Biallelic transcription was also detected in the transgenic *Drosophila* system, where further analysis indicated that the *H19* ICR transcripts induce gene silencing in an RNAi-independent manner (Schoenfelder et al., 2007). In transgenic *Drosophila*, mutations in RNAi genes failed to relieve reporter gene silencing, and no siRNAs were detected from the *H19* ICR. Furthermore, artificially producing *H19* ICR siRNAs resulted in a significant reduction of *H19* ICR transcripts, which was accompanied by a more than fivefold increase in mini-*white* expression. In their endogenous context, these ICR transcripts may be involved in forming a repressive chromatin structure and mediating *H19* repression on the paternal allele. This would be similar to the model for imprinting at the mammalian *Cdkn1c-Kcnq1* imprinted domain, where a noncoding RNA transcript is believed to mediate a repressive chromatin structure on the paternal allele (Umlauf et al., 2004). Noncoding RNA transcripts are similarly required to establish and maintain a heterochromatic structure at a ribosomal RNA gene cluster in mice (Mayer et al., 2006), and evidence suggests that the higher-order chromatin structure of mouse pericentric heterochromatin involves an RNA component (Maison et al., 2002). On the maternal allele, CTCF binding may prevent the repressive effect of the *H19* ICR transcripts, or alternatively, the ICR transcripts could serve a different functional role that is undetected in the transgenic *Drosophila* system, which does not imprint and acts most similarly to the silenced paternal allele.

The central A6-A4 region—also termed the *centrally conserved domain*—is located between the two genes and is unmethylated, DNAseI-hypersensitive, and GC-rich (Koide et al., 1994). Within this region, two subregions show a high level of homology between humans and mice. Region 1 is necessary for maintaining repression of *Igf2* from the maternal allele in skeletal muscle (Ainscough et al., 2000b), while region 2 appears to be an enhancer for *Igf2* expression in the choroid plexus, where it is normally expressed biallelically (Jones et al., 2001). Analysis of transgenic *Drosophila* containing the central A6-A4 region (Erhardt et al., 2003) may provide additional insight into the mechanism of repression within this region. Transgenic flies containing this region adjacent to mini-*white* and *lacZ* show overall silencing of both reporter genes, as well as eye-pigment variegation in some lines, indicating the formation of compact chromatin domains (Erhardt et al., 2003). Silencing increased in *Enhancer of Zeste* [*E(z)*] mutant flies and decreased in *Posterior Sex Combs* (*Psc*) mutant flies, which were also observed to bind to the transgene integration site (Erhardt et al., 2003). Both E(z) and Psc are highly conserved proteins involved in chromatin remodeling and maintaining silenced and/ or active gene states (LaJeunesse and Shearn, 1996), suggesting that similar genes or the mammalian gene homologues may be involved in modulating the chromatin structure at the A6-A4 region in mice.

ADDITIONAL REGULATORY REGIONS

In addition to the ICR and downstream enhancers that are essential for normal imprinted expression of both genes, several other sequences are required for appropriate tissue-specific expression and repression of *H19* and *Igf2* (Figure 4.2A). The central A6-A4 region has been discussed above. Two conserved sequences upstream of *H19* and the ICR, HUC1 and HUC2, are biallelically transcribed in both mice and humans and appear to be mesoderm-specific enhancers (Drewell et al., 2002). Additional differentially methylated regions (DMRs) surround the *Igf2* gene and affect its expression. In mice, this location contains three DMRs: DMR0, DMR1, and DMR2. DMR0 encompasses the promoter region of a placenta-specific transcript (P0) and is maternally hypermethylated in the placenta but biallelically methylated in the fetus (Moore et al., 1997). Both DMR1 and DMR2 are methylated on the active paternal allele, but they function oppositely. DMR1 is a mesodermal repressor located upstream of the Igf2 gene. Deletion of DMR1 results in biallelic expression of *Igf2* in several mesoderm-derived tissues (Constancia et al., 2000). Conversely, DMR2 is an *Igf2* enhancer located in the sixth exon, and while deletion does not affect imprinting, it results in reduced *Igf2* expression from the paternal allele (Murrell et al., 2001). In vitro experiments confirm that the methylation status of these two DMRs is important for their function and is conducive to paternal *Igf2* expression. Methylation of DMR1 causes a loss of reporter gene silencing (Eden et al., 2001), while in vitro methylation of DMR2 increases reporter gene expression (Murrell et al., 2001). The *IGF2* region of humans contains only DMR0 and DMR2; however, recent analysis indicates that DMR0 is methylated on the active paternal allele in all tissues and may function similarly to mouse DMR1 (Murrell et al., 2008).

CHROMATIN LOOPING

Further analysis in mice demonstrated that physical long-range interactions between the ICR and *Igf2* DMRs likely establish parent-of-origin-specific chromatin loops. On the maternal chromosome, the ICR physically interacts with both DMR1 and a matrix attachment region (MAR) located 3' to the *Igf2* gene, termed MAR3,

A

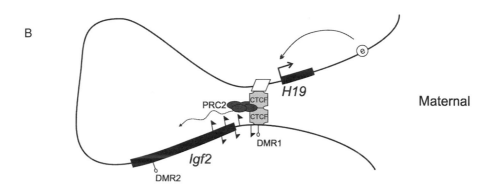

B

Maternal

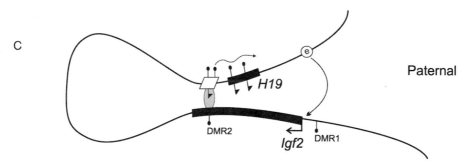

C

Paternal

FIGURE 4.2 A model for imprinting of the mammalian *H19* and *Igf2* genes. (A) Relative positions of the imprinted *Igf2* and *H19* genes, the imprint control region (ICR), downstream enhancers, and additional regulatory regions. The shared ICR is located upstream of *H19*. The positions of the downstream enhancers required for imprinted expression of both genes are given relative to the *H19* transcriptional start site. Sites of differential methylation are indicated with lollipops. (B) On the maternally inherited allele, CTCF binds to the unmethylated ICR and orchestrates chromatin looping and CTCF binding to DMR1 of *Igf2*, potentially through CTCF homodimerization. CTCF recruits PRC2 to the *Igf2* promoters via its interaction with Suz12. The PRC2 complex mediates H3K27 methylation throughout DMR1 and the *Igf2* promoters and gene region (indicated with filled flags and dotted arrow). This repressive histone modification leads to maternal *Igf2* silencing. The downstream enhancers interact with the *H19* promoters, leading to maternal-specific *H19* expression. (C) On the paternal chromosome, the ICR is methylated (filled lollipops) in sperm, which prevents binding of the insulator protein CTCF. During embryogenesis, this methylation spreads to encompass the *H19* promoter and gene region (dotted arrow). The ICR, *H19* promoter, and *H19* gene region also contain repressive histone modifications (filled flags) that likely contribute to repression. DMR1 and DMR2 are methylated (filled lollipops), and the ICR can interact with DMR2 (dotted ellipse), which may contribute to the formation of a chromatin loop that facilitates enhancer access to the *Igf2* promoters, resulting in paternal-specific *Igf2* expression.

while on the paternal chromosome the ICR interacts with DMR2 (Kurukuti et al., 2006; Murrell et al., 2004). CTCF binding to the ICR is necessary to mediate the higher-order chromatin structure on the maternal allele, and mutation of the CTCF-binding sites abolishes these physical interactions and causes the region to adopt the paternal chromatin structure (Kurukuti et al., 2006). Elimination of CTCF binding to the maternal ICR also causes a loss of CTCF binding within DMR1, as well as de novo methylation of DMR1 and DMR2. It thus appears that CTCF is recruited to DMR1 through the physical interaction with the ICR and that this recruitment protects the region from the acquisition of methylation on the maternal allele (Kurukuti et al., 2006).

Differential interactions between the gene promoters and the shared downstream enhancers have also been detected (Yoon et al., 2007). On the paternally inherited chromosome from which Igf2 is normally expressed, physical interactions between the Igf2 promoters and the downstream enhancers are observed. Conversely, on the maternally silenced chromosome, the Igf2 promoters physically interact with the ICR and ICR–enhancer interactions are also detected (Yoon et al., 2007). The maternal chromosome also exhibits physical interactions between the H19 promoter and the downstream enhancers, consistent with its expression of this gene (Yoon et al., 2007).

HISTONE MODIFICATIONS

Using chromatin immunoprecipitation, CTCF was found to bind to the Igf2 promoter region on the maternal chromosome, beginning at DMR1 and exhibiting the strongest binding at the two major promoters, P2 and P3. This region was also found to be hypermethylated at lysine 27 of histone H3 (H3K27), a modification mediated by the Polycomb repressive complex 2 (PRC2) in mammals. Consistent with this, CTCF was found to directly interact with Suz12, an essential component of PRC2. RNAi knockdown of Suz12 resulted in hypomethylation of H3K27 at the Igf2 promoters of the maternal allele and

biallelic Igf2 expression (Li et al., 2008). CTCF binding to the unmethylated maternal ICR therefore mediates long-range interactions with the Igf2 promoter and subsequently recruits PRC2, which results in histone methylation and repression of Igf2 on the maternal allele only.

Differential histone modifications have also recently been detected throughout the rest of the Igf2/H19 imprinted domain (Han et al., 2008; Verona et al., 2008). In addition to H3K27 methylation, the silenced maternal Igf2 region is enriched for repressive methylation at lysine 9 of histone H3 (H3K9) and the heterochromatic histone variant macroH2A1 (Han et al., 2008). Activating histone marks, including histone acetylation and histone H3 lysine 4 (H3K4) methylation, are predominant on the maternal chromosome at the ICR and H19 promoter–gene region and on the paternal chromosome at the Igf2 promoter–gene region (Han et al., 2008). Both ICRs contain H3K27 methylation, and the paternal ICR and H19 gene are also strongly enriched for H3K9 methylation and macroH2A1 (Han et al., 2008; Verona et al., 2008). H3K27 methylation at the H19 promoter and gene is unclear and is either enriched on the paternal allele (Han et al., 2008) or equivalent on the maternal and paternal alleles (Verona et al., 2008). Abolishing CTCF binding to the ICR caused the maternal chromosome to adopt the normal paternal histone composition throughout both Igf2 and H19, suggesting that CTCF is essential for organizing the maternal chromatin structure (Han et al., 2008). Transcription of H19 from the maternal allele also appears to be required for establishing H3K4 methylation and histone acetylation throughout this region as the maternal allele will lose these active chromatin modifications in cells with an ICR deletion where H19 is not expressed but not in cells with the same ICR deletion where H19 is expressed (Verona et al., 2008).

Although much of H19/Igf2 imprinting research has focused on the role of differential DNA methylation in establishing and maintaining imprinted expression of these two genes, it is now clear that *H19/Igf2* imprinting is much more complex. Given that DNA methylation, histone modifications, higher-order chromatin structure, and RNA-directed modifications are often mutually reinforcing epigenetic processes, it is not surprising that recent results have indicated that all of these processes are essential in *H19/Igf2* imprinting. A model for *Igf2* and *H19* imprinting is illustrated in Figure 4.2B.

GENOMIC IMPRINTING IN INSECTS

Although much of the recent research in the field of genomic imprinting has focused on mammals, imprinting was first described in two insect systems: *Sciara* (black fungus gnats) and coccids (scale insects). The process of imprinting in these insects, as well as in the model organism *Drosophila melanogaster*, is the same as that in mammals. The imprint is differentially established depending on the sex of the germ line, maintained throughout embryonic development, and then erased in gametogenesis so that the adult organism properly transmits the appropriate imprint to the progeny. However, while most documented examples of imprinting in mammals affect individual genes or several genes grouped in a large cluster, imprinting in *Sciara* and coccids results in a parent-of-origin effect on whole chromosomes. This can lead to elimination or heterochromatinization of chromosomes based strictly on whether they were inherited through the male or female germ line.

Imprinting in *Drosophila* has been observed for marker genes on rearranged chromosomes as well as for transgenes inserted at heterochromatic positions (Lloyd, 2000; Maggert and Golic, 2002). Paternal-specific chromosomal loss has also been observed in *Drosophila* in the presence of certain mutations, indicating that the chromosomes must carry a parent-of-origin-specific imprint, despite this imprint not normally causing an obvious consequence on chromosomal behavior. Together this evidence indicates that *Drosophila* is fully capable of im-

printing and can generate both smaller imprinted domains, similar to mammalian imprinting centers, that may result in imprinted expression of nondevelopmentally essential genes and imprints that can cause the loss of whole chromosomes, similar to those observed in other insects.

SCIARA

The term *imprint* was first used in describing the complex process of chromosome elimination in *Sciara* (Crouse, 1960). Following the observation that the developing *Sciara* embryo specifically eliminated chromosomes of paternal origin, it was hypothesized that the transmission of chromosomes through the male and female germ lines resulted in an "imprint" that marked the chromosomes based on their inheritance. This imprint was concluded to be unrelated to the genetic content of the chromosome and, therefore, determined solely by the sex of the parent that had transmitted it.

The complex process of chromosome elimination in *Sciara* occurs in three distinct elimination events (reviewed in Goday and Esteban, 2001). *Sciara* embryos inherit three X chromosomes: one maternally and two paternally. During early embryonic development, either one or both of the paternal X chromosomes are eliminated from somatic cells, depending on whether the sex of the embryo is female or male, respectively. The germ line retains all three X chromosomes until later in embryonic development, when a single paternal X chromosome is eliminated from the germ nuclei of both males and females by expulsion through the nuclear membrane and into the cytoplasm, where it is degraded. Female meiosis then proceeds normally. However, during male meiosis a third elimination event occurs. This elimination discards all remaining paternal chromosomes, including the autosomes, into a cytoplasmic bud that is extruded from the developing sperm nuclei. Nondisjunction of the maternal X chromosome during meiosis II results in the inclusion of both maternal X chromatids into a single mature sperm cell. Male *Sciara* therefore produce

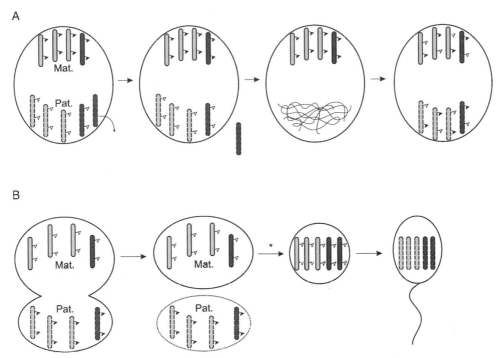

FIGURE 4.3 Genomic imprinting and chromosome elimination in the *Sciara* germ line (adapted from Greciano and Goday, 2006). Maternally inherited chromosomes are outlined with a solid line, and paternally inherited chromosomes are outlined with a dashed line. X chromosomes are filled with dark gray and autosomal chromosomes, with light gray. (A) In the early germ nucleus, paternal and maternal chromosomes are separated into distinct nuclear locations. Maternally inherited chromosomes exhibit histone hypermethylation (filled flags), and paternally inherited chromosomes exhibit histone hyperacetylation (white flags), with the exception of one paternally inherited X chromosome, which is neither hyperacetylated nor hypermethylated. This X chromosome is eliminated from the early germ nucleus of both sexes. Following this elimination, the paternal chromosomes undergo decondensation. This decondensation is maintained until just prior to gonial mitosis, at which point all chromosomes are equally condensed, methylated, and acetylated. Female meiosis then proceeds normally. (B) During male meiosis, the maternally inherited chromosomes are hyperacetylated and the paternally inherited chromosomes are hypermethylated. The paternally inherited chromosomes are eliminated during meiosis I. During meiosis II (*), nondisjunction of the maternal X chromosome occurs. The mature sperm nucleus contains the maternally inherited autosomes and two copies of the maternal X chromosome, which are re-marked as paternal so that they are recognized as being inherited from a male in the next generation.

sperm containing only the chromosomes that they inherited maternally (Figure 4.3). It is important to note that both males and females must reimprint their chromosomes at some point during germ-cell development so that their chromosomes are properly recognized as either "maternal" or "paternal" in the next generation.

Early cytological analysis provided evidence that maternal and paternal chromosomes exhibit distinct characteristics in the early germ line. Following elimination of the paternal X from the early germ nuclei, the paternal chromosomes appear unraveled and lightly stained, while the maternal chromosomes appear condensed and darkly stained (Berry, 1941; Rieffel and Crouse, 1966). This difference is evident until just prior to gonial mitosis, at which point the two chromosome sets appear equally condensed. Maternal and paternal chromosomes have also been observed to occupy distinct nuclear compartments in developing germ cells until meiosis, resulting in physical separation of the two chromosome sets within the nucleus (Kubai, 1987). The unique behaviors of the X chromosome, including both elimination and nondisjunction, require a distinct controlling element that maps to a block of X-chromosome heterochromatin (Crouse, 1960) that also contains ribosomal RNA genes (Crouse, 1979; Crouse et al., 1977).

HISTONE MODIFICATIONS

Molecular analysis of maternal and paternal chromatin modifications during germ-cell development discovered several differences in histone acetylation (Goday and Ruiz, 2002) and histone methylation (Greciano and Goday, 2006) during the chromosome elimination process. In early germ cells, prior to elimination, the paternal chromosomes are highly acetylated on histones H3 and H4, with the exception of the paternal X that will be eliminated. The paternal X that will be eliminated and the entire maternal chromosome complement are hypoacetylated (Goday and Ruiz, 2002). In addition, the maternal chromosomes exhibit H3K4 methylation, while the H3 histones of the paternal chromosomes are unmethylated (Greciano and Goday, 2006). The paternal X chromosome that is eliminated is therefore the only chromosome that is both unmethylated and hypoacetylated, which may be the distinguishing factor that identifies it for elimination. Hypoacetylation in particular is hypothesized to be required for its interaction with inner nuclear membrane proteins and subsequent elimination from the nucleus (Goday and Ruiz, 2002).

Histone acetylation differences are maintained during the X-chromosome elimination and subsequent decondensation of the remaining paternal chromosomes. Postdecondensation, the maternal chromosomes acquire histone acetylation, rendering both chromosome sets equally acetylated (Goday and Ruiz, 2002). At this stage, the H3 histones of the two chromosome sets are also equivalently methylated (Greciano and Goday, 2006). However, during male meiosis, the paternal chromosome set becomes hypoacetylated and hypermethylated. Interestingly, both the acetylation and methylation differences are reversed from those observed in early germ nuclei, where the maternal chromosomes were hypoacetylated and hypermethylated. In addition to methylation at H3K4, the paternal chromosomes acquire methylation at histone H4 lysine 20 (H4K20) during male meiosis (Greciano and Goday, 2006).

Localization of the maternal and paternal chromosome sets to distinct nuclear compartments is likely an essential component of the chromosome modification and elimination processes. These nuclear compartments may be associated with the activity of specific histone acetyltransferases, deacetylases, methyltransferases, or demethylases, resulting in histone modifications for one parental chromosome set but not the other. These modifications may then, in turn, participate in localizing or identifying the chromosomes for elimination or retention. The roles of other core epigenetic processes, such as RNA-mediated modifications and DNA methylation, have not yet been investigated. The current model of *Sciara* chromosome imprinting, which incorporates histone modifications, chromosome distribution, and elimination, is diagrammed in Figure 4.3.

COCCIDS

Imprinting of chromosomes has also been studied in coccids (superfamily Coccoidea), a group of insects that includes the Pseudococcidae family of mealybugs and the Diaspididae family of armored scale insects. Three complex genetic systems involving imprinting and chromosome elimination or inactivation have been studied in the coccid insects (reviewed in Khosla et al., 2006). Of these, the lecanoid chromosome system exhibited by a diverse group of Coccoidea families, including the mealybugs, has been the most thoroughly investigated.

Sex chromosomes are absent in lecanoid coccids, and thus, the chromosomal complement of all zygotes is initially identical. However, in male-determined embryos, a full haploid chromosome set is inactivated via heterochromatinization during embryogenesis. This chromosomal inactivation is nonrandom; it is consistently the paternally inherited chromosome set that becomes heterochromatic (Brown and Nelson-Rees, 1961). These heterochromatic chromosomes are transcriptionally inactive (Brown and Nelson-Rees, 1961), and thus, males are functionally haploid, with the exception of a few tissues that exhibit reversal of heterochromatinization (Nur, 1967). In spermatogenesis of lecanoid coccids, the paternally inherited heterochromatic chromosome set disintegrates; thus, males transmit only the chromosomes that were inherited maternally. Imprinting in diaspidoid coccids is similar, but in this case the entire paternally inherited haploid genome is eliminated early in male development, rather than inactivated in heterochromatin (reviewed in Khosla et al., 2006).

DNA METHYLATION

Analysis of the role of DNA methylation in imprinting in mealybugs has produced conflicting results. In analyzing methylation at CCGG sequences in the mealybug *Planococcus citri*, one study determined that the paternally inherited chromosomes are hypomethylated in both males and females (Bongiorni et al., 1999). Thus, methylation could serve as a mark that distinguishes the parental origin of the chromosomes but would probably not contribute directly to the silencing (Bongiorni and Prantera, 2003). However, a second study found no significant difference in the methylation of paternally and maternally inherited chromosomes (Buglia et al., 1999). Sequence-specific analysis of CpG methylation in the mealybug *Planococcus lilacinus* found that male-specific methylation occurs more frequently than female-specific methylation (Mohan and Chandra, 2005). In addition, these sex-specific methylated sequences were associated with transcriptionally silent chromatin but only in the sex exhibiting methylation, which may suggest a direct link between sex-specific DNA methylation and transcriptional silencing in mealybugs (Mohan and Chandra, 2005). A higher frequency of 5-methylcytosine in males compared to females was also found in two additional species of mealybugs, although only one was deemed statistically significant (Scarbrough et al., 1984). Interestingly, mealybugs also exhibit a significant amount of 5-methylcytosine in other dinucleotide combinations, and some species have been shown to also contain a high frequency of the normally rare 6-methyladenosine and 7-methylguanosine modified bases (Achwal et al., 1983; Deobagkar et al., 1982). The characterization of an active CpA methylase that methylates both CpG and CpA dinucleotides (Devajyothi and Brahmachari, 1992) confirms that mealybugs have the capacity for DNA methylation. However, the exact role of DNA methylation in imprinting in coccids remains to be elucidated.

CHROMATIN STRUCTURE AND HISTONE MODIFICATIONS

Chromatin analysis of the genomes of male and female mealybugs with micrococcal nuclease, an enzyme impeded by condensed chromatin, demonstrated that approximately 5%–10% of the genome is organized into nuclease-resistant chromatin in males but not females (Khosla et al., 1996, 1999). The nuclease-resistant chromatin sequences were found to be associated with the nuclear matrix (Khosla et al., 1996) and include unique sequences as well as middle-repetitive sequences distributed throughout the genome (Khosla et al., 1999). Detailed analysis of two middle-repetitive sequences found that these genomewide sequences are enriched within the nuclease-resistant chromatin of male mealybugs and exhibit different chromatin organization between males and females and within male nuclei (Khosla et al., 1999). The specialized organization of these sequences into nuclease-resistant chromatin is therefore likely a characteristic of the condensed paternal chromosomes, consistent with their cytologically visible heterochromatic structure. As only 10% of the male genome is organized into this nuclease-resistant chromatin and not 50% as would be expected if it were a property of the entire heterochromatinized chromosome set, it has been hypothesized that these sequences may function as initiation centers for heterochromatinization

(Khosla et al., 1996, 1999). Similarly, they may mediate differential organization of homologous chromosomes within the nucleus, which could serve as an epigenetic mark that distinguishes the genomes and triggers the heterochromatinization of the paternal genome (Khosla et al., 1999).

A protein with similarity to *Drosophila* Heterochromatin Protein 1 (HP1) is encoded by the *pchet2* gene in mealybugs (Bongiorni et al., 2007; Epstein et al., 1992). In males, this protein accumulates at the distinct chromocenter that contains the heterochromatic paternal chromosomes, with little binding elsewhere (Bongiorni et al., 2001). The heterochromatic chromocenter is also strongly enriched for H3K9 (Cowell et al., 2002) and H4K20 methylation (Kourmouli et al., 2004). In females, PCHET2 protein, H3K9 methylation, and H4K20 methylation exhibit a scattered distribution over all of the chromosomes. The heterochromatic paternal genome in male mealybugs was also found to be hypoacetylated at histone H4 compared to the euchromatic maternal genome, with an increase in acetylation accompanied by a decrease in condensation (Ferraro et al., 2001).

During male embryogenesis, a dense PCHET2 signal was evident prior to the formation of the chromocenter, indicating that PCHET2 accumulation precedes and likely contributes to heterochromatinization (Bongiorni et al., 2001). Consistent with this, a knockdown of *pchet2* was accompanied by a decondensation of the paternal chromosomes, a loss of H4K20 methylation, and overall genome instability (Bongiorni et al., 2007). In mammals, constitutive heterochromatin formation is thought to require HP1 binding to H3K9 methylation, which subsequently recruits H4K20 methyltransferases (Schotta et al., 2004). The facultative heterochromatinization of paternal chromosomes observed in male mealybugs appears to be consistent with this model. Analysis of cells undergoing reversal of heterochromatinization found that H3K9 methylation remains associated with the decondensing paternal chromosomes, while H4K20 methylation is lost and PCHET2 becomes dispersed (Bongiorni et al., 2007). H3K9 methylation may therefore be the primary epigenetic modification of the paternal chromosomes and may be the "imprint" carried by the paternal chromosomes, leading to their heterochromatinization (Bongiorni et al., 2007). The role of noncoding RNAs in coccid imprinting has not yet been studied.

DROSOPHILA MELANOGASTER

D. melanogaster is a model organism widely used to study gene expression. While endogenous imprinted genes remain to be identified, imprinted expression of transgenes, or marker genes on rearranged chromosomes, indicates that the capacity for differential gene expression based on parental inheritance is certainly present and mechanistically possible. Of the imprinting examples studied in *Drosophila*, all are associated with gene-poor regions of constitutive heterochromatin (reviewed in Lloyd, 2000). For example, imprinting has been observed on rearranged chromosomes when a chromosomal breakage results in the juxtaposition of a euchromatic marker gene with a region of broken heterochromatin. While this type of a disruption frequently results in variegation of the marker gene due to its new position adjacent to heterochromatin, imprinting of the marker gene is observed in only a small number of cases, suggesting it is a unique characteristic of only certain heterochromatin regions or segments. Imprinting is identified when transmission of the rearranged chromosome through one parent causes a significantly different level of marker gene expression from that transmitted through the other parent. In these cases, the imprint's origin appears to involve discrete regions at which the key epigenetic processes act, similar to mammalian imprint centers.

Imprinting has also been observed at a high frequency for transgenes inserted into the heterochromatic Y chromosome (Golic et al., 1998; Maggert and Golic, 2002). While most imprinted transgene insertions on the Y chromosome exhibit increased silencing when transmitted paternally, the reverse has also been observed, with increased silencing following maternal transmission. Furthermore, as with the imprinted domains in centric heterochromatin, some transgenes exhibit opposite imprinting of the two marker genes or imprinting of only one of the two marker genes, despite their insertion at the same genomic position. Imprinting in *Drosophila* may therefore include reciprocal imprinting of closely linked genes or differential gene response to an imprinting center, similar to many imprinting clusters in mammals (Lloyd, 2000; Maggert and Golic, 2002). The sequestering of imprinted domains to heterochromatic regions of the genome with low gene density appears to be both mechanistic, with heterochromatic repeat sequences nucleating the imprint, and a result of selection against the inclusion of too many genes in the imprinted domain (Anaka et al., 2009).

The observation of chromosomal loss in the presence of mutations in the *paternal loss inducer* (*pal*) gene provides additional evidence of imprinting in *Drosophila*. In these cases, chromosomal loss is not random, but instead it specifically affects chromosomes that were paternally inherited (Baker, 1975; Fitch et al., 1998). The *pal* gene acts exclusively in males and is hypothesized to encode a sperm-specific protein that could distinguish the paternal from the maternal chromosomes in the zygote after fertilization. *Drosophila* chromosomes therefore likely carry an imprint that distinguishes them based on their parent-of-origin, despite not normally causing chromosomal loss or inactivation, as in coccids or *Sciara*.

HISTONE MODIFICATIONS

The best-studied example of imprinting in *Drosophila* is the *Dp(1;f)LJ9* mini-X chromosome (Lloyd et al., 1999). Among the genes imprinted on this mini-X chromosome is the easily observable eye color gene *garnet*. Imprinting of such an easily monitored gene allowed for identification of genes involved in *Drosophila* imprinting (Joanis and Lloyd, 2002). Mutations in several *Suppressor of variegation* [*Su(var)*]

genes resulted in a loss of the maintenance of the paternal imprint. These included the well-characterized *Su(var)3-9* and *Su(var)2-5* (Joanis and Lloyd, 2002). *Su(var)3-9* encodes a histone methyltransferase that contributes to heterochromatin formation by catalyzing H3K9 methylation (Schotta et al., 2002), a mark that is recognized and bound by the *Su(var)2-5* gene product Heterochromatin Protein 1 (HP1) (Lachner et al., 2001). Mutations in the gene *Su(var)3-3*, which encodes a histone demethylase that associates with prospective heterochromatic regions and removes the H3K4 methylation mark that is normally associated with active chromatin (Rudolph et al., 2007), also resulted in a loss of paternal silencing (Joanis and Lloyd, 2002). Mutations in two *trithorax* group genes, *trithorax* and *brahma*, exhibited the opposite effect on the paternal imprint (Joanis and Lloyd, 2002), consistent with the role of these two proteins in complexes that participate in the formation and maintenance of active chromatin via activating histone modifications and chromatin remodeling (Simon and Tamkun, 2002). Overall, these results indicate that imprinting in *Drosophila* is likely accomplished by histone modifications that mediate the formation of a repressive heterochromatin structure upon passage through one germ line but not the other. The role of DNA methylation and antisense or noncoding RNA is under active investigation; preliminary results indicate that these epigenetic processes are also involved (MacDonald and Lloyd, 2004; Maggert and Golic, 2004).

GENOMIC IMPRINTING IN PLANTS

Several endogenous plant genes in both *Arabidopsis thaliana* (thale cress) and *Zea mays* (maize) are imprinted in the endosperm, a triploid tissue formed by the fusion of a haploid sperm from a pollen grain with the diploid central cell in the ovule. While the endosperm does not contribute its genome to the next generation, it nourishes the developing embryo and is an essential component of an angiosperm seed. Four of the five known imprinted genes in *Arabidopsis* and five of the six known imprinted genes in maize are expressed maternally and silenced paternally in the endosperm. The thorough examination of several endosperm-specific imprinted genes does not preclude the existence of nondevelopmentally essential imprinted genes in the embryo or adult plant. Imprinting of such genes could be specific to certain tissues or developmental stages or could involve partial, rather than complete, silencing of one parental allele. In support of this hypothesis, several paternally inherited genes and transgenes have been shown to be downregulated or silenced in the early embryo in *Arabidopsis*, providing evidence that the maternal and paternal genomes are nonequivalent during early embryogenesis (Baroux et al., 2001; Vielle-Calzada et al., 2000).

MEDEA IMPRINTING IN ARABIDOPSIS

Arabidopsis MEDEA (MEA) is an imprinted gene that encodes a SET domain–containing Polycomb group protein homologous to *Drosophila* Enhancer of Zeste [E(z)] (Grossniklaus et al., 1998). Polycomb group proteins function in multimeric protein complexes that maintain transcriptional repression by modifying chromatin structure (Orlando, 2003). E(z) and E(z) homologues are members of the Polycomb Repressive Complex 2 (PRC2), which exhibits histone methyltransferase activity through the SET domain of *E(z)* (Muller et al., 2002). In *Arabidopsis*, many of the core components of PRC2 are represented by small gene families rather than single-copy genes, and it is hypothesized that diversification of the ancestral PRC2 complex has led to multiple distinct PRC2 complexes that target different genes for repression (Chanvivattana et al., 2004). *MEA* is one of three *E(z)* homologues that has been identified, each of which has at least partially diverged in gene-expression pattern and protein function (Chanvivattana et al., 2004).

The *Arabidopsis* Polycomb group complex that contains MEA also includes the proteins Fertilization-Independent Endosperm (FIE) and Multicopy Suppressor of Ira 1 (MSI1), which are homologues of *Drosophila* Extra Sex Combs (Esc) and p55, respectively. This complex is hypothesized to also include Fertilization Independent Seed2 (FIS2), a homologue of *Drosophila* Suppressor of Zeste 12 (Chanvivattana et al., 2004; Kohler et al., 2003a). These four genes are members of the *Arabidopsis* Fertilization Independent Seed (FIS) group, a class of genes that are characterized by a mutant phenotype that includes seed development in the absence of fertilization. This Polycomb group complex is therefore also termed the "FIS complex." Interestingly, both *FIS2* and *PHERES1*, a known target of the FIS complex, are also imprinted in *Arabidopsis*. Maternal inheritance of a *mea* mutation results in seed abortion, aberrant proliferation of the central cell in the absence of fertilization, and overproliferation of the endosperm following fertilization (Kiyosue et al., 1999). The normal development of seeds inheriting a *mea* mutation paternally provided early evidence of a parent-of-origin effect (Grossniklaus et al., 1998; Kiyosue et al., 1999), which was later shown to be a result of *MEA* imprinting (Kinoshita et al., 1999). Paternal inheritance of a nonfunctional *mea* allele has no effect because the imprinted *MEA* gene is not normally expressed from the paternal allele. Similar asymmetrical consequences resulting from the inheritance of mutant alleles have led to the discovery of several mammalian imprinted genes.

DNA METHYLATION

Like other imprinted genes identified in plants, imprinting of *MEA* occurs only in the endosperm, where it is expressed from the maternal allele only. Conversely, biallelic expression of *MEA* occurs in the embryo and other tissues of the adult plant (Kinoshita et al., 1999). Early genetic analysis revealed that imprinting of *MEA* requires a maternal copy of the *DEMETER* (*DME*) gene, a DNA glycosylase that is primarily expressed in the central cell prior to fertilization. Maternal *dme* mutations resulted in a lack of maternal *MEA* expression in the central cell and endosperm, while ectopic *DME* expression in the endosperm resulted in *MEA* expression from the paternal allele (Choi et al., 2002). The role of DNA glycosylases in removing mismatched or altered bases from DNA and the nicks discovered at the *MEA* promoter upon ectopic expression of *DME* led to the hypothesis that DME may contribute to *MEA* imprinting by excising DNA methylation at the maternal allele.

This hypothesis was supported by the discovery that a maternal mutation in the DNA maintenance methyltransferase *met1* could suppress the *dme* mutant phenotype if a wild-type maternal *MEA* allele was also present (Xiao et al., 2003). The combination of maternal *met1* and *dme* mutations restored *MEA* expression to normal levels from the maternal allele in the endosperm, indicating that MET1 and DME act antagonistically in controlling *MEA* imprinting (Xiao et al., 2003). Three regions of methylation were detected in the *MEA* promoter, with a decrease in methylation detected in the presence of a *met1* mutation (Xiao et al., 2003). In the endosperm, the maternal alleles were found to be hypomethylated compared to the paternal allele, with maternal methylation increasing in *dme* mutant seeds (Gehring et al., 2006). Thus, expression of *DME* in the central cell prior to fertilization appears to remove *MEA* methylation on the maternal allele and establish a hypomethylated state that is required for its expression. Consistent with this, DME was found to excise 5-methylcytosines in vitro (Gehring et al., 2006).

Intriguingly, DME-mediated hypomethylation appears to be a unique requirement for *MEA* expression in the central cell and early endosperm. Hypomethylation of *MEA* is not required in the embryo, where *MEA* is biallelically expressed but exhibits methylation comparable to the silenced paternal allele in the endosperm. Furthermore, in *dme* mutants, the maternal allele is expressed late in endosperm development despite being hypermethylated (Gehring et al., 2006). Why then is DME required to establish hypomethylation of *MEA* in the central cell? Removal of maternal methylation may be a prerequisite for additional modifications required for expression in that environment, such

as the removal of histone methylation or changes in chromatin structure. Alternatively, the DME enzyme may directly mediate removal of both DNA and histone methylation at the *MEA* promoter, or it may activate *MEA* indirectly by removing methylation on an additional gene that, in turn, further modifies and activates the *MEA* locus (Jullien et al., 2006a).

HISTONE MODIFICATIONS

Hypomethylation of the paternal allele does not result in its expression in the endosperm, indicating that DNA methylation does not directly maintain paternal *MEA* silencing. Instead, the FIS Polycomb group complex containing MEA itself was found to be essential for this repression. Maternal mutations in *mea, fie, fis2,* or *msi1* resulted in expression from the paternal allele (Gehring et al., 2006; Jullien et al., 2006a), and chromatin immunoprecipitation analysis confirmed that MEA can physically interact with its own promoter (Baroux et al., 2006). Furthermore, the paternal allele was found to be enriched in H3K27 methylation (Gehring et al., 2006; Jullien et al., 2006a), a repressive histone mark that is well characterized in *Drosophila* and mammals, where it is catalyzed by the *MEA* homologues E(z) and E(z)H2 (Cao et al., 2002; Muller et al., 2002). The function of MEA and H3K27 methylation appears to be conserved in *Arabidopsis*, and a maternal *mea* mutation resulted in a decrease in H3K27 methylation at the paternal *MEA* allele and a loss of silencing (Gehring et al., 2006). *MEA* is therefore a gene that controls its own imprinting, with the maternally expressed protein contributing to the silencing of the paternal allele. The observation that a paternal mutation in *fie*, a single-copy gene essential for all known *Arabidopsis* Polycomb complexes, resulted in *MEA* expression from the paternal allele in the endosperm suggests that *MEA* silencing must also be maintained by a Polycomb group complex during male gametogenesis for successful imprinting in the endosperm (Jullien et al., 2006a). The current model of *MEA* imprinting in *Arabidopsis* is summarized in Figure 4.4.

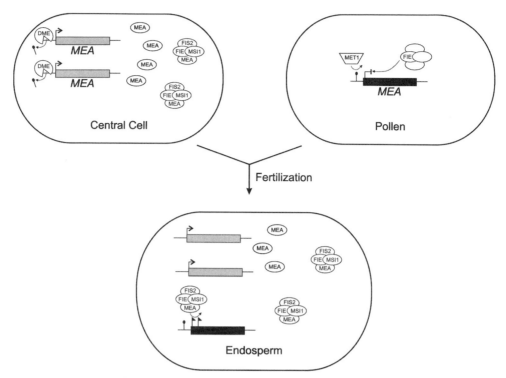

FIGURE 4.4 Imprinting of the *MEA* gene in *Arabidopsis*. In the central cell prior to fertilization, DEMETER (DME) removes DNA methylation (filled lollipops) from the maternal *MEA* alleles (light gray). The *MEA* gene is expressed and produces MEA protein, which assembles into the FIS Polycomb group complex. In the male pollen, DNA methylation is maintained by MET1, and a polycomb group complex containing FIE is necessary to maintain *MEA* silencing. Following fertilization, silencing of the paternal *MEA* allele in the endosperm is maintained by the FIS Polycomb group complex that contains the maternally expressed MEA protein. This complex catalyzes H3K27 methylation (filled flags) at the paternal *MEA* locus, a modification that inhibits paternal *MEA* expression via the formation of a repressive chromatin structure.

IMPRINTING OF OTHER PLANT GENES

Imprinting of the other known imprinted genes in *Arabidopsis* is accomplished using similar mechanisms to those acting at the *MEA* locus. The maternally expressed imprinted genes *FWA*, *FIS2*, and *MPC* require the maintenance methyltransferase MET1 to maintain methylation and paternal allele repression and require maternal DME to excise methylation and activate expression of the maternal allele (Jullien et al., 2006b; Kinoshita et al., 2004; Tiwari et al., 2008). The *PHERES1* gene is currently the only known imprinted gene in *Arabidopsis* that is paternally expressed in the endosperm. *PHERES1* expression requires DNA methylation at a region located 3′ to the gene. Conversely, repression of *PHERES1* requires both hypomethylation of the 3′ region and the *Arabidopsis* FIS

Polycomb complex, which catalyzes H3K27 methylation at the *PHERES1* promoter (Kohler et al., 2003b; Makarevich et al., 2006, 2008). Repression of the maternal allele may require hypomethylation of the 3′ sequence in order to facilitate the binding of a methylation-sensitive chromatin-binding protein. Such a protein could mediate the formation of a repressive chromatin loop and/or the recruitment of the FIS Polycomb complex to the *PHERES1* promoter. This would be similar to the mammalian *Igf2/H19* imprint center, where CTCF binds to the unmethylated ICR and orchestrates a chromatin loop that recruits PRC2 to *Igf2*, leading to its repression.

DE NOVO DNA METHYLATION AND SMALL RNAS

Thus far, de novo methylation has not been observed in imprinting in *Arabidopsis*. Instead, methylation must be maintained and then selectively removed in the endosperm. As the endosperm is terminally differentiated and does not contribute its genome to the next generation, remethylation is not required. However, there is evidence that de novo methylation may play a role in imprinting in maize. In maize, two orthologues of the *Arabidopsis FIE* gene, *fie1* and *fie2*, are imprinted and expressed from the maternal alleles during endosperm development (Danilevskaya et al., 2003; Gutierrez-Marcos et al., 2003). Similar to imprinted genes in *Arabidopsis*, *fie1* is hypermethylated on the silenced paternal allele in the endosperm, with methylation also detected in the sperm, egg, and embryo but not in the central cell that will contribute to the endosperm (Gutierrez-Marcos et al., 2006). It is thus likely that *fie1* methylation is actively removed from the maternal alleles in the central cell prior to fertilization. Examination of *fie1* histone modifications has demonstrated that the silenced paternal allele is enriched for H3K27 methylation, while the expressed maternal allele is enriched for H3 and H4 acetylation and H3K4 methylation, consistent with the histone patterns of many other imprinted genes (Haun and Springer, 2008). While the status of histone modifications at *fie2* is unknown, its methylation pattern is currently unique among the known imprinted genes in plants. Methylation is absent from *fie2* in the sperm, egg, embryo, and central cell; but the silenced paternal *fie2* allele exhibits hypermethylation in the endosperm (Gutierrez-Marcos et al., 2006). The imprinted paternal *fie2* allele must therefore acquire de novo methylation in the endosperm. This observation also further implies that the paternal *fie2* allele carries a non-DNA methylation–based imprint that identifies it for hypermethylation and repression in the endosperm (Scott and Spielman, 2006).

There is considerable evidence that de novo methylation in plants, mediated by the de novo methyltransferases Domains Rearranged Methyltransferase 1 (DRM1) and Domains Rearranged Methyltransferase 2 (DRM2), is directed by small RNAs. While this pathway has been extensively studied for genes and transgenes that are methylated and silenced in the adult plant, it also has applicability to any imprinted genes that may acquire de novo methylation, such as the *fie2* gene in maize. The *FWA* gene that is imprinted in the *Arabidopsis* endosperm is biallelically silenced in the embryo and adult plant and provides an excellent system for methylation studies as it exhibits CPG and non-CPG methylation at direct repeats located 5′ to the gene (Soppe et al., 2000). *FWA* transgenes have been observed to acquire de novo methylation and silencing in wild-type plants but not in *drm1/drm2* mutants (Cao and Jacobsen, 2002b). Interestingly, de novo methylation of *FWA* transgenes also requires a functional RNAi pathway. Mutations in several genes in the siRNA-generating pathway cause a similar loss of de novo methylation and silencing of the *FWA* transgenes (Chan et al., 2004). In the *drm1/drm2* and RNAi mutants, non-CPG methylation at the endogenous *FWA* locus was also lost (Cao and Jacobsen, 2002a; Chan et al., 2004), suggesting that the de novo methyltransferase and RNAi pathways may be required to maintain these modifications, in addition to establishing de novo methylation. The observation that the paternally inherited *fie2* allele is extensively methylated at CPG and non-CPG sites in the endosperm, whereas *fie1* methylation is almost entirely restricted to CPG sites (Gutierrez-Marcos et al., 2006), likely indicates that a similar RNA-mediated methylation pathway directs de novo methylation of the imprinted paternal *fie2* allele in maize.

There is also evidence that trans-communication between alleles is important in the methylation and silencing processes and may similarly be important in plant imprinting. Introduction of an *FWA* transgene into *fwa-1* mutant plants, in which the endogenous *FWA* gene is hypomethylated and expressed, occasionally results in methylation and silencing of the endogenous *fwa-1* mutant allele and rescue of the mutant phenotype (Chan et al., 2006; Soppe et al., 2000). Furthermore, while an *FWA* transgene very consistently acquires methylation and silencing when introduced into wild-type plants, in *fwa-1* plants methylation and silencing of the introduced *FWA* transgene is inefficient (Chan et al., 2006). The epigenetic status of the endogenous *FWA* locus can therefore influence that of the *FWA* transgene and vice versa. Given that siRNAs accumulate equally in wild-type and *fwa-1* mutant plants, additional chromatin or DNA modifications are likely required for efficient RNA-directed DNA methylation and epigenetic silencing (Chan et al., 2006). A similar mechanism involving RNA-mediated allelic communication and chromatin modifications has been proposed to function in maize *b1* paramutation, an epigenetic process related to imprinting that produces meiotically stable changes in chromatin structure and gene expression (Chandler, 2007). Overall, this evidence indicates that all three core epigenetic processes—DNA methylation, histone modifications, and small RNA–mediated modifications—underlie genomic imprinting in plants.

EVOLUTIONARY CONSERVATION OF GENOMIC IMPRINTING

Examination of genomic imprinting in organisms as diverse as mammals, insects, and plants suggests that imprinting is accomplished using phylogenetically conserved epigenetic mechanisms. Histone modifications are a common theme in genomic imprinting and are utilized in all species examined in order to establish higher-order chromatin structures that contribute to the imprinting of whole chromosomes, single genes, or gene clusters. At the mammalian *H19/Igf2* imprint center, the maternal chromosome is enriched for repressive histone modifications at *Igf2* and activating histone modifications at *H19*. Conversely, and consistent with the gene-expression patterns, the paternal chromosome is enriched for activating histone modifications at *Igf2* and repressive histone modifications at *H19*. A similar role for histone modifications is observed in insect imprinting. In *Sciara*, unique

patterns of histone acetylation and methylation contribute to chromosome elimination. The inactive paternal chromosomes in coccids are associated with repressive histone methylation and histone H4 hypoacetylation. On the *Drosophila Dp(1;f)LJ9* imprinted chromosome, the silenced paternal imprint requires a protein that catalyzes repressive histone methylation and a protein that removes activating histone methylation. The importance of histone modifications is also demonstrated for imprinted genes in plants, in which the silenced alleles are associated with repressive histone modifications and the expressed alleles are associated with activating histone modifications. In addition, both plants and mammals utilize a homologous Polycomb complex, the PRC2 complex in mammals and the FIS complex in plants, to catalyze repressive H3K27 methylation at silenced imprinted alleles.

DNA methylation is another common theme in plant and mammalian imprinting, and may also play a role in insect imprinting systems. Noncoding and small RNAs exhibit conserved functional roles in catalyzing heterochromatin formation in mammals, insects, and plants; it is therefore likely that these will be demonstrated to be important in many imprinting systems in different species. In mammals, noncoding RNAs are very common in imprinting clusters and both are imprinted and regulate imprinted expression of other genes. Within the *H19/Igf2* imprint center, for example, the *H19* gene is an imprinted noncoding RNA, the ICR is biallelically transcribed and produces sense and antisense transcripts, and the HUC1 and HUC2 sequences are biallelically transcribed. In plants, de novo DNA methylation can be guided by small RNAs; thus, imprinted plant genes that exhibit de novo methylation, such as *fie2* in maize, likely use this RNA-mediated mechanism. Given that DNA methylation, histone modifications, chromatin structure, and noncoding RNAs are frequently interconnected and mutually reinforcing, it is not surprising that all of these conserved mechanisms have been implicated in genomic imprinting.

In addition to the similarities apparent in the mechanisms underlying imprinting, the broad consequences of imprinting are similar across this diverse group of organisms. In both plants and mammals, inactivation of one parental allele via imprinting is common. There is accumulating evidence that this type of imprinting also occurs in *Drosophila*. In addition, the imprinted inactivation of whole chromosomes that are paternal in origin is not unique to insects. In marsupials, the paternal X chromosome is inactivated in females as a method of dosage compensation (VandeBerg et al., 1987). In eutherian mammals, X-chromosome inactivation in the somatic tissues of females is random; however, in the extraembryonic tissue of mice, the paternal X chromosome is imprinted and preferentially inactivated (Takagi and Sasaki, 1975). In marsupials and in the extraembryonic tissues of mice, the inactivated X is hypoacetylated at histone H4 (Wakefield et al., 1997). This mark is similarly associated with the inactive paternal chromosomes in coccids and the eliminated paternal chromosomes in *Sciara*.

The evolutionary conservation of the epigenetic mechanisms that underlie imprinting is further exemplified by the results of transgenic *Drosophila* experiments. Silencer elements from both the mouse and human *H19/Igf2* ICRs and the mouse A6-A4 region silence reporter genes in transgenic *Drosophila*. The mouse ICR is even biallelically transcribed and produces noncoding RNAs in *Drosophila*, as it does at its endogenous locus. Similarly, the human Prader-Willi imprint center also functions as a silencer in *Drosophila* (Lyko et al., 1998). Transgenic organisms are therefore valuable tools in studying genomic imprinting as they can be used to analyze the function of genetic sequences and epigenetic processes, as well as the conservation of epigenetic mechanisms. The fact that mammalian ICRs frequently function as epigenetic silencers but do not imprint in *Drosophila* may indicate that a silenced epigenetic state is often the default. Silencing may use core epigenetic mechanisms that are highly conserved from one species to another, while imprinting is a more divergent, gamete-specific modification of these conserved silencing processes.

Discordances in imprinting within the mammalian lineage may indicate that imprinting of genes can evolve rapidly and, thus, that genes that are subject to imprinting may differ in imprinted status, expression pattern, and regulatory sequences from one species to another. Nevertheless, the evidence presented here indicates that at the core genomic imprinting occurs via exploiting conserved epigenetic silencing mechanisms in order to establish distinct patterns of epigenetic gene regulation.

REFERENCES

Achwal, C.W., C.A. Iyer, and H.S. Chandra. 1983. Immunochemical evidence for the presence of 5mC, 6mA and 7mG in human, *Drosophila* and mealybug DNA. *FEBS Lett* 158:353–8.

Ainscough, J.F., L. Dandolo, and M.A. Surani. 2000a. Appropriate expression of the mouse *H19* gene utilises three or more distinct enhancer regions spread over more than 130 kb. *Mech Dev* 91:365–8.

Ainscough, J.F., R.M. John, S.C. Barton, and M.A. Surani. 2000b. A skeletal muscle-specific mouse *Igf2* repressor lies 40 kb downstream of the gene. *Development* 127:3923–30.

Alleman, M., and J. Doctor. 2000. Genomic imprinting in plants: Observations and evolutionary implications. *Plant Mol Biol* 43:147–61.

Anaka, M., A. Lynn, P. McGinn, and V. Lloyd. 2009. Genomic imprinting in *Drosophila* has properties of both mammalian and insect imprinting. *Dev Genes Evol* 219:59–66.

Arney, K.L., E. Bae, C. Olsen, and R.A. Drewell. 2006. The human and mouse *H19* imprinting control regions harbor an evolutionarily conserved silencer element that functions on transgenes in *Drosophila*. *Dev Genes Evol* 216:811–19.

Baker, B.S. 1975. *Paternal loss (pal)*: A meiotic mutant in *Drosophila melanogaster* causing loss of paternal chromosomes. *Genetics* 80:267–96.

Bao, N., K.W. Lye, and M.K. Barton. 2004. MicroRNA binding sites in *Arabidopsis* class III HD-ZIP mRNAs are required for methylation of the template chromosome. *Dev Cell* 7:653–62.

Baroux, C., R. Blanvillain, and P. Gallois. 2001. Paternally inherited transgenes are down-regulated but retain low activity during early embryogenesis in *Arabidopsis*. *FEBS Lett* 509:11–16.

Baroux, C., V. Gagliardini, D. R. Page, and U. Grossniklaus. 2006. Dynamic regulatory interactions of Polycomb group genes: *MEDEA* autoregulation is required for imprinted gene expression in *Arabidopsis*. *Genes Dev* 20:1081–6.

Bartolomei, M. S., A. L. Webber, M. E. Brunkow, and S. M. Tilghman. 1993. Epigenetic mechanisms underlying the imprinting of the mouse *H19* gene. *Genes Dev* 7:1663–73.

Bartolomei, M. S., S. Zemel, and S. M. Tilghman. 1991. Parental imprinting of the mouse *H19* gene. *Nature* 351:153–5.

Bean, C. J., C. E. Schaner, and W. G. Kelly. 2004. Meiotic pairing and imprinted X chromatin assembly in *Caenorhabditis elegans*. *Nat Genet* 36:100–5.

Beaudet, A. L., and Y. H. Jiang. 2002. A rheostat model for a rapid and reversible form of imprinting-dependent evolution. *Am J Hum Genet* 70: 1389–97.

Bell, A. C., and G. Felsenfeld. 2000. Methylation of a CTCF-dependent boundary controls imprinted expression of the *Igf2* gene. *Nature* 405:482–5.

Bernstein, E., and C. D. Allis. 2005. RNA meets chromatin. *Genes Dev* 19:1635–55.

Berry, R. O. 1941. Chromosome behavior in the germ cells and development of the gonads in *Sciara ocellaris*. *J Morphol* 68:547–83.

Bongiorni, S., O. Cintio, and G. Prantera. 1999. The relationship between DNA methylation and chromosome imprinting in the coccid *Planococcus citri*. *Genetics* 151:1471–8.

Bongiorni, S., M. Mazzuoli, S. Masci, and G. Prantera. 2001. Facultative heterochromatization in parahaploid male mealybugs: Involvement of a heterochromatin-associated protein. *Development* 128:3809–17.

Bongiorni, S., B. Pasqualini, M. Taranta, P. B. Singh, and G. Prantera. 2007. Epigenetic regulation of facultative heterochromatinisation in *Planococcus citri* via the Me(3)K9H3-HP1-Me(3)K20H4 pathway. *J Cell Sci* 120:1072–80.

Bongiorni, S., and G. Prantera. 2003 Imprinted facultative heterochromatization in mealybugs. *Genetica* 117:271–9.

Brown, S. W., and W. A. Nelson-Rees. 1961. Radiation analysis of a lecanoid genetic system. *Genetics* 46:983–1007.

Buglia, G., V. Predazzi, and M. Ferraro. 1999. Cytosine methylation is not involved in the heterochromatization of the paternal genome of mealybug *Planococcus citri*. *Chromosome Res* 7:71–3.

Buhler, M., A. Verdel, and D. Moazed. 2006. Tethering RITS to a nascent transcript initiates RNAi-and heterochromatin-dependent gene silencing. *Cell* 125:873–86.

Cao, R., L. Wang, H. Wang, L. Xia, H. Erdjument-Bromage, P. Tempst, R. S. Jones, and Y. Zhang. 2002. Role of histone H3 lysine 27 methylation in Polycomb-group silencing. *Science* 298:1039–43.

Cao, X., and S. E. Jacobsen. 2002a. Locus-specific control of asymmetric and CpNpG methylation by the *DRM* and *CMT3* methyltransferase genes. *Proc Natl Acad Sci USA* 99 (Suppl 4): 16491–8.

Cao, X., and S. E. Jacobsen. 2002b. Role of the *Arabidopsis* DRM methyltransferases in de novo DNA methylation and gene silencing. *Curr Biol* 12:1138–44.

Cattanach, B. M., and M. Kirk. 1985. Differential activity of maternally and paternally derived chromosome regions in mice. *Nature* 315:496–8.

Chan, S. W., X. Zhang, Y. V. Bernatavichute, and S. E. Jacobsen. 2006. Two-step recruitment of RNA-directed DNA methylation to tandem repeats. *PLoS Biol* 4:e363.

Chan, S. W., D. Zilberman, Z. Xie, L. K. Johansen, J. C. Carrington, and S. E. Jacobsen. 2004. RNA silencing genes control de novo DNA methylation. *Science* 303:1336.

Chandler, V. L. 2007. Paramutation: From maize to mice. *Cell* 128:641–5.

Chanvivattana, Y., A. Bishopp, D. Schubert, C. Stock, Y. H. Moon, Z. R. Sung, and J. Goodrich. 2004. Interaction of Polycomb-group proteins controlling flowering in *Arabidopsis*. *Development* 131:5263–76.

Choi, Y., M. Gehring, L. Johnson, M. Hannon, J. J. Harada, R. B. Goldberg, S. E. Jacobsen, and R. L. Fischer. 2002. DEMETER, a DNA glycosylase domain protein, is required for endosperm gene imprinting and seed viability in *Arabidopsis*. *Cell* 110:33–42.

Constancia, M., W. Dean, S. Lopes, T. Moore, G. Kelsey, and W. Reik. 2000. Deletion of a silencer element in *Igf2* results in loss of imprinting independent of *H19*. *Nat Genet* 26:203–6.

Cowell, I. G., R. Aucott, S. K. Mahadevaiah, P. S. Burgoyne, N. Huskisson, S. Bongiorni, G. Prantera, et al. 2002. Heterochromatin, HP1 and methylation at lysine 9 of histone H3 in animals. *Chromosoma* 111:22–36.

Crouse, H. V. 1960. The controlling element in sex chromosome behavior in *Sciara*. *Genetics* 45: 1429–43.

Crouse, H. V. 1979. X heterochromatin subdivision and cytogenetic analysis in *Sciara coprophila* (Diptera, Sciaridae). *Chromosoma* 74:219–39.

Crouse, H. V., S. A. Gerbi, C. M. Liang, L. Magnus, and I. M. Mercer. 1977. Localization of ribosomal DNA within the proximal X heterochromatin of

Sciara coprophila (Diptera, Sciaridae). *Chromosoma* 64:305–18.

Danilevskaya, O.N., P. Hermon, S. Hantke, M.G. Muszynski, K. Kollipara, and E.V. Ananiev. 2003. Duplicated *fie* genes in maize: Expression pattern and imprinting suggest distinct functions. *Plant Cell* 15:425–38.

Davies, K., L. Bowden, P. Smith, W. Dean, D. Hill, H. Furuumi, H. Sasaki, B. Cattanach, and W. Reik. 2002. Disruption of mesodermal enhancers for *Igf2* in the minute mutant. *Development* 129:1657–68.

Davies, W., A. Isles, R. Smith, D. Karunadasa, D. Burrmann, T. Humby, O. Ojarikre, et al. 2005. Xlr3b is a new imprinted candidate for X-linked parent-of-origin effects on cognitive function in mice. *Nat Genet* 37:625–9.

Day, T., and R. Bonduriansky. 2004. Intralocus sexual conflict can drive the evolution of genomic imprinting. *Genetics* 167:1537–46.

DeChiara, T.M., E.J. Robertson, and A. Efstratiadis. 1991. Parental imprinting of the mouse *insulin-like growth factor II* gene. *Cell* 64:849–59.

Deobagkar, D., K. Muralidharan, S. Devare, K. Kalghatgi, and H. Chandra. 1982. The mealybug chromosome system I: Unusual methylated bases and dinucleotides in DNA of a *Planococcus* species. *J Biosci* 4:513–26.

Devajyothi, C., and V. Brahmachari. 1992. Detection of a CpA methylase in an insect system: Characterization and substrate specificity. *Mol Cell Biochem* 110:103–11.

Drewell, R.A., K.L. Arney, T. Arima, S.C. Barton, J.D. Brenton, and M.A. Surani. 2002. Novel conserved elements upstream of the *H19* gene are transcribed and act as mesodermal enhancers. *Development* 129:1205–13.

Drewell, R.A., J.D. Brenton, J.F. Ainscough, S.C. Barton, K.J. Hilton, K.L. Arney, L. Dandolo, and M.A. Surani. 2000. Deletion of a silencer element disrupts *H19* imprinting independently of a DNA methylation epigenetic switch. *Development* 127:3419–28.

Eden, S., M. Constancia, T. Hashimshony, W. Dean, B. Goldstein, A.C. Johnson, I. Keshet, W. Reik, and H. Cedar. 2001. An upstream repressor element plays a role in *Igf2* imprinting. *EMBO J* 20:3518–25.

Engel, N., J.L. Thorvaldsen, and M.S. Bartolomei. 2006. CTCF binding sites promote transcription initiation and prevent DNA methylation on the maternal allele at the imprinted *H19/Igf2* locus. *Hum Mol Genet* 15:2945–54.

Epstein, H., T.C. James, and P.B. Singh. 1992. Cloning and expression of *Drosophila* HP1 homologs from a mealybug, *Planococcus citri*. *J Cell Sci* 101 (Pt 2): 463–74.

Erhardt, S., F. Lyko, J.F. Ainscough, M.A. Surani, and R. Paro. 2003. Polycomb-group proteins are involved in silencing processes caused by a transgenic element from the murine imprinted *H19/Igf2* region in *Drosophila*. *Dev Genes Evol* 213:336–44.

Ferraro, M., G.L. Buglia, and F. Romano. 2001. Involvement of histone H4 acetylation in the epigenetic inheritance of different activity states of maternally and paternally derived genomes in the mealybug *Planococcus citri*. *Chromosoma* 110:93–101.

Fitch, K.R., G.K. Yasuda, K.N. Owens, and B.T. Wakimoto. 1998. Paternal effects in *Drosophila*: Implications for mechanisms of early development. *Curr Top Dev Biol* 38:1–34.

Fuks, F. 2005. DNA methylation and histone modifications: Teaming up to silence genes. *Curr Opin Genet Dev* 15:490–5.

Gehring, M., J.H. Huh, T.F. Hsieh, J. Penterman, Y. Choi, J.J. Harada, R.B. Goldberg, and R.L. Fischer. 2006. DEMETER DNA glycosylase establishes *MEDEA* polycomb gene self-imprinting by allele-specific demethylation. *Cell* 124:495–506.

Goday, C., and M.R. Esteban. 2001. Chromosome elimination in sciarid flies. *Bioessays* 23:242–50.

Goday, C., and M.F. Ruiz. 2002. Differential acetylation of histones H3 and H4 in paternal and maternal germline chromosomes during development of sciarid flies. *J Cell Sci* 115:4765–75.

Golic, K.G., M.M. Golic, and S. Pimpinelli. 1998. Imprinted control of gene activity in *Drosophila*. *Curr Biol* 8:1273–6.

Greciano, P.G., and C. Goday. 2006. Methylation of histone H3 at Lys4 differs between paternal and maternal chromosomes in *Sciara ocellaris* germline development. *J Cell Sci* 119:4667–77.

Grewal, S.I., and D. Moazed. 2003. Heterochromatin and epigenetic control of gene expression. *Science* 301:798–802.

Grossniklaus, U., J.P. Vielle-Calzada, M.A. Hoeppner, and W.B. Gagliano. 1998. Maternal control of embryogenesis by *MEDEA*, a polycomb group gene in *Arabidopsis*. *Science* 280:446–50.

Gutierrez-Marcos, J.F., L.M. Costa, M.D. Pra, S. Scholten, E. Kranz, P. Perez, and H.G. Dickinson. 2006. Epigenetic asymmetry of imprinted genes in plant gametes. *Nat Genet* 38:876–8.

Gutierrez-Marcos, J.F., P.D. Pennington, L.M. Costa, and H.G. Dickinson. 2003. Imprinting in the endosperm: A possible role in preventing wide hybridization. *Philos Trans R Soc Lond B Biol Sci* 358:1105–11.

Han, L., D. H. Lee, and P. E. Szabo. 2008. CTCF is the master organizer of domain-wide allele-specific chromatin at the *H19/Igf2* imprinted region. *Mol Cell Biol* 28:1124–35.

Hark, A. T., C. J. Schoenherr, D. J. Katz, R. S. Ingram, J. M. Levorse, and S. M. Tilghman. 2000. CTCF mediates methylation-sensitive enhancer-blocking activity at the *H19/Igf2* locus. *Nature* 405:486–9.

Haun, W. J., and N. M. Springer. 2008. Maternal and paternal alleles exhibit differential histone methylation and acetylation at maize imprinted genes. *Plant J* 56:903–12.

Holliday, R. 1990. Genomic imprinting and allelic exclusion. *Development* (Suppl): 125–9.

Irvine, D. V., M. Zaratiegui, N. H. Tolia, D. B. Goto, D. H. Chitwood, M. W. Vaughn, T. L. Joshua, and R. A. Martienssen. 2006. Argonaute slicing is required for heterochromatic silencing and spreading. *Science* 313:1134–7.

Iwasa, Y., and A. Pomiankowski. 1999. Sex specific X chromosome expression caused by genomic imprinting. *J Theor Biol* 197:487–95.

Joanis, V., and V. K. Lloyd. 2002. Genomic imprinting in *Drosophila* is maintained by the products of *Suppressor of variegation* and *trithorax* group, but not *Polycomb* group, genes. *Mol Genet Genomics* 268:103–12.

Jones, B. K., J. Levorse, and S. M. Tilghman. 2001. Deletion of a nuclease-sensitive region between the *Igf2* and *H19* genes leads to *Igf2* misregulation and increased adiposity. *Hum Mol Genet* 10:807–14.

Jullien, P. E., A. Katz, M. Oliva, N. Ohad, and F. Berger. 2006a. Polycomb group complexes self-regulate imprinting of the Polycomb group gene *MEDEA* in *Arabidopsis*. *Curr Biol* 16:486–92.

Jullien, P. E., T. Kinoshita, N. Ohad, and F. Berger. 2006b. Maintenance of DNA methylation during the *Arabidopsis* life cycle is essential for parental imprinting. *Plant Cell* 18:1360–72.

Kaffer, C. R., M. Srivastava, K. Y. Park, E. Ives, S. Hsieh, J. Batlle, A. Grinberg, S. P. Huang, and K. Pfeifer. 2000. A transcriptional insulator at the imprinted *H19/Igf2* locus. *Genes Dev* 14:1908–19.

Khatib, H., I. Zaitoun, and E. S. Kim. 2007. Comparative analysis of sequence characteristics of imprinted genes in human, mouse, and cattle. *Mamm Genome* 18:538–47.

Khosla, S., M. Augustus, and V. Brahmachari. 1999. Sex-specific organisation of middle repetitive DNA sequences in the mealybug *Planococcus lilacinus*. *Nucleic Acids Res* 27:3745–51.

Khosla, S., P. Kantheti, V. Brahmachari, and H. S. Chandra. 1996. A male-specific nuclease-resistant chromatin fraction in the mealybug *Planococcus lilacinus*. *Chromosoma* 104:386–92.

Khosla, S., G. Mendiratta, and V. Brahmachari. 2006. Genomic imprinting in the mealybugs. *Cytogenet Genome Res* 113:41–52.

Kinoshita, T., A. Miura, Y. Choi, Y. Kinoshita, X. Cao, S. E. Jacobsen, R. L. Fischer, and T. Kakutani. 2004. One-way control of *FWA* imprinting in *Arabidopsis* endosperm by DNA methylation. *Science* 303:521–23.

Kinoshita, T., R. Yadegari, J. J. Harada, R. B. Goldberg, and R. L. Fischer. 1999. Imprinting of the *MEDEA* polycomb gene in the *Arabidopsis* endosperm. *Plant Cell* 11:1945–52.

Kiyosue, T., N. Ohad, R. Yadegari, M. Hannon, J. Dinneny, D. Wells, A. Katz, et al. 1999. Control of fertilization-independent endosperm development by the *MEDEA* polycomb gene in *Arabidopsis*. *Proc Natl Acad Sci USA* 96:4186–91.

Kloc, A., M. Zaratiegui, E. Nora, and R. Martienssen. 2008. RNA interference guides histone modification during the S phase of chromosomal replication. *Curr Biol* 18:490–5.

Kohler, C., L. Hennig, R. Bouveret, J. Gheyselinck, U. Grossniklaus, and W. Gruissem. 2003a. *Arabidopsis* MSI1 is a component of the MEA/FIE Polycomb group complex and required for seed development. *EMBO J* 22:4804–14.

Kohler, C., L. Hennig, C. Spillane, S. Pien, W. Gruissem, and U. Grossniklaus. 2003b. The Polycomb-group protein MEDEA regulates seed development by controlling expression of the MADS-box gene *PHERES1*. *Genes Dev* 17:1540–53.

Koide, T., J. Ainscough, M. Wijgerde, and M. A. Surani. 1994. Comparative analysis of *Igf-2/H19* imprinted domain: Identification of a highly conserved intergenic DNase I hypersensitive region. *Genomics* 24:1–8.

Kono, T. 2006. Genomic imprinting is a barrier to parthenogenesis in mammals. *Cytogenet Genome Res* 113:31–5.

Kourmouli, N., P. Jeppesen, S. Mahadevhaiah, P. Burgoyne, R. Wu, D. M. Gilbert, S. Bongiorni, et al. 2004. Heterochromatin and tri-methylated lysine 20 of histone H4 in animals. *J Cell Sci* 117:2491–501.

Kubai, D. F. 1987. Nonrandom chromosome arrangements in germ line nuclei of *Sciara coprophila* males: The basis for nonrandom chromosome segregation on the meiosis I spindle. *J Cell Biol* 105:2433–46.

Kurukuti, S., V. K. Tiwari, G. Tavoosidana, E. Pugacheva, A. Murrell, Z. Zhao, V. Lobanenkov, W. Reik, and R. Ohlsson. 2006. CTCF binding at the *H19* imprinting control region mediates maternally inherited higher-order chromatin conformation to restrict enhancer access to *Igf2*. *Proc Natl Acad Sci USA* 103:10684–9.

Lachner, M., D. O'Carroll, S. Rea, K. Mechtler, and T. Jenuwein. 2001. Methylation of histone H3 lysine 9 creates a binding site for HP1 proteins. *Nature* 410:116–20.

LaJeunesse, D., and A. Shearn. 1996. *E(z)*: A polycomb group gene or a trithorax group gene? *Development* 122:2189–97.

Leighton, P.A., J.R. Saam, R.S. Ingram, C.L. Stewart, and S.M. Tilghman. 1995. An enhancer deletion affects both *H19* and *Igf2* expression. *Genes Dev* 9:2079–89.

Li, T., J.F. Hu, X. Qiu, J. Ling, H. Chen, S. Wang, A. Hou, T.H. Vu, and A.R. Hoffman. 2008. CTCF regulates allelic expression of *Igf2* by orchestrating a promoter-polycomb repressive complex-2 intrachromosomal loop. *Mol Cell Biol* 28:6473–82.

Lippman, Z., and R. Martienssen. 2004. The role of RNA interference in heterochromatic silencing. *Nature* 431:364–70.

Lloyd, V. 2000. Parental imprinting in *Drosophila*. *Genetica* 109:35–44.

Lloyd, V.K., D.A. Sinclair, and T.A. Grigliatti. 1999. Genomic imprinting and position-effect variegation in *Drosophila melanogaster*. *Genetics* 151:1503–16.

Lyko, F., J.D. Brenton, M.A. Surani, and R. Paro. 1997. An imprinting element from the mouse *H19* locus functions as a silencer in *Drosophila*. *Nat Genet* 16:171–3.

Lyko, F., K. Buiting, B. Horsthemke, and R. Paro. 1998. Identification of a silencing element in the human 15q11-q13 imprinting center by using transgenic *Drosophila*. *Proc Natl Acad Sci USA* 95:1698–1702.

MacDonald, W.A., and V.K. Lloyd. 2004. Genomic imprinting in *Drosophila* and the role of noncoding RNAs. Presented at the 45th Annual *Drosophila* Research Conference, Washington DC. Abstract 323B.

Maggert, K.A., and K.G. Golic. 2002. The Y chromosome of *Drosophila melanogaster* exhibits chromosome-wide imprinting. *Genetics* 162:1245–58.

Maggert, K.A., and K.G. Golic. 2004. DNA methylation affects genomic imprinting. Presented at the 45th Annual Drosophila Research Conference, Washington DC. Abstract 133.

Maison, C., D. Bailly, A.H. Peters, J.P. Quivy, D. Roche, A. Taddei, M. Lachner, T. Jenuwein, and G. Almouzni. 2002. Higher-order structure in pericentric heterochromatin involves a distinct pattern of histone modification and an RNA component. *Nat Genet* 30:329–34.

Makarevich, G., O. Leroy, U. Akinci, D. Schubert, O. Clarenz, J. Goodrich, U. Grossniklaus, and C. Kohler. 2006. Different Polycomb group complexes regulate common target genes in *Arabidopsis*. *EMBO Rep* 7:947–52.

Makarevich, G., C.B. Villar, A. Erilova, and C. Kohler. 2008. Mechanism of *PHERES1* imprinting in *Arabidopsis*. *J Cell Sci* 121:906–12.

Martin, C.C., and R. McGowan. 1995. Genotype-specific modifiers of transgene methylation and expression in the zebrafish, *Danio rerio*. *Genet Res* 65:21–8.

Matzke, M.A., and J.A. Birchler. 2005. RNAi-mediated pathways in the nucleus. *Nat Rev Genet* 6:24–35.

Mayer, C., K.M. Schmitz, J. Li, I. Grummt, and R. Santoro. 2006. Intergenic transcripts regulate the epigenetic state of rRNA genes. *Mol Cell* 22:351–61.

Mohan, K.N., and H.S. Chandra. 2005. Isolation and analysis of sequences showing sex-specific cytosine methylation in the mealybug *Planococcus lilacinus*. *Mol Genet Genomics* 274:557–68.

Moore, T., M. Constancia, M. Zubair, B. Bailleul, R. Feil, H. Sasaki, and W. Reik. 1997. Multiple imprinted sense and antisense transcripts, differential methylation and tandem repeats in a putative imprinting control region upstream of mouse *Igf2*. *Proc Natl Acad Sci USA* 94:12509–14.

Moore, T., and D. Haig. 1991. Genomic imprinting in mammalian development: A parental tug-of-war. *Trends Genet* 7:45–9.

Morison, I.M., J.P. Ramsay, and H.G. Spencer. 2005. A census of mammalian imprinting. *Trends Genet* 21:457–65.

Muller, J., C.M. Hart, N.J. Francis, M.L. Vargas, A. Sengupta, B. Wild, E.L. Miller, M.B. O'Connor, R.E. Kingston, and J.A. Simon. 2002. Histone methyltransferase activity of a *Drosophila* Polycomb group repressor complex. *Cell* 111:197–208.

Murrell, A., S. Heeson, L. Bowden, M. Constancia, W. Dean, G. Kelsey, and W. Reik. 2001. An intragenic methylated region in the imprinted *Igf2* gene augments transcription. *EMBO Rep* 2:1101–6.

Murrell, A., S. Heeson, and W. Reik. 2004. Interaction between differentially methylated regions partitions the imprinted genes *Igf2* and *H19* into parent-specific chromatin loops. *Nat Genet* 36:889–93.

Murrell, A., Y. Ito, G. Verde, J. Huddleston, K. Woodfine, M.C. Silengo, F. Spreafico, et al. 2008. Distinct methylation changes at the *IGF2-H19* locus in congenital growth disorders and cancer. *PLoS ONE* 3:e1849.

Nur, U. 1967. Reversal of heterochromatization and the activity of the paternal chromosome set in the male mealy bug. *Genetics* 56:375–89.

Ohlsson, R., A. Paldi, and J. A. Graves. 2001. Did genomic imprinting and X chromosome inactivation arise from stochastic expression? *Trends Genet* 17:136–41.

Okamura, K., and T. Ito. 2006. Lessons from comparative analysis of species-specific imprinted genes. *Cytogenet Genome Res* 113:159–64.

Orlando, V. 2003. Polycomb, epigenomes, and control of cell identity. *Cell* 112:599–606.

Peters, J., and C. Beechey. 2004. Identification and characterisation of imprinted genes in the mouse. *Brief Funct Genomic Proteomic* 2:320–33.

Plagge, A., A. R. Isles, E. Gordon, T. Humby, W. Dean, S. Gritsch, R. Fischer-Colbrie, L. S. Wilkinson, and G. Kelsey. 2005. Imprinted *Nesp55* influences behavioral reactivity to novel environments. *Mol Cell Biol* 25:3019–26.

Rieffel, S. M., and H. V. Crouse. 1966. The elimination and differentiation of chromosomes in the germ line of *Sciara*. *Chromosoma* 19:231–76.

Rudolph, T., M. Yonezawa, S. Lein, K. Heidrich, S. Kubicek, C. Schafer, S. Phalke, et al. 2007. Heterochromatin formation in *Drosophila* is initiated through active removal of H3K4 methylation by the LSD1 homolog SU(VAR)3-3. *Mol Cell* 26:103–15.

Scarbrough, K., S. Hattman, and U. Nur. 1984. Relationship of DNA methylation level to the presence of heterochromatin in mealybugs. *Mol Cell Biol* 4:599–603.

Schoenfelder, S., G. Smits, P. Fraser, W. Reik, and R. Paro. 2007. Non-coding transcripts in the *H19* imprinting control region mediate gene silencing in transgenic *Drosophila*. *EMBO Rep* 8:1068–73.

Schotta, G., A. Ebert, V. Krauss, A. Fischer, J. Hoffmann, S. Rea, T. Jenuwein, R. Dorn, and G. Reuter. 2002. Central role of *Drosophila* SU(VAR)3-9 in histone H3-K9 methylation and heterochromatic gene silencing. *EMBO J* 21:1121–31.

Schotta, G., M. Lachner, K. Sarma, A. Ebert, R. Sengupta, G. Reuter, D. Reinberg, and T. Jenuwein. 2004. A silencing pathway to induce H3-K9 and H4-K20 trimethylation at constitutive heterochromatin. *Genes Dev* 18:1251–62.

Scott, R. J., and M. Spielman. 2006. Deeper into the maize: New insights into genomic imprinting in plants. *Bioessays* 28:1167–71.

Simon, J. A., and J. W. Tamkun. 2002. Programming off and on states in chromatin: Mechanisms of Polycomb and trithorax group complexes. *Curr Opin Genet Dev* 12:210–18.

Soppe, W. J., S. E. Jacobsen, C. Alonso-Blanco, J. P. Jackson, T. Kakutani, M. Koornneef, and A. J. Peeters. 2000. The late flowering phenotype of *fwa* mutants is caused by gain-of-function epigenetic alleles of a homeodomain gene. *Mol Cell* 6:791–802.

Srivastava, M., S. Hsieh, A. Grinberg, L. Williams-Simons, S. P. Huang, and K. Pfeifer. 2000. *H19* and *Igf2* monoallelic expression is regulated in two distinct ways by a shared cis acting regulatory region upstream of *H19*. *Genes Dev* 14:1186–95.

Suzuki, S., M. B. Renfree, A. J. Pask, G. Shaw, S. Kobayashi, T. Kohda, T. Kaneko-Ishino, and F. Ishino. 2005. Genomic imprinting of *IGF2*, *p57(KIP2)* and *PEG1/MEST* in a marsupial, the tammar wallaby. *Mech Dev* 122:213–22.

Szabo, P. E., S. H. Tang, F. J. Silva, W. M. Tsark, and J. R. Mann. 2004. Role of CTCF binding sites in the *Igf2/H19* imprinting control region. *Mol Cell Biol* 24:4791–800.

Takagi, N., and M. Sasaki. 1975. Preferential inactivation of the paternally derived X chromosome in the extraembryonic membranes of the mouse. *Nature* 256:640–2.

Thorvaldsen, J. L., K. L. Duran, and M. S. Bartolomei. 1998. Deletion of the *H19* differentially methylated domain results in loss of imprinted expression of *H19* and *Igf2*. *Genes Dev* 12:3693–702.

Tiwari, S., R. Schulz, Y. Ikeda, L. Dytham, J. Bravo, L. Mathers, M. Spielman, et al. 2008. *MATERNALLY EXPRESSED PAB C-TERMINAL*, a novel imprinted gene in *Arabidopsis*, encodes the conserved C-terminal domain of polyadenylate binding proteins. *Plant Cell* 20:2387–98.

Tremblay, K. D., K. L. Duran, and M. S. Bartolomei. 1997. A 5′ 2-kilobase-pair region of the imprinted mouse *H19* gene exhibits exclusive paternal methylation throughout development. *Mol Cell Biol* 17:4322–9.

Tremblay, K. D., J. R. Saam, R. S. Ingram, S. M. Tilghman, and M. S. Bartolomei. 1995. A paternal-specific methylation imprint marks the alleles of the mouse *H19* gene. *Nat Genet* 9:407–13.

Umlauf, D., Y. Goto, R. Cao, F. Cerqueira, A. Wagschal, Y. Zhang, and R. Feil. 2004. Imprinting along the *Kcnq1* domain on mouse chromosome 7 involves repressive histone methylation and recruitment of Polycomb group complexes. *Nat Genet* 36:1296–1300.

Vaissiere, T., C. Sawan, and Z. Herceg. 2008. Epigenetic interplay between histone modifications and DNA methylation in gene silencing. *Mutat Res* 659:40–8.

VandeBerg, J. L., E. S. Robinson, P. B. Samollow, and P. G. Johnston. 1987. X-linked gene expression and X-chromosome inactivation: Marsupials, mouse, and man compared. *Isozymes Curr Top Biol Med Res* 15:225–53.

Varmuza, S., and M. Mann. 1994. Genomic imprinting—defusing the ovarian time bomb. *Trends Genet* 10:118–23.

Verona, R. I., J. L. Thorvaldsen, K. J. Reese, and M. S. Bartolomei. 2008. The transcriptional status but not the imprinting control region determines allele-specific histone modifications at the imprinted *H19* locus. *Mol Cell Biol* 28:71–82.

Vielle-Calzada, J. P., R. Baskar, and U. Grossniklaus. 2000. Delayed activation of the paternal genome during seed development. *Nature* 404:91–4.

Wakefield, M. J., A. M. Keohane, B. M. Turner, and J. A. Graves. 1997. Histone underacetylation is an ancient component of mammalian X chromosome inactivation. *Proc Natl Acad Sci USA* 94:9665–8.

Wood, A. J., and R. J. Oakey. 2006. Genomic imprinting in mammals: Emerging themes and established theories. *PLoS Genet* 2:e147.

Xiao, W., M. Gehring, Y. Choi, L. Margossian, H. Pu, J. J. Harada, R. B. Goldberg, R. I. Pennell, and R. L. Fischer. 2003. Imprinting of the *MEA* Polycomb gene is controlled by antagonism between MET1 methyltransferase and DME glycosylase. *Dev Cell* 5:891–901.

Yoon, Y. S., S. Jeong, Q. Rong, K. Y. Park, J. H. Chung, and K. Pfeifer. 2007. Analysis of the *H19*ICR insulator. *Mol Cell Biol* 27:3499–3510.

Zaratiegui, M., D. V. Irvine, and R. A. Martienssen. 2007. Noncoding RNAs and gene silencing. *Cell* 128:763–76.

5

Methylation Mapping in Humans

Christoph Grunau

CONTENTS

The importance of DNA methylation mapping in humans was rapidly recognized once it became apparent that 5-methylcytosine (5mC) is not only an exotic and negligible modification of the DNA but an important carrier of epigenetic information. In particular the finding that cancer is characterized by aberrant methylation increased the interest of the scientific community. Many laboratories attempted to characterize these changes in methylation and to use them as biomarkers for the diagnosis of disease. It had been known since the 1950s (Wyatt, 1951) that human DNA contains 5mC (later determined to be roughly 1%), and the initial experiments of Vanyushin and his colleagues (1973) and other laboratories showed that DNA of different tissues and different developmental stages can actually differ considerably (Table 5.1).

Enzymatic digest indicated that in mammals (e.g., calf thymus) the majority of 5mC is followed in the 5′–3′ direction by a guanine (Sinsheimer, 1954). It is today assumed that human

TABLE 5.1

Examples for 5mC Content in Different Human Tissues

TISSUE	5MC CONTENT OF TOTAL DNA (MOL%)	REFERENCE
Cultured fibroblasts	0.57[a] (2.80 ± 0.3 mol% of C)	Wilson et al. (1986), Wilson and Jones (1983)
Lymphocytes	0.96 ± 0.010	Ehrlich et al. (1982)
Total blood	1.22[a] (5.96 ± 0.22 mol% of C)	Corvetta et al. (1991)
Liver	1.47 ± 0.05	Tawa et al. (1992)
Spleen	1.67 ± 0.08	Tawa et al. (1992)

[a] Recalculated from original data based on genomewide average GC content of 41%.

DNA contains ~1% 5mC, corresponding to methylation of ~5% of cytosines and ~70% of CpG pairs. Soon, the underrepresentation of such CpG pairs was noticed in bovine tissue (Swartz et al., 1962). A milestone, which could also be considered the starting point of genome-wide methylation mapping, was the discovery by Adrian Bird's laboratory that a small fraction (<2%) of the genome of humans and other vertebrates can actually be digested by the methylation-sensitive restriction enzyme *Hpa*II (Cooper et al., 1983). While treatment with *Hpa*II leaves the major part of human DNA undigested because of the presence of 5mC in the inner C of recognition site CCGG, a small portion of tiny fragments is generated that can be visualized after separation through migration in agarose gels by the incorporation of radioactive deoxycytidine triphosphate (dCTP). These pieces of DNA were first called *Hpa*II tiny fragments (HTF), and later it became clear that they correspond to unmethylated regions in the genome where CpG pairs occur with statistically expected frequency. These small genomic regions are now known as "CpG islands" (CGIs). The features of a "canonical" CGI are a GC content of 55%–70% (bulk DNA ~40%), a ratio of observed to expected CpG of 0.6–0.7, and a length of 200–1,000 bp (Gardiner-Garden and Frommer, 1987; Takai and Jones, 2002). Early attempts to characterize DNA methylation in genes and repetitive sequences relied entirely on the ability of restriction enzymes to distinguish between methylated and unmethylated sites. Genomic DNA was digested with pairs of enzymes that possess identical restriction sites (*isoschizomer*), but while one of these enzymes does cut regardless of methylation, the other is blocked by methylation in the restriction site. Fragments are then separated by gel electrophoresis, and regions of interest are revealed by Southern blotting. The resolution of this type of analysis is, of course, limited by the frequency of restriction sites. Since several micrograms of DNA are required, only large cell populations could be studied. Nevertheless, these experiments showed that important differences in DNA methylation exist between individuals but also between different cell types and in tissue of the same individual. Later, the development of bisulfite sequencing technology by Marianne Frommer and colleagues (1992) allowed for DNA methylation analysis in single-base and single-molecule resolution. The technique relies on the chemical desamination of unmethylated cytosine into uracil and subsequent amplification of the region of interest by polymerase chain reaction (PCR). In the PCR products, uracil is replaced by thymine and 5mC by cytosine. Comparison of the sequence of the PCR products with the unconverted genomic sequence delivers the exact position of methylated cytosines. With the first results of this new technique it became apparent that not only does each individual or each tissue have its proper DNA methylation profile but each individual cell has its own methylation pattern, its proper *methylome*. In comparison, the genetic information remains relatively invariant within a given person. With the completion of the Human Genome Project in 2001 the mapping of DNA methylation in humans on a large scale could be envisaged. However, given the differences in DNA methylation in different cell types, the complexity of this task is orders of magnitude larger than that of genome-sequencing projects, where DNA from a single cell type provides information about the genome of all other cells. Not only must the position of methylated sites in a model cell population be identified, but this task must be performed for different tissues and cell types and in a way that allows for quantitation of the degree of methylation in every single C site. Even if recent years have seen huge progress in mapping and sequencing technology, DNA methylation studies are forced to reduce the complexity of the task: Either a subset of the genome must be preselected for analysis or compromises in mapping resolution must be made. Our knowledge about DNA methylation depends therefore on the experimental approach that was used to generate the data. Before attempting to resume what is currently known about DNA methylation of the human genome, we will therefore go through the major techniques that are employed to study 5mC distribution on a larger scale.

DNA METHYLATION MAPPING TECHNIQUES

DNA-mapping techniques that reduce the amount of sequence to be analyzed to a subgenomic scale can roughly be divided into four basic principles: (1) those relying on interaction with a binding molecule to 5mC (such as antibodies against 5mC or 5mC-binding proteins), or moieties that do not bind to methylated DNA but to unmethylated cytosines (such as CXXC); (2) enzymes that digest either methylated or unmethylated DNA and that can be used to enrich

these genomic fractions; (3) bisulfite treatment and PCR of the regions of interest; and finally (4) bioinformatics tools. Many of these techniques can be combined.

AFFINITY PURIFICATION OF METHYLATED OR UNMETHYLATED DNA FRAGMENTS

Since only a small fraction of the genome shows a high density of unmethylated cytosines, the enrichment of these unmethylated regions was one of the first attempts to reduce the amount of sequence data to be analyzed. Sally Cross and colleagues (1994) used a two-step procedure and divided DNA from human peripheral blood cells into methylated and unmethylated fractions by affinity chromatography using the methyl-CpG binding domain (MBD) of the rat MeCP2 protein attached to a solid support: The coding region of MBD was PCR-amplified and cloned and the recombinant protein was attached to a nickel-agarose matrix via a polyhistidine tag. Genomic DNA was fragmented with a restriction enzyme whose restriction site is rare within CGIs (*Mse*I, restriction site TTAA). Methylated fragments were removed from the DNA pool by a passage through the MBD column, and the remaining unmethylated DNA was methylated in vitro with CpG methylase M.*Sss*I. By this, fragments with unmethylated CGIs were converted into high-affinity molecules. Finally, DNA was passed through the MBD column and washed and fragments with high affinity were eluted with a high salt concentration. This fraction contained DNA that strongly binds to MBD, that is, contains (in vitro methylated but in vivo unmethylated) CGIs (Cross et al., 1994). This unmethylated DNA fraction was cloned and later sequenced several times, for instance, by the CGI tagging project (available for download at www.sanger.ac.uk/HGP/cgi.shtml) and by He and colleagues (2008). The sequences were mapped to the human genome when it became publicly available. Depending on assembly strategies, 14,000 (Negre and Grunau, 2006) to 20,000 (He et al., 2008) unmethylated regions were reconstructed, many of them corresponding to CGIs.

The experiment provided the first library of experimentally confirmed unmethylated CGIs and became the basis for a number of subsequent analyses. Recently, a complementary approach was used to enrich the same unmethylated fraction of the human genome (Illingworth et al., 2008). This time, the immobilized CXXC domain of the mouse Mbd1 protein, which binds specifically to nonmethylated CpG pairs, was used. Again, DNA from pooled peripheral blood cells from three male individuals was used and digested by MseI to cut AT-rich DNA into small fragments that contain too few CpGs to be retained by the CXXC matrix. In this so-called bulk genomic DNA, CpG pairs are roughly fivefold underrepresented (~0.8 per 100 bp) compared to the statistically expected frequency (Bird, 1980) and compared to CGIs (~1 CpG per 10 bp). Methylated DNA and CpG-poor DNA were eluted in low-salt conditions, and nonmethylated CpG-rich DNA was eluted with high salt, rechromatographed, cloned into plasmids, sequenced, and assembled. The final data set contains about 17,500 CGIs and is available as DAS source "CPG island clones" from the Ensembl Genome browser maintained by the European Bioinformatics Institute and the Wellcome Trust Sanger Institute (http://www.ensembl.org). Finally, unmethylated and methylated fractions of the genome can be separated by immunoprecipitation with antibodies against 5mC (known as MeDIP, mDIP, or mCIP). DNA is purified, either digested with restriction enzymes or sheared by sonication, incubated with the antibody, and finally separated into the (methylated) bound fraction and the (unmethylated) unbound fraction by incubation with protein A– or G–coated sepharose beads and centrifugation (Weber et al., 2005, 2007). Potential caveats of this technique are that DNA needs to be completely denatured before immunoprecipitation to make the 5mC moiety accessible to the antibody and that binding of the antibody appears to be influenced by CpG density. The high GC content of CGIs makes these regions notoriously difficult to denature completely. The principal advantage of MeDIP is that existing tilling microarrays can be used to hybridize the immunoprecipitated DNA and compare to signals obtained with total DNA. As long as enough DNA is available (~4 μg), this technique can be applied for a genomewide screen of cell type–specific DNA methylation or methylation of DNA from different individuals. Until recently it was, however, not possible to determine absolute methylation levels with these techniques. A recently developed bioinformatics tools (Down et al., 2008) appears to have solved this issue, and the technique was applied to determine tissue-specific methylation (Rakyan et al., 2008).

METHYLATION MAPPING WITH RESTRICTION ENZYMES

Restriction enzymes have been used for more than 30 years to differentiate methylated and unmethylated DNA (Bird and Southern, 1978). They are part of the bacterial defense system against foreign DNA and allow bacteria to "restrict" the growth of bacteriophages. Foreign DNA is digested by bacterial endonucleases. In the noncompartmented bacterial cell, the bacterial DNA is protected by chemical modifications from the action of these restriction enzymes. One of these modifications can be the methylation of cytosine, identical to the one found in humans. The exhaustive screening of bacterial species for such endonucleases has provided biologists with a rich repertoire of restriction enzymes that are sensitive or insensitive to methylation and that by chance possess the same recognition site (isoschizomeres). However, not all isoschizomeric enzyme pairs are commercially available. One of the widely used enzyme pairs is HpaII and its methylation-insensitive isoschizomere MspI (restriction site CCGG). Other frequently used CpG methylation-sensitive endonucleases are TaiI (ACGT), BstUI (CGCG), HhaI (GCGC), and AciI (CCGC). In combination with Southern blotting, these enzymes were initially applied to the analysis of individual loci and repetitive sequences; but recent years have seen the development of new methods for methylation mapping on a

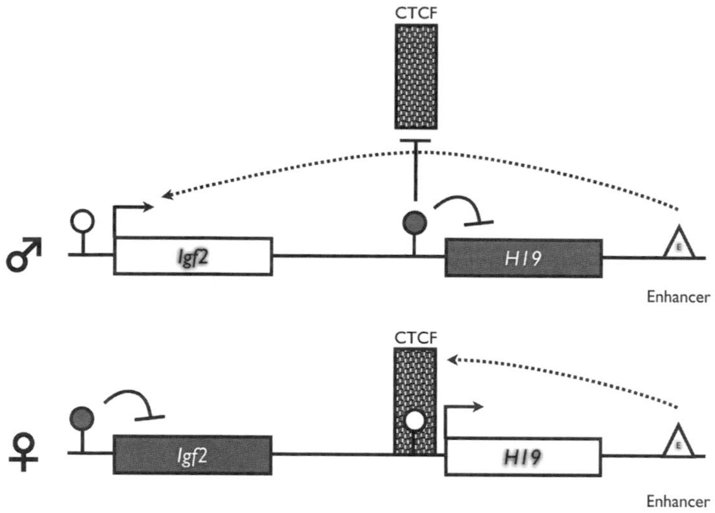

FIGURE 5.2 Schematic representation of DNA methylation of imprinted genes using the *Igf2/H19* locus as an example. (Empty circles) Unmethylated differentially methylated region (DMR), (full circles) methylated DMR. Paternal chromosome on top, maternal on bottom. Not to scale.

deoxyuracil glycosylase system (reviewed in Krokan et al., 2002). It is still in discussion to what extent this deamination rate is dependent on local sequence composition, in particular the GC content.

IMPRINTED GENES

Imprinted genes are a class of genes whose expression status depends on the parent—either father or mother—from whom they were inherited. It is the parental origin that strictly determines which of the two alleles is silenced and which is active. Imprinted genes illustrate the fact that information inherited through the paternal line and that inherited through the maternal line is not identical, even if the genetic information (i.e., the DNA sequence) was absolutely the same. The information about the parental origin is coded by epigenetic information carriers and, in particular, by DNA methylation. The imprinted gene catalogue (www.otago.ac.nz/IGC) lists currently about 60 imprinted genes in humans (in mice about 100). Loss of imprinting (LOI) leads to (or accompanies) severe developmental diseases or cancerogenesis. Most imprinted genes are organized in clusters, and transcription is controlled by DNA elements that are called "imprinting centers" (ICs). Another feature of these clusters is the presence of differentially methylated regions

(DMRs, or differentially methylated domains [DMDs]) with parent-specific methylation on the different alleles. A classic example is the *IGF2* (Insulin-like growth factor 2)/*H19* gene pair located on 11p15. *IGF2* coding for a fetal growth factor is in most tissues expressed only from the paternal allele. Its transcription is under the control of enhancers downstream of the *H19* locus (Figure 5.2). In contrast to *IGF2*, *H19* is transcribed only from the maternal allele. In the *IGF2* gene there are two DMRs. DMR0 is in intron 2 of *IGH2*, and DMR2 overlaps intron 8 and exon 9. Both DMRs show methylation on the maternal allele. Upstream of *H19*, the unmethylated maternal allele has binding sites for CTCF, an insulator element that prevents access of the enhancers to *IGF2*. LOI of *IGF2* in Wilms tumor (a tumor of the kidneys that can occur during childhood) and Beckwith-Wiedemann syndrome (a disease characterized by overgrowth and susceptibility to tumor formation) is associated with transcriptional repression and hypermethylation of the maternal *H19* allele.

PSEUDOGENES AND DUPLICATED GENES

In mammals, despite several large-scale epigenetic studies, relatively little is known about the epigenetic state of duplicated loci. This stems from the experimental procedures that were used in most studies and that were not suitable for the analysis of very similar DNA sequences. Essentially, duplicated loci are carefully avoided in these studies. Nevertheless, for a few duplicated loci, the epigenetic status is known: In humans, the pericentromeric region is highly enriched in long stretches of segmental duplications (Bailey et al., 2002), and for a number of human genes in these regions (*SLC6A8, ANKRD21, BAGE, TPTE*) we have analyzed the methylation status. We have shown that DNA of duplicated loci that reside in this region is highly methylated and the loci are in general transcriptionally silent. In contrast, the original copy is hypomethylated and can be transcribed (Grunau et al., 2000, 2006). Only during cancerogenesis and spermatogenesis do the duplicated genes lose methylation (i.e., the heterochromatic compartment becomes transcriptionally competent) The creatine-transporter gene *SLC6A8* is located in Xq28. The gene is expressed tissue-specifically, and the associated CGI is not methylated. The duplicated autosomal copy (ψ*SLC6A8*) has a similarity of 95% with *SLC6A8*, resides in the pericentromeric region of chromosome 16, and is not transcribed in somatic tissue. During spermatogenesis, the highly methylated CGI becomes hypomethylated and the duplicated locus is activated (Grunau et al., 2000). The next example is the *BAGE* family of genes. *BAGE* (B melanoma antigen) genes were generated by duplication and translocation of part of the *MLL3* gene on chromosome 7 to the pericentromeric heterochromatic compartment of chromosome 21 or 13, exon shuffling, and further duplication and translocation events to pericentromeric regions of other chromosomes (Ruault et al., 2003). CGIs of *BAGE* genes are highly methylated in all analyzed loci (Grunau et al., 2006). Another example is *TPTE*. Copies of human *TPTE* (coding for a putative transmembrane phosphatase) map to the pericentromeric regions of several human chromosomes. There is only one orthologous copy in the mouse genome, mapping to a region that is syntenic to human 13q14.2-q21. This euchromatic chromosomal region con-

tains a divergent *TPTE* sequence that represents probably the ancestral locus from which the other copies arose through duplication and translocation events (Tapparel et al., 2003). Except for the testis, human *TPTE* loci are highly methylated (Grunau et al., 2006). Finally, members of the *POTE* gene family (e.g., *ANKRD21*) map to the pericentromeric regions of many human chromosomes and to the long arm of chromosome 2 (2q21). Phylogenetic studies have identified the original locus on the short arm of chromosome 10 (*ANKRD26*) from which the *POTE* loci were generated by duplication and translocation (Hahn et al., 2006). All pericentromeric copies are transcriptionally silent. We have shown (Grunau et al., 2006) that the locus on chromosome 21 is methylated, which is probably true for all pericentromeric copies. The *POTE* gene family can be subdivided into three groups based on sequence similarity. The members of group 3, POTE-2α, 2β, and 2γ are located on the long arm of chromosome 2 (2q21); and in contrast to the pericentromeric copies, these loci are transcribed in other tissues than the testis (prostate, ovary, and placenta) (Bera et al., 2006). The long arm of chromosome 2 arose from a telomere-to-telomere fusion of two ancestral ape-type chromosomes in 2q12-14 (Kasai et al., 2000). Phylogenetic analysis of the *POTE* family of genes indicated that group 3 was formed before the separation of apes and Old World monkeys (macaque and baboon). By in situ hybridization, under low-stringency conditions with two alphoid DNA probes, Avarello et al. (1992) detected signals on the long arm of chromosome 2 at approximately q21.3-q22.1. These findings indicate that the ancestral centromere was located here and that the entire pericentromeric region or parts of it (including *POTE-2*) underwent euchromatization after the fusion event. Using data from the Human Epigenome Project (Eckhardt et al., 2006), Cortese et al. (2008) analyzed methylation of the *PLG* and *TBX* gene families and in 17 processed and 15 unprocessed pseudogenes for which the original gene had tissue-specific methylation. The *PLG* sequences of POTE2β, -γ, and -δ; POTE14α and -β; and POTE22 are nearly identical (≥96%). However, CGIs of these loci are highly divergent due to VNTR detectable as length polymorphism.

DNA METHYLATION IN CENTROMERES, IN PERICENTROMERIC AND SUBTELOMERIC REGIONS, AND ON THE INACTIVE X CHROMOSOME

Microscopic studies of human metaphase chromosomes using immunofluorescence and antibodies against 5mC have shown that the DNA component of the constitutive heterochromatin of centromeres, pericentromeric regions, and subtelomeric regions is densely methylated. The telomeric DNA sequence $(TTAGGG)_n$ itself does not contain CpG and is probably unmethylated. It is not entirely clear to what extent the pericentromeric regions carry DNA methylation. We have shown for chromosome 21 that the region that contains methylated CGIs coincidence with the region that is composed of large segmental duplications. These stretches of several hundred base pairs that exist in multiple copies in the genome can be found in the pericentromeric region of every human chromosome (Bailey et al., 2002). It remains to be investigated if high DNA methylation is a general characteristic of these duplicated regions. Unfortunately, they are in general avoided in large-scale studies that rely on hybridization or short sequencing stretches for practical experimental reasons (the duplicated sequences cannot be distinguished). It could be that the pericentromeric chromosomal region is methylated and therefore heterochromatic and that gene duplications are tolerated here since expression of potentially detrimental transcripts is less likely than in other regions. It could also be that the dense DNA methylation is a result of the duplication event. In mammals, females carry two X chromosomes, while males have only one. To assure that in both males and females the same number of transcripts from the genes encoded on the X chromosome are produced, dose compensation is required. One X chromosome in females is inactivated in a process that involves specific expression of the Xist RNA from the inactivated chromosome, histone deacetylation, and dense DNA methylation of the inactivated chromosome. Marsupial mammals undergo nonrandom X-chromosome inactivation: It is the paternal X chromosome that is preferentially inactivated (*imprinted X-inactivation*). In eutherians (placental mammals), random X-inactivation occurs in the somatic cells and either the paternal or the maternal X chromosome is silenced.

DATABASES

Catalogues and databases have always been indispensable tools for biologists. Not surprisingly, databases have also been developed to store DNA methylation data. Most large-scale studies have made their data available through genome browsers that display the degree of methylation along the analyzed DNA stretches. Examples are the Methylation Landscape of the Human Genome (http://epigenomics.cu-genome.org/html/meth_landscape/index.html#results) (Rollins et al., 2006; accessible through the genome browser developed by the University of Southern California) and MeDIP-chip data (Rakyan et al., 2008; accessible as DAS source through the Ensembl genome browser maintained at the European Bioinformatics Institute, http://www.ensembl.org/Homo_sapiens/). However, still most DNA methylation data are produced by small- to medium-scale gene-by-gene and tissue-by-tissue studies. These data are published, but in general, they are not available in a database. The often very different experimental approaches make it also difficult to store and to compare the data. The DNA Methylation Database (http://www.methdb.net) is a curated database that attempts to store and to standardize these literature data. The database can be linked to a genome browser through the DAS protocol. Additional databases have been developed that are based on automatic scanning of the literature, such as PubMeth (http://www.pubmeth

.org/) (Ongenaert et al., 2008) and Meth CancerDB (http://www.methcancerdb.net/methcancerdb/) (Lauss et al., 2008). An interesting approach is the MethyCancer database of the Chinese Epigenome project (He et al., 2008): There, data have been loaded from other databases, were obtained through automatic literature searches, and can be queried through a common search form (http://methycancer.psych.ac.cn/). The name *MethyCancer* is slightly misleading since methylation in healthy tissue is also included. When data are mirrored this way, there is a considerable risk of redundancy with discrepancy through update lags. Nevertheless, redundancy helps also to maintain the information flow in case a database becomes temporarily or permanently unavailable. An example is the database of the first large-scale DNA methylation study (Human Epigenome Project, http://www.sanger.ac.uk/Post Genomics/epigenome/mvpviewer.shtml) that currently has been inactivated, but the data are still available in MethyCancer.

During the last 40 years our knowledge of DNA methylation in human DNA has been expanding rapidly. We know now that the majority of DNA is methylated in CpG pairs and that only small (<2kb), precisely defined promoter-near CGIs are free of methylation. If these CGIs are methylated, the associated gene becomes silenced; however, if the CGI is unmethylated, its gene is not necessarily active. Only for a small portion of genes (probably less than 10%) is strict correlation between tissue-specific methylation and tissue-specific expression observed. CGIs that are not associated with promoters can be either hypo- or hypermethylated. Highly repetitive sequences are always methylated, even if they contain CGIs. The methylation status of duplicated sequences (2–20 copies) such as gene–pseudogene pairs depends on the local chromatin context and probably the time since duplication occurred. Each individual cell possesses a different sequence of methylated CpG pairs, but these methylation patterns are similar in similar cell types. Important regular differences exist between the DNA methylation profiles of somatic tissue and reproductive tissue such as testis, sperm cells, oocytes, and placenta. Even moderate changes in DNA methylation can have profound effects on the phenotype of the concerned cells (e.g., cancerogenesis), underlining the importance of this carrier of epigenetic information.

REFERENCES

Amoreira, C., W. Hindermann, and C. Grunau. 2003. An improved version of the DNA methylation database (MethDB). *Nucleic Acids Res* 31:75–7.

Avarello, R., A. Pedicini, A. Caiulo, O. Zuffardi, and M. Fraccaro. 1992. Evidence for an ancestral alphoid domain on the long arm of human chromosome 2. *Hum Genet* 89:247–9.

Bailey, J. A., Z. Gu, R. A. Clark, K. Reinert, R. V. Samonte, S. Schwartz, M. D. Adams, E. W. Myers, P. W. Li, and E. E. Eichler. 2002. Recent segmental duplications in the human genome. *Science* 297:1003–7.

Batzer, M. A., and P. L. Deininger. 2002. Alu repeats and human genomic diversity. *Nat Rev Genet* 3:370–9.

Benson, G. 1999. Tandem repeats finder: A program to analyze DNA sequences. *Nucleic Acids Res* 27:573–80.

Bera, T. K., A. Saint Fleur, Y. Lee, A. Kydd, Y. Hahn, N. C. Popescu, D. B. Zimonjic, B. Lee, and I. Pastan. 2006. POTE paralogs are induced and differentially expressed in many cancers. *Cancer Res* 66:52–6.

Bird, A. P. 1980. DNA methylation and the frequency of CpG in animal DNA. *Nucleic Acids Res* 8:1499–1504.

Bird, A. P., and E. M. Southern. 1978. Use of restriction enzymes to study eukaryotic DNA methylation: I. The methylation pattern in ribosomal DNA from *Xenopus laevis*. *J Mol Biol* 118:27–47.

Boissinot, S., and A. V. Furano. 2005. The recent evolution of human L1 retrotransposons. *Cytogenet Genome Res* 110:402–6.

Brock, G. J., and A. Bird. 1997. Mosaic methylation of the repeat unit of the human ribosomal RNA genes. *Hum Mol Genet* 6:451–6.

Burden, A. F., N. C. Manley, A. D. Clark, S. M. Gartler, C. D. Laird, and R. S. Hansen. 2005. Hemimethylation and non-CpG methylation levels in a promoter region of human LINE-1 (L1) repeated elements. *J Biol Chem* 280:14413–19.

Cokus, S. J., S. Feng, X. Zhang, Z. Chen, B. Merriman, C. D. Haudenschild, S. Pradhan, et al. (2008) Shotgun bisulphite sequencing of the

Arabidopsis genome reveals DNA methylation patterning. *Nature* 452:215–19.

Cooper, D.N., M.H. Taggart, and A.P. Bird. 1983. Unmethylated domains in vertebrate DNA. *Nucleic Acids Res* 11:647–58.

Cortese, R., M. Krispin, G. Weiss, K. Berlin, and F. Eckhardt. 2008. DNA methylation profiling of pseudogene–parental gene pairs and two gene families. *Genomics* 91:492–502.

Corvetta, A., R. Della Bitta, M.M. Luchetti, and G. Pomponio. 1991. 5-Methylcytosine content of DNA in blood, synovial mononuclear cells and synovial tissue from patients affected by autoimmune rheumatic diseases. *J Chromatogr* 566: 481–91.

Cross, S.H., J.A. Charlton, X. Nan, and A.P. Bird. 1994. Purification of CpG islands using a methylated DNA binding column. *Nat Genet* 6:236–44.

Das, R., N. Dimitrova, Z. Xuan, R.A. Rollins, F. Haghighi, J.R. Edwards, J. Ju, T.H. Bestor, and M.Q. Zhang. 2006. Computational prediction of methylation status in human genomic sequences. *Proc Natl Acad Sci USA* 103:10713–16.

Down, T.A., V.K. Rakyan, D.J. Turner, P. Flicek, H. Li, E. Kulesha, S. Gräf, et al. 2008. A Bayesian deconvolution strategy for immunoprecipitation-based DNA methylome analysis. *Nat Biotechnol* 26:779–85.

Eckhardt, F., J. Lewin, R. Cortese, V.K. Rakyan, J. Attwood, M. Burger, J. Burton, et al. 2006. DNA methylation profiling of human chromosomes 6, 20 and 22. *Nat Genet* 38:1378–85.

Ehrlich, M. 2002. DNA methylation in cancer: Too much, but also too little. *Oncogene* 21:5400–13.

Ehrlich, M., M.A. Gama-Sosa, L.H. Huang, R.M. Midgett, K.C. Kuo, R.A. McCune, and C. Gehrke. 1982. Amount and distribution of 5-methylcytosine in human DNA from different types of tissues of cells. *Nucleic Acids Res* 10:2709–21.

Frommer, M., L.E. McDonald, D.S. Millar, C.M. Collis, F. Watt, G.W. Grigg, P.L. Molloy, and C.L. Paul. 1992. A genomic sequencing protocol that yields a positive display of 5-methylcytosine residues in individual DNA strands. *Proc Natl Acad Sci USA* 89:1827–31.

Futscher, B.W., M.M. Oshiro, R.J. Wozniak, N. Holtan, C.L. Hanigan, H. Duan, and F.E. Domann. 2002. Role for DNA methylation in the control of cell type specific maspin expression. *Nat Genet* 31:175–9.

Gardiner-Garden, M., and M. Frommer. 1987. CpG islands in vertebrate genomes. *J Mol Biol* 196:261–82.

Grunau, C., W. Hindermann, and A. Rosenthal. 2000. Large-scale methylation analysis of human genomic DNA reveals tissue-specific differences between the methylation profiles of genes and pseudogenes. *Hum Mol Genet* 9:2651–63.

Grunau, C., J. Buard, M.E. Brun, and A. De Sario. 2006. Mapping of the juxtacentromeric heterochromatin–euchromatin frontier of human chromosome 21. *Genome Res* 16:1198–1207.

Hahn, Y., T.K. Bera, I.H. Pastan, and B. Lee. 2006. Duplication and extensive remodeling shaped POTE family genes encoding proteins containing ankyrin repeat and coiled coil domains. *Gene* 366:238–45.

Hansen, R.S., C. Wijmenga, P. Luo, A.M. Stanek, T.K. Canfield, C.M. Weemaes, and S.M. Gartler. 1999. The DNMT3B DNA methyltransferase gene is mutated in the ICF immunodeficiency syndrome. *Proc Natl Acad Sci USA* 96:14412–17.

He, X., S. Chang, J. Zhang, Q. Zhao, H. Xiang, K. Kusonmano, L. Yang, Z.S. Sun, H. Yang, and J. Wang. 2008. MethyCancer: The database of human DNA methylation and cancer. *Nucleic Acids Res* 36:D836–41.

Hellmann-Blumberg, U., M.F. Hintz, J.M. Gatewood, and C.W. Schmid. 1993. Developmental differences in methylation of human Alu repeats. *Mol Cell Biol* 13:4523–30.

Holliday, R. 1987. The inheritance of epigenetic detects. *Science* 238:163–70.

Illingworth, R., A. Kerr, D. DeSousa, H. Jørgensen, P. Ellis, J. Stalker, D. Jackson, et al. 2008. A novel CpG island set identifies tissue-specific methylation at developmental gene loci. *PLoS Biol* 6:e22.

Kasai, F., E. Takahashi, K. Koyama, K. Terao, Y. Suto, K. Tokunaga, Y. Nakamura, and M. Hirai. 2000. Comparative FISH mapping of the ancestral fusion point of human chromosome 2. *Chromosome Res* 8:727–35.

Khulan, B., R.F. Thompson, K. Ye, M.J. Fazzari, M. Suzuki, E. Stasiek, M.E. Figueroa, et al. 2006. Comparative isoschizomer profiling of cytosine methylation: The HELP assay. *Genome Res* 16:1046–55.

Krokan, H.E., F. Drablos, and G. Slupphaug. 2002. Uracil in DNA—occurrence, consequences and repair. *Oncogene* 21:8935–48.

Lander, E.S., L.M. Linton, B. Birren, C. Nusbaum, M.C. Zody, J. Baldwin, K. Devon, et al. 2001. Initial sequencing and analysis of the human genome. *Nature* 409:860–921.

Lauss, M., I. Visne, A. Weinhaeusel, K. Vierlinger, C. Noehammer, and A. Kriegner. 2008. MethCancerDB—aberrant DNA methylation in human cancer. *Br J Cancer* 98:816–17.

Li, L.C., and R. Dahiya. 2002. MethPrimer: Designing primers for methylation PCRs. *Bioinformatics* 18:1427–31.

Lo, H. S., Z. Wang, Y. Hu, H. H. Yang, S. Gere, K. H. Buetow, and M. P. Lee. 2003. Allelic variation in gene expression is common in the human genome. *Genome Res* 13:1855–62.

Marques, C. J., P. Costa, B. Vaz, F. Carvalho, S. Fernandes, A. Barros, and M. Sousa. 2008. Abnormal methylation of imprinted genes in human sperm is associated with oligozoospermia. *Mol Hum Reprod* 14:67–74.

Meklat, F., Z. Li, Z. Wang, Y. Zhang, J. Zhang, A. Jewell, and S. H. Lim. 2007. Cancer-testis antigens in haematological malignancies. *Br J Haematol* 136:769–76.

Negre, V., and C. Grunau. 2006. The MethDB DAS server: Adding an epigenetic information layer to the human genome. *Epigenetics* 1:101–5.

Ohno, S., U. Wolf, and N. B. Atkin. 1968. Evolution from fish to mammals by gene duplication. *Hereditas* 59:169–87.

Ongenaert, M., L. Van Neste, T. De Meyer, G. Menschaert, S. Bekaert, and W. Van Criekinge. 2008. PubMeth: A cancer methylation database combining text-mining and expert annotation. *Nucleic Acids Res* 36:D842–6.

Pfeifer, G. P. 2006. Mutagenesis at methylated CpG sequences. *Curr Top Microbiol Immunol* 301:259–81.

Rakyan, V., T. A. Down, N. P. Thorne, P. Flicek, E. Kulesha, S. Gräf, E. M. Tomazou, et al. 2008. An integrated resource for genome-wide identification and analysis of human tissue-specific differentially methylated regions (tDMRs). *Genome Res* 18:1518–29.

Rakyan, V. K., T. Hildmann, K. L. Novik, J. Lewin, J. Tost, A. V. Cox, T. D. Andrews, et al. 2004. DNA methylation profiling of the human major histocompatibility complex: A pilot study for the human epigenome project. *PLoS Biol* 2:e405.

Rodin, S. N., and A. D. Riggs. 2003. Epigenetic silencing may aid evolution by gene duplication. *J Mol Evol* 56:718–29.

Rollins, R. A., F. Haghighi, J. R. Edwards, R. Das, M. Q. Zhang, J. Ju, and T. H. Bestor. 2006. Large-scale structure of genomic methylation patterns. *Genome Res* 16:157–63.

Ruault, M., M. Ventura, N. Galtier, M. E. Brun, N. Archidiacono, G. Roizes, and A. De Sario. 2003. BAGE genes generated by juxtacentromeric reshuffling in the Hominidae lineage are under selective pressure. *Genomics* 81:391–9.

Rubin, C. M., C. A. VandeVoort, R. L. Teplitz, and C. W. Schmid. 1994. Alu repeated DNAs are differentially methylated in primate germ cells. *Nucleic Acids Res* 22:5121–7.

Saxonov, S., P. Berg, and D. L. Brutlag. 2006. A genome-wide analysis of CpG dinucleotides in the human genome distinguishes two distinct classes of promoters. *Proc Natl Acad Sci USA* 103:1412–17.

Shen, L., Y. Kondo, Y. Guo, J. Zhang, L. Zhang, S. Ahmed, J. Shu, X. Chen, R. A. Waterland, and I. P. Issa. 2007. Genome-wide profiling of DNA methylation reveals a class of normally methylated CpG island promoters. *PLoS Genet* 3:2023–36.

Sinsheimer, R. L. 1954. The action of pancreatic desoxyribonuclease. I. Isolation of mono- and dinucleotides. *J Biol Chem* 208:445–59.

Strichman-Almashanu, L. Z., R. S. Lee, P. O. Onyango, E. Perlman, F. Flam, M. B. Frieman, and A. P. Feinberg. 2002. A genome-wide screen for normally methylated human CpG islands that can identify novel imprinted genes. *Genome Res* 12:543–54.

Sutherland, E., L. Coe, and E. A. Raleigh. 1992. McrBC: A multisubunit GTP-dependent restriction endonuclease. *J Mol Biol* 225:327–48.

Swartz, M. N., T. A. Trautner, and A. Kornberg. 1962. Enzymatic synthesis of deoxyribonucleic acid. XI. Further studies on nearest neighbor base sequences in deoxyribonucleic acids. *J Biol Chem* 237:1961–7.

Takai, D., and P. A. Jones. 2002. Comprehensive analysis of CpG islands in human chromosomes 21 and 22. *Proc Natl Acad Sci USA* 99:3740–5.

Takai, D., and P. A. Jones. 2003. The CpG island searcher: A new WWW resource. *In Silico Biol* 3:235–40.

Tapparel, C., A. Reymond, C. Girardet, L. Guillou, R. Lyle, C. Lamon, P. Hutter, and S. E. Antonarakis. 2003. The TPTE gene family: Cellular expression, subcellular localization and alternative splicing. *Gene* 323:189–99.

Tawa, R., S. Ueno, K. Yamamoto, Y. Yamamoto, K. Sagisaka, R. Katakura, T. Kayama, T. Yoshimoto, H. Sakurai, and T. Ono. 1992. Methylated cytosine level in human liver DNA does not decline in aging process. *Mech Ageing Dev* 62:255–61.

Taylor, K. H., R. S. Kramer, J. W. Davis, J. Guo, D. J. Duff, D. Xu, C. W. Caldwell, and H. Shi. 2007. Ultradeep bisulfite sequencing analysis of DNA methylation patterns in multiple gene promoters by 454 sequencing. *Cancer Res* 67:8511–18.

Vanyushin, B. F., A. L. Mazin, V. K. Vasilyev, and A. N. Belozersky. 1973. The content of 5-methylcytosine in animal DNA: The species and tissue specificity. *Biochim Biophys Acta* 299:397–403.

Waechter, D. E., and R. Baserga. 1982. Effect of methylation on expression of microinjected genes. *Proc Natl Acad Sci USA* 79:1106–10.

Weber, M., J. J. Davies, D. Wittig, E. J. Oakeley, M. Haase, W. L. Lam, and D. Schubeler. 2005. Chromosome-wide and promoter-specific analyses identify sites of differential DNA methylation in normal and transformed human cells. *Nat Genet* 37:853–62.

Weber, M., I. Hellmann, M. B. Stadler, L. Ramos, S. Paabo, M. Rebhan, and D. Schubeler. 2007. Distribution, silencing potential and evolutionary impact of promoter DNA methylation in the human genome. *Nat Genet* 39:457–66.

Wilson, V. L., and P. A. Jones. 1983. DNA methylation decreases in aging but not in immortal cells. *Science* 220:1055–7.

Wilson, V. L., R. A. Smith, H. Autrup, H. Krokan, D. E. Musci, N. N. Le, J. Longoria, D. Ziska, and C. C. Harris. 1986. Genomic 5-methylcytosine determination by ^{32}P-postlabeling analysis. *Anal Biochem* 152:275–84.

Woodcock, D. M., C. B. Lawler, M. E. Linsenmeyer, J. P. Doherty, and W. D. Warren. 1997. Asymmet-

ric methylation in the hypermethylated CpG promoter region of the human L1 retrotransposon. *J Biol Chem* 272:7810–16.

Wyatt, G. R. 1951. Recognition and estimation of 5-methylcytosine in nucleic acids. *Biochem J* 48:581–4.

Yamada, Y., H. Watanabe, F. Miura, H. Soejima, M. Uchiyama, T. Iwasaka, T. Mukai, Y. Sakaki, and T. Ito. 2004. A comprehensive analysis of allelic methylation status of CpG islands on human chromosome 21q. *Genome Res* 14:247–66.

Yamada, Y., T. Shirakawa, T. D. Taylor, K. Okamura, H. Soejima, M. Uchiyama, T. Iwasaka, et al. 2006. A comprehensive analysis of allelic methylation status of CpG islands on human chromosome 11q: Comparison with chromosome 21q. *DNA Seq* 17:300–306.

6

Asexuality and Epigenetic Variation

Root Gorelick, Manfred Laubichler, and Rachel Massicotte

CONTENTS

Epigenetic processes are of fundamental importance for all living organisms as an individual phenotype is shaped by both its genome and its epigenome (Richards, 2006; Bossdorf et al., 2008). By *epigenetic*, we mean all molecular signals that are literally on top of DNA, such as cytosine methylation, chromatin marks, histone modification, and RNAi (Allis et al., 2007), many of which are responsible for classical developmental processes, as seen by embryologists. Collectively, we refer to all of these molecular epigenetic signals as the *epigenome* (Suzuki and Bird, 2008). Like Holliday and Pugh (1975), we consider these molecular epigenetic signals to be the nuts and bolts underlying classic epigenesis *sensu* Waddington. The developmental programs that lead to differentiated cell phenotypes are based on the interaction of genomic control mechanisms with programmed epigenetic signals that are established very precisely in space and time at the scale of an individual (Bird, 2002; Meissner et al., 2008). The integration of these intrinsic epigenetic signals by the genome enables cellular differentiation of all multicellular eukaryotic organisms (Jaenisch and Bird, 2003; Holliday, 2006; Allis et al., 2007). While this is true for any individual of any taxon, there are evidently some important differences among the epigenomes of different taxa (Suzuki and Bird, 2008); but there might even be some epigenetic variation among individuals of the same taxon that is not necessarily caused by genetic variation (Richards, 2006, 2008).

These two interacting variables, the genome and epigenome, though connected during development and differentiation can vary more or

Epigenetics: Linking Genotype and Phenotype in Development and Evolution, ed. Benedikt Hallgrímsson and Brian K. Hall.

less independently from one another; and as a result, we need to try to disentangle genetic from epigenetic variation in the shaping of an individual phenotype (Gorelick, 2004a, 2005; Bossdorf et al., 2008; Richards, 2008). Furthermore, a growing body of literature shows that organisms can have heritable epigenetic variation despite little or no heritable variation in DNA nucleotides (Cubas et al., 1999; Rakyan et al., 2003; Chong and Whitelaw, 2004; Weaver et al., 2004; Blewitt et al., 2006; Manning et al., 2006; Richards, 2006; Whitelaw and Whitelaw, 2006; Crews et al., 2007; Vaughn et al., 2007; Kucharski et al., 2008; Jablonka and Raz, 2009). While these phenomena are particularly difficult to investigate for organisms that reproduce sexually (with constant genetic mixing during meiosis and syngamy), asexual organisms are a better model system to study epigenetic variation (Massicotte and et al., in press).

In this chapter we first demonstrate how asexual lineages allow us to measure epigenetic variation that is distinct from genetic variation in natural populations. Then, we show how epigenetic variation can be used to demarcate generations—and thus to define individuals and heritability—in asexual organisms. While epigenetic signals often vary over the course of development, the same epigenetic signals can in some instances be more or less immutable from one generation to the next. Such constancy of epigenetic signals probably also exists in asexual lineages. Epigenetic signals are thus a convenient measurement "device" that allows us to define individuals, generations, heritability, and even species in asexual taxa.

By *asexual*, here, we mean those eukaryotic taxa that never engage in *amphimixis*, i.e., genetic mixing. Asexual taxa can include lineages that rely on hybridogenesis, gynogenesis, parthenogenesis, or autogamy, including complete automixis, restitutional automixis, and even apomixes (Gorelick and Carpinone, 2009) (see Table 6.1). With *autogamy*, each individual produces both eggs and sperm, which then fuse with one another to form a zygote. With *complete* automixis, females undergo meiosis, but then two egg nuclei from the same meiosis fuse with one another to form a zygote. Complete automixis includes many forms of parthenogenesis and possibly gynogenesis. With *restitutional automixis*, including premeiotic doubling, females undergo meiosis, but diploidy is restored only by the egg cell or its mitotic progeny spontaneously duplicating all chromosomes (i.e., endomitosis or endoreduplication).

Epigenetic signals are generally more readily alterable than are DNA nucleotides (Gorelick, 2005; Angers et al., 2010). Heritable and inducible epigenetic variability may thus allow asexual taxa to succeed without genetic mixing of DNA nucleotides. The frozen niche variation (FNV) model (Vrijenhoek, 1984) suggests that for a population composed of multiple different clones, each clonal lineage uses only a fraction of the total niche, reducing competition among individuals and allowing efficient resource utilization. This model can be extended to variation among the epigenomes of individuals from the same clonal lineage, thus creating epiclonal lineages. According to the FNV model, selection should favor epiclonal lineages that have minimal overlap and, as a result, enhance the ecological success of a clonal lineage. However, epimutations typically are less heritable than changes in nucleotide sequence because of the necessity of the epigenetic reset prior to the initiation of development. Epigenetic reset can be seen in the classical case of resetting a mature multicellular adult composed of highly differentiating cells to a small (almost unicellular) zygote or preembryo that is composed of totipotent cells. Epigenetic reset could also be seen in molecular signals, such as cytosine methylation or telomere degradation, which are reset to "juvenile" levels following meiosis and syngamy (Gorelick and Carpinone, 2009). It is not clear yet whether the balance between these two factors—epimutations and epigenetic reset—will cause evolution of obligate asexual taxa to be more or less affected by epimutations than by DNA nucleotide mutations.

TABLE 6.1
Nomenclature of Asexuality

TERM	DEFINITION	EXAMPLES
Amphimixis	Outcrossing Reduction division + gametes + syngamy	Most plants and animals, any taxon with male individuals
Complete automixis	Both gametic nuclei are products of the same meiotic division Reduction division + syngamy (either with or without gametes)	Stick insect (*Bacillus atticus*), dewberry (*Rubus caesius*), *Paramecium aurelia*
Restitutional automixis	Meiosis but no gametes and no syngamy Endomitosis in lieu of syngamy Reduction division but no syngamy	Lumbricid earthworms (e.g., *Octolasion cyaneum*), garlic chives (*Allium tuberosum*), probably *Giardia intestinalis*
Parasex	No discrete reduction division or gametes Syngamy + reduction	*Aspergillus nidulans*, possibly some heliozoans
Apomixis	No meiosis or syngamy	Possibly bdelloid rotifers and orabatid mites, although these may be cryptically automictic
Autogamy	Self-fertilization	Mangrove killifish (*Kleptolebias* [*Rivulus*] *marmoratus*), many flowering plants
Endomitosis	Duplication of chromosomes without nuclear division (aka endoploidy or endoreduplication)	Probably all eukaryotes
Fertilization	Fusion of gametic cell membranes (aka plasmogamy)	All amphimictic and completely automictic taxa
Syngamy	Fusion of gametic nuclei or pronuclei and subsequent mixing (i.e., decondensation and unpairing) of homologous chromosomes (aka karyogamy)	All amphimictic and completely automictic taxa
Parthenogenesis	Reproduction without sperm	Stick insect (*Bacillus atticus*), dewberry (*Rubus caesius*), *Paramecium aurelia*, hammerhead shark (*Sphyrna tiburo*), teiid lizard (*Aspidoscelis* [*Cnemidophorus*] *tesselatus*)
Gynogenesis	Reproduction in which sperm induce development of the egg to form an embryo or preembryo, but the sperm nucleus or pronucleus does not fuse with the egg nucleus or pronucleus	Summer flounder (*Paralichthys dentatus*), Amazon molly (*Poecilia formosa*), diploid hybrid *Chrosomus* [*Phoxinus*] *eosneogaeus* complex
Hybridogenesis	Apparent amphimixis but eggs comprised only of maternal chromosomes	Hemiclonal *Poeciliopsis monachalucida*, hybrid frog *Rana esculenta* complex

If nothing makes sense except in light of population genetics (Lynch, 2007), then we need to know how the four horsemen of evolution—selection, mutation, gene flow, and drift—are affected by lineages being sexual versus asexual. Incidentally, such a broader understanding of the roles of different evolutionary factors in sexual versus asexual lineages, including the role of epigenetic variation, will also be relevant to one of the most fundamental puzzles of all evolutionary theory, the evolution of sex (Gorelick and Heng, 2011). Selection should have roughly equal effects on obligate asexual lineages and related amphimictic taxa. Gene flow should also be roughly the same in obligately asexual and amphimictic lineages. It is difficult to generalize whether sexual and asexual taxa will have different mutation rates. The only instances when their mutations rates should clearly differ are when asexuals evolved via polyploidy and geographical parthenogenesis. Neoploids have higher proportions of cytosine methylation and, hence, higher mutation rates than their diploid ancestors (Adams et al., 2003; Rapp and Wendel, 2005; Salmon et al., 2005). This also results in higher epimutation rates, due to loss of methylation following mismatch repair of deaminated 5-methylcytosine (Gorelick, 2003). Geographical parthenogens tend to live in harsher environments and, hence, might have higher mutation rates than amphimictic sister taxa (Vandel, 1928; Lynch, 1984). Only with drift do we unequivocally expect higher rates of evolution with asexual lineages than with sister sexual lineages. First, however, we must digress.

Genetic drift is defined for any locus, including many different loci that contain epigenetic signals. A *locus* is a specific location on a eukaryotic or prokaryotic chromosome. A location on a map, including a chromosomal map, is not inherently genetic or epigenetic. What matters are the items that reside at that location. By analogy, on a map of Canada, the *location* of the province of Quebec has no linguistic attributes but can be considered francophone only when we ask who resides there. A location on a chromosome can be considered genetic if we ask which DNA nucleotides reside at that locus or epigenetic if we ask about methylation at that locus. Similarly, a locus could be considered epigenetic if we ask whether the polymorphism observed at a given histone is methylated, unmethylated, acetylated, deacetylated, phosphorylated, or not phosphorylated. As a final example, we could ask how many cytosines are methylated at a given promoter. Drift at any of these loci could be considered epigenetic drift, although we are reluctant to distinguish epigenetic from genetic (Gorelick and Laubichler, 2008). Effective population size of obligate asexual lineages is one. Therefore, evolution of obligate asexual lineages should be dominated by genetic drift of epigenetic alleles (Richards, 2008) and, as we saw in the previous paragraph, to a lesser degree by mutation and epimutation, which could also be considered synonymous.

Because population genetics was developed before the elucidation of DNA as the carrier of genetic information, within the conceptual framework of population genetics, mutation can mean DNA point mutations as well as epimutations (Gorelick and Laubichler, 2008). We have drift not only at DNA loci but also at epigenetic loci. Thus, when studying epigenetics and asexuality, we are justified in focusing on epigenetic drift and epimutations and, of course, their impacts on additive (epi)genetic variation (Gorelick, 2005).

In this context, studying epigenetic variation of asexual taxa has several advantages. First, asexuality makes the investigation of epigenetic variation not related to genetic variation much easier and allows us to quantitatively measure degrees of epigenetic variation. As epigenetic variation can potentially influence phenotypes, influence sustained phenotypic plasticity, and/or be considered as a phenotype by itself, such measurements are obviously important. Furthermore, epigenetic variation helps to demarcate clones (individuals) and epigenetic reset helps to define generations, thus allowing us to apply the traditional conceptual apparatus of evolutionary theory to these taxa.

EPIGENETIC VARIATION • *What Can Asexual Organisms Tell Us?*

A growing body of literature suggests that epigenetic mechanisms might be of particular importance in driving microevolutionary processes, including speciation, mostly because subtle variation in gene expression can strongly influence the phenotypic outcome even without having to modify the underlying DNA sequence. In that sense, heritable epigenetic variation and its effects on the regulation of gene expression can be see as a logical extension of the paradigm of regulatory evolution.

Epigenetic variation has many sources. One can classify epigenetic variation into three classes, reflecting its dependence upon underlying DNA nucleotides (Richards, 2006; Bird, 2007). Genetic information can control epigenetic marks via the interaction in *cis* and *trans*. This represents the obligate epigenetic variation, which can be modeled as epistatic interactions between DNA sequences and epigenetic markers (Gorelick, 2004b). However, epigenetic variation is not exclusively linked to DNA sequence variation. Facilitated—semi-independent, e.g., *agouti* locus (Morgan et al., 1999)—and pure—fully independent, e.g., monozygotic twins (Fraga et al., 2005)—epigenetic variation have also been observed. The independence of epigenotype and DNA sequences highlights the potential of individuals with identical DNA sequences to express different phenotypes. The amount of variation of a phenotype that is linked to facilitated or pure epimutation can therefore be a function of stochastic events and/or environmental influence (Angers et al., 2010).

Before trying to reach conclusions about the effects of epigenetic variation in the evolution of any organisms, it is important to look at the epigenetic variation in natural populations (Kalisz and Purugganan, 2004; Richards, 2008). It is particularly important to define how much epigenetic variation in natural population is not strictly related to DNA sequence variation, i.e., how much facilitated and pure epigenetic variation exists (Massicotte et al., in press). To do so,

we need to be able to isolate the proportion of the phenotype that is not encoded by DNA. Some examples can be found in the literature (we know that there is some variation in natural populations), but we do not yet have much of a picture of how much facilitated and pure epigenetic variation exists, except perhaps for studies of monozygotic twins (largely human; Rakyan et al., 2004). An equally important problem is to determine how much of the existing epigenetic variation (of any kind) is heritable. This is especially relevant in medical contexts as a growing body of evidence suggests that some environmentally induced variation in the epigenome can indeed be passed on to future generations (Pembrey et al., 2006; Hitchins et al., 2007).

The question is, How can we isolate the effect of epigenetic (facilitated and pure epigenetic variation) versus DNA-induced genetic and epigenetic effects (obligatory epigenetic variation) in order to highlight the phenotypic variation that is the result of individuals having different epialleles? The easiest solution would be to control for DNA-based variation. How can we do so and still investigate the epigenetic variation in natural populations? Naturally occurring populations of clones are found in organisms that reproduce asexually via parthenogenesis, gynogenesis, and androgenesis; (Bird, 2007; Avise, 2008; Scali and Milani, 2009). Individuals that belong to the same clonal lineage are all genetically identical. Populations are composed of many individuals from a single lineage, unlike with monozygotic twin studies in which we can investigate variation in a sample size of only two, albeit across many sets of twins in a population. We can also borrow ideas from quantitative epigenetics and try to measure what proportion of additive genetic variance is due to heritability of some life-history traits that are related to an epigenetic state (Rutherford and Henikoff, 2003; Gorelick 2005).

PARTITIONING EPIGENETIC VARIATION

Because epigenetic marks are reversible, they are more susceptible to being altered than are

DNA nucleotides. The modification of the original epigenetic pattern that controls gene expression is called an *epimutation* (Jeggo and Holliday, 1986). Epimutations can become established over the course of development and/or at maturity. As an example, the removal of an epigenetic mark, such as DNA methylation, can lead to improper gene expression in space and/or in time at the scale of an individual and, consequently, modify the phenotype. Epimutations can be the result of stochastic events, i.e., the inability to establish or to maintain the programmed epigenetic pattern through cell multiplication (mitosis). Such epigenetic variation among individuals can lead to the formation of epialleles (Kalisz and Purugganan, 2004). Naturally occurring epialleles have been described, mostly in plants, e.g., *Linaria* floral symmetry (Cubas et al., 1999) and tomato ripening (Manning et al. 2006). More interestingly and in contrast to the genome, the environment can influence the epigenome and lead to epimutations (Jaenisch and Bird, 2003). For example, Pembrey et al. (2006) showed that smoking by teenage human males adversely affected their offspring and grand-offspring, even though the males quit smoking long before they fathered offspring. The consequences of prescribing 5-azacytidine for treating human ailments are particularly insidious because this chemical inhibits maintenance methylation throughout the genome, effectively removing methylation from cytosines. Furthermore, much of the 5-azacytidine is urinated into the water supply, wreaking environmental damage that is currently unregulated (Gorelick, 2005). Epimutations are far more common than DNA mutations over the course of an individual's life span. Numerous examples of the integration of extrinsic signals and the subsequent response of shaping of the epigenome have been observed: temperature (Sheldon et al., 2002), diet (Feil, 2006), and chemicals (Crews et al., 2007) being three examples. Both the stochastic events and the integration of the extrinsic signal represent ways by which the phenotype can be modified without changing the underlying DNA se-

quence, resulting in more phenotypic plasticity (Angers et al., 2010).

Epigenetic variation can thus occur at different levels and scales: (1) within an individual, (2) among individuals in a population, and (3) between populations.

Epigenetic variation that possibly modifies the phenotype can occur at the scale of an individual (in addition to integration of intrinsic signals related to the developmental program, which technically are also epigenetic signals). The inability to maintain the initial epigenetic state at any given locus in some cells leads to variation among cells of the same tissue. This process is called *variegation*. An example of a variegated phenotype is the *agouti* locus that controls coat color in mice. This system illustrates well the metastability concept of epialleles (Rakyan et al., 2002, 2003). The development of cancer clones is another example of how epigenetic modifications can have phenotypic consequences during the life cycle of an individual.

There will, of course, also be a certain amount of epigenetic variability among individuals of the same population (variable expressivity) (Rakyan et al., 2002). This level of variation is integrative of the stochastic events that occur at the scale of an individual and/or to the effect of the environment. Environmental influence on the shaping of the epigenome is a well-known process (Jaenisch and Bird, 2003). Individuals experiencing similar environmental pressures should have more similar epigenetic profiles, while individuals experiencing different environmental pressures (inhabiting different areas, eating different food, higher exposition to contaminants, etc.) should have more dissimilar epigenetic profiles. As a result, epigenetic variation at the scale of a population is a function of stochastic events and of the heterogeneity of the environment (both in space, allowing individuals inhabiting different parts of an environment to have different epigenetic profiles, and in time, allowing modification of the epigenetic profile of individuals over the course of development).

Epigenetic differentiation of populations will be dependent on heterogeneity of environmental conditions. Two recently separated populations under similar environmental pressures should have low epigenetic differentiation. Over time, these two populations should evolve divergent epigenetic signatures due to experiencing different environmental pressures, as well as epimutations, epigenetic drift, and selection pressure. How should we measure this epigenetic divergence? Genetic tools for measuring differentiation, such as Wright's F statistics, all have heritability of the variation as a premise. The good news is that Wright did not presume the molecular mode of inheritance—DNA versus epigenetic—insofar as population genetics was largely formulated prior to the realization that DNA was the primary carrier of the genetic code (Gorelick and Laubichler, 2008). Therefore, the conceptual apparatus of heritability measurements and the estimation of genetic variation can easily be adopted to quantify epigenetic variation and heritability.

DEFINING GENERATIONS AND INDIVIDUALS • *What Can Epigenetic Reset Tell Us?*

In order to define development or evolution in eukaryotes, one has to demarcate generations, which is ordinarily a seemingly trivial task when the alternation of haploid and diploid generations is bookmarked by meiosis and syngamy. However, how do we define development or evolution in asexual eukaryotes, such as with parthenogenesis (strictly clonal), gynogenesis (where the sperm of parental species is needed only to trigger embryogenesis but sperm DNA is not incorporated into the zygote, possibly with or possibly without egg meiosis), or restitutional automixis (endoreduplication and no syngamy), in all of which meiosis and/or syngamy are lacking? In asexual taxa, we have an intuitive notion of what constitutes generations, especially by comparison with amphimictic sister taxa. Here, we propose that the start of a generation in all multicellular eukaryotes can be demarcated by epigenetic reset, such as of cytosine methylation or chromatin marks, which are clearly needed for development (Santos and Dean, 2004; Gorelick and Carpinone, 2009). Although this might make quantitative geneticists cringe, epigenetic resets can provide a sufficient demarcation of generations to rigorously define heritability in all eukaryotes, including asexual ones. The term *genetic* should be synonymous with *heritable* (a synonymy that we have reluctantly not invoked in this chapter) (Gorelick and Laubichler, 2008); hence, the fidelity of epigenetic reset provides a measure of heritability and, consequently, implicitly provides a definition of generations. Epigenetic reset also defines individuals in such so-called asexual lineages, demarcating the end of one individual and the start of the next. Also, due to the relative fluidity of epigenetic signals, at least when compared with DNA nucleotides, asexual individuals may well possess different epigenetic profiles (variable expressivity).

THE PROBLEM OF DEFINING GENERATIONS AND INDIVIDUALS IN ASEXUAL TAXA

Is epigenetic reset needed to define generations and individuals in asexual taxa? This depends upon how one defines *epigenetic*. Contrary to popular misconceptions (pardon the pun), meiotic females of most taxa do not produce single-celled gametes with only a single haploid nucleus. Furthermore, most (all?) zygotes are not single-celled with a single diploid nucleus and alternation (see next paragraph for details). Moreover, ploidy is fuzzier than usually believed. Thus, we need something other than return to a single-celled stage to demarcate generations in multicellular organisms. Regardless, in multicellular eukaryotes, reset of development from a large multicellular stage to a small unicellular or oligocellular stage would provide a demarcation between generations and individuals; but this is epigenetic *sensu* Waddington and all biologists before the molecular era. Our contribution here is to make this definition of generations and individuals more molecular and, in many ways, more quantifiable.

Instead of demarcating generations by meiosis and syngamy, we could try defining generation by reversion to a single-celled state with either a single haploid or diploid nucleus, as apparently occurs during meiosis and syngamy. However, this single-celled state is often illusory. The product of meiosis in many females is a single cell with either two or four haploid nuclei. See, for example, bisporic and tetrasporic megagametogenesis in angiosperms (Klekowski, 1988). The so-called polar bodies are often not jettisoned or digested until after the sperm fertilizes the egg. In many organisms, such as humans (Austin, 1965), most stages of female meiosis are not even initiated until the egg, which is still really diploid, is fertilized by the sperm. Thus, in human females there is no true haploid phase. Things get even crazier in some plants, such as *Gnetum*, where the egg nucleus is embedded in a single huge cell with thousands of haploid nuclei (Friedman and Carmichael, 1996). However, it cannot be universally true that sperm are needed for female meiosis. Female coral and sea urchins complete meiosis before fertilization (Austin, 1965; Longo, 1973). Moreover, automixis occurs in many parthenogens without sperm. With these parthenogens, does a second egg pronucleus trigger meiosis? While male gametes almost always go through a single-celled stage with a single haploid nucleus, females seldom do. Females are the more fundamental sex, at least if we accept the paradigm that anisogamy evolved from isogamy (Bell, 1978). Thus, for organisms in which males do not exist, we need something other than gametes with single nuclei to demarcate generations.

Contrary to what is in most textbooks, most outcrossing sexual taxa also do not go through a single-celled stage with only a single diploid nucleus. In most organisms, including humans, gametic pronuclei duplicate all haploid chromosomes prior to the two pronuclear envelopes dissolving to form a zygote (Austin, 1965; Gwatkin, 1977; Veeck, 1999; Gorelick and Carpinone, 2009). Thus, in diploid taxa, zygotes have four copies of each homologous chromosome (4C). The first stage in which cells each contain a single diploid nucleus is usually the two-cell stage. For sexual taxa, we could demarcate the start of a generation as the production of a 4C zygote or the subsequent two-celled 2C (diploid) stage.

For the foregoing reasons, we cannot rely on reversion to a single-celled haploid or diploid state to demarcate generations in most organisms, regardless of whether they are sexual or asexual. For sexual (outcrossing) lineages, we can simply demarcate generations by when genetic mixing occurs. How, though, can we demarcate generations for asexual taxa? The answer here depends on the form of asexuality, automixis versus apomixis, i.e., with or without meiosis.

Can we demarcate generations by changes in ploidy? This might provide a surrogate for single-celled stages (Kondrashov, 1994), but, as we discuss in the next three paragraphs, this approach has a few problems. First, it cannot account for endoploidy. Second, it cannot account for lineages that alternate between more than two ploidy levels, i.e., more than just haploid and diploid. Third, and perhaps most fundamentally, ploidy level has no obvious connection with development and epigenesis. We therefore, in the next section, propose an epigenetic demarcation of generations.

Should we define individuals of new ploidy as new individuals and new generations? Does it make sense to define each sperm as a separate haploid individual? What about each egg cell, especially if it never undergoes an unambiguous haploid state? For unicellular taxa, alternation of ploidy may be the best we can do for defining individuals and generations, but we can do better for multicellular taxa. Alternation of haploid and diploid generations seems like a contrived definition of generations, especially with endoploidy rampant in many eukaryotes (Cavalier-Smith, 1995). Human heart and liver cells are highly polyploid (Anatskaya and Vinogradov, 2004), but this should not raise them to the level of another individual (possibly parasitic) in your body. Endoploidy also exists in invertebrates (Johnston et al., 2004; Mello, 2005). There is some evidence that endoploidy is common in cells with high metabolic demand in animals (Vinogradov et al., 2001; Anatskaya and Vinogradov, 2002), such as cardiac or flight muscles. Endoploidy is also common in plants, although it is not obvious whether ploidy levels are correlated with metabolic demand in plants (De Rocher et al., 1990; Palomino et al., 1999). Nonetheless, changes in ploidy, without other changes, do not seem to warrant the demarcation between two generations or between two individuals.

Endoploidy probably has interesting epigenetic effects in its own right. From a developmental perspective, endoploid cells appear to be terminal, the zenith of ontogeny. At least in animals, endoploid cells are usually (always?) highly specialized and probably incapable of further mitotic or meiotic divisions. From a molecular perspective, we suspect that endoploid cells have a disproportionate number of their regulatory loci methylated and are highly heterochromatic.

How do we define generations and individuals in taxa that, instead of alternating only between haploid and diploid stages (for the moment, ignoring how fuzzy these stages are, especially for female gametes), cycle between haploid, diploid, tetraploid, and perhaps higher ploidies and then undergo the decreasing cavalcade of ploidies: tetraploid, diploid, haploid? This occurs in some members of the genera *Polysiphonia* (red alga), *Ectocarpus* (brown alga), *Pyrsonympha* (Excavata, oxymonad), and *Giardia* (Excavata, diplomonad), some of which even have higher ploidy levels in this cycle (Müller, 1967; Hollande and Carruette-Valentin, 1970; Goff and Coleman, 1986). Each change in ploidy should constitute a new generation if and only if there is an associated epigenetic reset. We do not have a good answer here if there appears to be one extended epigenetic rest spanning all increases in ploidy and a second extended epigenetic reset spanning all decreases in ploidy (cf. Davis et al., 2000; Farthing et al., 2008).

EPIGENETIC DEMARCATION OF GENERATIONS

We propose that the crux of what constitutes a new generation and a new individual is epigenetic reset. Heuristically, we want to demarcate a generation by the *abrupt shift* from (1) a huge, complex, multicellular organism with many specialized cell types, many of which are incapable of further cell divisions, to (2) a small, unicellular, or oligocellular organism, with one type of totipotent cell. The one or two totipotent cells then divide mitotically, *gradually differentiating* into the complex multicellular individual. This is classic epigenesis. Notice that this heuristic definition does not mention ploidy levels. While some taxa, such as bryophytes, pteridophytes, and many algae, have complex multicellular haploid and diploid stages; others, such as most animals, do not. Thus, we are left with the uncomfortable situation that gametes and gametophytes may or may not be considered separate haploid individuals. This seems peculiar insofar as many of these haploid entities undergo extensive development. Even animal sperm undergo extensive development, from a prototypical spherical cell to an elongated cell virtually devoid of cytoplasm. We thus need a more precise—less heuristic—definition of epigenetic reset to demarcate individuals and generations.

We thus propose that epigenetic reset of various molecular epigenetic signals provides a definitive demarcation of individuals and generations in all eukaryotes, including autogamic, automictic, and apomictic taxa. We simply have to plot a time series of some molecular epigenetic signal and then use time series methods to detect discrete jumps in the signal. Epigenetic signals, such as cytosine methylation and chromatin modification, are known to take discrete jumps during or immediately following meiosis and syngamy in outcrossing sexual taxa (El-Maarri et al., 2001; Santos and Dean, 2004; Ruiz-García et al., 2005). Why not use discrete jumps in these same molecular epigenetic signals to define individuals and generations in self-sexual and asexual taxa? Why not use these signals to determine whether endoploid tissues constitute a separate (parasitic) individual, albeit one that is an evolutionary dead end? Why not use these signals to determine whether the 1N-2N-4N-8N-16N-8N-4N-2N-1N ploidy cycle is composed of eight generations? Or is it fewer? For unicellular eukaryotes, do molecular epigenetic signals allow us to define individuals and generations despite a lack of development in either haploid or diploid stage? Why not use these signals to determine whether there are interesting asymmetries between female and male gametes, possibly providing clues as to causes of differential genomic imprinting? While we do not believe that such inquiries should influence existing abortion and contraception debates, it would be fascinating if these signals helped to inform us whether egg or sperm cells are ever really individuals.

Epigenetic reset provides virtually the same demarcation of generations as existed for outcrossing sexual organisms that used genetic mixing as the demarcation. The epigenetic reset demarcation provides the same number of generations, although the timing of the start of generations may be slightly different from when genetic mixing occurs. In outcrossing sexual taxa, molecular epigenetic signals are usually thought to have a sawtooth pattern over multiple generations, with the period of the signal being one generation (Gorelick and Carpinone, 2009). Over the course of diploid development, the frequency of a molecular epigenetic signal over all or a portion of the genome (epigenome) changes gradually and monotonically. For example, telomeres gradually degrade, while cytosine methylation at regulatory loci gradually increases. There is a rapid (albeit not instantaneous) shift in these epigenetic signals during gametogenesis (Farthing et al., 2008). This epigenetic reset is a return to levels that existed at the start of the previous haploid generation but not necessarily to levels at the start of the diploid generation. Molecular epigenetic changes probably then change gradually and monotonically over the course of haploid development, although data are not available to corroborate this. However, during or immediately following

syngamy, there is a second rapid shift in these epigenetic signals to those levels that existed at the start of the previous diploid generation. Outcrossing sexual taxa have two distinct types of generation, haploid and diploid. Over the course of a generation, the time history of molecular epigenetic signals in outcrossing sexual eukaryotes undergoes two gradual monotonic periods, interspersed with two discrete jumps, with the pattern repeating in subsequent generations.

In the hypothetical situation in which all epigenetic variation (marks associated with the developmental program and epimutations) is heritable, the offspring epigenome would be identical to the parental epigenome and strict sense heritability would be equal to one. This situation is highly improbable because the epigenetic reset is needed to initiate proper development of individuals of the next generation, at least in mammals. As a result, most of the epigenetic marks are cleared each generation. However, the epigenetic marks associated with developmental programs are reestablished each generation with a high fidelity to enable proper development of organisms (epigenetic signals associated with developmental programs are highly heritable because they are genetically determined, i.e., obligate epigenetic variation). Furthermore, the epigenetic resets of meiosis and syngamy might not be perfect, allowing transgenerational epigenetic inheritance of some epimutations, thus leading to a certain level of additive epigenetic variation. If both epigenetic resets were perfect, we would expect to see no facilitated or pure epigenetic component of additive genetic variance because all changes in epigenetic signatures would be reset (assuming the low probability of the appearance of the same epimutation in the next generation). Although the epigenetic mark itself might not be present in the germ line, some particular molecular mechanisms have been proposed to lead to faithful reestablishment of the marks by passing on small interfering RNA via the cytoplasm of gametes (Chandler, 2007). There is thus a tension or trade-off between the two parts of this chapter: Epigenetic resets make transgenerational inheritance of epimutations more difficult, but epigenetic reset allows demarcation of generations and individuals. We can have our cake and eat it too because epigenetic reset can sometimes be imperfect, although exactly how imperfect is an outstanding empirical question that may vary across taxa, as we discuss in the next section. Among individuals, epigenetic variation also helps to define individuals, as we discuss at the end of this chapter.

Notice that epigenetic reset demarcates generations even in obligately self-fertilizing lineages, for which there is never genetic mixing. This includes lineages with autogamy, complete automixis, and restitutional automixis. The question, however, remains whether an epigenetic reset demarcation of generations works for forms of ploidy cycling lacking any evidence of meiosis, such as the putatively apomictic gynogenetic fishes like *Poecilia* and *Chromosomus* [*Phoxinus*] *eos-neogaeus*. For apomictic gynogens, sperm somehow seem to provide the epigenetic signal that resets development and presumably also resets cytosine methylation signatures.

Because epigenetic resets are imperfect, time series analysis may be needed to detect these resets. Time series may also be needed because epigenetic resets are not instantaneous. Continuous periodic time histories are often analyzed with Fourier series. Sawtooth waves, however, are a mix of discrete and continuous variables. Therefore, the orthogonal basis used to estimate the molecular epigenetic signals should ideally contain discrete and continuous functions, e.g., sinusoids and step functions. We therefore suggest using Walsh-Hadamard series to estimate where the discrete jumps occur, which demarcate generations (Elliott and Rao, 1982).

Asexual or self-sexual individuals may not undergo meiosis and/or syngamy but should still undergo molecular epigenetic resets once or twice each generation. Otherwise, they will have no way of resetting diploid and/or haploid development each generation. Taxa with autogamy or complete automixis should undergo two molecular epigenetic resets each generation because they still undergo both meiosis and syngamy. With many instances of parthenogenesis, standard meiosis almost certainly occurs after premeiotic endomitosis; there is no syngamy and, thus, only one epigenetic reset (Dawley, 1989; Gorelick & Carpinone 2009). Apomictic gynogenesis is unusual insofar as the epigenetic reset is initiated by something other than meiosis and syngamy. Only one reset is important, the one before the initiation of the developmental program. The sperm that triggers development may serve as the signal for the epigenetic reset, although the sperm genome (i.e., sperm nuclear DNA) is not incorporated. Taxa with restitutional automixis, such as lumbricid earthworms (e.g., *Octolasion cyaneum*) and garlic chives (*Allium tuberosum*), may undergo only one molecular epigenetic reset each generation because they have meiosis but no syngamy. Their endomitotic duplication resembles that of endoploidy in outcrossing sexual taxa and, therefore, may not count as a generation. Using the demarcation of epigenetic reset, we may be unable to distinguish haploid and diploid generations in taxa with restitutional automixis.

We should also point out that there has been contemporary debate about when generations start in amphimictic (outcrossing) lineages, especially in humans, largely due to political tensions between in vitro fertilization researchers and antiabortion activists (Spallone, 1996). The terms *conceptus* and *preembryo* describe the state between fertilization and blastocyst (aka embryo) stage, with the subtle message that generations and individuals are not clearly demarcated in humans. Preembryos are composed of many cells that will not form the next generation but instead contribute to the placenta. Preembryos can be split into monozygotic twins and, hence, represent an indeterminate number of individuals. Just as epigenetic signals are not reset instantaneously, epigenesis from a fertilized egg to a blastocyst with a primitive streak also does not happen instantaneously.

Parasex refers to life cycles with syngamy but without a traditional reduction division. Instead, these organisms go from a diploid to a haploid state by jettisoning one homologous chromosome at a time, going through a succession of aneuploid states until only one copy of each homologue remains (Pontecorvo, 1956). Recent work shows that this process is a highly modified form of meiosis (Forche et al., 2008). However, it is not obvious whether epigenetic reset occurs during or following these successive aneuploid events. Thus, the epigenetic time history of parasexual lineages may be like restitutional automicts (one epigenetic reset per generation) or like amphimicts and complete automicts (two epigenetic resets per generation). Nevertheless, we should still be able to use epigenetic resets to demarcate generations in the few lineages of fungi that are parasexual.

Other than with apomictic gynogenesis, in which sperm pronuclei (products of meiosis!) trigger epigenetic reset, we hypothesize that *obligate* apomictic reproduction does not exist in multicellular eukaryotes because there is nothing akin to meiosis or syngamy to induce the needed epigenetic resets. This needs to be tested in two parallel ways. First, we need to look for cryptic meiosis in putatively obligately apomictic lineages (Solari, 2002). Second, we need to see whether epigenetic reset occurs each generation in these lineages. By epigenetic reset, we refer to the molecular signals, such as cytosine methylation, because it is already obvious that developmental reset *sensu* Waddington (1940, 1957) occurs.

Having defined generations for asexual organisms, conventional definitions of heritability apply. We can apply parent–offspring regression or full-sib (but not half-sib) analysis.

DEFINING INDIVIDUALS AND HERITABILITY

Once we define generations for asexual taxa, it becomes easier to define individuals. If two organisms are separated by an epigenetic reset, i.e., a generation, then they must be different individuals. The only question remaining is whether two individuals in the same generation are the same individual (Smith et al., 1992; Scrosati, 2002). Here, we need to rely again on epigenetic variation, as we did in defining asexual species. Epigenetic reset is not perfect and is, in fact, less heritable than DNA nucleotides. There are small stochastic differences in epigenetic reset from generation to generation as well as between individuals within a generation. In meiotic taxa, each gamete should have slightly different molecular epigenetic resets, allowing us to distinguish individuals even if these are not outcrossing taxa, as with automixis or autogamy. If eukaryotes still exist without meiosis, there will probably still be molecular epigenetic reset occurring during ploidy cycling. Stochastic differences in this epigenetic reset should allow us to distinguish individuals following ploidy cycling. If two organisms differ in their epigenetic marks at more than a certain number of loci, then call the two organisms different individuals. If the two differ in fewer than that predetermined number of epigenetic loci, then say that they are clones, i.e., the same individual. The only thing arbitrary about this definition is setting the threshold number of epigenetic loci that differ. The same tack has been taken to define prokaryotic species (Moreno, 1997; Stackebrandt et al., 2002).

Having rigorously defined generations and individuals for obligately asexual lineages, defining heritabilities of any signals is now trivial, whether the signals or traits are ultimately caused by DNA nucleotides, epigenetic marks, or both. Simply apply Lush's classic definition of heritability: additive genetic variance divided by phenotypic variance (Lush, 1937). Being able to demarcate individuals and generations allows us to apply conventional quantitative genetic techniques, such as parent–offspring and full-sib analysis, as well as more sophisticated methods, such as restricted maximum likelihood (Knott et al., 1995). The measured phenotype can be anything from color of a maize kernel or *agouti* mouse to amount of cytosine methylation present on a stretch of DNA.

CONCLUDING REMARKS

Asexual lineages can illuminate the relative importance of epigenetic signals in evolution. Conversely, epigenetic signals can be used to define generations, individuals, and even species in asexual lineages. How well we can accomplish both of these goals depends on the extent of epigenetic variation in natural populations, and we still have much work to do in order to evaluate this, including how imperfect epigenetic reset is during meiosis, syngamy, or more generally ploidy cycling.

REFERENCES

Adams, K.L., R. Cronn, R. Percifield, and J.F. Wendel. 2003. Genes duplicated by polyploidy show unequal contributions to the transcriptome and organ-specific reciprocal silencing. *Proc Natl Acad Sci USA* 100(8):4649–54.

Allis, C.D., T. Jenuwein, and D. Reinberg. 2007. Overview and concepts. In *Epigenetics*, ed. C.D. Allis, T. Jenuwein, D. Reinberg, and M.-L. Caparros, 23–61. Woodbury, NY: Cold Spring Harbor Laboratory Press.

Anatskaya, O.V., and A.E. Vinogradov. 2002. Myocyte ploidy in heart chambers of birds with different locomotor activity. *J Exp Zool* 293(4):427–41.

Anatskaya, O.V., and A.E. Vinogradov. 2004. Heart and liver as developmental bottlenecks of mammal design: Evidence from cell polyploidization. *Biol J Linn Soc* 83(2):175–86.

Angers, B., E. Castonguay, and R. Massicotte. 2010. Environmentally induced phenotype and DNA methylation: How to deal with unpredictable conditions till the next generation, and after. *Mol Ecol* 19(7):1283–95.

Austin, C.R. 1965. *Fertilization*. Edgewood Cliffs, NJ: Prentice Hall.

Avise, J.C. 2008. *Clonality: The Genetics, Ecology, and Evolution of Sexual Abstinence in Vertebrate Animals*. Oxford: Oxford University Press.

Bell, G. 1978. Evolution of anisogamy. *J Theor Biol* 73(2):247–70.

Bird, A. 2002. DNA methylation patterns and epigenetic memory. *Genes Dev* 16(1):6–21.

Bird, A. 2007. Perceptions of epigenetics. *Nature* 447(7143):396–8.

Blewitt, M.E., N.K. Vickaryous, A. Paldi, H. Koseki, and E. Whitelaw. 2006. Dynamic reprogramming of DNA methylation at an epigenetically sensitive allele in mice. *PLoS Genet* 2(4):399–405.

Bossdorf, O., C.L. Richards, and M. Pigliucci. 2008. Epigenetics for ecologists. *Ecol Lett* 11(2):106–15.

Cavalier-Smith, T. 1995. Cell cycles, diplokaryosis and the Archezoan origin of sex. *Arch Protist* 145(3–4):189–207.

Chandler, V.L. 2007. Paramutation: From maize to mice. *Cell* 128(4):641–45.

Chong, S.Y., and E. Whitelaw. 2004. Epigenetic germline inheritance. *Curr Opin Genet Dev* 14(6):692–6.

Crews, D., A.C. Gore, T.S. Hsu, N.L. Dangleben, M. Spinetta, T. Schallert, M.D. Anway, and M.K. Skinner. 2007. Transgenerational epigenetic imprints on mate preference. *Proc Natl Acad Sci USA* 104(14):5942–6.

Cubas, P., C. Vincent, and E. Coen. 1999. An epigenetic mutation responsible for natural variation in floral symmetry. *Nature* 401(6749):157–61.

Davis, T.L., G.J. Yang, J.R. McCarrey, and M.S. Bartholomei. 2000. The H19 methylation imprint is erased and re-established differentially on the parental alleles during male germ cell development. *Hum Mol Genet* 9(19):2885–94.

Dawley, R.M. 1989. An introduction to unisexual vertebrates. In *Evolution and Ecology of Unisexual Vertebrates*, ed. R.M. Dawley and J.P. Bogart, 1–18. Albany: New York State Museum.

De Rocher, E.J., K.R. Harkins, D.W. Galbraith, and H.J. Bohnert. 1990. Developmentally regulated systemic endopolyploidy in succulents with small genomes. *Science* 250(4977):99–101.

Elliott, D.F., and K.R. Rao. 1982. *Fast Transforms: Algorithms, Analyses, Applications*. New York: Academic Press.

El-Maarri, O., K. Buiting, E.G. Peery, P.M. Kroisel, B. Balaban, K. Wagner, B. Urman, et al. 2001. Maternal methylation imprints on human chromosome 15 are established during or after fertilization. *Nat Genet* 27(3):341–4.

Farthing, C.R., G. Ficz, R.K. Ng, C.-F. Chan, S. Andrews, W. Dean, M. Hemberger, and W. Reik. 2008. Global mapping of DNA methylation in mouse promoters reveals epigenetic reprogramming of pluripotency genes. *PLoS Genet* 4(6) e1000116.

Feil, R. 2006. Environmental and nutritional effects on the epigenetic regulation of genes. *Mutat Res Fundam Mol Mech Mutagen* 600(1–2):46–57.

Forche, A., K. Alby, D. Schaefer, A.D. Johnson, J. Berman, and R.J. Bennet. 2008. The parasexual cycle in *Candida albicans* provides an alternative pathway to meiosis for the formation of recombinant strains. *PLoS Biol* 6(5):1084–97.

Fraga, M.F., E. Ballestar, M.F. Paz, S. Ropero, F. Setien, M.L. Ballestart, D. Heine-Suñer, et al.

2005. Epigenetic differences arise during the lifetime of monozygotic twins. *Proc Natl Acad Sci USA* 102(30):10604–9.

Friedman, W. E., and J. S. Carmichael. 1996. Double fertilization in Gnetales: Implications for understanding reproductive diversification among seed plants. *Int J Plant Sci* 157(6):S77–94.

Goff, L. J., and A. W. Coleman. 1986. A novel pattern of apical cell polyploidy, sequential polyploidy reduction and intercellular nuclear transfer in the red alga *Polysiphonia*. *Am J Bot* 73(8):1109–30.

Gorelick, R. 2003. Evolution of dioecy and sex chromosomes via methylation driving Muller's ratchet. *Biol J Linn Soc* 80(2):353–68.

Gorelick, R. 2004a. Neo-Lamarckian medicine. *Med Hypotheses* 62(2):299–303.

Gorelick, R. 2004b. *Evolutionary epigenetic theory.* PhD thesis, Arizona State University.

Gorelick, R. 2005. Environmentally-alterable additive genetic variance. *Evol Ecol Res* 7(3):371–9.

Gorelick, R., and J. Carpinone. 2009. Origin and maintenance of sex: The evolutionary joys of self sex. *Biol J Linn Soc* 98:707–28.

Gorelick, R., and H. H. Q. Heng. 2011. Sex reduces genetic variation: a multidisciplinary review. *Evolution* [DOI:10.1111/j.1558-5646.2010.01173.x].

Gorelick, R., and M. D. Laubichler. 2008. Genetic = heritable (genetic ≠ DNA). *Biol Theory* 3(1): 79–84.

Gwatkin, R. B. L. 1977. *Fertilization Mechanisms in Man and Mammals.* New York: Plenum Press.

Hitchins, M. P., V. Ap Lin, A. Buckle, K. Cheong, N. Halani, S. Ku, C. T. Kwok, et al. 2007. Epigenetic inactivation of a cluster of genes flanking MLH1 in microsatellite-unstable colorectal cancer. *Cancer Res* 67(19):9107–16.

Hollande, A., and J. Carruette-Valentin. 1970. Appariement chromosomique et complexes synaptonematiques dans les noyaux en cours de depolyploidisation chez Pyrsonymphaflagellata: Le cycle evolutif des Pyrsonymphines symbiontes de *Reticulitermes lucifugus*. *C R Acad Sci Paris* 270: 2550–3.

Holliday, R. 2006. Dual inheritance. In *DNA Methylation: Basic Mechanisms*, ed. W. Doerfler and P. Böhm, 243–56. Berlin: Springer-Verlag.

Holliday, R., and J. E. Pugh. 1975. DNA modification mechanisms and gene activity during development. *Science* 187(4173):226–32.

Jablonka, E., and G. Raz. 2009. Transgenerational epigenetic inheritance: Prevalence, mechanisms, and implications for the study of heredity and evolution. *Q Rev Biol* 84(2):131–76.

Jaenisch, R., and A. Bird. 2003. Epigenetic regulation of gene expression: How the genome integrates intrinsic and environmental signals. *Nat Genet* 33:245–54.

Jeggo, P. A., and R. Holliday. 1986. Azacytidine-induced reactivation of a DNA repair gene in Chinese hamster ovary cells. *Mol Cell Biol* 6(8):2944–9.

Johnston, J. S., L. D. Ross, L. Beani, D. P. Hughes, and J. Kathirithamby. 2004. Tiny genomes and endoreduplication in *Strepsiptera*. *Insect Mol Biol* 13(6):581–5.

Kalisz, S., and M. D. Purugganan. 2004. Epialleles via DNA methylation: Consequences for plant evolution. *Trends Ecol Evol* 19(6):309–14.

Klekowski, E. J. 1988. *Mutation, Developmental Selection, and Plant Evolution.* New York: Columbia University Press.

Knott, S. A., R. M. Sibly, R. H. Smith, and H. Møller. 1995. Maximum likelihood estimation of genetic parameters in life history studies using the "animal model." *Funct Ecol* 9(1):122–6.

Kondrashov, A. S. 1994. The asexual ploidy cycle and the origin of sex. *Nature* 370(6486):213–16.

Kucharski, R., J. Maleszka, S. Foret, and R. Maleszka. 2008. Nutritional control of reproductive status in honeybees via DNA methylation. *Science* 319(5871):1827–30.

Longo, F. J. 1973. Fertilization: A comparative ultrastructural review. *Biol Reprod* 9(2):149-215.

Lush, J. L. 1937. *Animal Breeding Plans.* Ames: Iowa State College Press.

Lynch, M. 1984. Destabilizing hybridization, general-purpose genotypes and geographic parthenogenesis. *Q Rev Biol* 59(3):257–90.

Lynch, M. 2007. The frailty of adaptive hypotheses for the origins of organismal complexity. *Proc Natl Acad Sci USA* 104:8597–604.

Manning, K., M. Tor, M. Poole, Y. Hong, A. J. Thompson, G. J. King, J. J. Giovannoni, and G. B. Seymour. 2006. A naturally occurring epigenetic mutation in a gene encoding an SBP-box transcription factor inhibits tomato fruit ripening. *Nat Genet* 38(8):948–52.

Massicotte, R., E. Whitelaw, and B. Angers. in press. DNA methylation: a source of random variations in random populations. *Epigenetics.*

Meissner, A., T. S. Mikkelsen, H. C. Gu, M. Wernig, J. Hanna, A. Sivachenko, X. L. Zhang, et al. 2008. Genome-scale DNA methylation maps of pluripotent and differentiated cells. *Nature* 454(7205): 766–91.

Mello, M. L. S. 2005. Apoptosis in polyploid cells of the blood-sucking hemipteran, *Triatoma infestans* Klug. *Caryologia* 58(3):281–7.

Moreno, E. 1997. In search of a bacterial species definition. *Rev Biol Trop* 45(2):753–71.

Morgan, H. D., H. G. E. Sutherland, D. I. K. Martin, and E. Whitelaw. 1999. Epigenetic inheritance at the *agouti* locus in the mouse. *Nat Genet* 23(3): 314–18.

Müller, D. G. 1967. Culture experiments on life cycle nuclear phases and sexuality of brown alga *Ectocarpus siliculosus*. *Planta* 75(1):39–54.

Palomino, G., J. Doležel, R. Cid, I. Brunner, I. Méndez, and A. Rubluo. 1999. Nuclear genome stability of *Mammillaria san-angelensis* (Cactaceae) regenerants induced by auxins in long-term in vitro culture. *Plant Sci* 141(2):191–200.

Pembrey, M. E., L. O. Bygren, G. Kaati, S. Edvinsson, K. Northstone, M. Sjostrom, and J. Golding. 2006. Sex-specific, male-line transgenerational responses in humans. *Eur J Hum Genet* 14(2):159–66.

Pontecorvo, G. 1956. Parasexual cycle in fungi. *Annu Rev Microbiol* 10:393–400.

Rakyan, V. K., M. E. Blewitt, R. Druker, J. I. Preis, and E. Whitelaw. 2002. Metastable epialleles in mammals. *Trends Genet* 18(7):348–51.

Rakyan, V. K., S. Chong, M. E. Champ, P. C. Cuthbert, H. D. Morgan, K. V. K. Luu, and E. Whitelaw. 2003. Transgenerational inheritance of epigenetic states at the murine *Axin*^Fu^ allele occurs after maternal and paternal transmission. *Proc Natl Acad Sci USA* 100(5):2538–43.

Rakyan, V. K., T. Hildmann, K. L. Novik, J. Lewin, J. Tost, A. V. Cox, T. D. Andrews, et al. 2004. DNA methylation profiling of the human major histocompatibility complex: A pilot study for the Human Epigenome Project. *PLoS Biol* 2(12): 2170–82.

Rapp, R. A., and J. F. Wendel. 2005. Epigenetics and plant evolution. *New Phytol* 168(1):81–91.

Richards, E. J. 2006. Inherited epigenetic variation: Revisiting soft inheritance. *Nat Rev Genet* 7: 395–401.

Richards, E. J. 2008. Population epigenetics. *Curr Opin Genet Dev* 18(2):221–6.

Ruiz-García, L., M. T. Cervera, and J. M. Martínez-Zapater. 2005. DNA methylation increases throughout *Arabidopsis* development. *Planta* 222(2):301–6.

Rutherford, S. L., and S. Henikoff. 2003. Quantitative epigenetics. *Nat Genet* 33(1):6–8.

Salmon, A., M. L. Ainouche, and J. F. Wendel. 2005. Genetic and epigenetic consequences of recent hybridization and polyploidy in Spartina (Poaceae). *Mol Ecol* 14(4):1163–75.

Santos, F., and W. Dean. 2004. Epigenetic reprogramming during early development in mammals. *Reproduction* 127(6):643–51.

Scali, V., and L. Milani. 2009. New *Clonopsis* stick insects from Morocco: the amphigonic *C. felicitatis* sp.n., the parthenogenetic *C. soumiae* sp.n., and two androgenetic taxa. *Ital J Zool* 76(3): 291–305.

Scrosati, R. 2002. An updated definition of genes applicable to clonal seaweeds, bryophytes, and vascular plants. *Basic Appl Ecol* 3(2):97–9.

Sheldon, C. C., A. B. Conn, E. S. Dennis, and W. J. Peacock. 2002. Different regulatory regions are required for the vernalization-induced repression of flowering locus C and for the epigenetic maintenance of repression. *Plant Cell* 14(10): 2527–37.

Smith, M. L., J. N. Bruhn, and J. B. Anderson. 1992. The fungus *Armillaria bulbosa* is among the largest and oldest living organisms. *Nature* 356 (6368):428–31.

Solari, A. J. 2002. Primitive forms of meiosis: The possible evolution of meiosis. *BioCell* 26(1):1–13.

Spallone, P. 1996. The salutory tale of the preembryo. In *Between Monsters, Goddesses and Cyborgs: Feminist Confrontations with Science, Medicine and Cyberspace*, ed. N. Lykke and R. Braidotti, 207–26. Atlantic Highlands, NJ: Zed Books.

Stackebrandt, E., W. Frederiksen, G. M. Garrity, P. A. D. Grimont, P. Kämpfer, M. C. J. Maiden, X. Nesme, et al. 2002. Report of the ad hoc committee for the re-evaluation of the species definition in bacteriology. *Int J Syst Evol Microbiol* 52:1043–7.

Suzuki, M. M., and A. Bird. 2008. DNA methylation landscapes: Provocative insights from epigenomics. *Nat Rev Genet* 9(6):465–76.

Vandel, A. 1928. La parthenogenese geographique: Contribution a l'etude biologique et cytologique de la parthenogenese naturelle. *Bull Biol France Belg* 62:164–281.

Vaughn, M. W., M. Tanurdzic, Z. Lippman, H. Jiang, R. Carrasquillo, P. D. Rabinowicz, N. Dedhia, et al. 2007. Epigenetic natural variation in *Arabidopsis thaliana*. *PLoS Biol* 5(7):1617–29.

Veeck, L. L. 1999. *An Atlas of Human Gametes and Conceptuses: An Illustrated Reference for Assisted Reproductive Technology*. New York: Pantheon Publishing.

Vinogradov, A. E., O. V. Anatskaya, and B. N. Kudryavtsev. 2001. Relationship of hepatocyte ploidy levels with body size and growth rate in mammals. *Genome* 44(3):350–60.

Vrijenhoek, R. C. 1984. Ecological differentiation among clones: The frozen niche variation model. In *Population Biology and Evolution*. ed. K. Wohrmann and V. Loeschcke, 217–31. Heidelberg: Springer-Verlag.

Waddington, C. H. 1940. *Organizers and Genes*. Cambridge: Cambridge University Press.

Waddington, C. H. 1957. *The Strategy of the Genes: A Discussion of Some Aspects of Theoretical Biology.* London: George Allen & Unwin.

Weaver, I. C. G., N. Cervoni, F. A. Champagne, A. C. D'Alessio, S. Sharma, J. R. Seckl, S. Dymov, M. Szyf, and M. J. Meaney. 2004. Epigenetic programming by maternal behavior. *Nat Neurosci* 7(8):847–54.

Whitelaw, N. C., and E. Whitelaw. 2006. How lifetimes shape epigenotype within and across generations. *Hum Mol Genet* 15:R131–7.

7

Epigenesis, Preformation, and the Humpty Dumpty Problem

Ellen W. Larsen and Joel Atallah

CONTENTS

Humpty Dumpty, of nursery rhyme fame, was an egg that fell off a wall and "all the King's horses and all the King's men, couldn't put Humpty together again." If we think of an incubated fertilized hen's egg maintained under proper conditions of temperature, humidity, and egg rolling, a chick will develop in a few weeks. Try the same thing with a lightly scrambled egg, and . . . ? Why, when all the material, including the genome, exists in the bowl, do we not expect a downy hatchling to develop? To understand why embryo content alone is insufficient for normal development, it is necessary to formulate a model of *epigenesis*, the gradual un-folding of structure and function during development. This contribution contains a framework for exploring several issues in understanding this process. We start by addressing the issue that biological systems are unusual, in an otherwise self-organizing world, in having a crucial preformed element, the genome. We present a conceptual solution to the potential paradox of epigenetic development having both preformed and self-organizing elements by arguing that the most pervasive preformed element, the genome, evolves to be reactive to epigenetic context. We show that the Humpty Dumpty problem can be resolved if we appreciate that lower levels of organization can produce emergent phenomena at higher levels. We discuss reasons for believing that an increased focus on dynamics, rather than concentrating almost exclusively on content, may provide important insights into the emergent properties of the epigenetic process.

PREFORMATION AND EPIGENESIS

It may be argued that preformation is currently in the ascendancy as a general conceptual framework for understanding development and its evolution. This is not, of course, the crude preformationism of the ovists or spermists who envisioned a homunculus curled inside a gamete that had only to grow in size to display its inherent morphology. No one today doubts the physical changes that occur in embryos as forms develop in the miraculous unfolding of development as the zygote becomes an adult. Today's preformation rests not on fanciful depictions of gametes but on solid evidence of genes (Hall, 1998), whose molecular biology and regulation materialistically underpin morphological development. Not only does the morphology of an embryo unfold in a predictable sequence but so too, it is envisioned, does an underlying genetic program. Although it is now recognized that genes are highly conserved among metazoa, and hence cannot be solely responsible for variation, the regulatory networks controlling gene expression are thought to be "the essence of animal development" (Carroll, 2008). The implication is that ontogeny is the unfolding of developmental programs inherent in a preformed gamete. While the homunculus had merely to grow to produce species-specific form, genomes are envisioned as being decoded, sometimes explicitly, as in a genetic code for protein structure and somewhat less concretely in the invocation of cascades and networks controlling gene regulation. This position is arguably the worldview many developmental biologists have when they approach epigenesis.

An alternative, more inclusive model is that morphogenesis is the outcome of dynamically changing interactions between the genome, the developing embryo, and the external environment (Gilbert, 2003). If this position has merit, it is essential to explicitly formulate a conceptual framework integrating these three components. As C. H. Waddington noted (1969), our worldview influences the questions we ask and ultimately the observations we make and the experiments we devise. If we conceive of developmental change solely in terms of a programmed, spatiotemporal sequence of gene expression, our goal will be to describe the gene cascades and networks responsible. If we consider the genomic elements participating in development to be part of a context in which development occurs, then we will search for the elements of "context" and for the interactions between that context and "preformed" genetic elements.

When explaining most aspects of the cosmos, it is not necessary to appeal to an independent program. Instead, the universe is seen as a self-organized system in which patterns emerge on the basis of physicochemical properties (Stent, 1982). A large-scale example would be the regularity with which a planet revolves around a star. This emergent behavior is a consequence of the mass and position of the two objects and the laws of gravitation; there is no program or blueprint that encodes the details of a specific planetary trajectory. Likewise, Figure 7.1 shows a morphogenetic change occurring over a period of 5 minutes. This change in pattern was wrought not with cells but with sand rearranged on a New Zealand beach by a freshwater rivulet flowing into the ocean. On a more micro scale, the precise base-pairing of separated, complementary DNA helices is attributed to hydrogen bonding, not to an independent program.

Although the world is a self-organized place, "preformed" genomes are a sine qua non of living organisms as we know them. This distinguishes the biological world from the rest of natural phenomena and explains why some of the finest physicists, including Schrodinger (1967), were inspired to think deeply

FIGURE 7.1 (A) Sand under a rivulet of water on a beach on Stewart Island, New Zealand. (B) The same area 5 minutes later. Note the change in the morphology of the pattern.

about genetics. In biology, a description of the development and evolution of organisms would be incomplete without considering the role of the genome.

Evidence will be presented here that epigenesis must be understood as a dynamic interaction between the genome and internal and external factors, one of the important internal factors being the process of development itself. The practical result of this general view is that to understand developmental mechanisms we must consider not only the ways in which the genome and its products drive development but also how genes are reactive to developmental context. We will consider aspects of self-organization in development as an example of epigenesis that is context-dependent and can drive gene expression. The flexibility of this type of epigenetic system will also be explored in the context of a genome supporting developmental *plasticity* (production of environmentally spe-

cific morphology) and *canalization* (producing the same phenotype over a variety of environmental conditions). Finally, a sample of epigenetic strategies will be discussed, and the suggestion will be put forward that a greater emphasis on developmental dynamics (rather than content) is going to be necessary to help us understand the relationship between epigenetic processes at multiple levels of organization.

EPIGENESIS AND THE COORDINATION OF DEVELOPMENT

Epigenesis involves those processes by which developmental structures and functions unfold during development. As such, the terms *epigenesis* and *development* may sometimes be used interchangeably in what follows. Epigenesis of the early amphibian nervous systems would include both induction of neural ectoderm and the physical processes involved in neural plate delineation, the subsequent thickening of its edges to form the neural folds and the elevation of these folds followed by their union at the midline. Even with such a general idea of epigenesis, it is intuitively reasonable that if complicated changes occur in a regular sequence, there has to be a high level of coordination so that cells divide, grow, change, shape, move (in the case of animal cells), die, and secrete molecules at the proper place and time. Morphological evolution may be seen as a change in the coordination of basic cell behaviors in producing different structures, and this change has a hereditary basis. One way of exploring the role of genes in development is to observe how they are affected by evolution. To what degree are evolutionary changes in development the result of new genes as opposed to coordinating the expression of old genes in new ways? The best answer, at present, seems to be that while in some cases single allelic differences are probably important for changing morphologies (Colosimo et al., 2005; Barret et al., 2008), "new genes" unrelated by gene duplication to currently available gene families are probably rare, and most changes appear to occur using old genes in new ways (Nilsson, 1996). The similarity of gene families across multicellular animal groups and plants described in the dozens of organisms whose genomes have been sequenced has led to the idea of a common toolkit, shared widely, often even across kingdom lines (Carroll, 2005, chap. 3). By the time the human genome was sequenced, about 99% of the genes annotated had putative homologues in other eukaryotes and of the remaining 1%, most, if not all, are now thought to have been bacterial in origin (International Human Genome Sequencing Consortium, 2001). Thus, at the cellular and molecular level, a restricted set of modules (cell behavior, genes, and their RNA and protein products) provides a basis for biodiversity and its evolution. The evolutionary conservation of cell behaviors and genes suggests that morphological novelty arises not so much by new material content but by the way conserved material and processes are "used" in time and space.

EPIGENESIS AND REACTIVE VS. INSTRUCTIVE GENES

There is a prominent school of thought that genes and genetic programs drive developmental dynamics. It is often said that a gene is instructive for a particular cell fate or fates if it can be shown that expression of that gene in a particular group of cells is correlated with a particular cell fate and, conversely, if absence of that gene product is associated with loss of a cell fate. It may be argued that the same type of data may lead to the opposite conclusion: that the effect of a gene product is context-dependent since it is often shown that ubiquitous expression of a gene is able to influence cell fates only in particular cells at particular times. For example, *Antennapedia* is a gene in the Antennapedia complex of flies which, when expressed everywhere under the influence of a heat shock promoter in fly larvae, initiates ectopic leg development in the presumptive antenna only during a certain period of time (Gibson and Gehring, 1988). In no other regions are ectopic legs produced, suggesting that these regions lack the

contextual cues for initiating leg development in the presence of Antennapedia protein. Furthermore, different bristles and structures on the transformed antennal leg have their own periods of sensitivity and quantitative thresholds to Antennapedia protein (Gibson and Gehring, 1988; Scanga et al., 1995; Larsen et al., 1996). If generalized, these phenomena suggest that only at certain places and times in development are cells competent to respond to particular levels of gene products and that without the proper context a gene product is unable to induce cell or morphological fates.

It may be difficult to think that the genome, the very basis of inherited difference between taxa and the material repository of evolutionary change, can be strongly context-dependent in its function. Any system depending on a code, however, must be decoded; and thus, the system's "meaning" is dependent on the context of the decoding process. Consider the word *sin*. In Spanish it means "without," in English it is a deed one must not commit, in geometry it is the ratio of particular sides of a right triangle, and in Canada it signifies "Social Insurance Number." Context is essential for decoding the word *sin*, and context is similarly necessary to convert DNA base-pair sequences into "meaning" in organisms.

EPIGENESIS AND DEVELOPMENTAL PLASTICITY

With the resurgence of evolutionary developmental biology, Darwin's insight that morphological diversity is the result of hereditary changes influencing development has again become widely appreciated. Although we may think of development as a sequence of stereotyped changes underlain by a stereotyped pattern of gene expression producing a standard phenotype, in practice, even if two individuals are genetically similar and develop in the same environment, they often demonstrate phenotypic variation. Not all zygotes reach maturity, and those that do differ to a smaller or greater extent. Perhaps the most readily observed phenotypic variation within a species is sexual dimorphism. While this often has a well-understood gene or chromosomal basis, nonbiological factors, such as temperature, can play a role. Various taxa among fish, amphibians, and reptiles show temperature-dependent sex determination (Valenzuela and Lance, 2004) (leading to concern that global warming may occur too rapidly to prevent extinction from lack of adaptation to new temperature regimes). Condition-dependent sex determination may also involve factors other than temperature. In a bizarre example, *Bonellia*, a marine worm, the sex-indeterminate larva becomes a female if it settles on the sea bottom and a male if it settles on a female (Jaccarini et al., 1983). The female grows to be a visible 10 cm or so adult, while the male becomes a tiny parasite on the female, being reduced to essentially a sac of sperm. The fact that such a fundamental phenotypic difference as the sex of the individual may be dependent on the environment rather than genetics shows the limitations of a model that views genes as the only driving force behind development.

Many other polymorphisms correlated with external factors have been described (West-Eberhard, 2003). Not surprisingly, in view of the importance of photosynthesis, the botanical literature is particularly rich in examples where light plays a role in morphogenesis. *Heterophylly* is a term describing individual plants with at least two distinctive leaf morphologies. It has been studied in pond plants with differing morphologies for submerged and aerial leaves. Submerged leaves face different physiological problems compared to aerial leaves, based on lower light and reduced access to carbon dioxide for photosynthesis; and they also face different physical stresses of water vs. air flow. These different environments may have created selective advantages for leaves of different shapes and sizes on the same plant. Current mechanistic explanations for leaf shape suggest that changes in the synthesis and/or transport of the plant hormone abscisic acid are associated with the environment-dependent change from one leaf type to another (Smith and Hake, 1992).

Recent work has clarified the interaction of gibberellic acid (another plant hormone) with light in affecting cell elongation during development. Cell elongation apparently depends on the concentration of a protein which is both degraded by light and transcriptionally upregulated by gibberellic acid (Feng et al. 2008). Therefore, endogenous and exogenous contexts collaborate in regulating an important cell behavior with morphogenetic consequences.

In animals as well as plants, hormones can mediate effects between environmental conditions and morphology in producing developmental plasticity. Termite colonies can produce reproductive, worker, and soldier castes. The production of soldiers from workers appears to be influenced by nutrition and temperature, with consequent effects of juvenile hormone control and its downstream effects (Scharf et al., 2007, fig. 6). Once more, environmental factors, interact with endogenous metabolic machinery to create a flexible morphogenetic response (as well as a behavioral one).

EPIGENESIS AND DEVELOPMENTAL BUFFERING

The flip side of developmental plasticity is *canalization*, the ability of developmental systems to buffer genetic or environmental variations and produce a "wild-type" phenotype. Near the beginning of the twentieth century, it was thought that the normal or wild-type phenotype of a species was the result of similar heredity. By the 1970s, so much evidence for genetic variation within populations had been found that the very maintenance of all that variation became an unsolved issue (Franklin and Lewontin, 1970)—and that was before we were capable of detecting single-nucleotide polymorphisms! How can all this genetic diversity and environmental variation be "ignored" in thinking about the fact that members of a species, though subject to these kinds of variations, still develop species-specific phenotypes? C. H. Waddington, with typical prescience, brought this issue to the fore with beautiful experiments showing how "abnormal" phenotypes could be produced at environmental extremes. The sensitivity to environmental extremes could be selected for so that eventually, even without the environmental insult, some of the selected strain would produce the abnormality (Waddington , 1953, 1961). The ability to destabilize the buffered phenotype through selection suggests that developmental stability, as well as instability, is genetically controlled. One could infer from this that there is no single genetic network to produce a "normal" wing or thorax in nature; a basic scaffolding of gene interactions, involving genetic pathways such as the *Drosophila* genes *engrailed* and *wingless*, may be conserved, but in each habitat we can expect this scaffold to interact with numerous other environmentally sensitive pathways to produce an adapted phenotype. Selection can operate on a very small scale. Weber (1992) showed that selection is possible for minute differences in wing vein length, on the order of a few tenths of millimeters, demonstrating that the genetic fine-tuning of morphology may act very locally.

Buffering development is inherent in some of the processes of "normal" development and may consist of feedback controls to regulate, for example, sizes of organs relative to each other. There is also buffering under conditions of stress in which different genotypes cope differently in the degree to which they produce the wild-type phenotype. As Waddington showed, under stress, mutational or external, otherwise hidden genetic variation may be expressed that is important for producing the wild-type phenotype. When this type of buffering fails, abnormal phenotypes may be produced; and within a few generations of selection, the originally abnormal phenotype may be produced even without stress (Waddington, 1953, 1961). The threshold between the breakdown and reestablishment of developmental buffering under selection provides a potentially important mechanism for rapid evolution and links genetic variation to the evolution of epigenetic processes under stress.

EPIGENESIS AND SELF-ORGANIZATION

Many developmental phenomena are considered self-organized in that they are triggered by stochastic processes, physical forces, or previously developed internal conditions. Stochastic events are among the most pervasive phenomena to trigger change (Larsen, 2005, and references therein). The polarization of budding yeast cells in determining a new bud site is one example (Altschuler et al., 2008). In a model of cell polarization, a signaling molecule randomly recruited to the membrane from a cytoplasmic pool can initiate cell polarization if there is a positive feedback whereby other molecules from the pool are recruited to join membrane-bound molecules. In the model, the ratio of the rates of molecules entering and leaving the membrane as well as the number of molecules in the cytoplasmic pool are predictive of whether polarized localization will occur. Experimental data of membrane recruitment of fluorescently tagged Cdc42 molecules fit model expectations quite well. In this situation, a stochastic process of recruitment of molecules to a membrane results in a qualitatively different membrane structure depending on kinetics and pool size. In this case, cell membrane polarity is an emergent property of the kinetics of recruitment and of protein concentration.

Mechanical forces can also be important triggers for emergent behavior in development. A dramatic example is that of heart looping in embryonic zebrafish. Hove et al. (2002) discovered that only when there is stress on the developing heart tube from blood flow does looping occur. Here, the physics of an unfolding epigenetic trajectory can signal the next change in morphology. While genes were involved in creating the context in which the shear forces are thought to act, the driving force for morphological change, in this case, was the physical interactions of the developing tissue.

For mechanical forces to have morphogenetic effects, it is assumed that the mechanical signal is transduced to affect specific gene expression. This has been described in the

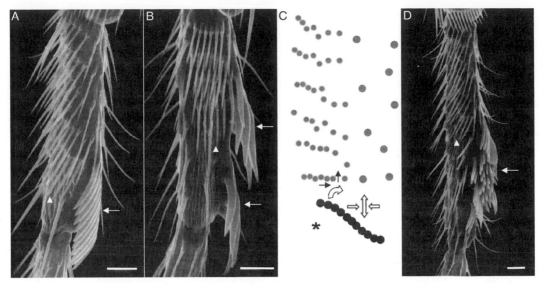

FIGURE 7.2 (A, B) *Drosophila* sex combs (arrows) are always found anterior to (to the right of) neighboring contiguous transverse bristle rows (arrowheads), despite the considerable differences in sex comb morphology between species such as *D. melanogaster* (A) and *D. biarmipes* (B). (C) In *D. melanogaster* and many other species, the sex comb is initially transverse (horizontal) and rotates during metamorphosis. The distal foreleg first tarsal segment (TS1) of a developing pupal *D. melanogaster* male is depicted in the diagram, at a time when the presumptive sex comb teeth (filled black circles) have joined together into a tight, contiguous formation and the structure is about mid-way through the rotation. The presumptive transverse row bristles (light gray circles on the left) are in the process of joining together; most of the distal transverse row (lower left) is contiguous, while in other rows many of the bristles are separated by intervening cells. Part of the sex comb (left, next to the curvilinear arrow) overlaps with the transverse rows in the proximodistal direction (vertical axis in the diagram) at this stage and moves during development in an anterior direction (to the right) relative to these rows. We showed (Atallah, 2008) that sex comb rotation is a consequence of male-specific convergent extension (block arrows) in the distal TS1, which results in proximodistal tissue elongation (denoted by the vertical double-headed block arrow). As elongation proceeds, cells are able to pass between spaces separating nonadjoining bristles (vertical black arrow), but they are forced to move around contiguous bristle formations (horizontal black arrow) like the sex comb and most of the distal transverse row. This leads to a rapid rotation (curvilinear arrow) and anterior displacement of the portion of the sex comb posterior to this transverse row. The position of the sex comb, similar across lineages, is thus largely a consequence of the cell dynamics in the distal TS1, which are constrained by the tendency of contiguous bristle arrangements to act as barriers to cell movement, an emergent property of these multibristle configurations. (D) The biased position of the sex comb relative to the transverse rows is also evident in this mutant, *eyeless-Dominant* (eyD). Anterior is to the right and distal is down in all panels. Scale bars = 20 µm.

initiation of gastrulation in the embryonic fruit fly gut (Desprat et al., 2008). Using fluorescently tagged molecules in a variety of experiments, it was concluded that mechanical stimulation induced by germ-band extension during development causes translocation of Armadillo protein to the nucleus, which in turn activates the *twist* gene, which is associated with invagination. Here again, the developmental process itself, in terms of germ-band extension, creates a signal resulting in another morphogenetic process, invagination. Mechanical stress is so pervasive in organisms that it has long been recognized as important for morphogenesis

(Muller and Streicher, 1989), and we are now beginning to unravel the details of the gene activation it signals as well as the cytoskeletal changes required to effect processes such as invagination (Tamada et al., 2004).

Stable patterns of differentiated elements may also result from a combination of stereotyped cell behavior and existing physical constraints in the developing structure. Atallah (2008) noted that in all fly species with male sex combs, the comb shows an anterior displacement relative to a transverse bristle row adjacent or just proximal to it (Figure 7.2A,B). He showed that this displacement may be the natural outcome of the dynamics in the fruit fly tarsus during metamorphosis, where cells, during a process of male-specific convergent extension, must navigate around contiguous formations of presumptive bristles (Figure 7.2C). By preventing cells from passing between developing bristles within a tandem row, multibristle configurations act as a barrier, an emergent property which is not apparent in the hypothesized evolutionarily basal condition in which bristles are always separated by one or more cell diameters. Interestingly, the positional bias of sex combs relative to transverse row bristles is apparent even in mutants with bizarre and highly variable sex comb patterns (Figure 7.2D), suggesting that the phenotype is an inherent property of the system. Since there are over 200 species of flies with sex combs, as well as a developmentally well-characterized species (*Drosophila melanogaster*) with dozens of mutations affecting sex comb patterning, this is an excellent system in which to study not only the evolution of development (True, 2008) but also the role of physical biases in epigenesis and how they can be modified genetically.

The examples above show that physical and geometric conditions may profoundly influence morphogenesis and pattern formation. The evolution of development probably incorporated physical and chemical principles frequently. Tuning the dynamics of reactants may be required for new morphologies initiated by stochastic processes, and such tuning may be hypothesized to require that fewer genes be recruited; an additional "savings" is that the stability of morphogenesis (canalization) can be the outcome of cellular processes constrained by local geometries. Implications arising from this type of epigenesis are that (1) preexisting states, both genetic and physical, determine future states and (2) knowing the genome or the gene-activation status of a cell or tissue may not be predictive of the next phase of morphogenesis because of the central importance of context. One could therefore argue that a simple mapping of genotype to phenotype, such as exists for the relationship between DNA and the primary structure of proteins, will not be generally possible for epigenesis.

EPIGENESIS AND MORPHOGENETIC FIELDS

The type of constraint proposed to explain sex comb position is a type of self-organization likely to be sui generis in that much morphogenetic change will have local determinants. There is, however, ample evidence of generic self-organizing properties in development, one of the most prominent being morphogenetic fields (Gilbert et al., 1996; Carlson, 2003; Atallah et al., 2004). Classically, morphogenetic fields were operationally defined as regions fated to become a particular structure, which were autonomous when transplanted to an ectopic site. Portions of a field could be severed, and the partial field could subsequently reconstitute part or all of the fated structure. The organ field had regions of highest potency as measured by the probability of reconstituting a new structure from that region, and this suggested that gradients of some type were involved. The field concept seemed to lose favor with the rise of developmental genetics and molecular biology, but several highly influential theoretical models in the late 1960s and 1970s, including the positional information (Wolpert, 1969) and clock (French et al., 1976) models, dealt with problems associated with morphogenetic field phenomena. Since it appears that after early embryogenesis, a minimum number of cells (from about 10 in hydra to about 50 for limb fields in vertebrates) is required to form a field, the field must be viewed as a collective, higher-order phenomenon based on a population of cells, rather than the individual property of a single cell. One could therefore hypothesize that intercellular communication is essential for field properties and that the loss of this communication provides the signal for the formation of new fields. This loss of communication may occur naturally (as when the single eye field in vertebrate embryos separates into two) or may be surgically induced. It has been shown that perturbing gene expression by creating *engrailed*-deficient clones in the fly wing imaginal disc can induce a supernumerary wing and, hence, appear to have initiated a new wing field (Tabata et al., 1995). Such findings suggest that morphogenetic fields might be defined in molecular terms (Gilbert et al., 1996). It is unlikely, however, that particular molecules will be of universal significance in understanding field phenomena because morphogenetic fields are described in such diverse contexts as plant meristems and the protozoan cortex (Frankel, 1997). Thus, field principles transcend the effects of specific gene products, despite the importance of these products in any particular context.

Recently, a new theoretical attempt to capture the salient aspects of morphogenetic fields has been published. The approach differs from that of Wolpert (1969), in which it was envisioned that fields contained a (universal) morphogenetic gradient that was interpreted differently depending on cellular context. Jaeger et al. (2008) reviewed the literature on early fly development and suggested that field boundaries are constantly changing as development proceeds. They proposed that morphogenetic field properties will be better understood if the positional information (morphogenetic gradient) is not considered static but if the gradient and its interpretation are dynamically coupled and change over time. Work described above (Larsen et al. 1996), in which the competence to transform an antenna field into a leg field varied spatially and temporally, lends credence to this perspective.

Dynamic aspects of development have not been given the same attention as the cataloging of genes and molecules (Keller, 2005). It is worth recalling that in the 1940s Richard Goldschmidt (1946) suggested that gene function should be considered from a rate of production viewpoint: If the rates were above a given time-dependent threshold, the "normal" phenotype would emerge. Today, we have a superabundance of information about the players relative to their dynamics, despite the fact that dynamics

are an inherent aspect of a developing system. What makes dynamics particularly important is that the molecular players, in morphogenesis are several orders of magnitude lower on the level of organization scale than the morphological phenotype. As Herbert Simon (1973) pointed out, the salient difference between objects at different levels of organization is their dynamics. According to Simon's model, the dynamics at every level of organization that we recognize must be one or two orders of magnitude slower than that at the next lower level. Molecules, for example, move and change their state at much faster rates than cells. It will be fascinating to see how dynamic descriptions of epigenesis will allow us to understand the emergent properties inherent in morphogenesis and morphogenetic fields, and this should be a priority in exploring epigenesis.

EPIGENESIS AND INTERGENOMIC INTERACTIONS

To date, most of the cellular and genetic studies of morphogenesis have focused on organisms derived from single genotypes. There is growing awareness that some morphologies are the result of interactions of cells with very different genomes and that analysis of these systems may allow us to distinguish aspects of morphogenesis that are dependent on the emergent properties of interactions rather than the readout of a single "genetic program." A fascinating example is that of light organs in the squid *Euprymna scolopes*. These squid have a lantern structure containing the luminescent bacterium *Vibrio fischeri*. McFall-Ngai (1999) has shown that if young squid are deprived of these bacteria, the lantern is not formed. Once formed, however, the squid is able to control the population density of the bacteria by expelling excess numbers every few days. This may seem like an exceptional phenomenon, but many interactions between different kinds of organisms produce morphologies neither biont would be capable of on its own. This is illustrated by arthropods that utilize the same species of goldenrod plants for part of their life cycle by stimulating different types of gall formation, each arthropod species being identified by the morphology of the induced gall (Darling and Gibson, 2000). Figure 7.3 shows an elliptical, a spherical, and a bunch gall stimulated by a moth, fly, and midge, respectively. A difficulty in studying such interactions is that two organisms have to be maintained and their life cycles and special requirements may not lend themselves to laboratory culture. Furthermore, it would be desirable to study systems that are tractable from a molecular as well as cellular point of view.

As a system for studying morphology produced through interactions of two different kinds of bionts, lichens have distinct advantages. They are composed of a fungus and a photobiont (which is usually algal). Many lichen species have been separated into their fungal and algal components; and although each can be propagated separately, neither on its own will produce the morphologies found in combination (Brodo et al., 2001). Some startling changes have been noted between the free-living and lichenized fungi of at least one lichen species, with fungal DNA methylation (commonly referred to as an epigenetic change) being much more pronounced in the lichenized form (Armaleo and Miao, 1999), suggesting that molecular changes during lichenization may yield insight into processes required for the cooperation of two different cell types in morphogenesis. Because genome sequences of fungi and algae are already available and lichen genome projects are under way in several laboratories, lichens might be ideal symbiotic systems to study developmentally. A formidable problem, however, is that because lichens develop and grow relatively slowly, no model lichen system for developmental studies is yet available, although the work of Sanders and Lucking (2002) suggests that tropical leaf-dwelling lichens which readily grow on coverslips may have a sufficiently rapid life cycle to surmount this obstacle. One of us (E. W. L.) has pursued this possibility, finding that in the tropics initiation of development from asexual propagules of one species may take less than 2.5 months to produce a new structure containing asexual propagules (Larsen, 2010). Whether these in situ findings can be replicated in laboratory growth chambers remains to be seen.

CONCLUSION

Epigenesis, the gradual unfolding of developmental trajectories in which succeeding changes depend on preceding ones, has many sources of coordination. Self-organizing properties of physical and geometric constraints, stochastic processes, morphogenetic fields, and external environments all contribute to epigenesis, along with genomes containing encoded properties. The relationship of genotype and phenotype during development is presented here as a reciprocal, dynamic interaction in which gene expression depends on developmental context and developmental context is shaped by gene expression. Without a genome, development would grind to a halt; without developmental context, the genome is quite powerless to affect form. Epigenesis as a conceptual framework provides a natural approach to describing and analyzing dynamic developmental processes at many levels of organization, without an implicit bias as to what factors are driving the process. Epigenetic phenomena highlight the challenge ahead and demonstrate that our quest for a genotype-to-phenotype "map" will require more than correlations of coding motifs to structures. We will need to ask the question of what else, in addition to a genotype, is required to arrive at a particular phenotype and to consider whether, in the origination of biological form, adaptive features that were not initially genetically encoded may have been "captured" by genes to increase the frequency with which those forms appeared (Muller and Newman, 2003; Newman and Muller, 2000). The best systems to address these issues may yet prove to be those in which the interaction of more than one genome yields an emergent phenotype.

FIGURE 7.3 Galls (arrows) on three plants of the goldenrod *Solidago canadensis*. (A) The elliptical stem gall caused by the larva of the moth *Epiblema scudderiana*. (B) The spherical stem gall of the fly larva *Eurosta solidaginis*. (C) The bunch gall associated with the immature midge of *Rhopalomyia solidaginis*, which distorts leaf growth at the plant apex.

ACKNOWLEDGMENTS

Support for this work was provided by a fellowship from the Konrad Lorenz Institute to E. W. L. Funding for research in the authors' laboratory was provided by the Natural Sciences and Engineering Research Council (Canada) and Self-Funded grants (University of Toronto) to E. W. L.

REFERENCES

Altschuler, S. J., S. B. Angenent, Y. Wang, and L. F. Wu. 2008. On the spontaneous emergence of cell polarity. *Nature* 454:886–9.

Armaleo, D., and V. Miao. 1999. Symbiosis and DNA methylation in the *Cladonia* lichen fungus. *Symbiosis* 26:143–63.

Atallah, J., I. Dworkin, U. Cheung, A. Greene, B. Ing, L. Leung, and E. Larsen. 2004. The environmental and genetic regulation of *obake* expressivity: Morphogenetic fields as evolvable systems. *Evol Dev* 6:114–22.

Atallah, J. 2008. The development and evolution of complex patterns: The *Drosophila* sex comb as a model system. PhD diss., University of Toronto.

Barrett, R. D. H., S. M. Rogers, and D. Schluter. 2008. Natural selection on a major armor gene in three-spine stickleback. *Science* 322:255–7.

Brodo, I., S. D. Sharnoff, and S. Sharnoff. 2001. *Lichens of North America*. New Haven, CT: Yale University Press.

Carlson, B. M. 2003. Developmental mechanisms, animal. In *Keywords & Concepts in Evolutionary Developmental Biology*, ed. B. K. Hall and W. M. Olson, 74–83. Cambridge, MA: Harvard University Press.

Carroll, S. 2008. Evo–devo and an expanding evolutionary synthesis. A genetic theory of morphological evolution. *Cell* 134:25–36.

Carroll, S. B. 2005. *Endless Forms Most Beautiful.* New York: W. W. Norton.

Colosimo, P. F., K. E. Hosemann, S. Balabhadra, G. Villarreal, Jr., M. Dickson, J. Grimwood, J. Schmutz, R. M. Myers, D. Schluter, and D. M. Kingsley. 2005. Widespread parallel evolution in sticklebacks by repeated fixation of ectodysplasin alleles. *Science* 307:1928–33.

Darling, C. D., and C. Gibson. 2000. Life and death on the Leslie Street Spit. *Rotunda* 32:24–9.

Desprat, N., W. Supatto, P.-A. Pouille, E. Beaurepaire, and E. Farge. 2008. Tissue deformation modulates Twist expression to determine anterior midgut differentiation in *Drosophila* embryos. *Dev Cell* 15:470–7.

Feng, S., C. Martinez, G. Gusmaroli, Y. Wang, J. Zhou, F. Wang, L. Chen, et al. 2008. Coordinated regulation of *Arabidopsis thaliana* development by light and gibberellins. *Nature* 451:475–9.

Frankel, J. 1997. Is spatial pattern formation homologous in unicellular and multicellular organisms? In *Physical Theory in Biology. Studies of Nonlinear Phenomena in Life Sciences*, vol. 4, ed. C. J. Lumsden, W. A. Brandts, and L. E. H. Trainor, 245–62. Singapore: World Science.

Franklin, I., and R. C. Lewontin. 1970. Is the gene the unit of selection? *Genetics* 65:707–34.

French, V., P. G. Bryant, and S. V. Bryant. 1976. Pattern regulation in epimorphic fields. *Science* 193:969–81.

Gibson, G., and W. Gehring. 1988. Head and thoracic transformations caused by ectopic expression of Antennapedia during *Drosophila* development. *Development* 102:657–75.

Gilbert, S. F. 2003. The reactive genome. In *Origination of Organismal Form*. The Vienna Series in Theoretical Biology. ed. G. B. Muller and S. A. Newman, 87–101. Cambridge, MA: MIT Press.

Gilbert, S. F., J. M. Opitz, and R. A. Raff. 1996. Resynthesizing evolutionary and developmental biology. *Dev Biol* 173:357–72.

Goldschmidt, R. B. 1946. An empirical evolutionary generalization viewed from the standpoint of phenogenetics. *Am Nat* 80:305–17.

Hall, B. K. 1998. *Evolutionary Developmental Biology.* Dordrecht, the Netherlands: Kluwer Academic.

Hove, J. R., R. W. Köster, A. S. Forouhar, G. Acevedo-Bolton, S. E. Fraser, and M. Gharib. 2002. Intracardiac fluid forces are an essential epigenetic factor for embryonic cardiogenesis. *Nature* 421:172–7.

International Human Genome Sequencing Consortium. 2001. Initial sequencing and analysis of the human genome. *Nature* 409:860–921.

Jaccarini, V., L. Agius, P. J. Schembri, and M. Rizzo. 1983. Sex determination and larval sexual interaction in *Bonellia viridis* Rolando (Echiura, Bonelliidae). *J Exp Mar Biol Ecol* 66:25–40.

Jaeger, J., D. Irons, and N. Monk. 2008. Regulative feedback in pattern formation, towards a general relativistic theory of positional information. *Development* 135:3175–83.

Keller, E. F. 2005. The century beyond the gene. *J Biosci* 30:3–10.

Larsen, E. 2005. Developmental origins of variation. In *Variation*, ed. B. Hallgrimsson and B. K. Hall, 113–29. Boston: Elsevier Academic Press.

Larsen, E. 2010. Progress in culturing foliicolous lichens on coverslips. *Bibliotheca Lichendogica* 105:17–23.

Larsen, E. W., T. Lee, and N. Glickman. 1996. Antenna to leg transformation: Dynamics of developmental competence. *Dev Genet* 19: 333–9.

McFall-Ngai, M. 1999. Consequences of evolving with bacterial symbionts: Insights from squid–vibrio associations. *Annu Rev Ecol Syst* 30: 235–56.

Muller, G. B., and S. A. Newman. 2003. Origination of organismal form: The forgotten cause. In *Origination of Organismal Form*. The Vienna Series in Theoretical Biology. ed. G. B. Muller and S. A. Newman, 87–101. Cambridge, MA: MIT Press.

Muller, G. B., and J. Streicher. 1989. Ontogeny of the syndesmosis tibiofibularis and the evolution of the bird hind limb: A caenogenetic feature triggers phenotypic novelty. *Anat Embryol* 179: 327–39.

Newman, S. A., and G. B. Muller. 2000. Epigenetic mechanisms of character origination. *J Exp Zoolog B Mol Dev Evol* 288:304–17.

Nilsonn, D. E. 1996. Eye ancestry: Old genes for new eyes. *Curr Biol* 6:39–42.

Sanders, W. B., and R. Lucking. 2002. Reproductive strategies, relichenization and thallus development observed in situ in leaf-dwelling lichen communities. *New Phytol* 155:425–35.

Scanga, S., A. Manoukian, and E. Larsen. 1995. Time- and concentration-dependent response of the *Drosophila* antenna imaginal disc to Antennapedia. *Dev Biol* 169:673–82.

Scharf, M. E., C. E. Buckspandagger, T. L. Grzymala, and X. Zhou. 2007. Regulation of polyphenic caste differentiation in the termite *Reticulitermes flavipes* by interaction of intrinsic and extrinsic factors. *J Exp Biol* 210:4390–8.

Schrodinger, E. 1967. *What Is Life?* Cambridge: Cambridge University Press.

Simon, H. A. 1973. The organization of complex systems. In *Hierarchy Theory*, ed. H. H. Patee, 1–27. New York: George Braziller.

Smith, L. G., and S. Hake. 1992. The initiation and determination of leaves. *Plant Cell* 4:1017–27.

Stent, G. 1982. Comments. In *A Dahlem Conference*, ed. J. T. Bonner, 110–13. Dahlem, Germany: Springer Verlag.

Tabata, T., C. Schwartz, E. Gustavson, Z. Ali, and T. Kornberg. 1995. Creating a *Drosophila* wing de novo: The role of engrailed, and the compartment border hypothesis. *Development* 121:3359–69.

Tamada, M., M. P. Sheetz, and Y. Sawada. 2004. Activation of a signaling cascade by cytoskeleton stretch. *Dev Cell* 7:709–18.

True, J. R. 2008. Combing evolution. *Evol Dev* 10: 400–2.

Valenzuela, N., and V. A. Lance. 2004. *Temperature-Dependent Sex Determination in Vertebrates*. New York: W. W. Norton.

Waddington, C. H. 1953. Genetic assimilation of an acquired character. *Evolution* 7:118–26.

Waddington, C. H. 1961. Genetic assimilation. *Adv Genet* 10:257–93.

Waddington, C. H. 1969. The practical consequences of metaphysical ideas on a biologists work. An autobiographical note. In *Towards a Theoretical Biology*, vol. 2, ed. C. H. Waddington, 72–81. Edinburgh: Edinburgh University Press.

Weber, K. E. 1992. How small are the smallest selectable domains of form? *Genetics* 130:345–53.

West-Eberhard, M. J. 2003. *Developmental Plasticity and Evolution*. New York: Oxford University Press.

Wolpert, L. 1969. Positional information and the spatial pattern of differentiation. *J Theor Biol* 25:1–47.

8

A Principle of Developmental Inertia

Alessandro Minelli

Self-organization . . . is frequently overlooked . . . in biology . . . partly due to a lack of understanding about how self-organization integrates with modern evolutionary synthesis, particularly natural selection. . . . From the smallest of scales to the largest, self-organization is present at most (if not all) levels of biology. . . . This parallels the ubiquity of self-organization in the non-biological world.

HALLEY AND WINKLER (2008, 148–9)

CONTENTS

Epigenetics: Linking Genotype and Phenotype in Development and Evolution, ed. Benedikt Hallgrímsson and Brian K. Hall.

A NEED FOR THEORY

Biology has long suffered from comparison with physics—biology is largely concerned with the description of historically determined phenomena, rather than with eternal laws. Recently, however, the status of biology has increased substantially, largely because of its success in studying genes and their expression and because of the widely appreciated importance of life sciences for the welfare of humanity and to safeguard our living space on Earth.

To some extent, progress in the life sciences has involved technical, rather than conceptual, improvements in the way we study life phenomena. We are still largely confronted with experimental results obtained from a handful of model systems, rather than with generalizations rooted in universal principles. Thus, most of what we know today is still the embryology of the worm or the developmental genetics of the fruit fly—with the proviso, however, that simply moving from *Caenorhabditis elegans* to *C. briggsae* would mean writing a somewhat different worm story and choosing a *Drosophila* species other than the popular *D. melanogaster* would possibly give us a fruit fly with a *Hox* gene cluster with a different arrangement on the chromosome. Admittedly, the use of mathematics in biology is more frequent currently than in the past, and progress in understanding biological mechanisms is increasingly based on explicit, testable, and eventually tested hypotheses. Thus, why worry about possible shortcomings in the current trend of a triumphantly expanding science such as today's biology?

Difficulties, indeed, emerge at the level of systematization of knowledge. It is often repeated that the only theory in biology is the theory of evolution, and the latter in turn is often equated to a population genetics–based, neo-Darwinian reading of the history of life. To be sure, one may argue that the theoretical basis of biology is much larger and includes, for example, principles about individuality, sex, reproduction, and development that go beyond their strict relevance for our understanding of evolution, not to mention properties that are perhaps most conspicuous in living systems but not exclusive to them, like robustness, hierarchical organization, and the like. However, as soon as we try to articulate this theoretical background of biology in explicit and detailed terms, we are generally confronted with unsatisfactory results, either pushing us astray toward purely mathematical formulations of limited use in advancing knowledge of biological systems or forcing us to abandon hope of universality because our most cherished biological concepts— e.g., individual, gene, reproduction, sex—turn out to function only within some strict set of conditions that correspond to the properties of selected kinds of organism.

However, this is not a good reason to abandon the effort. The problem is how to find a workable strategy. An obvious starting point is to enquire why and how other disciplines were more successful at developing a heuristically useful corpus of theory. To be sure, I do not dare to speculate about a general theory of biological systems. What I will advance are suggestions as to an approach to the study of development that is able to generate sensible questions about the evolution of development. I will use here and there the language of physics, but the approach I will suggest is basically rooted in evo–devo.

Despite its growing popularity and the obvious wealth of experimental results produced and advertised under its name, evo–devo is still a discipline in search of its identity (Müller, 2008). To many (e.g., Carroll, 2005), it is just developmental genetics writ comparative. To others, it is a scientific discipline with an autonomous research program, one of its main targets being arguably the study of evolvability (Hendrikse et al., 2007). Eventually, we should realize that evo–devo deserves to be recognized as an independent and "healthy" biological discipline only if it covers questions, or approaches, that cannot be specifically addressed under either of the parent disciplines taken alone and if it manages to integrate evolutionary biology and developmental biology on a fundamental, theoretical level. In my opinion, a large part of current "evo–devo" research has such a marginal evolutionary content as to adequately fit into the traditionally circumscribed limits of developmental biology. Less frequent, but far from rare, is the case of evo–devo studies where development is not really addressed to such an extent as to justify moving a conventional evolutionary biology paper into the province of evo–devo. Nevertheless, a real evo–devo biology is now growing in extent and importance; but integration between the two disciplinary components is still basically fought on the battlefield, case by case.

I remarked years ago (Minelli, 2003) that in order to promote the dialogue between developmental biology and evolutionary biology we must frame developmental and evolutionary questions in conceptually comparable terms. Here, I will further elaborate that idea.

A NULL MODEL FOR DEVELOPMENTAL BIOLOGY

In the past, the metaphor of the tree was used to express *genealogy* (e.g., these two recent species derive from that Miocene species), but cladistics has taught us to use it instead to express *relationships* (e.g., this recent species and that Miocene species share a common ancestor that a third recent or extinct species does not share with them). Today, evolutionary biology is less and less concerned with origins as such and even less so with the destiny of lineages or features. Its subject is, rather, the change of populations through time. This remains true irrespective of the different perspectives adopted in any given research in evolutionary biology, with focus on genes or morphology, behavior or development.

What about developmental biology? In respect to evolutionary biology, the difference in the way the temporal dimension of phenomena is approached is dramatic. We take for granted that individual development has a recognizable origin (usually in an egg, seed, or spore) and an end point (the adult or mature animal or the seed-ripe plant). In agreement with this perspective, it seems also appropriate to say that the egg, the seed, or the spore contains the program for the adult or, in other terms, that development is nothing but the deployment of that program.

When putting together these seemingly unrelated perspectives on life, rooted in evolutionary biology and developmental biology, respectively, most researchers in evolutionary developmental biology are ready to regard evo–devo as the study of how developmental programs change in evolutionary time. Different views are expressed by those who are unhappy with a strict genetic determinism as *the* explanation of development. This reaction to the notion of the genetic program is a sensible one, but it is not sufficient to bring evolutionary biology closer to developmental biology.

No reason to despair. A way to move ahead can be found by a closer inspection of the history of evolutionary biology. By disposing of the traditional framing of questions in terms of origins and progress and finally emerging as a real "science of change," evolutionary biology has been forced to identify the "zero condition" of its subject phenomena. This is somehow similar to what happened in mechanics, which was eventually developed starting from a principle of inertia, i.e., that in the absence of forces applied to it, a material point will maintain forever its condition of rest or of uniform rectilinear motion. Of course, although the material point in inertial conditions is the system onto which mechanics has been developed, our real interest in it begins only at the time its inertia is disturbed.

Similarly, in a population genetic–centered view of evolution, we could say that the "inertial system" is represented by a population in Hardy-Weinberg equilibrium (see Gayon, 1992). Such a population is an elementary study object of evolutionary biology, although our real interest in it will begin only at the time the population moves away from the equilibrium. A critically important feature of this conceptual perspective is that the *inertial conditions* (of either the material point in mechanics or the population in evolutionary biology) *do not represent origin*, in the ordinary meaning of the word, but only a convenient "zero" term of comparison for the study of something that happens in time—a segment of history, without prejudice of what happened before it. Speaking of evolution, this translates into something like "let's study the modifications affecting a lineage of common descent along a given time interval."

This is the background against which we should be prepared to adopt, in developmental biology, a corresponding "principle of inertia." In the following pages I will try to demonstrate how profitable it can be to approach the wealth of developmental processes occurring in the most different kinds of living beings as deviations from a *local self-perpetuation of cell-level dynamics* to which we can give the name *developmental inertia*.

CELL PROLIFERATION AS INERTIAL CONDITION IN DEVELOPING SYSTEMS

In many multicellular organisms, short phases of intense cell proliferation are usually followed by a long-term or even definitive arrest of mitotic activity, often but not necessarily coupled with terminal cell differentiation. Cell proliferation is largely or even completely confined to the embryonic segment of ontogeny in different, often miniature-size animals such as mites, chaetonotid gastrotrichs, and many nematodes. In other groups, such as arrowworms, the cessation of mitosis is limited mainly to the epidermal cells (Shinn, 1997); the same is true of a great many flatworms (Littlewood et al., 1999) but with exceptions (Drobysheva, 2003).

It is reasonable to accept the concept that the "inertial" condition of all cells is proliferation (Soto and Sonnenschein, 2004) rather than the lack of mitosis or, still worse, apoptosis (a view defended, e.g., by Meier et al., 2000). This is not to negate that many cell types, in metazoans, are dependent for their survival on a regular supply of specific signals; but this is clearly a derived condition. Mitotic arrest is arguably a derived condition, sometimes correlated with gross morphological changes in the epidermal cells, e.g., with the production of cilia, whose presence is incompatible with the formation of a mitotic spindle.

Textbook examples of "diffuse inertial multiplicity" are provided by the archaeocytes of sponges and by the interstitial cells of hydra, but it is worth adding that in adult planarians neoblasts can comprise up to 30% of the total number of cells (Ellis and Fausto-Sterling, 1997).

Taking mitotic activity as a cell's inertial feature seems sensible with respect to the generalized, although far from universal, downregulation of mitosis during major morphogenetic changes, such as gastrulation (Foe et al., 1993) or the dorsal closure of a *Drosophila* embryo.

Indeterminate growth is widespread in plants but is also recorded from many animals, including annelids (*Pristina*), molluscs (several bivalves), sea urchins (*Strongylocentrotus*), cladoceran and decapod crustaceans, and vertebrates (many bony fishes, reptiles, and a few mammals, e.g., bison, giraffe, and elephant) (Karkach, 2006).

If the inertial condition for a metazoan cell is proliferation, what requires explanation, both in mechanistic and in evolutionary terms, is the cessation of mitosis, which is so generally associated with the change to a differentiated state and so often remains as an irreversible acquired condition. Still more in need of explanation is apoptosis, something we cannot accept as a default condition, as has been sometimes contended (Raff et al., 1993; Meier et al., 2000). It is quite possible that apoptosis first evolved as a way for multicellular organisms to create a sheltering barrier of dead cells from which they obtained protection against pathogens (Biella et al., 2002) or to limit the danger potentially growing from the proliferation of cancerous cells (Krakauer and Plotkin, 2002). This hypothesis could apply to multicellular organisms as different as animals, fungi, and plants.

SINGULARITIES WITHIN SYNCYTIA

Similar to the perfect immobility of the material point or the conditions of Hardy-Weinberg equilibrium of a Mendelian population, developmental inertia is a theoretical "null" condition that real phenomena can only approximate to some extent. Deviations from it will occur everywhere, and it is to them that we shall direct our attention.

Syncytial systems are worth a glance in this respect. In principle, a *syncytium* is a system within which functional conditions are expected to be largely averaged over a distance larger, or much larger, than the usual size of one cell and possibly to a better degree than in a cluster of cells of equivalent overall size. However, some degree of compartmentalization and, hence, of deviation from developmental inertia is to be expected in syncytia too. A good example is provided by the urochordate *Oikopleura*, where the female germ line is represented by multinucleate (i.e., syncytial) coenocysts, within which the individual nuclei undergo meiosis asynchronously (Ganot et al., 2007).

In a cellular environment, one of the most obvious signs of breaking down of developmental inertia is the different pace at which mitosis goes on or, more conspicuously, the local stop given to mitosis. A more "courageous" proof will be the breaking down of physical connectivity among the cells in a cluster, as in the splitting off of a gamete, a spore, or a multicellular propagule. In this sense, reproduction is an example of a biological phenomenon where a system moves away from its state of developmental inertia.

EVERYTHING EVERYWHERE

Multiplicity of centers of local developmental dynamics, as opposed to global control over the whole of the multicellular system, is the most obvious trait of developmental inertia, of which examples are abundant.

Cell proliferation is only the most widespread and the most obvious of the many processes that can run in parallel in an unlimited number of foci within a multicellular system, so far as no internal or external agents show up to locally displace a cell, or a cluster of cells, from an otherwise inertial dynamics.

SYMMETRY

Evidence of multiple equivalent foci of development is provided by symmetry, including that peculiar kind of translational symmetry we describe in metazoans as "segmentation." If this is actually a proof of autonomous developmental dynamics, we must expect some lack of precision, that is, some degree of fluctuating asymmetry to occur. This is indeed what we see, transiently, in some embryos. For example, in the amphipod crustacean *Orchestia cavimana*, from the eight-cell stage on, two types of embryos are recognizable, which are mirror images of each other, but the asymmetry eventually disappears at more advanced stages (Scholtz and Wolff, 2002). Another example is provided by bird feathers. At the beginning of their growth, these are often distinctly and increasingly asymmetric in length, but they will eventually symmetrize when approaching complete development. Aparicio's (1998) experiments do not suggest the existence of an internal comparison between the two sides, followed by compensatory growth: More likely, the feathers on the two sides follow a similar but independent trajectory of "targeted growth" toward their maximal potential size.

Counterintuitively perhaps, the most highly symmetrical systems are also the most random (Ball, 1999), that is, the closest to an inertial condition. What requires explanation as a developmental phenomenon is thus directional asymmetry. In other terms, there are not genes, arguably, for body symmetry, whereas genes involved in breaking the inertial symmetry or in elaborating over a first lateralization of a developing system do obviously exist. The growing list of genes that have been identified as responsible for body asymmetry in vertebrates includes *nodal* (e.g., Levin et al., 1995; Blum et al., 1999; Yost, 2001) and *Pitx2* (e.g., Ryan et al., 1998). Their homologues have been recently found to be involved in establishing the chirality of snail torsion (Grande and Patel, 2009). A corresponding role, although through very different mechanisms, is taken in *Drosophila* by *Myo ID* (Hozumi et al., 2006; Speder et al., 2006; Speder and Noselli, 2007; Coutelis et al., 2008).

The usefulness of the notion of developmental inertia as a parsimonious perspective on the production of multiple (two, at least) equivalent parts in symmetrical organisms is also shown by the lawfulness of monsters with duplicated body parts. Symmetry rules over the double heads of bicephalous lambs and cattle not too rarely recorded by the chronicles, as it also rules in the bifurcated or trifurcated legs and antennae occasionally found in the fruit fly bottles or among field-collected insects (Balazuc, 1948).

The multiple parts we can parsimoniously interpret as a product of developmental inertia are not necessarily the left and right halves of a bilaterally symmetrical animal because the same interpretation applies to the many examples of translational symmetry found in nature, i.e., the repetition of segments along an animal's main body axis or the sequence of node + leaf + internode modules of plants (Minelli and Fusco, 2004; Fusco, 2005).

Some mutations apparently release a latent potentiality for the multiplication of developmental foci. For example, mice mutants for *jaws* develop ectopic interphalangeal joints (Sohaskey et al., 2008) and the *Drosophila* mutant *spiny legs* develops up to eight tarsal elements, instead of the normal five (Held et al., 1986). In most of these *Drosophila* mutants, the extra elements have inverted polarity, that is, in their tarsus normal and inverted articles alternate, demonstrating the lawfulness of local duplications with mirror symmetry effect (Held et al., 1986). Another interesting feature of this mutation is that along the axis of the leg it gives rise to a multiplication of equivalent (and, by pairs, symmetrical) developmental foci, in exactly the same way that alternating normal and extra segment boundaries develop in another *Drosophila* mutant, the embryonic lethal *patched*, along the main body axis (Ingham, 1991).

SEGMENTS

The latter examples lead us straight to another class of multiple units within a developing body, segments.

Segments have been described as parallel worksites (Minelli, 2009b) where similar or even virtually identical sequences of events give rise, either simultaneously or in regular temporal progression, to similar or virtually identical body parts. Actually, this description would better apply to the serially repeated elements of any one of the several components which are likely to develop as distinctly modules, only to be eventually integrated at a later stage into a unitary morphological and functional unit deserving the name of "segment." For example, in an arthropod segment there is no evidence of a strict and original developmental integration between what will eventually emerge as the segmental unit of the nervous system (the neuromere) and the segmentally repeated sclerites, or the muscles ensuring their articulation in respect to the preceding and the following segments. Even in animals with more overtly expressed segmentation the identification of segments as unambiguously integrated units is not possible, due to a gross mismatch between different series of structural elements. The most popular example is provided by millipedes, where most of the trunk is articulated in rings, each of which corresponds to two pairs of legs as well as to pairs of ganglia in the ventral nerve cord; but similar mismatches are widespread in arthropods generally (Minelli, 2003; Minelli and Fusco, 2004; Fusco, 2005). In the pill millipede *Glomeris marginata*, Janssen et al. (2004) have demonstrated that different genes are involved in establishing the dorsal and the ventral series of sclerites, and the genes involved in both sets of elements do not necessarily have an identical role.

It is reasonable to hypothesize that in the course of evolution a diversity of serially repeated features have been progressively added together. The starting point was probably a regularly patterned series of neuromeres (Scholtz, 2003). In this context (and as a further example of the inertial principle that a developmental process tends to happen in parallel wherever nothing forbids it), the development of the ventral nervous cord of nematodes offers an interesting example. In *Caenorhabditis elegans*, this cord derives from a set of 13 regularly spaced pioneer motoneurons that will eventually divide, in parallel, in a stereotyped and broadly identical way (Sulston and Horvitz, 1977; Walthall, 1995).

GROWING OUT OF MULTIPLE INDEPENDENT CENTERS

Another aspect of developmental inertia is the spatial delocalization of developmental foci. While plants would offer a virtually inexhaustible number of examples, animals are arguably less rich of evidence; but there are outstanding exceptions. One is the Hydrozoa, where buds can sprout out in the most diverse parts of the body and in virtually all stages of the conventional life cycle (Bouillon et al., 2006). Budding polyps are more popular; buds, however, are also produced by medusas, and these buds can give rise either to other medusas or to gonophores (specialized branches where gametes will eventually mature) or even to polypoid structures, as observed in *Bougainvillia platygaster* and other species. More precisely, buds are produced virtually everywhere on a budding medusa, including the manubrium and the radial canals.

A peculiar example of multiple focal points of cell proliferation is the tree-like mother sporocyst of the digenean flatworm *Lechriorchis primus*, each of whose branches has its own center of proliferation of germinal cells (Galaktionov and Dobrovolskij, 2003).

Multiplicity of developmental centers does not negate the eventually integrated nature of animal development, even in cases of temporary physical independence of the developing centers that invite us to think otherwise. Some of these phenomena are long known. An example is the so-called blastomere anarchy first described by Hallez (1887) from freshwater planarians of the genus *Dendrocoelum*. In the embryo of these free-living flatworms, an apparently "normal" eight-cell stage disintegrates into separate cells that migrate away from each other within the mass of yolk cells—only at a later stage will their progeny gather together again into a single embryo. A similar behavior is known for the stenolaematan bryozoans (Zimmer, 1997). More impressive are, however, the embryos of a freshwater fish (*Cynolebias*) where a similar split followed by later rejoining of parts happens at a quite later stage. Here, in fact, the zygote often gives rise to two separate blastoderms, which develop into two distinct advanced blastulae that reaggregate at gastrulation (Carter and Wourms, 1993). Also, freshwater sponges (*Ephydatia* and *Spongilla*) are known to occasionally form from the fusion of two or more larvae (Brien, 1973).

POLYEMBRYONY

Reciprocally, parts of a prospective embryo may become centers of parallel development of as many independent embryos. This is the phenomenon known as *polyembryony*, by which more than one embryo is eventually obtained from a single zygote.

The occurrence of polyembryony across the metazoans is scattered among six different phyla (Zhurov et al., 2007). Strictly speaking, polyembryony means that two or more embryos derive from the splitting of the embryo originating from an egg. This is known for the rhabditophoran flatworm *Gyrodactylus elegans*, the stenolaematan bryozoans, several tiny hymenopterans belonging to different families (Encyrtidae: *Copidosoma*, Platygasteridae: *Platygaster*, Braconidae: *Macrocentrus*, Dryinidae: *Aphelopus*), and some armadillos (*Dasypus*). In other metazoans it is the larva that splits into two or more individuals. This is known for representatives of the hydrozoans (the trachyline medusas of the genera *Pegantha*, *Cunina*, and *Cunochtantha* and the curious parasite *Polypodium hydriforme*), the tapeworms (*Echinococcus*), the flukes (*Schistosoma*), plus the parasitic crustacean (rhizocephalan) *Loxothylacus panopaei*, a sea star of the genus *Luidia*, and the brittle sea star *Ophiopluteus opulentus*.

Interestingly, in the never-ceasing confrontation between the inertial development of virtually identical parallel modules and the possibility to turn them into eventually different systems, sometimes as the effect of seemingly minimal point differences, polyembryony does not necessarily give rise to only one kind of embryos. In the parasitic wasp *Copidosoma*, the splitting of a large morula-like embryo gives rise to a huge number of larvae, of which there will be two different kinds: Some (those that will inherit a germ-line determinant in the form of *vasa* transcripts; Zhurov et al., 2004) will turn into larvae developing into adults, while the others (those that will not receive the germ-line determinant) develop into "soldier larvae" that will not metamorphose to adults. In many respects, the very different developmental fate of these two kinds of larva is comparable to the plurality of morphologically different castes generally obtained among social insects by alternative canalization of development as a result of environmental manipulation.

SPATIOTEMPORAL SEQUENCES

The equivalence of multiple local foci of growth and developmental dynamics is usually masked by the spatial progression that accompanies ontogeny. In plants, spatial progression is most conspicuously embodied in the activity of the apical meristematic cell of ferns or in the equivalent though more diffuse proliferation of the multicellular apical shoot meristem of flowering plants.

In animals, growth and differentiation are usually dominated by an anteroposterior trend, which is obvious, e.g., in the somites serially produced by an elongating vertebrate embryo or in the segments sprouting from the lower half of a polychaete trochophora. However, what is spread along the anteroposterior axis is, in a sense, a wave that progressively sows new centers of growth and differentiation. This is very clear in the case of the clitellate embryos, such as the well-studied leech *Helobdella*, where the progenitors of ectodermal and mesodermal tissues of each segment are serially generated by the proliferating activity of a small number of teloblasts; in this way, all segments receive an identical set of precursors that will soon start dividing, and their progeny will differentiate in broadly similar ways all along the 32 segments of the leech. From the same perspective, the difference between a short germ-band insect like the cricket and a long germ-band insect like *Drosophila* is basically in the way, and the temporal sequence, the cellular material is distributed that will form largely equivalent foci of later differentiation.

Starting with a null model of developmental inertia, we should have no problem regarding isometric growth as a kind of default condition for growth. Rather, we should not take the obviously widespread anteroposterior progression in growth and development as a necessary rule. This rule, indeed, is conspicuously negated by the inertial intercalary growth that allows a juvenile *Ascaris* worm to grow perhaps 10-fold in length following the last moult and by the frequent deviations from the rule of anteroposterior growth and differentiation found in many crustaceans (or, equivalently, from the proximodistal growth and differentiation of their appendages): For example, in some stomatopods, the appendages of the *pleon* (the posterior body region) differentiate before those of the *pereion* (the middle body region) (Schram, 1986).

THE STARTING POINT, IF ANY—IS THIS THE EGG?

One of the major difficulties in acknowledging the heuristic usefulness of a principle of developmental inertia as a "default model" of developmental processes is likely due to the traditional view of development as a sequence of events starting—as a rule—with an egg (or a seed or a spore).

However, the metazoan egg has little to share with the ideal "inertial cell." An egg's ability to undergo mitosis must not obscure its conspicuously derived features, e.g., its enormous size, its yolk content, its specialized envelopes. A much more sensible candidate to represent the default metazoan cell is a stem cell or one of those blastomeres which, in "regulative" embryos such as a sea urchin's, can eventually give rise to a normal individual if isolated from an early cleavage stage's embryo.

In evolutionary terms, the hen's egg is not necessarily older than the hen. Out of metaphor, eggs are far from "primitive," or generalized, cells. To equate an egg to the closest unicellular ancestors of metazoans would contradict phylogeny in the grossest way. It is even possible that the earliest metazoans were devoid of sexuality, as apparently are the choanoflagellates, their closest unicellular relatives. That the egg is one of the most specialized cell types evolved in the animal lineage (Boyden and Shelswell, 1959) is something we can understand not only from the organization of a typical mature egg but also from the profiles of gene expression during oogenesis. Furthermore, egg cleavage mitoses do not have the usual inertial character typical of self-perpetuating processes, as later cell divisions have, by which cell size is often kept essentially constant; to the contrary, cleavage divisions produce cells of rapidly decreasing size, something often required for subsequent developmental steps to go on smoothly.

Finally, animal development does not start necessarily from an egg. This shows how inadequate is the narrow view of development as the execution of a program stored in the fertilized egg. The nonprogrammed component of developmental control is exactly the epigenetic aspect of ontogenetic canalization, onto which gene expression often appears as a kind of optional overdetermination. There are examples of vegetative reproduction among representatives of groups as diverse as the "Porifera," Placozoa, Ctenophora, Cnidaria, Acoela, Catenulida, Rhabditophora, Cycliophora, Bryozoa, Entoprocta, Dicyemida, Nemertea, Annelida, Sipuncula, Arthropoda, Xenoturbellida, Echinodermata, Enteropneusta, Pterobranchia, Urochordata, and even Vertebrata (Minelli, 2009a).

REGENERATION

Physical damage to a full-grown organism (or one sensibly advanced along its usual developmental schedule) is often followed by regeneration, a set of processes often entailing local restitution to inertial conditions of growth and development. Regeneration is known in "Porifera," Ctenophora, Cnidaria, Acoela, Gastrotricha (*Turbanella* only), Catenulida, Rhabditophora (especially in Macrostomorpha and Tricladida, much less in other groups), Bryozoa, Entoprocta, Nemertea, Phoronozoa, Mollusca, Annelida (but not, e.g., in Hirudinea), Sipuncula, Arthropoda (appendages only), Echinodermata, Enteropneusta, Urochordata, and Vertebrata (summarized in Minelli, 2009a).

In a recent review on regeneration, Birnbaum and Sánchez Alvarado (2008, 707) ask "how does the disruption of homeostatic mechanisms caused by injury canalize development in plants and animals to regenerate anatomically precise and functionally integrated tissues from such unpredictable starting points?" If we accept a principle of developmental inertia as a default condition of developing systems, there is nothing obscure in this state of affairs. Regeneration is not a peculiar phenomenon, or strategy, that arose independently in a diversity of lineages but a jump to conditions closer to those of developmental inertia, which in different organisms takes the idiosyncratic features of the specific lineage involved—no wonder, if life expectancy is prolonged in regenerating organisms with respect to intact ones. For example, in the flatworm *Macrostomum lignano* repeated regeneration following repeated amputation causes life to be prolonged beyond normal expectance (Egger et al., 2006). We used to say that many organisms regenerate in order to live longer, but I would rather say that those organisms live longer because they are able to regenerate. The problem is, however, whether, or how far, we can regard the regenerated organism as "the same" as the original one; in *Botryllus*, for example, regeneration entails a complete turnover of the original components (Lauzon et al., 2002).

If we regard regeneration as a return to inertial, generic conditions of development, rather than as a peculiar, newly discovered mechanism for coping with the consequences of physical damage to a developmental system, we obtain a more parsimonious and more realistic view of the phenomenon. First of all, there is no more need to describe regeneration in finalistic terms, as a solution eventually adopted by organisms in order to erase the consequences of damage and to restore conditions that will improve the chance of survival and reproduction. Second, we will not insist on searching for a specific innovation responsible for regeneration, while a developmental repertoire is already available and needs only to be used again. Even if regeneration does not closely replicate a segment of embryonic development, its mechanisms are mainly picked out of an already available series of developmental activities, which is (re)activated outside its usual context. Thus, it makes little sense to ask what a newt and a stick insect have in common, only with reference to the circumstance that both of them, but not many of either's close relatives, can regenerate a lost leg.

There is abundant experimental evidence showing the common background of embryogenesis, asexual reproduction, and regeneration. One of the most illuminating studies is Cardona et al.'s (2005) comparison between embryogenesis and regeneration in planarians. In these flatworms there is no clear evidence of germ layers or of well-defined organ primordia. Therefore, their embryos are more similar to regenerating blastemas than other embryos would be. Embryogenesis will nevertheless differ from regeneration in these animals too because cells of different origin are involved and because regeneration can take advantage of patterning cues provided by already differentiated structures, while the embryo is bound to rely on maternal determinants only. However, in either case the formation of a provisional epidermis is followed by the activation of neoblasts, after whose proliferation the definitive epidermal cells will be produced. Parallel between embryogenesis and regeneration are the subsequent differentiation of the muscle tissue, the formation of ventral nerve cord pioneers—which will eventually differentiate into the ventral nerve cords—and the processes of brain precursor condensation and nerve cord development.

EVERYTHING AT ALL SCALES

The parallel, inertial progress of identical developmental foci at multiple sites within a developing organisms is likely prone to repeated reactivation, with the consequent generation of a recursive, fractal-like pattern. Examples of structures likely generated this way are the compound leaves of many ferns.

We must be cautious, however, before claiming recursivity of mechanisms, that is, a plain application of the principle of developmental inertia, whenever we are confronted with similarity of patterns at different size scales. That these are actually produced by the uniform operation of the same inertial morphogenetic processes is only a null hypothesis to be tested case by case. Development, indeed, is often less "elegant" than we would expect. In generating the primary striped pattern that will eventually give rise to a segmented *Drosophila* embryo, each transversal stripe is actually controlled separately from the other, eventually equivalent stripes. A condition intermediate between "elegant" inertial iteration of branching and "inelegant" specific control for each branching step has been recently suggested for the development of the mouse lung (Metzger et al., 2008).

DIFFERENT BUT EVENTUALLY EQUIVALENT DEVELOPMENTAL PATHWAYS

Developmental inertia suggests that the history of a developmental system can be irrelevant, to a degree, as a determinant of its current and future behavior. This is what we observe in regeneration, if compared to "normal" development. However, this is also demonstrated by the nearly identical course of development and the nearly identical results obtained by embryogenesis and *blastogenesis* (asexual reproduction from a bud), alternative routes to reproduction and development, both of which are available to some metazoans.

In colonial ascidians like *Botryllus*, nearly identical zooids are formed either by embryogenesis, starting from a fertilized egg, or by blastogenesis, starting from a multicellular bud. Also interesting is the case of *Prorhynchus stagnalis*, a freshwater flatworm belonging to the clade of the Lecithoepitheliata, whose embryo can progress through two alternative modes of cleavage, with the eight-cell stage represented either by eight blastomeres of equal size, as expected in animals with radial cleavage, or by four macromeres and four micromeres, as typical of spiral cleavage (Steinböck and Ausserhofer, 1950). Again, in the medusa *Aurelia aurita* embryos may gastrulate in different ways, but this has no consequence on the eventual organization of the adult (Franc, 1993).

METAGENESIS VS. METAMORPHOSIS

Developmental inertia can help us to understand why the divide between metagenesis and metamorphosis is often far from neat, and sometimes an arbitrary one. The distinction between the two classes of phenomena was first proposed by Steenstrup (1845), following his pioneering studies of cnidarian life cycles. Steenstrup characterized metamorphosis as implying changes within the same individual, while in metagenesis, or alternation of generation, changes are spread over subsequent generations of distinct individuals.

Problems with this distinction, however, are numerous (Minelli, 2009a). Difficulties already emerge when describing cnidarian life cycles including a polyp stage and a medusa stage because Steenstrup's concept of alternating generations does not apply to cubozoans. In this class, the polyp does not give rise to the medusa by a process of vegetative reproduction (strobilation or budding), that is, by a mechanism that preserves the parental identity of the polyp as distinct from the identity of the medusa to which it gives rise. In this group, the polyp does not generate a medusa, in the conventional sense of the term, but literally metamorphoses into a medusa. In this case, we should perhaps regard the polyp as a larva and describe the life cycle as including one generation only. However (Minelli, 2009a), if we describe the cubozoan polyp as a larva because it turns integrally into a medusa, where shall we actually fix the divide between metagenesis and metamorphosis? One tentative solution could be to frame the question in terms of numbers. In cubozoans, one individual (polyp) turns into a different but still single individual (a medusa); but in scyphozoans or hydrozoans, the medusas budding off from the parent polyp increase the total number of individuals in the system, thus offering an instance of reproduction, rather than metamorphosis. Again, the divide is not a clear-cut one. If a hydroid polyp produces one medusa only, is this really different from the "catastrophic" metamorphoses of many invertebrate larvae, where most of the larval body is discarded after a few days or weeks of independent life, while the juvenile is formed from an originally inconspicuous cluster of set-aside cells?

A most puzzling borderline case is offered by *Mutela bourguignati*, a freshwater bivalve mollusc with alternation of generations. As soon as the lasidium larva of this mussel is attached to its fish host, it metamorphoses into a parasitic haustorial larva, the latter eventually producing a bud fated to give rise to a juvenile mussel (Fryer, 1961). One may argue that this mollusc has a metagenetic cycle, while its closest relatives, even those in the same family Mutelidae, "simply" undergo metamorphosis. The two phenomena are thus arguably mechanistically quite close and together show that development (or reproduction?) occurs wherever nothing deviates its inertial course elsewhere.

EMERGING SINGULARITIES

Fungi, with the amazing diversity of shapes of their fruiting bodies made of a loose meshwork of hyphae whose "social life" is arguably very limited, will be a rewarding subject for evo–devo studies. The same is true of the thallus of lichenizing fungi, where "Fungal hyphae and algal cells or filaments . . . initially independent . . . join together in development to produce the lichen thallus; the morphology of this secondary structure is . . . an emergent property distinct from that of its components. The surfaces and volumes of the lichen thallus are assembled de novo in the course of development, and show no continuity with those of the primary cellular elements" (Sanders, 2006, 97). I would bet that a better knowledge of these systems will reveal a diversity of forms that development can produce through mostly inertial dynamics. This will help us to set in a more adequate context the phenomena we usually describe in terms of cell–cell competition, something that obviously becomes important as soon as cells are closely packed together as in animals.

BREAKING THE SYMMETRY: COMPETITION AND CELLULAR IDENTITY

Molecularly marking their own identity is a way by which a cell mass may obtain advantage in terms of perpetuation of its inertial conditions. This is relevant with respect to biological phenomena ordinarily treated as quite separate items such as the differentiation of germ layers, the production of set-aside cells and imaginal discs, the origin of tissues, and the evolution of histocompatibility genes.

Let us begin with the latter topic, about which we have learnt much from grafting experiments with colonial ascidians. For example, Laird et al. (2005) fused together the vascular systems of genetically distinct colonies of *Botryllus schlosseri*, thus releasing phenomena of cellular parasitism involving both somatic tissues and gametes, which induced the authors to formulate the hypothesis that histocompatibility genes evolved in these tunicates as a way to protect the body from parasitic stem cells that usurp asexual or sexual inheritance.

Competition among neighbors or, more generally, among lineages of inertially proliferating cells can be avoided, reduced, or balanced. Establishing a physical barrier to strong competitors or invaders is only one of the possible solutions. A different one is to take divergent routes of differentiation so that two neighboring groups of cells will eventually form two metabolically distinct compartments. Another, nearly opposite strategy is to completely dispose of the boundaries between cells, thus pooling their otherwise contrasting needs. This can be the result of incomplete citodieresis or of cytoplasmic fusion, as in syncytia, or of complete fusion including karyogamy, as in fertilization. The least obvious way cells can escape from disruptive competition, but possibly the most widespread one, is by acquiring metabolic conditions very similar to those of their nearest neighbors. In this way, competition is not canceled, but it can reach a stable equilibrium.

Specifically, there are many ways by which developmental inertia can be established, thus providing the structural and metabolic uniformity within a set of neighboring cells we describe as belonging to the same tissue. Paraphrasing Gurdon (1992), we can say that these cells are similar because they have acquired similar instructions either from their ancestors or from their contemporaries. Held (2002) has rephrased this by saying that a cell can get specific cues, in the form of diffusible ligands, from its more or less proximate neighbors but can also inherit a specific epigenetic state from its mother. However, it is also possible that uniformity of epigenetic and metabolic states is produced by the sustained, two-way exchange of chemical or electrical inputs between neighboring cells, a local inertial phenomenon whose "synchronizing" consequences can spread over a wide range, similar to what local coupling produces in a diversity of phenomena, including human choral singing (Minelli, 2009a).

DEVELOPMENTAL MODULES

To some extent, there are important points in common between the notions of developmental inertia and developmental module, extensively floated in recent literature (e.g., Schlosser, 2002; Schlosser and Wagner, 2004; Callebaut and Rasskin-Gutman, 2005). However, the growing body of literature about developmental

uncritically taken as primitive turn out to be secondarily simplified. This is relevant to this chapter's main argument because features we would like to interpret as "default morphology" produced by developmental inertia may turn out to be instead the result of a secondary simplification, if not even the sophisticated product of complex control systems.

Addressing the study of development from a parsimonious perspective as hypothesizing developmental inertia as the default status of developmental systems can anyway help to avoid preconceived views of character evolution, based either on functional properties or on abstract notions of structural complexity. One example is provided by sponges. In terms of gross morphology, three sponge models are traditionally distinguished: ascon (a simple sac whose internal wall is covered with choanocytes), sycon (a more complex sac, with choanocytes lining layers of flagellated chambers in the sac's wall), and leucon (a principally undefined three-dimensional meshwork of flagellated chambers variously interconnected through narrow channels). In terms of efficiency in obtaining food by filtering water, a progression from ascon to sycon to leucon seems to be warranted as the way for a sponge to achieve larger size and eventually to develop a diversity of growth forms, but a perspective based on developmental inertia would suggest otherwise. Within the diffuse organization of a leucon-type sponge, every little piece comprised of a flagellated chamber and its connecting channels is, in principle, an exact replica of any other similar piece and their reciprocal topographic distribution seems to respond to generic principles of spatial adjustments of equivalent units rather than to an overall shape control depending on the expression of pattern-controlling genes. To the contrary, the pretty regular shape of a syconoid sponge, with a main body axis going through the internal cavity and its apical outflow opening (the osculum), signals a strictly controlled symmetry breaking imposed over an inertial diffuse system. This revised polarity of changes among basic kinds of body organization fits much better than its tra-

ditional alternative within the newly emerging (although still somehow unresolved) phylogenetic relationships among the main groups of sponges (Minelli, 2009a).

EPIGENETICS AND DEVELOPMENTAL INERTIA

In mechanics, inertial conditions are not necessarily those of rest but, alternatively, those of uniform rectilinear motion. Similarly, in development, inertial conditions do not necessarily imply a lack of change, as exemplified by a growing mass of dividing cells. Moreover, as in mechanics, the lack of motion is sometimes the result of the equilibrium of contrasting forces acting on the same material object, so in development a lack of change, or *morphostasis*, is not necessarily a symptom of inertia but possibly the result of a dynamic confrontation, that is, a developmental process per se (Wagner and Misof, 1993; Wagner, 1994; Minelli, 2009a). Accordingly, a potential list of inertial aspects of development has little in common with a list of conditions or developmental phases where little seems to happen besides basic metabolism.

Stasis and change: What is *explanans* and what the *explanandum*? As in many other contexts, much depends on the perspective from which we address the problem.

On the one hand, inertial features of development, such as an indefinitely repeating cycle of mitosis or the generation of multiple identical units like body segments along the temporal and spatial axes of development, are often the mere starting point for more exciting studies of the mechanisms by which the basic iterativity or symmetry is broken.

On the other hand, the conditions, or processes, we can otherwise take as inertial with respect to the more complex phenomena we are interested in studying are nevertheless in need of mechanistic explanation themselves.

ACKNOWLEDGMENTS

I am grateful to the editors for their welcome invitation and their insightful comments on a first draft.

REFERENCES

Aparicio, J. M. 1998. Patterns of fluctuating asymmetry in developing primary feathers: A test of the compensatorial growth hypothesis. *Proc R Soc Lond B Biol Sci* 265:2353–7.

Baguñà, J., P. Martinez, J. Paps, and M. Riutort. 2008. Unravelling body-plan and axial evolution in the Bilateria with molecular phylogenetic markers. In *Evolving Pathways: Key Themes in Evolutionary Developmental Biology*, ed. A. Minelli and G. Fusco, 217–28. Cambridge: Cambridge University Press.

Balazuc, J. 1948. La tératologie des Coléoptères et expériences de transplantation sur *Tenebrio molitor* L. *Mem Mus Natl Hist Nat Paris (N. S.)* 25:1–293. (Orig. pub. 1947.)

Ball, P. 1999. *The Self-Made Tapestry: Pattern Formation in Nature.* Oxford: Oxford University Press.

Berking, S., and K. Herrmann. 2007. Compartments in Scyphozoa. *Int J Dev Biol* 51:221–8.

Biella, S., M. L. Smith, J. R. Aist, P. Cortesi, and M. G. Milgroom. 2002. Programmed cell death correlates with virus transmission in a filamentous fungus. *Proc R Soc Lond B Biol Sci* 269:2269–76.

Birnbaum, K. D., and A. Sánchez Alvarado. 2008. Slicing across kingdoms: Regeneration in plants and animals. *Cell* 132:697–710.

Blackstone, N. W., and A. M. Ellison. 2000. Maximal indirect development, set-aside cells, and levels of selection. *J Exp Zoolog B Mol Dev Evol* 288: 99–104.

Blum, M., H. Steinbeisser, M. Campione, and A. Schweickert. 1999. Vertebrate left–right asymmetry: Old studies and new insights. *Cell Mol Biol* 45:505–16.

Bouillon, J., C. Gravili, F. Pagès, J.-M. Gili, and F. Boero. 2006. *An Introduction to Hydrozoa.* Mémoires du Muséum national d'Histoire naturelle, 194. Paris: Muséum national d'Histoire naturelle.

Boyden, A., and E. M. Shelswell. 1959. Prophylogeny: Some considerations regarding primitive evolution in lower Metazoa. *Acta Biotheor* 13: 115–30.

Brien, P. 1973. Les démosponges. Morphologie et reproduction. In *Traité de Zoologie*, vol. 3(1), ed. P. P. Grassé, 133-461. Paris: Masson.

Callebaut, W., and D. Rasskin-Gutman, eds. 2005. *Modularity. Understanding the Development and Evolution of Natural Complex Systems.* Cambridge, MA: MIT Press.

Cardona, A., V. Hartenstein, and R. Romero. 2005. The embryonic development of the triclad *Schmidtea polychroa. Dev Genes Evol* 215:109–31.

Carroll, S. B. 2005. *Endless Forms Most Beautiful: The New Science of Evo Devo and the Making of the Animal Kingdom.* New York: Norton.

Carter, C. A., and J. P. Wourms. 1993. Naturally occurring diblastodermic eggs in the annual fish *Cynolebias*: Implications for developmental regulation and determination. *J Morphol* 215:301–12.

Chagas-Junior, A., G. D. Edgecombe, and A. Minelli. 2008. Variability in trunk segmentation in the centipede order Scolopendromorpha: A remarkable new species of *Scolopendropsis* Brandt (Chilopoda: Scolopendridae) from Brazil. *Zootaxa* 1888:36–46.

Coutelis, J. B., A. G. Petzoldt, P. Spéder, M. Suzanne, and S. Noselli. 2008. Left–right asymmetry in *Drosophila. Semin Cell Dev Biol* 19:252–62.

Drobysheva, I. M. 2003. On mitosis in embryos and larvae of the polyclads *Cycloporus japonicus* and *Notoplana humilis* (Plathelminthes) with different types of development. *Zoolog Zh* 82:1292–9.

Eaves, A. A., and A. R. Palmer. 2003. Reproduction: Widespread cloning in echinoderm larvae. *Nature* 425:146.

Egger, B., P. Ladurner, K. Nimeth, R. Gschwentner, and R. Rieger. 2006. The regeneration capacity of the flatworm *Macrostomum lignano*—on repeated regeneration, rejuvenation, and the minimal size needed for regeneration. *Dev Genes Evol* 216:565–77.

Ellis, C. H., and A. Fausto-Sterling. 1997. Platyhelminths, the flatworms. In *Embryology: Constructing the Organism.* ed. S. F. Gilbert and A. M. Raunio, 115–30. Sunderland, MA: Sinauer Associates.

Foe, V. E. 1989. Mitotic domains reveal early commitment of cells in *Drosophila* embryos. *Development* 107:1–22.

Foe, V. E., G. M. Odell, and B. A. Edgar. 1993. Mitosis and morphogenesis in the *Drosophila* embryo: Point and counterpoint. In *The Development of Drosophila melanogaster*, ed. M. Bate and A. Martinez-Arias, 149–300. Woodbury, NY: Cold Spring Harbor Laboratory Press.

Franc, A. 1993. Classe des Scyphozoaires. In *Traité de Zoologie*, vol. 3(2), ed. P. P. Grassé, 597–884. Paris: Masson.

Fryer, G. 1961. The developmental history of *Mutela bourguignati* (Ancey) Bourguignat (Mollusca: Bivalvia). *Philos Trans R Soc Lond B Biol Sci* 244:259–98.

Fusco, G. 2005. Trunk segment numbers and sequential segmentation in myriapods. *Evol Dev* 7:608–17.

Galaktionov, K. V., and A. A. Dobrovolskij. 2003. *The Biology and Evolution of Trematodes. An Essay on the Biology, Morphology, Life Cycles, and Evolution* of *Digenetic Trematodes*. Dordrecht, the Netherlands: Kluwer Academic.

Ganot, P., T. Kallesøe, and E. M. Thompson. 2007. The cytoskeleton organizes germ nuclei with divergent fates and asynchronous cycles in a common cytoplasm during oogenesis in the chordate *Oikopleura*. *Dev Biol* 302:577–90.

Garcia-Bellido, A., P. Ripoll, and G. Morata. 1973. Developmental compartmentalisation of the wing disc of *Drosophila*. *Nat New Biol* 245:251–3.

Gayon, J. 1992. *Darwin et l'Après-Darwin: Une Histoire de l'Hypothèse de Sélection Naturelle*. Paris: Kimé.

Grande, C., and N. H. Patel. 2009. Nodal signalling is involved left–right asymmetry in snails. *Nature* 457:1007–11.

Gurdon, J. B. 1992. The generation of diversity and pattern in animal development. *Cell* 68:185–99.

Guthrie, S., V. Prince, and A. Lumsden. 1993. Selective dispersal of avian rhombomere cells in orthotopic and heterotopic grafts. *Development* 118:527–38.

Hall, B. K. 1998. Germ layers and the germ-layer theory revisited: Primary and secondary germ layers, neural crest as a fourth germ layer, homology, demise of the germ-layer theory. *Evol Biol* 30:121–86.

Hall, B. K. 1999. *The Neural Crest in Development and Evolution*. New York: Springer.

Halley, J. D., and D. A. Winkler. 2008. Critical-like self-organization and natural selection: Two facets of a single evolutionary process? *BioSystems* 92:148–58.

Hallez, P. 1887. *Embryogénie des Dendrocoeles d'Eau Douce*. Paris: Baillière.

Held, L. I. 2002. *Imaginal Discs: The Genetic and Cellular Logic of Pattern Formation*. Cambridge: Cambridge University Press.

Held, L. I., Jr., C. M. Duarte, and K. Derakhshanian. 1986. Extra tarsal joints and abnormal cuticular polarities in various mutants of *Drosophila melanogaster*. *Roux Arch Dev Biol* 195:145–57.

Hendrikse, J. L., T. E. Parsons, and B. Hallgrímsson. 2007. Evolvability as the proper focus of evolutionary developmental biology. *Evol Dev* 9:393–401.

Hozumi, S., R. Maeda, K. Taniguchi, M. Kanai, S. Shirakabe, T. Sasamura, P. Spéder, et al. 2006. An unconventional myosin in *Drosophila* reverses the default handedness in visceral organs. *Nature* 440:798–802.

Ingham, P. W. 1991. Segment polarity genes and cell patterning within the *Drosophila* body segment. *Curr Opin Genet Dev* 1:261–7, 417.

Janssen, R., N. M. Prpic, and W. G. M. Damen. 2004. Gene expression suggests decoupled dorsal and ventral segmentation in the millipede *Glomeris marginata* (Myriapoda: Diplopoda). *Dev Biol* 268:89–104.

Karkach, A. S. 2006. Trajectories and models of individual growth. *Demogr Res* 15:347–400.

Krakauer, D., and J. Plotkin. 2002. Redundancy, antiredundancy, and the robustness of genomes. *Proc Natl Acad Sci USA* 99:1405–9.

Laird, D. J., A. W. De Tomaso, and I. L. Weissman. 2005. Stem cells are units of natural selection in a colonial ascidian. *Cell* 123:1351–60.

Lauzon, R. J., K. J. Ishizuka, and I. L. Weissman. 2002. Cyclical generation and degeneration of organs in a colonial urochordate involves crosstalk between old and new: A model for development and regeneration. *Dev Biol* 249:333–48.

Levin, M., R. L. Johnson, C. D. Stern, M. Kuehn, and C. Tabin. 1995. A molecular pathway determining left–right asymmetry in chick embryogenesis. *Cell* 82:803–14.

Littlewood, D. T. J., K. Rohde, and K. A. Clough. 1999. The interrelationships of all major groups of Platyhelminthes: Phylogenetic evidence from morphology and molecules. *Biol J Linn Soc* 66:75–114.

Meier, P., A. Finch, and G. Evan. 2000. Apoptosis in development. *Nature* 407:796–801.

Metzger, R. J., O. D. Klein, G. R. Martin, and M. A. Krasnow. 2008. The branching programme of mouse lung development. *Nature* 453:745–50.

Michod, R. E., and D. Roze. 2001. Cooperation and conflict in the evolution of multicellularity. *Heredity* 86:1–7.

Minelli, A. 2000a. Holomeric vs. meromeric segmentation: A tale of centipedes, leeches, and rhombomeres. *Evol Dev* 2:35–48.

Minelli, A. 2000b. Limbs and tail as evolutionarily diverging duplicates of the main body axis. *Evol Dev* 2:157–65.

Minelli, A. 2001. A three-phase model of arthropod segmentation. *Dev Genes Evol* 211:509–21.

Minelli, A. 2003. *The Development of Animal Form: Ontogeny, Morphology, and Evolution*. Cambridge: Cambridge University Press.

Minelli, A. 2009a. *Perspectives in Animal Phylogeny and Evolution*. Oxford: Oxford University Press.

Minelli, A. 2009b. *Forms of Becoming*. Princeton, NJ: Princeton University Press.

Minelli, A., and S. Bortoletto. 1988. Myriapod metamerism and arthropod segmentation. *Biol J Linn Soc* 33:323–43.

Minelli, A., and G. Fusco. 2004. Evo–devo perspectives on segmentation: Model organisms, and beyond. *Trends Ecol Evol* 19:432–29.

Mitchell, S. D. 2006. Modularity—more than a buzzword? *Biol Theory* 1:98–101.

Müller, G. B. 2008. Evo–devo as a discipline. In *Evolving Pathways: Key Themes in Evolutionary Developmental Biology*, ed. A. Minelli and G. Fusco, 5–30. Cambridge: Cambridge University Press.

Omont, N., and F. Képès. 2005. Book review: *Modularity in Development and Evolution. BioEssays* 27:667–8.

Panganiban, G., S. M. Irvine, C. Lowe, H. Roehl, L. S. Corley, B. Sherbon, J. K. Grenier, et al. 1997. The origin and evolution of animal appendages. *Proc Natl Acad Sci USA* 94:5162–6.

Peterson, K. J., R. A. Cameron, and E. H. Davidson. 1997. Set-aside cells in maximal indirect development: Evolutionary and developmental significance. *BioEssays* 19:623–31.

Raff, M. C., B. A. Barres, J. F. Burne, H. S. Coles, Y. Ishizaki, and M. D. Jacobson. 1993. Programmed cell death and the control of cell survival: Lessons from the nervous system. *Science* 262:695–700.

Ryan, A. K., B. Blumberg, C. Rodriguez-Esteban, S. Yonei-Tamura, K. Tamura, T. Tsukui, J. de la Peña, et al. 1998. *Pitx2* determines left–right asymmetry of internal organs in vertebrates. *Nature* 394:545–51.

Sanders, W. B. 2006. A feeling for the superorganism: Expression of plant form in the lichen thallus. *Bot J Linn Soc* 150:89–99.

Schlosser, G. 2002. Modularity and the units of evolution. *Theory Biosci* 121:1–80.

Schlosser, G., and G. P. Wagner, eds. 2004. *Modularity in Development and Evolution*. Chicago: University of Chicago Press.

Scholtz, G. 2003. Is the taxon Articulata obsolete? Arguments in favour of a close relationship between annelids and arthropods. In *The New Panorama of Animal Evolution. Proceedings of the 18th International Congress of Zoology*, ed. A. Legakis, S. Sfenthourakis, R. Polymeni, and M. Thessalou-Legaki, 489–501. Moscow: Pensoft.

Scholtz, G., and C. Wolff. 2002. Cleavage pattern, gastrulation, and germ disc formation of the amphipod crustacean *Orchestia cavimana*. *Contrib Zool* 71:9–28.

Schram, F. R. 1986. *Crustacea*. Oxford: Oxford University Press.

Shinn, G. L. 1997. Chaetognatha. In *Microscopic Anatomy of Invertebrates*. Vol. 15 of *Hemichordata, Chaetognatha, and the Invertebrate Chordates*, ed. F. W. Harrison and E. E. Ruppert, 103–220. New York: Wiley-Liss.

Sohaskey, M. L., J. Yu, M. A. Diaz, A. H. Plaas, and R. M. Harland. 2008. JAWS coordinates chondrogenesis and synovial joint positioning. *Development* 135:2215–20.

Soto, A. M., and C. Sonnenschein. 2004. The somatic mutation theory of cancer: Growing problems with the paradigm? *BioEssays* 26:1097–1107.

Speder, P., G. Adam, and S. Noselli. 2006. Type ID unconventional myosin controls left–right asymmetry in *Drosophila. Nature* 440:803–7.

Speder, P., and S. Noselli. 2007. Left–right asymmetry: Class I myosins show the direction. *Curr Opin Cell Biol* 19:82–7.

Steenstrup, J. J. S. 1845. *On the Alternation of Generation or the Propagation and Development of Animals through Alternate Generations*. London: Ray Society.

Steinböck, O., and B. Ausserhofer. 1950. Zwei grundverschiedene Entwicklungsabläufe bei einer Art (*Prorhynchus stagnatilis* M. Sch., Turbellaria). *Arch Entwicklungsmech* 144:155–77.

Sulston, J., and H. R. Horvitz. 1977. Postembryonic cell lineages of the nematode *Caenorhabditis elegans. Dev Biol* 56:110–56.

Technau, U., and C. B. Scholz. 2003. Origin and evolution of endoderm and mesoderm. *Int J Dev Biol* 47:531–9.

Van Speybroeck, L. 2005. Review of Werner Callebaut & Diego Rasskin-Gutman (eds.), *Modularity. Understanding the Development and Evolution of Natural Complex Systems*, Cambridge: MIT Press, 2005. *Philosophica* 76:129–35.

Wagner, G. P. 1994. Homology and the mechanisms of development. In *Homology: The Hierarchical Basis of Comparative Biology*, ed. B. K. Hall, 273–99. San Diego: Academic Press.

Wagner, G. P., and B. Y. Misof. 1993. How can a character be developmentally constrained despite variation in developmental pathways? *J Evol Biol* 6:449–55.

Walthall, W. W. 1995. Repeating patterns of motoneurons in nematodes: The origin of segmentation? In *The Nervous System of Invertebrates: An Evolutionary and Comparative Approach*, ed. O. Breidbach and W. Kutsch, 61–75. Basel: Birkhäuser.

Yost, H. J. 2001. Establishment of left–right asymmetry. *Int Rev Cytol* 203:357–81.

Zhurov, V., T. Terzin, and M. Grbić. 2004. Early blastomere determines embryo proliferation and caste fate in a polyembryonic wasp. *Nature* 432:764–9.

Zhurov, V., T. Terzin, and M. Grbić. 2007. (In)discrete charm of the polyembryony: Evolution of embryo cloning. *Cell Mol Life Sci* 64:2790–8.

Zimmer, R. L. 1997. Phoronids, brachiopods, and bryozoans, the lophophorates. In *Embryology: Constructing the Organism*, ed. S. F. Gilbert and A. M. Raunio, 279–305. Sunderland, MA: Sinauer Associates.

PART THREE

Epigenetics of Vertebrate Organ Development

9

The Role of Epigenetics in Nervous System Development

Chris Kovach, Pierre Mattar, and Carol Schuurmans

CONTENTS

A fundamental question is how cells acquire their specific identities and functional properties during embryogenesis. The intricate molecular controls that guide progression from pluripotent stem cells, which make up the early embryo, to a differentiated cell with a unique identity have begun to be elucidated (Figure 9.1). They include extrinsic signals, such as secreted or transmembrane signaling molecules, as well as intrinsic cues within cells, primarily transcription factors. It is currently well established that transcription factors play essential roles in cell fate specification, acting as key regulators of most, if not all, developmental programs in both invertebrates and vertebrates. However, transcription factor function is highly dependent on cellular context. Hence, the presence of a particular transcription factor in the nucleus of a cell does not necessarily equate with its ability to activate all of its target genes. Several molecular events are known to spatially and temporally regulate transcriptional factor function, including direct posttranslational modifications and cofactor associations. In addition, the ability of transcriptional regulators to influence cell fate decisions is intimately tied to the epigenetic status of their target genes. How transcriptional regulation and epigenetic modifications are coordinated during embryonic development is currently a hot topic of research.

During mouse development, the embryo proper is derived from the inner cell mass (ICM) of blastocyst stage embryos (reviewed in Rossant, 2008). As development proceeds, the three germ layers (endoderm, mesoderm, ectoderm) and the main dorsal–ventral, anterior–posterior, and left–right body axes are specified. Accompanying the establishment of the body plan is a loss of cellular pluripotency as pluripotent stem cells within the epiblast (which will give rise to the embryo proper) become specified to acquire specific cellular fates. During organogenesis, the vertebrate central nervous system (CNS) is derived from the dorsal ectoderm through the process of neural induction, which involves several well-characterized signaling molecules (reviewed by Aboitiz and Montiel, 2007, and by Gaulden and Reiter, 2008). At early developmental stages, the neural plate and neural tube are initially comprised of neural stem cells, which are characterized by their ability to self-renew, as well as by their multipotency—they have the potential to give rise to neurons, astrocytes, and oligodendrocytes (Figure 9.1). As development proceeds, neural stem cells produce progenitors that may remain multipotent but are more limited in their developmental potential and self-renewal properties than neural stem cells (Figure 9.1). Finally, progenitor cells give rise to precursor cells that are committed to one or more neural cell fates and divide a limited number of times before exiting the cell cycle to undergo terminal differentiation (Figure 9.1). The role that epigenetic modifications play during these cellular transitions is just beginning to be appreciated. We review how epigenetic changes influence key cell fate decisions in the developing mammalian CNS as well as highlight the major open questions in the field. For the most part, this chapter focuses on the embryonic forebrain, although similar principles are known to, or are likely to apply to, other regions of the developing CNS.

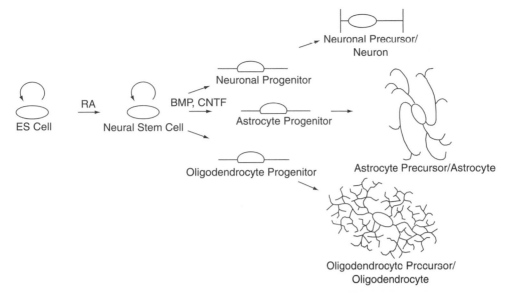

FIGURE 9.1 Neural lineage progression. Embryonic stem (ES) cells differentiate into neural stem cells in response to signals such as retinoic acid (RA). Neural stem cells are multipotent and can give rise to progenitor cells that are more limited in their developmental potential. Progenitor cells in turn give rise to committed precursors for neurons, astrocytes, or oligodendrocytes. BMP, bone morphogenetic protein; CNTF, ciliary neurotrophic factor.

A PRIMER ON EPIGENETIC MODIFICATIONS

In a eukaryotic cell, genomic DNA is tightly associated with histones, which together form compacted and highly organized chromatin domains. The basic chromatin unit is the *nucleosome*, which is comprised of a histone octamer (two copies each of the four core histones: H2A, H2B, H3, and H4) wrapped around 146 base pairs of DNA (Figure 9.2; reviewed in Fischle et al., 2003). The DNA that falls between the nucleosomes is bound by the linker histone H1. Both histones and genomic DNA can be post-translationally modified, influencing gene-expression levels by altering nucleosome arrangements and changing the accessibility of gene-regulatory regions to the transcriptional machinery. *Euchromatin* is the term used to define actively transcribed regions of the genome that are characterized by displaced nucleosomes. *Heterochromatin* refers to inactive regions of the genome where nucleosomes are more packed, preventing access of the transcriptional machinery to the chromatin. DNA and histone modifications are termed *epigenetic alterations* as they influence gene expression without altering the primary DNA code. The term *epigenetic* also implies that the changes are heritable during cell division, with histone and DNA modifications copied into daughter cells. This does not mean that all epigenetic modifications are permanent; enzymes that both add and remove chemical groups to DNA and histones have been identified, and post-translational chromatin modifications, often denoted as chromatin marks, are dynamic during development. Moreover, chromatin-remodeling complexes impart a higher-order organization to the genome, setting up active areas of the genome in a lineage-specific manner and imparting a relatively stable transcriptional signature to specific cell types. We briefly introduce the major epigenetic modifications, focusing on those molecules and mechanisms with a known role in neural lineage development

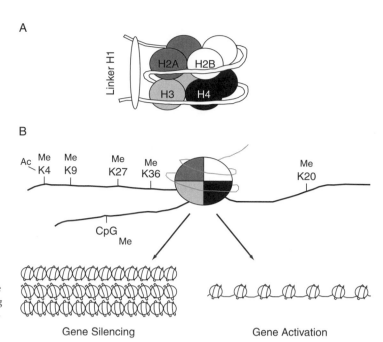

A

Linker H1

H2A H2B
H3 H4

B

Ac Me Me Me Me Me
K4 K9 K27 K36 K20

CpG
Me

FIGURE 9.2 Major chromatin modifications. Depicted are the major activating and repressing modifications to chromatin, including DNA and histones.

Gene Silencing   Gene Activation

(summarized in Figure 9.2 and reviewed in Hamby et al., 2008).

THE HISTONE CODE

HISTONE ACETYLATION

Several posttranslational modifications of histone tails have been identified, including acetylation, methylation, phosphorylation, adenosine diphosphate (ADP) ribosylation, ubiquitination, and sumoylation. The term *histone code* was originally coined to describe the overall pattern of histone modifications, which together dictate chromatin architecture and thus determine whether a gene is accessible or inaccessible to the transcriptional machinery (reviewed in Jenuwein and Allis, 2001; Strahl and Allis, 2000; Turner, 2000). One of the best-studied activating histone modifications is the transfer of acetyl groups from acetyl-coenzyme A (CoA) to ε-amino groups of lysine residues in the amino-terminal tails of histones, an event catalyzed by a group of enzymes known as histone acetyltransferases (HATs) (reviewed in Carrozza et al., 2003). Acetylation influences chromatin structure by decreasing the interaction between negatively charged DNA and positively charged histones, thereby destabilizing the formation of the most highly compact form of chromatin (i.e., 30 nm fibers) (Shogren-Knaak et al., 2006). Five subfamilies of HATs have been identified, with the two most commonly studied in neural lineage development including the CBP/p300—cAMP-responsive element binding protein (CREB) binding protein—and the Gcn5 (e.g., p300/CBP-associated factor, PCAF) families (reviewed in Carrozza et al., 2003; Roth et al., 2001). HAT activity allows for the recruitment of bromodomain-containing coactivators, which bind acetylated histones.

The acetylation of histones is reversible and, hence, dynamic. The removal of acetyl groups allows for tighter associations between histones and the DNA backbone and is therefore associated with transcriptional repression. Deacetylation is mediated by histone deacetylases (HDACs), which fall into four classes: class I (HDAC1–3, 8); class II (HDAC4–7, 9); class III (SIRT1–7), and class IV (HDAC11) (reviewed in Thiagalingam et al., 2003; Yang and Gregoire, 2005). The HDAC proteins form multiprotein corepressor complexes on promoters. The HDACs studied most extensively in the developing nervous system include the class I HDACs, which interact with proteins such as Sin3A, NuRD, Co-REST (Heinzel et al., 1997), and MeCP2 (reviewed in Tucker, 2001), as well as class II HDACs, which associate with the corepressors NCoR (nuclear receptor corepressor 1) and SMRT (silencing mediator of retinoid and thyroid hormone receptors) (Fischle et al., 2002).

HISTONE METHYLATION

The effects of histone methylation on transcriptional activity strictly depend on the specific lysine residue that is modified. One of the major transcriptional activating events is methylation of lysine 4 (K4) in histone H3, which can be mono- (K4me1), di- (K4me2), or tri- (K4me3) methylated. In general, H3K4me3 groups are clustered at transcriptional start sites of actively transcribed genes, while H3K4me2 groups predominate on genes primed for transcription (Bernstein et al., 2006; Pavri et al., 2006; Santos-Rosa et al., 2002; Sims and Reinberg, 2006). Another histone mark that is strongly associated with active transcription is H3K36me3, which is enriched throughout the actively transcribed regions of individual genes following the H3K4 modifications in the promoter sites (Mikkelsen et al., 2007). In contrast, the methylation of histone H3K9, H3K27, and H4K20 is considered a silencing modification as it negatively regulates transcription (Papp and Muller, 2006; Ringrose et al., 2004). Notably, histone methylation results in the recruitment of chromodomain-containing proteins that include other chromatin-modifying proteins and transcriptional regulators.

More than 50 putative histone methyltransferases (HMTases) have been identified in mammalian genomes based on the presence of a SET—Su(var)3-9, Enhancer of Zeste, Trithorax—domain, which is common to most, but not all, HMTases (reviewed in Lachner et al., 2003). SET-domain proteins are subdivided based on their HMTase substrate specificity, with distinct enzymes responsible for the methylation of different lysine residues in H2A, H3,

and/or H4 (Lachner et al., 2003). The best-studied HMTases are those associated with the trithorax (TrxG) and Polycomb (PcG) complexes, which act in a mitotically heritable and lineage-specific manner to activate or silence gene transcription, respectively (reviewed in Ringrose and Paro, 2007). In general, HMTases associated with TrxG complexes activate gene transcription by specifically methylating H3K4 residues. One exception is Ash1, a TrxG complex HMTase that can also modify H3K9 and H4K20, although it does not add these repressive marks to the promoters or coding sequences of actively transcribed genes (Beisel et al., 2002; Papp and Muller, 2006; Petruk et al., 2001; Smith et al., 2004). In contrast, the two major PcG complexes, PRC1 and PRC2, add repressive marks to genes that are actively suppressed. For instance, enhancer of zeste (Ezh)–Eed complexes specifically methylate H3K27, while G9a and the Suv39h HMTases methylate H3K9 in the promoter regions of genes that should be silenced (Rea et al., 2000; Rice et al., 2003).

More recently, the enzymes that remove methyl groups from histones have also been identified. Demethylases include lysine-specific histone demethylase 1 (LSD1) and jumonji-C-domain (JmjC-domain) histone demethylases (JHDMs). While demethylases such as LSD1 are thought to nonspecifically remove methyl groups from histone tails, JMJD3 is thought to be an H3K27me3-specific demethylase (Jepsen et al., 2007). In contrast, the lysine demethylase 5 (KDM5) family of four demethylases, which also contain Jmj domains, are responsible for the removal of H3K4me2 and H3K4me3 modifications (Christensen et al., 2007; Iwase et al., 2007; Lee et al., 2006; Yamane et al., 2007).

DNA METHYLATION

DNA can be methylated on cytosine position 5 in CpG dinucleotides to form 5-methylcytosine. It has been estimated that 60%–90% of CpG dinucleotides are methylated in the mammalian genome (reviewed in Tucker, 2001). However, methylated CpG dinucleotides are not randomly distributed and are, in general, scarce in

CpG islands found in the promoter regions of many genes. DNA methylation is associated with the silencing of gene expression at distinct developmental stages and/or in specific cellular lineages and is the main mechanism underlying gene silencing associated with maternal or paternal imprinting and X-chromosome inactivation (Li et al., 1993; Panning and Jaenisch, 1996). For imprinted genes, de novo methylation occurs very early in development, often prior to implantation, and is carried out by DNA methyltransferase (Dnmt) 3a and Dnmt3b, which methylate unmodified CpGs (reviewed in Goll and Bestor, 2005; Robertson and Wolffe, 2000). In contrast, Dnmt1 is responsible for maintenance methylation after cell division, modifying the new DNA strand in hemimethylated DNA (Bestor et al., 1988; Hirasawa et al., 2008).

Methylated CpG dinucleotides can suppress gene transcription in two ways: directly interfering with the ability of transcriptional regulators to bind to their target sequences or altering chromatin structure through the recruitment of methylated DNA–binding domain (MBD) proteins such as MECP2 (methyl CpG-binding protein 2) and MBD1 (Namihira et al., 2008). MBD1 shares sequence homology with MBD2–4, but not all of these family members have been shown to bind methylated DNA. The interaction of methylated CpG dinucleotides with MBD proteins allows a functional link with histone modifications as MECP2 can also recruit SIN3 and HDAC1 molecules to form a corepressor complex that induces a closed chromatin structure, leading to gene silencing (reviewed in Tucker, 2001). Notably, deacetylation mediated by a MECP2–SIN3A complex is often considered irreversible and is commonly seen in imprinted genes. This is in stark contrast to other alterations of chromatin, such as HDAC-mediated deacetylation in the absence of DNA methylation, which is often reversible.

CHROMATIN REMODELING

For active transcription to take place, in addition to acetylation events that decondense chroma-

tin, it is necessary to displace nucleosomes for promoter and enhancer regions to be accessible to the transcriptional machinery. Chromatin-remodeling complexes are responsible for unwinding DNA from histones so that the architecture of nucleosomes can be modified and moved to adjacent regions of the genome, opening up promoters and enhancers to allow access to the transcriptional machinery (Gutierrez et al., 2007). Chromatin remodeling occurs through an adenosine triphosphate (ATP)–dependent process mediated by large protein complexes that include SWI (switching) and SNF (sucrose nonfermenting) family ATPases, which were initially cloned in *Saccharomyces cerevisiae* (Cairns et al., 1994; Cote et al., 1994). To date, 29 SWI/SNF-like proteins have been identified in mammals, including Brahma (Brm) and Brahma-related protein 1 (Brg1), which bind to multiple Brg/Brm-associated factors (BAFs) to form distinct, lineage- and time-specific chromatin-remodeling complexes (Khavari et al., 1993; Kwon et al., 1994; Lemon et al., 2001; Lessard et al., 2007; Wang et al., 1996). A second family of chromatin-remodeling proteins is the imitation switch (ISWI) family, which can dislocate nucleosomes but has less ability to induce structural changes in the nucleosomes themselves (reviewed in Racki and Narlikar, 2008).

The accessibility of the transcriptional machinery to promoters and enhancers is also influenced by the tertiary structure of the genome. Chromatin is organized into functional units by the formation of loops that are anchored to the nuclear matrix at AT-rich sequences known as scaffold/matrix attachment regions (S/MAR) (Gohring and Fackelmayer, 1997). Proteins that bind to AT-rich S/MAR sequences influence higher-order chromatin structure and include the special AT-rich sequence-binding protein 1 (SATB1) and SATB2 (Cai et al., 2003; Gyorgy et al., 2008; Seo et al., 2005a; Yasui et al., 2002). Altering the tertiary structure of chromatin can either enhance or suppress transcription. Attachment of genomic DNA to the nuclear matrix can promote transcription by bringing distant enhancer elements more proximal to a gene's promoter sequences (Callinan and Feinberg, 2006). Alternatively, looping of genomic DNA through S/MAR sequences can block the association of distal enhancer elements with promoter sequences, thereby preventing transcription. SATB1 was initially shown to block transcription by binding to the nuclear matrix and preventing the attachment of transcription factors to gene-regulatory regions in the vicinity (Seo et al., 2005b, 2005c). SATB1 and SATB2 also negatively influence transcription by recruiting HDAC1 (Gyorgy et al., 2008; Yasui et al., 2002). Finally, SATB1 interacts with components of the nucleosome remodeling and deacetylase (NuRD) complex, which is considered to be a "landing platform" for several classes of molecules involved in chromatin remodeling and histone modifications and is recruited by transcriptional repressors (Gyorgy et al., 2008; Yasui et al., 2002).

EPIGENETIC MODIFICATIONS REGULATE STEM-CELL PROLIFERATION

HISTONE METHYLATION STATUS IS ASSOCIATED WITH THE PROLIFERATIVE CAPACITY OF STEM CELLS

ICM cells are pluripotent and can be isolated and propagated as embryonic stem (ES) cell lines. ES cells are used for *gene targeting*, which is the introduction of a site-specific mutation via homologous recombination. Mutated ES cells can be used to generate mutant mice by aggregating the modified ES cells with wild-type morula or injecting into wild-type blastocysts and then implanting the modified embryos into pseudopregnant females to generate chimeric embryos. If the mutation is passed through the germ line of the chimeric mouse, it is possible to generate a line of mutant mice. ES cell lines are also exploited in in vitro differentiation paradigms, allowing the molecular events accompanying the first steps in lineage commitment to be assessed. In general, ES cell differentiation is accompanied by a restriction

in developmental potential that is regulated in part by epigenetic mechanisms. To promote a neural cell identity, ES cells are cultured with retinoic acid (RA), generating cells that most closely resemble cortical progenitors based on molecular marker expression and their subsequent differentiation into glutamatergic neurons that are characteristic of the neocortex (Figure 9.1; and see Bibel et al., 2004, 2007). Most promoter sequences are unmethylated in pluripotent ES cells, allowing genes to be rapidly transcribed upon lineage commitment. As ES cells are induced to differentiate into neural lineages, a subset of promoters become methylated, including the *Nanog* and *Oct4* promoters, transcription factors that regulate the expression of genes associated with the maintenance of a pluripotent stem cell state (Fouse et al., 2008; Meissner et al., 2008; Mohn et al., 2008) (see Table 9.1).

HISTONE MARKS ARE DYNAMIC IN THE PROGRESSION FROM ES CELL TO NEURAL STEM CELL

Histone methylation patterns are also dynamically regulated during ES-cell differentiation. In ES cells, a bivalent chromatin signature consisting of both H3K4me3 (activating) and H3K27me3 (silencing) marks is associated with putative regulatory regions of key developmental genes, including many transcriptional regulators (Bernstein et al., 2006). This bivalent chromatin mark is thought to silence gene expression in ES cells, while simultaneously priming certain genes for activation upon lineage specification (Bernstein et al., 2006; Mohn et al., 2008). As ES cells differentiate, H3K27me3 marks are lost in the regulatory regions of genes that are required for the development of that specific lineage. Both PRC1 and PRC2 complexes, which are associated with H3K27 methylases, play essential roles in repressing gene expression in ES cells, with mutations in the core components of these complexes resulting in a loss of pluripotency (Boyer et al., 2006; O'Carroll et al., 2001). Notably, the H3K27me3 demethylase Jmjd3 is required for

TABLE 9.1
Common Epigenetic Modifications in Neural Lineages

SITE	MODIFICATION	EFFECT ON TRANSCRIPTION
H3,H4	Acetylation	Gene activation
H3K4	Methylation	Gene activation
H3K9	Methylation	Gene silencing
H3K27	Methylation	Gene silencing
H3K36	Methylation	Gene activation
H4K20	Methylation	Gene silencing

the switch from an ES-cell to a neural stem-cell identity (Burgold et al., 2008). In contrast, KDM5, an H3K4 demethylase, is required to maintain an ES-cell identity, with the overexpression of KDM5 in ES cells blocking neural differentiation (Dey et al., 2008).

EPIGENETIC MODIFIERS REGULATE NOTCH SIGNALING IN NEURAL STEM CELLS

Early in development, neural stem cells undergo self-renewing symmetric cell divisions that serve to expand the progenitor cell pool. This is followed by asymmetric, differentiative cell divisions that give rise to another neural stem cell and a postmitotic neuronal or glial cell. These cell-division patterns can be recapitulated to some extent in vitro. Neural stem cells isolated from the CNS can be expanded and maintained in culture in the presence of fibroblast growth factor 2 (FGF2), a key growth factor at early developmental stages—from embryonic day (E) 10—or epidermal growth factor (EGF), which more efficiently expands neural progenitor cells isolated at later developmental stages and from the adult (Li et al., 2008; Reynolds and Weiss, 1992; Tropepe et al., 1999).

INTRODUCTION TO NOTCH SIGNALING

One of the key signaling pathways implicated in maintaining a neural stem-cell state in vivo, especially early in development, is the Notch signaling pathway (Alexson et al., 2006; Chojnacki et al., 2003; Hitoshi et al., 2002; Mizutani et al.,

2007; Nagao et al., 2007; Nyfeler et al., 2005; Shimojo et al., 2008; Yoon et al., 2004). In the murine genome four single-pass transmembrane Notch receptors (1–4) are activated upon binding with transmembrane ligands of the Delta (Dll 1, 3, 4) or Jagged (Jag 1, 2) families, which are DSL family ligands (Figure 9.3, and reviewed in Bolos et al., 2007). Upon ligand binding, two sequential proteolytic processing events involving tumor necrosis factor–converting enzyme (TACE) and gamma-secretase lead to the release of the Notch intracellular domain (NICD) into the cytoplasm. The NICD then translocates to the nucleus, where it interacts with the transcription factor CSL—C-promoter binding factor 1 (CBF-1); suppressor of hairless (SuH), longevity assurance gene (LAG1)—to activate the transcription of Notch target genes. Notch target genes include members of the *Hes—hairy and enhancer of split, E(spl)*—gene family, which encode basic helix–loop–helix (bHLH) transcription factors. In the absence of nuclear NICD, CSL associates with corepressor proteins, such as SMRT or NCoR, which recruit class II HDACs, repressing the transcription of *Hes* and other Notch target genes.

In support of the idea that Notch signaling is required to maintain neural stem-cell identity, *Hes1* single and *Hes1;Hes5* double mutants undergo premature neurogenesis in several neural cell lineages, while misexpression of *Hes* genes prevents neuronal differentiation (Cau

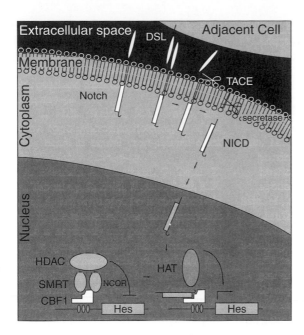

FIGURE 9.3 Notch signaling. DSL family ligands bind Notch receptors, which are proteolytically cleaved by TACE and g-secretase to give rise to the Notch intracellular domain (NICD). The NICD is translocated to the nucleus, where it associates with CBF1 proteins, changing CBF1 from a transcriptional repressor (associated with HDAC/SMRT/NCoR) to a transcriptional activator (associated with HAT activity). CBF1 regulates *Hes* gene expression.

et al., 2000; Ishibashi et al., 1995, 1994; Ohtsuka et al., 1999). Classically, the *Hes* genes are thought to repress neurogenesis by blocking the expression and/or function of the proneural or neural determination genes, which also encode bHLH transcription factors but have distinct functions—specifying both generic (i.e., pan-neuronal) and cell type–specific aspects of neuronal identity (reviewed in Bertrand et al., 2002). The mammalian proneural genes include *Ascl1*, a murine homologue of genes in the *achaete-scute* cluster (AS-C) in *Drosophila*, and Neurogenins (*NeuroG*) 1–3, which are more highly related to *Drosophila atonal*. Contrasting to the *Hes* genes, the loss of proneural gene function is associated with a block in neuronal differentiation in different neural cell lineages (Cau et al., 1997; Fode et al., 1998; Guillemot et al., 1993; Ma et al., 1996), while overexpression of proneural genes promotes premature neuronal differentiation (Britz et al., 2006; Mattar et al., 2008).

DYNAMIC PATTERNS OF HISTONE ACETYLATION MODULATE NOTCH SIGNALING

Epigenetic modifications regulate the transcriptional activities of Notch effector molecules in a temporally regulated manner. In neural stem cells, Hes proteins bind to N-box sequences (CACNAG) in the promoter regions of proneural genes, such as *Ascl1*, negatively regulating transcription (Takebayashi et al., 1994). Transcriptional repression by Hes proteins is mediated by their interactions with corepressors of the groucho (Gro)/transducin-like enhancer of split (TLE) family, which bind a TRPW C-terminal motif in Hes1 (reviewed in Cinnamon and Paroush, 2008). TLE proteins recruit HDACs, Sin3, and other proteins to promote the formation of a closed chromatin structure and repress transcription (Yochum and Ayer, 2001). However, TLE1 and Hes1 associate with the *Ascl1* promoter only in proliferating neural stem cells isolated from the cortex as the addition of platelet-derived growth factor (PDGF), which induces neuronal differentiation, causes Hes1 to instead recruit HAT coactivators (e.g., CBP), to *Ascl1* regulatory regions (Ju et al., 2004). Notably, the switch from a Hes1 repressor to activator complex is regulated by poly(ADP-ribose) polymerase 1 (PARP-1), a component of the Hes1 repressor complex that upon neuronal differentiation directly modifies Hes1 to catalyze the removal of TLE1 and the subsequent recruitment of coactivator proteins (Ju et al., 2004). Thus, by altering cofactor associations, Hes1 switches from a repressor to an activator, thereby controlling the onset of neurogenesis in cortical progenitor cells.

CSL interacts with class II HDACs and the corepressors NCoR and SMRT to repress Notch target gene transcription in the absence of Notch ligands (Fischle et al., 2002; Kao et al., 1998; Perissi et al., 2008, and references therein). The analysis of *SMRT* and *NCoR* knockout mice supports a role for these corepressors in maintaining neural progenitor cells in a proliferative state. In both *SMRT* and *NCoR* mutants, the proliferative capacity of cortical stem cells is reduced, with progenitor cells undergoing premature neuronal and astroglial differentiation in *SMRT* mutant cortices and premature astroglial differentiation in *NCoR* mutant cortices (Jepsen et al., 2007). *NCoR* mutant cortical progenitors are also not able to self-renew in the presence of FGF (Hermanson et al., 2002). An interesting feature of the *SMRT* mutant cortical progenitors is derepression of expression of JMJD3, an H3K27me3 demethylase, correlating with the aberrant expression of neuronal and astroglial differentiation genes (Jepsen et al., 2007). This study supports the view that H3K27me3 is a functional marker of developmental potential, as suggested by genomewide expression profiling (Mikkelsen et al., 2007). The altered expression of a demethylase following the mutation of a corepressor protein also highlights the numerous cross-interactions that occur between chromatin-modifying enzymes.

CONCLUSIONS AND FUTURE DIRECTIONS

In summary, the conversion of a pluripotent ES cell to a neural stem cell is accompanied by global alterations in the chromatin marks surrounding key developmental genes. The key challenge now is to determine how these dramatic changes in chromatin architecture are orchestrated. How are the activities of the chromatin- and histone-remodeling enzymes themselves regulated in such a precise and gene-specific manner during the differentiation process? It will also be important to examine whether Hes1's ability to change its association with corepressor and coactivator complexes is a unique feature of this protein or is common to other transcriptional regulators involved in the specification of a neural stem-cell fate.

EPIGENETIC MODIFICATIONS REGULATE NEUROGENESIS

RE1 SILENCING TRANSCRIPTION FACTOR NEGATIVELY REGULATES NEURONAL DIFFERENTIATION

Neuronal differentiation is promoted by proneural bHLH transcription factors, which initiate the transcription of downstream genes and/or transcriptional cascades (reviewed in Schuurmans and Guillemot, 2002). However, neuronal differentiation is also actively repressed in nonneural cells by RE1 silencing transcription factor (REST)/neuron restrictive silencing factor (NRSF) (Ballas et al., 2001; Chong et al., 1995; Schoenherr and Anderson, 1995; Schoenherr et al., 1996). REST/NRSF is expressed transiently in neural progenitor cells, but its expression is extinguished in terminally differentiated neurons (Chong et al., 1995; Schoenherr and Anderson, 1995). REST/NRSF encodes a zinc finger protein that binds to a specific DNA sequence called RE1, also known as neuron-restrictive silencer elements (NRSE). RE1 elements are present in at least 800 genes in the human genome, including neuronal genes encoding neurotransmitters, ion channels, and synaptic vesicle proteins (Belyaev et al., 2004). REST/NRSF is also bound to RE1 sites in some neuronal target genes in neural stem cells derived from the embryonic hippocampus (Greenway et al., 2007). The importance of REST/NRSF in silencing neuronal gene expression is at least partially born out by the phenotype of *REST/NRSF* knockout mice, which die in early embryogenesis with morphological defects and the ectopic expression of some, but surprisingly not the majority of, neuron-specific genes in

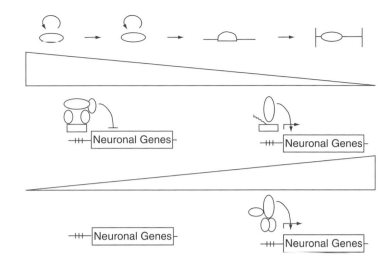

FIGURE 9.4 REST and bHLH gene regulation during neurogenesis. In embryonic stemm cells and neural stem cells, REST levels are high, blocking neuronal differentiation. As neurogenesis is initiated, REST levels decline, mediated in part by ncRNA, and proneural bHLH is able to induce neurogenesis in association with chromatin-remodeling factors.

nonneural lineages (Chen et al., 1998). Thus, REST/NRSF has a different capacity to regulate individual neuronal target genes.

Mechanistically, differences in the ability of REST/NRSF to regulate target genes have been attributed to its ability to recruit different corepressor complexes through two distinct repressor domains (Figure 9.4 and see Belyaev et al., 2004; Lunyak et al., 2002; Tapia-Ramirez et al., 1997). The N-terminal repression domain of REST/NRSF recruits SIN3–HDAC complexes (Grimes et al., 2000; Huang et al., 1999; Roopra et al., 2000), while the C-terminal repressor domain recruits a complex of proteins that includes CoREST, HDAC1, and MeCP2 (Andres et al., 1999; Ballas et al., 2001, 2005; Ballas and Mandel, 2005). By recruiting HDACs, REST/NRSF is able to reduce H3 and H4 acetylation levels in target genes, thereby generating a repressive transcriptional signature (Greenway et al., 2007). A more permanent repressive signature is conferred to a subset of genes when REST/NRSF associates with MeCP2 bound to methylated CpG.

REST/NRSF protein levels are also tightly regulated at the posttranscriptional level. In differentiating neurons, REST/NRSF transcript levels are high while protein levels are low (Kuwabara et al., 2004). REST/NRSF gene silencing in neuronal cells is mediated by a small noncoding (nc) RNA that acts as an antisense repressor (Kuwabara et al., 2004). Specifically, a 20-nucleotide double-stranded (ds) RNA molecule with sequence homology to RE1 was isolated from hippocampal neural stem cells (Kuwabara et al., 2004). The NRSE dsRNA molecule base pairs with RE1 sites in DNA, as well as interacting directly with REST/NRSF, converting this zinc finger protein to a transcriptional activator, in part by promoting new associations with HATs and disrupting HDAC associations (Kuwabara et al., 2004). The activity of REST/NRSF, and its association with chromatin modifying proteins, is thus a dynamic process in neuronal and nonneuronal cells.

CHROMATIN-REMODELING COMPLEXES REGULATE bHLH ACTIVITY DURING NEUROGENESIS

It has been proposed that the switch from neural stem-cell proliferation to neuronal differentiation is accompanied by chromatin remodeling on a global scale (Lessard et al., 2007). This idea is supported by biochemical analyses of Brg1/Brm-associated BAF complexes, which are comprised of distinct subunits in neuronal versus nonneuronal lineages (Olave et al., 2002) and in neural progenitor cells versus differentiated neurons (Lessard et al., 2007). Moreover, BAF subunits that are specifically complexed with Brg1/Brm in neural progenitors, such as BAF45a and BAF53a, are sufficient to promote the proliferation of neural stem cells (Lessard et al., 2007). Moreover, if BAF subunit switching during neural lineage progression is blocked, the ability of neural stem cells to undergo neuronal differentiation is severely hampered (Lessard et al., 2007).

Consistent with the idea that chromatin remodeling is important in the developing nervous system, tissue-specific deletion of *Brg1*, a SWI/SNF-family ATPase, in neural progenitors throughout the CNS results in severe neural tube defects (Lessard et al., 2007). *Brg1* mutants are characterized by a depletion of the neural progenitor cell pool, and consequently, fewer neuronal and glial cells differentiate (Lessard et al., 2007). Similarly, the knockdown of *Brg1* in neural progenitor cells cultured in vitro prevents neuronal differentiation (Seo et al., 2005c). It has been proposed that Brg1 regulates neurogenesis by forming obligate associations with bHLH proteins of the proneural (e.g., NeuroG1) and differentiation (e.g., NeuroD) classes, catalyzing the formation of active chromatin domains in NeuroG1/NeuroD target genes (Seo et al., 2005c). Another player in this pathway, Geminin (Gem), antagonizes Brg1 activity and prevents it from associating with bHLH-type transcriptional regulators (Seo et al., 2005b).

Additional chromatin-remodeling molecules have also been implicated in regulating neural development. For instance, a neural progenitor-specific knockout of α-thalassemia mental retardation, X-linked (*ATRX*), which encodes an SNF-family ATPase, results in reduced neuronal number due largely to increased apoptosis (Berube et al., 2005). Notably, mutations in *ATRX* are associated with mental retardation in humans, highlighting the importance of chromatin remodeling for nervous system functioning (Gibbons et al., 1995). Finally, ISWI chromatin remodelers also play a role in neural lineage development. The mammalian ISWI family ATPases Snf2h and Snf2l are highly expressed in dividing progenitors and differentiated neurons, respectively (Lazzaro and Picketts, 2001). A knockdown of ISWI ATPases in

Xenopus results in aberrant gastrulation and defects in neural tube closure that are associated with the deregulated expression of key signaling molecules, such as sonic hedgehog (Shh) and bone morphogenetic protein (BMP) 4 (Dirscherl et al., 2005). In addition, Cecr2 is a protein associated with Snf2l that is also highly expressed in the developing nervous system, the knockout of which results in severe neural tube defects (Banting et al., 2005).

CONCLUSIONS AND FUTURE DIRECTIONS

It is currently accepted that both negative (e.g., REST/NRSF) and positive (e.g., proneural genes and BAF complexes) regulators of neurogenesis are required during nervous system development. The use of both positive and negative effectors ensures that the timing of neurogenesis is precisely regulated in the developing embryo. However, it remains to be determined how these systems are coordinated so that the precise temporal and spatial patterns of neurogenesis are maintained during development. For instance, while the NRSE dsRNA molecule adds an extra layer of regulation onto REST/NRSF, it is unclear whether this mechanism contributes to regulating the precise order of gene expression in developing neural lineages. For instance, in the developing neocortex, the onset of gene expression is precisely coordinated, with genes expressed in the ventricular zone, subventricular zone, intermediate zone, and cortical plate in a precise temporal order. It will be interesting to determine whether each of these gene sets is removed from REST/NRSF repression at different times in development.

While chromatin-remodeling proteins clearly have essential roles in the developing nervous system and are known to associate with different bHLH regulators, the temporal regulation of these associations has yet to be assessed. This is an important issue given that the proneural genes *NeuroG2* and *Ascl1* have distinct functions at different stages of cortical development (Britz et al., 2006; Schuurmans et al., 2004). In contrast, the associations between chromatin modifiers and bHLH transcription factors have been more extensively studied during myogenesis (Palacios and Puri, 2006). In skeletal muscle, the bHLH protein MyoD is similar to NeuroG2 in that it does not activate target gene transcription in all cells where it is expressed. This has been explained in part by the observation that in proliferating myoblasts MyoD initially associates with HDACs, making promoters and enhancers inaccessible for transcription (Mal and Harter, 2003). Once differentiation is initiated, MyoD instead recruits HATs and SWI/SNF chromatin-remodeling complexes, opening the chromatin to make it accessible for transcription (Cao et al., 2006). We suggest that changes in NeuroG2 and/or Ascl1 cofactor binding could underlie temporal and spatial alterations in their activities in the developing cortex (dorsal telencephalon) and ventral telencephalon.

EPIGENETIC MODIFICATIONS REGULATE THE NEURONAL-TO-GLIAL TRANSITION

In the developing nervous system, progenitor cells change their cellular output over time, following a general temporal rule, which is that neurogenesis precedes astrogliogenesis which precedes oligodendrogliogenesis (Sauvageot and Stiles, 2002). Strikingly, the temporal order of neuronal followed by glial differentiation is recapitulated by cortical cells cultured at clonal cell densities in vitro, indicating that the timing of these cell fate switches is at least partially cell-intrinsic (Qian et al., 2000). The competence of cortical progenitors to respond to astrocytogenic signals also changes over time, with only later-stage progenitors competent to respond to gliogenic cytokines (Song and Ghosh, 2004). Astrocyte-inducing signals include cytokines of the interleukin-6 (IL-6) family, such as ciliary neurotrophic factor (CNTF) and leukemia inhibitory factor (LIF), which bind to a common gp130 receptor (Bonni et al., 1997; Nakashima et al., 1999a; Nakashima et al., 1999b; Nakashima et al., 1999c; Uemura et al., 2002; Yanagisawa et al., 2000). These cytokines signal through the Janus kinase (JAK) to phosphorylate and activate transcription factors of the signal transducers and activators of transcription (STAT) family. Another gliogenic signal is mediated by BMP2 and BMP4, which bind to the receptor serine/threonine kinases BMP receptors (BMPR) 1A/B and BMPR2, which heterodimerize and upon ligand binding signal through the Smad1/5/8 pathway (reviewed in Fukuda and Taga, 2005). The importance of epigenetic modifications in regulating the response of neural progenitors to gliogenic signals, and hence the decision to make neuronal or glial cells, has been extensively studied and involves epigenetic modifications.

DYNAMIC HISTONE ACETYLATION EVENTS ARE ASSOCIATED WITH ASTROCYTE DIFFERENTIATION

The choice by cortical progenitor cells to make either neuronal or glial cells is controlled in part by the proneural bHLH genes, which both promote a neuronal identity and actively inhibit an astrocytic fate (Nieto et al., 2001; Sun et al., 2001). Accordingly, in double mutants lacking the proneural genes NeuroG2 and Ascl1, neurogenesis is reduced in the neocortex, with progenitor cells instead undergoing premature gliogenesis (Nieto et al., 2001). Similar patterns of premature astrocytogenesis and reduced neurogenesis are observed in other regions of the CNS, including the hindbrain and midbrain, when the proneural genes Ascl1 and Math3 are both deleted (Tomita et al., 2000).

Mechanistically, proneural genes such as NeuroG1 and NeuroG2 promote neurogenesis in a DNA binding–dependent manner by directly initiating the transcription of downstream target genes and transcriptional cascades (Mattar et al., 2004, 2008; Sun et al., 2001). In contrast, the ability of NeuroG1 to inhibit gliogenesis is independent of DNA binding (Sun et al., 2001). NeuroG1 blocks gliogenesis by preventing cortical progenitors from responding to the gliogenic properties of BMP and cytokine signaling, interfering with both pathways at the level of their downstream effector molecules. During normal gliogenesis, BMP and cytokines (e.g., CNTF, cardiotrophin 1) cooperate to promote gliogenesis (Nakashima et al., 1999b, 1999c; Barnabe-Heider et al., 2005). STAT3-binding sites have been identified in the promoter regions of several glia-specific genes, such as glial fibrillary acidic protein (GFAP), suggesting that cytokine signaling directly regulates glia-specific gene transcription (Bonni et al., 1997; Nakashima et al., 1999a, 1999b, 1999c). STAT3–CBP complexes bound to their recognition sites then recruit Smad1 to the promoter complex in a non-DNA binding–dependent manner (Nakashima et al., 1999a, 1999b, 1999c). NeuroG1 interferes with this process by complexing with Smad1–CBP complexes and preventing their recruitment to gliogenic target genes (Sun et al., 2001). In addition, NeuroG1 can prevent cytokine-mediated phosphorylation of Y701/Y705 in STAT1/3, thereby blocking STAT activation (Sun et al., 2001).

These studies show that transcriptional regulators, such as NeuroG1, can regulate developmental decisions in a non-DNA-binding fashion in part by their ability to compete for HAT proteins, such as CBP. The importance of low histone acetylation levels for astrocyte differentiation has recently been appreciated. In general, astrocytic lineages display relatively low levels of histone H3 and H4 acetylation compared to neuronal lineages (Hsieh et al., 2004). Moreover, HDAC inhibitors such as valproic acid and trichostatin A, which effectively increase acetylation levels, induce hippocampal progenitor cells to undergo neurogenesis, even in the presence of gliogenic cytokines (Hsieh et al., 2004). The inhibition of a glial fate by HDAC inhibitors is in part due to the initiation of a neurogenic program that includes expression of the bHLH differentiation factor *NeuroD*.

ASTROCYTE COMPETENCE IS REGULATED BY SITE-SPECIFIC DNA AND HISTONE METHYLATION PATTERNS

The competence of cortical progenitors to undergo astrocytogenesis increases during development (Song and Ghosh, 2004). Epigenetic

modifications have been implicated in regulating the temporal response of cortical progenitors to gliogenic signals. Specifically, FGF2 potentiates the ability of CNTF to induce astrocyte differentiation by increasing H3K4me and reducing H3K9me marks at the STAT-binding site in the GFAP promoter (and not in other regions such as the TATA-binding sequence) (Song and Ghosh, 2004). The importance of these modifications was demonstrated by the use of 5'-methylthioadenosine (MTA), a general inhibitor of methyltransferase activity that prevents CNTF and FGF2 from inducing GFAP expression and glial differentiation. These studies support the idea that extracellular signals such as FGF2 can regulate cell fate decisions by controlling the access of transcription factors to promoters through alterations of chromatin state.

The role of DNA methylation in neural lineage development has been most extensively studied during the neuronal-to-astrocytic transition that occurs in later stages of development. Of the three known Dnmts, *Dnmt1* is ubiquitously expressed, while *Dnmt3b* and *Dnmt3a* show regionalized patterns of expression in the developing CNS, with *Dnmt3b* transcripts detected in neural progenitors between E10 and E14 and *Dnmt3a* expressed in postmitotic neurons (Feng et al., 2005). The importance of DNA methylation in development is highlighted by the severity of the knockout phenotypes of mutant mice, with both *Dnmt1* and *Dnmt3b* mutant mice dying embryonically, while *Dnmt3a* mutants die within 1 month after birth (Allen, 2008; Feng et al., 2007; Li et al., 1992, and references therein). Moreover, *Dnmt1* mutant mice display precocious astrocyte differentiation, due to global demethylation of promoters as well as an overall increase in JAK-STAT signaling levels (Fan et al., 2005). It has also been found that glia-specific genes, including *GFAP* and *S100β*, are highly methylated in ES cells, correlating with the inability of these cells to undergo glial differentiation in response to cytokine signaling (Shimozaki et al., 2005). There is also a general trend toward more demethylated CpG dinucleotides in the regulatory regions of astrocytic genes, such as *GFAP* and *S100β*, in later- (E14.5) compared to earlier- (E11.5) stage cortical progenitors, with the loss of methyl groups correlating with the period when astrocytogenesis begins (Hatada et al., 2008; Namihira et al., 2004; Takizawa et al., 2001; Teter et al., 1996). The demethylation of astroglial genes also regulates the competence of cortical progenitors to respond to astrogliogenic signals such as CNTF. Specifically, methylation of the STAT3-binding site in the *GFAP* promoter (TT**CG**GAGAA) blocks STAT3 binding and transcription of *GFAP* (Takizawa et al., 2001).

The MBD proteins that bind methylated CpG dinucleotides have also been implicated as negative regulators of gliogenesis. MBD proteins such as MeCP2 are not normally expressed in astrocytes and oligodendrocytes, whereas they are abundantly expressed in differentiating neurons (Kishi and Macklis, 2004; Setoguchi et al., 2006). Moreover, the ectopic expression of MBDs such as MeCP2 or MBD1 inhibits astrocyte differentiation in response to cytokine signaling (Setoguchi et al., 2006). Taken together these studies strongly support a role for DNA methylation in regulating the timing of astrocyte differentiation.

OLIGODENDROCYTE DIFFERENTIATION IS REGULATED BY HISTONE DEACETYLATION AND METHYLATION

Oligodendrocytes are the second type of glial cell in the CNS. While astrocytes are generated throughout the CNS, oligodendrocyte production is regionally restricted. In the telencephalon, long-term lineage-tracing studies using different cre drivers have revealed that oligodendrocytes are generated from three distinct regions of the forebrain in a temporally defined order (Kessaris et al., 2006). The first wave of oligodendrocyte progenitor cell (OPC) differentiation occurs in ventral telencephalic domains, including the medial ganglionic eminence (MGE) and anteropeduncular area (AEP), followed by the caudal (CGE) and lateral (LGE) ganglionic eminences. Ventrally derived oligo-

dendrocytes not only populate the ventral forebrain but also migrate tangentially to reach the neocortex. The last wave of OPC differentiation occurs in the dorsal telencephalon at early postnatal stages (Kessaris et al., 2006). In addition, several studies have shown that OPCs have intrinsic timing mechanisms that promote OPC differentiation at specific times in development (reviewed in Raff et al., 2001).

The idea that histone acetylation may regulate the timing of oligodendrocyte differentiation initially came from correlative studies that showed that postnatal cortical progenitors cultured in vitro and induced to undergo oligodendrocyte differentiation by the removal of mitogens showed a global decrease in histone acetylation levels (Marin-Husstege et al., 2002). A functional assay showed that the addition of HDAC inhibitors, such as TSA or sodium butyrate, also blocked oligodendrocyte differentiation in vitro (Marin-Husstege et al., 2002). Mechanistically, the transcription factor Yin Yang 1 (YY1) has been identified as a key recruiter of HDACs during oligodendrocyte differentiation (He et al., 2007a, 2007b). YY1 is a transcriptional repressor that binds HDACs, and YY1 consensus-binding sequences (NNCCATNN) are found in ~30% of the promoters for genes that are downregulated by histone deacetylation during oligodendrocyte differentiation (He et al., 2007a, 2007b). Moreover, conditional knockouts of YY1 show clear defects in the differentiation of oligodendrocytes and subsequent myelination (He et al., 2007a, 2007b). Thus, in the OPC lineage, a single transcriptional repressor and its HDAC cofactors are key regulators of the timing of oligodendrocyte differentiation.

The expression of Ezh2, an H3K27 HMTase and PcG protein, is also dynamically regulated during neural lineage progression, with high levels of Ezh2 detected in proliferating neural stem cells and in OPCs and reduced expression levels observed during neuronal and astrocytic differentiation (Sher et al., 2008). Moreover, the forced expression of Ezh2 in neural stem cells promotes OPC proliferation, resulting in the generation of supernumerary oligodendrocytes, while knockdown of Ezh2 reduced the number of oligodendrocytes generated (Sher et al., 2008). The methylation status of H3K27 is thus a key determinant of cell fate during neural lineage progression.

CHROMATIN REMODELING AND DEDIFFERENTIATION OF OLIGODENDROCYTE PRECURSORS TO A MULTIPOTENT STATE

Sox1, Sox2, and Sox3 are SRY-related, HMG-box-containing transcription factors that are expressed early in the developing neural tube and operate along with Notch signaling to maintain a multipotent stem cell-like fate (reviewed in Shi et al., 2008). *Sox1/2/3* expression levels must be downregulated for neuronal differentiation to proceed as these SoxB1 proteins directly interfere with proneural bHLH activity (Bylund et al., 2003; Holmberg et al., 2008). Conversely, it has recently been shown that the reexpression of Sox2 is associated with the conversion of differentiated type 2 astrocytes into a multipotent stem cell and, furthermore, that this conversion involves chromatin remodeling (Kondo and Raff, 2004). OPCs are induced to differentiate into oligodendrocytes or type 2 astrocytes by the addition of thyroid hormone (TH) or BMP2, respectively. BMP2 also induces the expression of some markers of a neural stem-like (NSL) fate, including *Sox2*. When BMP2-induced type 2 astrocytes are cultured with FGF2, they dedifferentiate into neural stem-like cells (NSLCs) in a Sox2-dependent manner. Interestingly, Brm, a SWI/SNF ATPase, is recruited to the *Sox2* enhancer upon exposure of OPCs to BMP2 and type 2 astrocytes to bFGF, and the *Sox2* enhancer displays increased acetylation of H3K9 and methylation of H3K4 (Kondo and Raff, 2004). Dedifferentiation events are therefore also associated with wide-scale chromatin remodeling that allows genes involved in maintaining a multipotent stem cell-like phenotype to be reexpressed.

CONCLUSIONS AND FUTURE DIRECTIONS

In summary, the ability of growth factors and cytokines to promote gliogenesis occurs at the level of gene transcription, as well as at the level of histone and DNA modifications. Strikingly, in some cases, the effects of the gliogenic signals have been tracked down to the level of chromatin modifications at individual promoter elements. Interactions between transcription factors (e.g., YY1, proneural bHLH) and chromatin-remodeling agents that influence gliogenesis have also been described, making the neuronal-to-glial transition one of the best-understood cell fate switches in the developing nervous system.

EPIGENETICS AND NEURONAL FATE SPECIFICATION IN THE NEOCORTEX

INTRODUCTION TO NEOCORTICAL DEVELOPMENT

Neurons acquire not only generic neuronal properties but also subtype-specific identities, including neurotransmitter phenotypes, morphologies, axonal projections, and molecular identities that reflect their regional, spatial, and temporal characteristics. The acquisition of cell type–specific identities is important as the formation of functional neural circuits requires that appropriate numbers of the correct types of neuronal cells are generated during development. Understanding how neurons acquire their appropriate identities is of particular relevance for the neocortex, the area of the brain that is responsible for higher cognitive functioning and perceptual processing and where defects in neuronal fate specification are associated with severe behavioral disorders, including bipolar disorder, schizophrenia, and autism. The neocortex is primarily comprised of excitatory projection neurons that are organized into six radial layers (i.e., layers I–VI), each distinguishable based on neuronal morphologies, molecular identities, and axonal projection patterns (Tamamaki et al., 1997). For instance, while layer V/VI neurons either project subcortically or to the contralateral hemisphere, neurons in upper layers II–IV project only within the cortex, either to the contralateral or the ipsilateral hemisphere.

SATB2 AND THE DIFFERENTIATION OF LAMINA-SPECIFIC PROPERTIES

The importance of chromatin remodeling in regulating lamina-specific neuronal differentiation has recently been appreciated based on the analysis of *Satb2* knockout mice (Alcamo et al., 2008; Britanova et al., 2008). The possibility that *Satb2* may be involved in cortical development was initially proposed based on its expression in the cortical subventricular zone, a secondary progenitor cell layer, and the progeny of these progenitors, neurons that occupy upper layers of the cortex (Britanova et al., 2005; Szemes et al., 2006). Strikingly, in *Satb2* knockout mice, upper-layer neurons aberrantly project subcortically along the corticospinal tract (Alcamo et al., 2008; Britanova et al., 2008). In addition, *Satb2* mutant upper-layer neurons ectopically express some (but not all) deep-layer markers, including Ctip2, a zinc finger transcription factor implicated in the establishment of a deep layer identity (Arlotta et al., 2005). Notably, Satb2 negatively regulates Ctip2 expression by binding to upstream S/MAR sequences in the *Ctip2* upstream regulatory regions, reducing H3K4 methylation and acetylation (Alcamo et al., 2008). Hence, in *Satb2* mutants, levels of H4 acetylation at the Ctip2 locus were increased (Britanova et al., 2008). Thus, chromatin remodeling at the level of DNA looping via nuclear matrix attachment is also essential for neuronal fate decisions to be correctly made during development.

CONCLUSIONS AND FUTURE DIRECTIONS

The phenotypic analysis of *Satb2* mutants highlights the importance of higher-order chromatin organization during neural development. An interesting feature of *Satb2* mutant cortical cells is that AUF1, another AT-rich binding protein, replaces Satb2 in a NuRD chromatin-remodeling complex (Gyorgy et al., 2008). However, AUF1 obviously cannot replace the function of Satb2, given the aberrant neuronal phenotype of mutant upper-layer neurons, suggesting that different AT-rich binding proteins confer different chromatin architectures. The extent to which the layer specification defects in *Satb2* mutant mice are due to the loss of Satb2 versus the addition of AUF1 remains to be addressed. It is also curious that while Satb2 affects the expression of layer-specific genes in postmitotic neurons, previous studies have suggested that layer fates are determined in part by temporal changes in the competence of cortical progenitors (Bohner et al., 1997; Desai and McConnell, 2000; Frantz and McConnell, 1996; McConnell, 1988; McConnell and Kaznowski, 1991). Given that Satb2 is not expressed at detectable levels in primary cortical progenitors (Alcamo et al., 2008), this raises the possibility that other chromatin-remodeling factors influence layer fate specification in the progenitor pool. What might these chromatin-remodeling factors be? In addition, do signaling molecules such as brain-derived neurotrophic factor (BDNF), which influence laminar fate specification (Fukumitsu et al., 2006), operate by altering the composition of NuRD complexes?

EPIGENETICS IN ADULT NEUROGENESIS

While the majority of neurogenesis is completed in the embryonic and early postnatal period, there are a few sites of adult neurogenesis. These include subgranular cells in the hippocampus, which give rise to neurons that repopulate the dentate gyrus, and cells lining the lateral ventricle of the forebrain, which differentiate and migrate along the rostral migratory stream to give rise to new olfactory bulb neurons (reviewed in Taupin and Gage, 2002). In general, the transcriptional mechanisms that regulate neurogenesis in the embryo are recapitulated in the adult. However, some chromatin-modifying agents appear to operate specifically during adult neurogenesis.

POLYCOMB GENES AND ADULT NEURAL STEM-CELL RENEWAL

The PcG genes form large, multimeric complexes that are in general involved in gene silencing through epigenetic mechanisms. A

R. Godbout, H. E. McDermid, and R. Shiekhattar. 2005. CECR2, a protein involved in neurulation, forms a novel chromatin remodeling complex with SNF2L. *Hum Mol Genet* 14:513–24.

Barnabe-Heider, F., J. A. Wasylnka, K. J. Fernandes, C. Porsche, M. Sendtner, D. R. Kaplan, and F. D. Miller. 2005. Evidence that embryonic neurons regulate the onset of cortical gliogenesis via cardiotrophin-1. *Neuron* 48:253–65.

Beisel, C., A. Imhof, J. Greene, E. Kremmer, and F. Sauer. 2002. Histone methylation by the *Drosophila* epigenetic transcriptional regulator Ash1. *Nature* 419:857–62.

Belyaev, N. D., I. C. Wood, A. W. Bruce, M. Street, J. B. Trinh, and N. J. Buckley. 2004. Distinct RE-1 silencing transcription factor–containing complexes interact with different target genes. *J Biol Chem* 279:556–61.

Bernstein, B. E., T. S. Mikkelsen, X. Xie, M. Kamal, D. J. Huebert, J. Cuff, B. Fry, et al. 2006. A bivalent chromatin structure marks key developmental genes in embryonic stem cells. *Cell* 125: 315–26.

Bertrand, N., D. S. Castro, and F. Guillemot. 2002. Proneural genes and the specification of neural cell types. *Nat Rev Neurosci* 3:517–30.

Berube, N. G., M. Mangelsdorf, M. Jagla, J. Vanderluit, D. Garrick, R. J. Gibbons, D. R. Higgs, R. S. Slack, and D. J. Picketts. 2005. The chromatin-remodeling protein ATRX is critical for neuronal survival during corticogenesis. *J Clin Invest* 115: 258–67.

Bestor, T., A. Laudano, R. Mattaliano, and V. Ingram. 1988. Cloning and sequencing of a cDNA encoding DNA methyltransferase of mouse cells. The carboxyl-terminal domain of the mammalian enzymes is related to bacterial restriction methyltransferases. *J Mol Biol* 203:971–83.

Bibel, M., J. Richter, E. Lacroix, and Y. A. Barde. 2007. Generation of a defined and uniform population of CNS progenitors and neurons from mouse embryonic stem cells. *Nat Protoc* 2: 1034–43.

Bibel, M., J. Richter, K. Schrenk, K. L. Tucker, V. Staiger, M. Korte, M. Goetz, and Y. A. Barde. 2004. Differentiation of mouse embryonic stem cells into a defined neuronal lineage. *Nat Neurosci* 7:1003–9.

Bohner, A. P., R. M. Akers, and S. K. McConnell. 1997. Induction of deep layer cortical neurons in vitro. *Development* 124:915–23.

Bolos, V., J. Grego-Bessa, and J. L. de la Pompa. 2007. Notch signaling in development and cancer. *Endocr Rev* 28:339–63.

Bonni, A., Y. Sun, M. Nadal-Vicens, A. Bhatt, D. A. Frank, I. Rozovsky, N. Stahl, G. D. Yancopoulos, and M. E. Greenberg. 1997. Regulation of gliogenesis in the central nervous system by the JAK-STAT signaling pathway. *Science* 278: 477–83.

Boyer, L. A., K. Plath, J. Zeitlinger, T. Brambrink, L. A. Medeiros, T. I. Lee, S. S. Levine, et al. 2006. Polycomb complexes repress developmental regulators in murine embryonic stem cells. *Nature* 441:349–53.

Britanova, O., S. Akopov, S. Lukyanov, P. Gruss, and V. Tarabykin. 2005. Novel transcription factor Satb2 interacts with matrix attachment region DNA elements in a tissue-specific manner and demonstrates cell-type-dependent expression in the developing mouse CNS. *Eur J Neurosci* 21:658–68.

Britanova, O., C. de Juan Romero, A. Cheung, K. Y. Kwan, M. Schwark, A. Gyorgy, T. Vogel, et al. 2008. Satb2 is a postmitotic determinant for upper-layer neuron specification in the neocortex. *Neuron* 57:378–92.

Britz, O., P. Mattar, L. Nguyen, L. M. Langevin, C. Zimmer, S. Alam, F. Guillemot, and C. Schuurmans. 2006. A role for proneural genes in the maturation of cortical progenitor cells. *Cereb Cortex* 16 (Suppl 1): i138–51.

Burgold, T., F. Spreafico, F. De Santa, M. G. Totaro, E. Prosperini, G. Natoli, and G. Testa. 2008. The histone H3 lysine 27-specific demethylase Jmjd3 is required for neural commitment. *PLoS ONE* 3:e3034.

Bylund, M., E. Andersson, B. G. Novitch, and J. Muhr. 2003. Vertebrate neurogenesis is counteracted by Sox1-3 activity. *Nat Neurosci* 6:1162–8.

Cai, S., H. J. Han, and T. Kohwi-Shigematsu. 2003. Tissue-specific nuclear architecture and gene expression regulated by SATB1. *Nat Genet* 34: 42–51.

Cairns, B. R., Y. J. Kim, M. H. Sayre, B. C. Laurent, and R. D. Kornberg. 1994. A multisubunit complex containing the SWI1/ADR6, SWI2/SNF2, SWI3, SNF5, and SNF6 gene products isolated from yeast. *Proc Natl Acad Sci USA* 91:1950–4.

Callinan, P. A., and A. P. Feinberg. 2006. The emerging science of epigenomics. Special issue, *Hum Mol Genet* 15:R95–101.

Cao, Y., R. M. Kumar, B. H. Penn, C. A. Berkes, C. Kooperberg, L. A. Boyer, R. A. Young, and S. J. Tapscott. 2006. Global and gene-specific analyses show distinct roles for Myod and Myog at a common set of promoters. *EMBO J* 25:502–11.

Carrozza, M. J., R. T. Utley, J. L. Workman, and J. Cote. 2003. The diverse functions of histone acetyltransferase complexes. *Trends Genet* 19:321–9.

Cau, E., G. Gradwohl, S. Casarosa, R. Kageyama, and F. Guillemot. 2000. *Hes* genes regulate sequential stages of neurogenesis in the olfactory epithelium. *Development* 127:2323–32.

Cau, E., G. Gradwohl, C. Fode, and F. Guillemot, 1997. Mash1 activates a cascade of bHLH regulators in olfactory neuron progenitors. *Development* 124:1611–21.

Chang, Q., G. Khare, V. Dani, S. Nelson, and R. Jaenisch. 2006. The disease progression of Mecp2 mutant mice is affected by the level of BDNF expression. *Neuron* 49:341–8.

Chen, R. Z., S. Akbarian, M. Tudor, and R. Jaenisch. 2001. Deficiency of methyl-CpG binding protein-2 in CNS neurons results in a Rett-like phenotype in mice. *Nat Genet* 27:327–31.

Chen, W. G., Q. Chang, Y. Lin, A. Meissner, A. E. West, E. C. Griffith, R. Jaenisch, and M. E. Greenberg. 2003. Derepression of BDNF transcription involves calcium-dependent phosphorylation of MeCP2. *Science* 302:885–9.

Chen, Z. F., A. J. Paquette, and D. J. Anderson. 1998. NRSF/REST is required in vivo for repression of multiple neuronal target genes during embryogenesis. *Nat Genet* 20:136–42.

Chojnacki, A., T. Shimazaki, C. Gregg, G. Weinmaster, and S. Weiss. 2003. Glycoprotein 130 signaling regulates Notch1 expression and activation in the self-renewal of mammalian forebrain neural stem cells. *J Neurosci* 23:1730–41.

Chong, J. A., J. Tapia-Ramirez, S. Kim, J. J. Toledo-Aral, Y. Zheng, M. C. Boutros, Y. M. Altshuller, M. A. Frohman, S. D. Kraner, and G. Mandel. 1995. REST: A mammalian silencer protein that restricts sodium channel gene expression to neurons. *Cell* 80:949–57.

Christensen, J., K. Agger, P. A. Cloos, D. Pasini, S. Rose, L. Sennels, J. Rappsilber, K. H. Hansen, A. E. Salcini, and K. Helin. 2007. RBP2 belongs to a family of demethylases, specific for tri-and dimethylated lysine 4 on histone 3. *Cell* 128:1063–76.

Cinnamon, E., and Z. Paroush. 2008. Context-dependent regulation of Groucho/TLE-mediated repression. *Curr Opin Genet Dev* 18:435–40.

Cote, J., J. Quinn, J. L. Workman, and C. L. Peterson. 1994. Stimulation of GAL4 derivative binding to nucleosomal DNA by the yeast SWI/SNF complex. *Science* 265:53–60.

Desai, A. R., and S. K. McConnell. 2000. Progressive restriction in fate potential by neural progenitors during cerebral cortical development. *Development* 127:2863–72.

Dey, B. K., L. Stalker, A. Schnerch, M. Bhatia, J. Taylor-Papidimitriou, and C. Wynder. 2008. The histone demethylase KDM5b/JARID1b plays a role in cell fate decisions by blocking terminal differentiation. *Mol Cell Biol* 28:5312–27.

Dirscherl, S. S., J. J. Henry, and J. E. Krebs. 2005. Neural and eye-specific defects associated with loss of the imitation switch (ISWI) chromatin remodeler in *Xenopus laevis*. *Mech Dev* 122:1157–70.

Fan, G., C. Beard, R. Z. Chen, G. Csankovszki, Y. Sun, M. Siniaia, D. Biniszkiewicz, et al. 2001. DNA hypomethylation perturbs the function and survival of CNS neurons in postnatal animals. *J Neurosci* 21:788–97.

Fan, G., K. Martinowich, M. H. Chin, F. He, S. D. Fouse, L. Hutnick, D. Hattori, et al. 2005. DNA methylation controls the timing of astrogliogenesis through regulation of JAK-STAT signaling. *Development* 132:3345–56.

Fasano, C. A., J. T. Dimos, N. B. Ivanova, N. Lowry, I. R. Lemischka, and S. Temple. 2007. shRNA knockdown of Bmi-1 reveals a critical role for p21-Rb pathway in NSC self-renewal during development. *Cell Stem Cell* 1:87–99.

Feng, J., H. Chang, E. Li, and G. Fan. 2005. Dynamic expression of de novo DNA methyltransferases Dnmt3a and Dnmt3b in the central nervous system. *J Neurosci Res* 79:734–46.

Feng, J., S. Fouse, and G. Fan. 2007. Epigenetic regulation of neural gene expression and neuronal function. *Pediatr Res* 61:58R–63R.

Fischle, W., F. Dequiedt, M. J. Hendzel, M. G. Guenther, M. A. Lazar, W. Voelter, and E. Verdin, 2002. Enzymatic activity associated with class II HDACs is dependent on a multiprotein complex containing HDAC3 and SMRT/N-CoR. *Mol Cell* 9:45–57.

Fischle, W., Y. Wang, and C. D. Allis. 2003. Histone and chromatin cross-talk. *Curr Opin Cell Biol* 15:172–83.

Fode, C., G. Gradwohl, X. Morin, A. Dierich, M. LeMeur, C. Goridis, and F. Guillemot. 1998. The bHLH protein NEUROGENIN 2 is a determination factor for epibranchial placode-derived sensory neurons. *Neuron* 20:483–94.

Fouse, S. D., Y. Shen, M. Pellegrini, S. Cole, A. Meissner, L. Van Neste, R. Jaenisch, and G. Fan. 2008. Promoter CpG methylation contributes to ES cell gene regulation in parallel with Oct4/Nanog, PcG complex, and histone H3 K4/K27 trimethylation. *Cell Stem Cell* 2:160–9.

Frantz, G. D., and S. K. McConnell. 1996. Restriction of late cerebral cortical progenitors to an upper-layer fate. *Neuron* 17:55–61.

Fukuda, S., and T. Taga. 2005. Cell fate determination regulated by a transcriptional signal network in the developing mouse brain. *Anat Sci Int* 80:12–18.

Fukumitsu, H., M. Ohtsuka, R. Murai, H. Nakamura, K. Itoh, and S. Furukawa. 2006. Brain-derived neurotrophic factor participates in determination of neuronal laminar fate in the developing mouse cerebral cortex. *J Neurosci* 26:13218–30.

Gaulden, J., and J. F. Reiter. 2008. Neur-ons and neur-offs: Regulators of neural induction in vertebrate embryos and embryonic stem cells. *Hum Mol Genet* 17:R60–6.

Gibbons, R. J., D. J. Picketts, L. Villard, and D. R. Higgs. 1995. Mutations in a putative global transcriptional regulator cause X-linked mental retardation with alpha-thalassemia (ATR-X syndrome). *Cell* 80:837–45.

Glinsky, G. V., O. Berezovska, and A. B. Glinskii. 2005. Microarray analysis identifies a death-from-cancer signature predicting therapy failure in patients with multiple types of cancer. *J Clin Invest* 115:1503–21.

Gohring, F., and F. O. Fackelmayer. 1997. The scaffold/matrix attachment region binding protein hnRNP-U (SAF-A) is directly bound to chromosomal DNA in vivo: A chemical cross-linking study. *Biochemistry* 36:8276–83.

Goll, M. G., and T. H. Bestor. 2005. Eukaryotic cytosine methyltransferases. *Annu Rev Biochem* 74:481–514.

Greenway, D. J., M. Street, A. Jeffries, and N. J. Buckley. 2007. RE1 silencing transcription factor maintains a repressive chromatin environment in embryonic hippocampal neural stem cells. *Stem Cells* 25:354–63.

Grimes, J. A., S. J. Nielsen, E. Battaglioli, E. A. Miska, J. C. Speh, D. L. Berry, F. Atouf, B. C. Holdener, G. Mandel, and T. Kouzarides. 2000. The co-repressor mSin3A is a functional component of the REST-CoREST repressor complex. *J Biol Chem* 275:9461–7.

Guillemot, F., L. C. Lo, J. E. Johnson, A. Auerbach, D. J. Anderson, and A. L. Joyner. 1993. Mammalian achaete-scute homolog 1 is required for the early development of olfactory and autonomic neurons. *Cell* 75:463–76.

Gutierrez, J., R. Paredes, F. Cruzat, D. A. Hill, A. J. van Wijnen, J. B. Lian, G. S. Stein, J. L. Stein, A. N. Imbalzano, and M. Montecino. 2007. Chromatin remodeling by SWI/SNF results in nucleosome mobilization to preferential positions in the rat osteocalcin gene promoter. *J Biol Chem* 282:9445–57.

Gyorgy, A. B., M. Szemes, C. de Juan Romero, V. Tarabykin, and D. V. Agoston. 2008. SATB2 interacts with chromatin-remodeling molecules in differentiating cortical neurons. *Eur J Neurosci* 27:865–73.

Hamby, M. E., V. Coskun, and Y. E. Sun. 2008. Transcriptional regulation of neuronal differentiation: The epigenetic layer of complexity. *Biochim Biophys Acta* 1779:432–7.

Hatada, I., M. Namihira, S. Morita, M. Kimura, T. Horii, and K. Nakashima. 2008. Astrocyte-specific genes are generally demethylated in neural precursor cells prior to astrocytic differentiation. *PLoS ONE* 3:e3189.

He, Y., J. Dupree, J. Wang, J. Sandoval, J. Li, H. Liu, Y. Shi, K. A. Nave, and P. Casaccia-Bonnefil, 2007a. The transcription factor Yin Yang 1 is essential for oligodendrocyte progenitor differentiation. *Neuron* 55:217–30.

He, Y., J. Sandoval, and P. Casaccia-Bonnefil, 2007b. Events at the transition between cell cycle exit and oligodendrocyte progenitor differentiation: The role of HDAC and YY1. *Neuron Glia Biol* 3:221–31.

Heinzel, T. R. M., Lavinsky, T. M. Mullen, M. Soderstrom, C. D. Laherty, J. Torchia, W. M. Yang, et al. 1997. A complex containing N-CoR, mSin3 and histone deacetylase mediates transcriptional repression. *Nature* 387:43–8.

Hermanson, O., K. Jepsen, and M. G. Rosenfeld. 2002. N-CoR controls differentiation of neural stem cells into astrocytes. *Nature* 419:934–9.

Hirasawa, R., H. Chiba, M. Kaneda, S. Tajima, E. Li, R. Jaenisch, and H. Sasaki. 2008. Maternal and zygotic Dnmt1 are necessary and sufficient for the maintenance of DNA methylation imprints during preimplantation development. *Genes Dev* 22:1607–16.

Hitoshi, S., T. Alexson, V. Tropepe, D. Donoviel, A. J. Elia, J. S. Nye, R. A. Conlon, T. W. Mak, A. Bernstein, and D. van der Kooy. 2002. Notch pathway molecules are essential for the maintenance, but not the generation, of mammalian neural stem cells. *Genes Dev* 16:846–58.

Holmberg, J. E. Hansson, M. Malewicz, M. Sandberg, T. Perlmann, U. Lendahl, and J. Muhr. 2008. SoxB1 transcription factors and Notch signaling use distinct mechanisms to regulate proneural gene function and neural progenitor differentiation. *Development* 135:1843–51.

Hsieh, J., K. Nakashima, T. Kuwabara, E. Mejia, and F. H. Gage. 2004. Histone deacetylase inhibition–mediated neuronal differentiation of multipotent adult neural progenitor cells. *Proc Natl Acad Sci USA* 101:16659–64.

Huang, Y., S.J. Myers, and R. Dingledine. 1999. Transcriptional repression by REST: Recruitment of Sin3A and histone deacetylase to neuronal genes. *Nat Neurosci* 2:867–72.

Ishibashi, M., S.L. Ang, K. Shiota, S. Nakanishi, R. Kageyama, and F. Guillemot. 1995. Targeted disruption of mammalian hairy and Enhancer of split homolog-1 (HES-1) leads to up-regulation of neural helix–loop–helix factors, premature neurogenesis, and severe neural tube defects. *Genes Dev* 9:3136–48.

Ishibashi, M., K. Moriyoshi, Y. Sasai, K. Shiota, S. Nakanishi, and R. Kageyama. 1994. Persistent expression of helix–loop–helix factor HES-1 prevents mammalian neural differentiation in the central nervous system. *EMBO J* 13:1799–1805.

Iwase, S., F. Lan, P. Bayliss, L. de la Torre-Ubieta, M. Huarte, H.H. Qi, J.R. Whetstine, A. Bonni, T.M. Roberts, and Y. Shi. 2007. The X-linked mental retardation gene *SMCX/JARID1C* defines a family of histone H3 lysine 4 demethylases. *Cell* 128:1077–88.

Jenuwein, T., and C.D. Allis. 2001. Translating the histone code. *Science* 293:1074–80.

Jepsen, K., D. Solum, T. Zhou, R.J. McEvilly, H.J. Kim, C.K. Glass, O. Hermanson, and M.G. Rosenfeld. 2007. SMRT-mediated repression of an H3K27 demethylase in progression from neural stem cell to neuron. *Nature* 450: 415–19.

Ju, B.G., D. Solum, E.J. Song, K.J. Lee, D.W. Rose, C.K. Glass, and M.G. Rosenfeld. 2004. Activating the PARP-1 sensor component of the groucho/TLE1 corepressor complex mediates a CaMKinase IIdelta-dependent neurogenic gene activation pathway. *Cell* 119:815–29.

Kao, H.Y., P. Ordentlich, N. Koyano-Nakagawa, Z. Tang, M. Downes, C.R. Kintner, R.M. Evans, and T. Kadesch. 1998. A histone deacetylase corepressor complex regulates the Notch signal transduction pathway. *Genes Dev* 12:2269–77.

Kessaris, N., M. Fogarty, P. Iannarelli, M. Grist, M. Wegner, and W.D. Richardson. 2006. Competing waves of oligodendrocytes in the forebrain and postnatal elimination of an embryonic lineage. *Nat Neurosci* 9:173–9.

Khavari, P.A., C.L. Peterson, J.W. Tamkun, D.B. Mendel, and G.R. Crabtree. 1993. BRG1 contains a conserved domain of the SWI2/SNF2 family necessary for normal mitotic growth and transcription. *Nature* 366:170–4.

Kishi, N., and J.D. Macklis. 2004. MECP2 is progressively expressed in post-migratory neurons and is involved in neuronal maturation rather than cell fate decisions. *Mol Cell Neurosci* 27:306–21.

Kondo, T., and M. Raff. 2004. Chromatin remodeling and histone modification in the conversion of oligodendrocyte precursors to neural stem cells. *Genes Dev* 18:2963–72.

Kuwabara, T., J. Hsieh, K. Nakashima, K. Taira, and F.H. Gage. 2004. A small modulatory dsRNA specifies the fate of adult neural stem cells. *Cell* 116:779–93.

Kwon, H., A.N. Imbalzano, P.A. Khavari, R.E. Kingston, and M.R. Green. 1994. Nucleosome disruption and enhancement of activator binding by a human SWI/SNF complex. *Nature* 370:477–81.

Lachner, M., R.J. O'Sullivan, and T. Jenuwein. 2003. An epigenetic road map for histone lysine methylation. *J Cell Sci* 116:2117–24.

Lazzaro, M.A., and D.J. Picketts. 2001. Cloning and characterization of the murine Imitation Switch (ISWI) genes: Differential expression patterns suggest distinct developmental roles for Snf2h and Snf2l. *J Neurochem* 77:1145–56.

Lee, M.G., C. Wynder, J. Norman, and R. Shiekhattar. 2006. Isolation and characterization of histone H3 lysine 4 demethylase-containing complexes. *Methods* 40:327–30.

Lemon, B., C. Inouye, D.S. King, and R. Tjian. 2001. Selectivity of chromatin-remodelling cofactors for ligand-activated transcription. *Nature* 414: 924–8.

Lessard, J., J.I. Wu, J.A. Ranish, M. Wan, M.M. Winslow, B.T. Staahl, H. Wu, R. Aebersold, I.A. Graef, and G.R. Crabtree. 2007. An essential switch in subunit composition of a chromatin remodeling complex during neural development. *Neuron* 55:201–15.

Leung, C., M. Lingbeek, O. Shakhova, J. Liu, E. Tanger, P. Saremaslani, M. Van Lohuizen, and S. Marino. 2004. Bmi1 is essential for cerebellar development and is overexpressed in human medulloblastomas. *Nature* 428:337–41.

Li, E., C. Beard, and R. Jaenisch. 1993. Role for DNA methylation in genomic imprinting. *Nature* 366:362–5.

Li, E., T.H. Bestor, and R. Jaenisch. 1992. Targeted mutation of the DNA methyltransferase gene results in embryonic lethality. *Cell* 69:915–26.

Li, X., B.Z. Barkho, Y. Luo, R.D. Smrt, N.J. Santistevan, C. Liu, T. Kuwabara, F.H. Gage, and X. Zhao. 2008. Epigenetic regulation of the stem cell mitogen FGF-2 by mbd1 in adult neural stem/progenitor cells. *J Biol Chem* 283:27644–52.

Lunyak, V.V., R. Burgess, G.G. Prefontaine, C. Nelson, S.H. Sze, J. Chenoweth, P. Schwartz, et al. 2002. Corepressor-dependent silencing of chromosomal regions encoding neuronal genes. *Science* 298:1747–52.

Ma, Q., C. Kintner, and D.J. Anderson. 1996. Identification of neurogenin, a vertebrate neuronal determination gene. *Cell* 87:43–52.

Mal, A., and M.L. Harter. 2003. MyoD is functionally linked to the silencing of a muscle-specific regulatory gene prior to skeletal myogenesis. *Proc Natl Acad Sci USA* 100:1735–9.

Marin-Husstege, M., M. Muggironi, A. Liu, and P. Casaccia-Bonnefil. 2002. Histone deacetylase activity is necessary for oligodendrocyte lineage progression. *J Neurosci* 22:10333–45.

Mattar, P., O. Britz, C. Johannes, M. Nieto, L. Ma, A. Rebeyka, N. Klenin, F. Polleux, F. Guillemot, and C. Schuurmans. 2004. A screen for downstream effectors of Neurogenin2 in the embryonic neocortex. *Dev Biol* 273:373–89.

Mattar, P., L.M. Langevin, K. Markham, N. Klenin, S. Shivji, D. Zinyk, and C. Schuurmans. 2008. Basic helix–loop–helix transcription factors cooperate to specify a cortical projection neuron identity. *Mol Cell Biol* 28:1456–69.

McConnell, S.K. 1988. Fates of visual cortical neurons in the ferret after isochronic and heterochronic transplantation. *J Neurosci* 8:945–74.

McConnell, S.K., and C.E. Kaznowski. 1991. Cell cycle dependence of laminar determination in developing neocortex. *Science* 254:282–5.

Meissner, A., T.S. Mikkelsen, H. Gu, M. Wernig, J. Hanna, A. Sivachenko, X. Zhang, et al. 2008. Genome-scale DNA methylation maps of pluripotent and differentiated cells. *Nature* 454: 766–70.

Mikkelsen, T.S., M. Ku, D.B. Jaffe, B. Issac, E. Lieberman, G. Giannoukos, P. Alvarez, et al. 2007. Genome-wide maps of chromatin state in pluripotent and lineage-committed cells. *Nature* 448: 553–60.

Mizutani, K., K. Yoon, L. Dang, A. Tokunaga, and N. Gaiano. 2007. Differential Notch signalling distinguishes neural stem cells from intermediate progenitors. *Nature* 449:351–5.

Mohn, F., M. Weber, M. Rebhan, T.C. Roloff, J. Richter, M.B. Stadler, M. Bibel, and D. Schubeler. 2008. Lineage-specific polycomb targets and de novo DNA methylation define restriction and potential of neuronal progenitors. *Mol Cell* 30:755–66.

Molofsky, A.V., S. He, M. Bydon, S.J. Morrison, and R. Pardal. 2005. Bmi-1 promotes neural stem cell self-renewal and neural development but not mouse growth and survival by repressing the p16Ink4a and p19Arf senescence pathways. *Genes Dev* 19:1432–7.

Molofsky, A.V., R. Pardal, T. Iwashita, I.K. Park, M.F. Clarke, and S.J. Morrison. 2003. Bmi-1 dependence distinguishes neural stem cell self-renewal from progenitor proliferation. *Nature* 425:962–7.

Nagao, M., M. Sugimori, and M. Nakafuku. 2007. Cross talk between notch and growth factor/cytokine signaling pathways in neural stem cells. *Mol Cell Biol* 27:3982–94.

Nakashima, K., S. Wiese, M. Yanagisawa, H. Arakawa, N. Kimura, T. Hisatsune, K. Yoshida, T. Kishimoto, M. Sendtner, and T. Taga. 1999a. Developmental requirement of gp130 signaling in neuronal survival and astrocyte differentiation. *J Neurosci* 19:5429–34.

Nakashima, K., M. Yanagisawa, H. Arakawa, N. Kimura, T. Hisatsune, M. Kawabata, K. Miyazono, and T. Taga. 1999b. Synergistic signaling in fetal brain by STAT3-Smad1 complex bridged by p300. *Science* 284:479–82.

Nakashima, K., M. Yanagisawa, H. Arakawa, and T. Taga. 1999c. Astrocyte differentiation mediated by LIF in cooperation with BMP2. *FEBS Lett* 457:43–46.

Namihira, M., J. Kohyama, M. Abematsu, and K. Nakashima. 2008. Epigenetic mechanisms regulating fate specification of neural stem cells. *Philos Trans R Soc Lond B Biol Sci* 363:2099–2109.

Namihira, M., K. Nakashima, and T. Taga. 2004. Developmental stage dependent regulation of DNA methylation and chromatin modification in a immature astrocyte specific gene promoter. *FEBS Lett* 572:184–8.

Nieto, M., C. Schuurmans, O. Britz, and F. Guillemot. 2001. Neural bHLH genes control the neuronal versus glial fate decision in cortical progenitors. *Neuron* 29:401–13.

Nyfeler, Y., R.D. Kirch, N. Mantei, D.P. Leone, F. Radtke, U. Suter, and V. Taylor. 2005. Jagged1 signals in the postnatal subventricular zone are required for neural stem cell self-renewal. *EMBO J* 24:3504–15.

O'Carroll, D., S. Erhardt, M. Pagani, S.C. Barton, M.A. Surani, and T. Jenuwein. 2001. The polycomb-group gene *Ezh2* is required for early mouse development. *Mol Cell Biol* 21:4330–6.

Ohtsuka, T., M. Ishibashi, G. Gradwohl, S. Nakanishi, F. Guillemot, and R. Kageyama. 1999. Hes1 and Hes5 as notch effectors in mammalian neuronal differentiation. *EMBO J* 18:2196–2207.

Olave, I., W. Wang, Y. Xue, A. Kuo, and G.R. Crabtree. 2002. Identification of a polymorphic, neuron-specific chromatin remodeling complex. *Genes Dev* 16:2509–17.

Palacios, D., and P. L. Puri. 2006. The epigenetic network regulating muscle development and regeneration. *J Cell Physiol* 207:1–11.

Panning, B., and R. Jaenisch. 1996. DNA hypomethylation can activate Xist expression and silence X-linked genes. *Genes Dev* 10:1991–2002.

Papp, B., and J. Muller. 2006. Histone trimethylation and the maintenance of transcriptional ON and OFF states by trxG and PcG proteins. *Genes Dev* 20:2041–54.

Pavri, R., B. Zhu, G. Li, P. Trojer, S. Mandal, A. Shilatifard, and D. Reinberg. 2006. Histone H2B monoubiquitination functions cooperatively with FACT to regulate elongation by RNA polymerase II. *Cell* 125:703–17.

Perissi, V., C. Scafoglio, J. Zhang, K. A. Ohgi, D. W. Rose, C. K. Glass, and M. G. Rosenfeld. 2008. TBL1 and TBLR1 phosphorylation on regulated gene promoters overcomes dual CtBP and NCoR/SMRT transcriptional repression checkpoints. *Mol Cell* 29:755–66.

Petruk, S., Y. Sedkov, S. Smith, S. Tillib, V. Kraevski, T. Nakamura, E. Canaani, C. M. Croce, and A. Mazo. 2001. Trithorax and dCBP acting in a complex to maintain expression of a homeotic gene. *Science* 294:1331–4.

Qian, X., Q. Shen, S. K. Goderie, W. He, A. Capela, A. A. Davis, and S. Temple. 2000. Timing of CNS cell generation: A programmed sequence of neuron and glial cell production from isolated murine cortical stem cells. *Neuron* 28:69–80.

Racki, L. R., and G. J. Narlikar. 2008. ATP-dependent chromatin remodeling enzymes: Two heads are not better, just different. *Curr Opin Genet Dev* 18:137–44.

Raff, M., J. Apperly, T. Kondo, Y. Tokumoto, and D. Tang. 2001. Timing cell-cycle exit and differentiation in oligodendrocyte development. *Novartis Found Symp* 237:100–7; discussion 107–12, 158–63.

Rea, S., F. Eisenhaber, D. O'Carroll, B. D. Strahl, Z. W. Sun, M. Schmid, S. Opravil, et al. 2000. Regulation of chromatin structure by site-specific histone H3 methyltransferases. *Nature* 406:593–9.

Reynolds, B. A., and S. Weiss. 1992. Generation of neurons and astrocytes from isolated cells of the adult mammalian central nervous system. *Science* 255:1707–10.

Rice, J. C., S. D. Briggs, B. Ueberheide, C. M. Barber, J. Shabanowitz, D. F. Hunt, Y. Shinkai, and C. D. Allis. 2003. Histone methyltransferases direct different degrees of methylation to define distinct chromatin domains. *Mol Cell* 12:1591–8.

Ringrose, L., H. Ehret, and R. Paro. 2004. Distinct contributions of histone H3 lysine 9 and 27 methylation to locus-specific stability of polycomb complexes. *Mol Cell* 16:641–53.

Ringrose, L., and R. Paro. 2007. Polycomb/Trithorax response elements and epigenetic memory of cell identity. *Development* 134:223–32.

Robertson, K. D., and A. P. Wolffe. 2000. DNA methylation in health and disease. *Nat Rev Genet* 1:11–19.

Roopra, A., L. Sharling, I. C. Wood, T. Briggs, U. Bachfischer, A. J. Paquette, and N. J. Buckley. 2000. Transcriptional repression by neuron-restrictive silencer factor is mediated via the Sin3-histone deacetylase complex. *Mol Cell Biol* 20:2147–57.

Rossant, J. 2008. Stem cells and early lineage development. *Cell* 132:527–31.

Roth, S. Y., J. M. Denu, and C. D. Allis. 2001. Histone acetyltransferases. *Annu Rev Biochem* 70:81–120.

Santos-Rosa, H., R. Schneider, A. J. Bannister, J. Sherriff, B. E. Bernstein, N. C. Emre, S. L. Schreiber, J. Mellor, and T. Kouzarides. 2002. Active genes are tri-methylated at K4 of histone H3. *Nature* 419:407–11.

Sauvageot, C. M., and C. D. Stiles. 2002. Molecular mechanisms controlling cortical gliogenesis. *Curr Opin Neurobiol* 12:244–9.

Schoenherr, C. J., and D. J. Anderson. 1995. The neuron-restrictive silencer factor (NRSF): A coordinate repressor of multiple neuron-specific genes. *Science* 267:1360–3.

Schoenherr, C. J., A. J. Paquette, and D. J. Anderson. 1996. Identification of potential target genes for the neuron-restrictive silencer factor. *Proc Natl Acad Sci USA* 93:9881–6.

Schuurmans, C., O. Armant, M. Nieto, J. M. Stenman, O. Britz, N. Klenin, C. Brown, et al. 2004. Sequential phases of cortical specification involve Neurogenin-dependent and -independent pathways. *EMBO J* 23:2892–2902.

Schuurmans, C., and F. Guillemot. 2002. Molecular mechanisms underlying cell fate specification in the developing telencephalon. *Curr Opin Neurobiol* 12:26–34.

Seo, J., M. M. Lozano, and J. P. Dudley. 2005a. Nuclear matrix binding regulates SATB1-mediated transcriptional repression. *J Biol Chem* 280:24600–9.

Seo, S., A. Herr, J. W. Lim, G. A. Richardson, H. Richardson, and K. L. Kroll. 2005b. Geminin regulates neuronal differentiation by antagonizing Brg1 activity. *Genes Dev* 19:1723–34.

Seo, S., G. A. Richardson, and K. L. Kroll. 2005c. The SWI/SNF chromatin remodeling protein Brg1 is

In the blue angelfish, on the other hand, new stripes are added as the fish grows. Thus, a fish 2 cm in length has three vertical yellow stripes. When the fish is 4 cm long, new stripes have formed between the old ones; and when the fish has doubled in length again, the number of stripes has also doubled again (Figure 10.3). In this way, the distances between the vertical stripes remain constant. The model could mimic this pattern formation in great detail by producing a stationary wave pattern, which retains the distance between stripes (Figure 10.3).

Even more support for their model was gained by Kondo and Asai when they explained the formation of the color pattern in a related fish with the same type of model. In this species adult fish have horizontal stripes which run from head to (caudal) fin. While the fish grows, the stripes split so that the distances between them stay the same. One stripe splits in two in the same manner as a zipper is unfastened (Figure 10.4). The activator–inhibitor model shows exactly the same behavior. Even the rather complicated changes to the pattern which take place at the uppermost and lowermost parts of the fish are in agreement with the model.

Impressive though this is, the definitive proof is lacking that Turing's morphogens actually exist. It is important to note that the "diffusion" of morphogens in reaction–diffusion models need not correspond to diffusing chemicals in the biological system. They can be replaced by, e.g., a relay of direct cell–cell or cell–ECM interactions (Kondo, 2002). We do not know if *Pomacanthus* fish actually use activators and inhibitors to make their color patterns, nor do we know which genes could be involved in coding for this pattern formation process.

TRAVELING WAVES ON A STRIPED MOUSE

The same research group around Dr. Kondo has turned to a spontaneous mouse mutant to obtain insight into the molecular genetics of stripe formation. This mouse mutant has a splicing defect in a gene, *FoxN1*, which is important for proper hair follicle development. In the mutant,

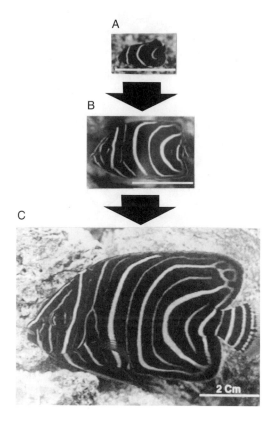

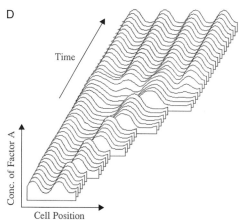

FIGURE 10.3 Rearrangement of the stripe pattern of the growing *Pomacanthus semicirculatus* and its computer simulation. (A–C) Juveniles 2, 6, and 12 months old. When the distance between the vertical lines increases in size, new vertical lines are formed between them so that the wavelength oscillates back to it original value. (D) Computer simulation of the reaction–diffusion wave on a growing one-dimensional array of cells. One of the cells duplicates periodically (once in 100 iterations). The concentration of activator A is represented by the vertical height. For details about the model, see Kondo and Asia (1995). Reprinted by permission from Macmillan Publishers Ltd: *Nature* 376:765–8, copyright © (1995).

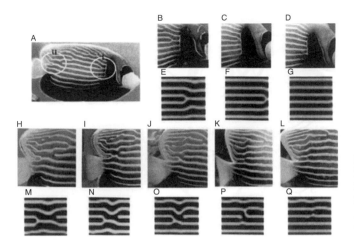

FIGURE 10.4 *Pomacanthus.* Kondo and Asia (1995, fig. 3). Reprinted by permission from Macmillan Publishers Ltd: *Nature* 376:765–8, copyright © (1995).

development stops shortly after melanin has started to accumulate in the hair follicles. After a while, follicles are replaced by new hair follicles, which in their turn die and are replaced by the next generation of hair follicles, and so on. This does not sound very spectacular, but the effect is that the resulting color pattern on the mouse undergoes dramatic changes. Newborn mice are uniformly dark, but after about 1 month, a pattern of traveling dark bands starts to appear, becoming more and more prominent as the mouse grows. Older mice have thinner bands, which travel in wave after wave over the skin. The waves emanate in the area around the armpits and spread from there over the entire animal (Figure 10.5A–D). Traveling waves are well known from oscillating chemical reactions of the reaction–diffusion type predicted by Alan Turing (such as the Beloussov-Zhabotinski reaction, see Figure 10.5E), but these mutant mice are a stunning example of the existence of traveling waves in a vertebrate. How common this mechanism is for creating pigment patterns in animals is unknown. Kondo and his coauthors (Suzuki et al., 2003) point out that a pattern found in many mammals—that hairs divide into alternating dark and light bands—might well be caused by traveling waves of pigment synthesis. If a growing hair is repeatedly struck by a wave of pigment synthesis, the combination of hair growth and the repeated formation

of pigment in the area where hair growth is taking place will lead to exactly the pattern of alternating pigmented and unpigmented bands we can observe in many mammals, ranging from domestic cats to porcupines.

Thus, more and more evidence accumulates that supports the notion that Alan Turing was on the right track when he developed his ideas on pattern formation more than 50 years ago. We now turn to the case of pigment pattern formation in salamanders, where a different formalism, cellular automata, has been used to model pigment pattern formation. However, first is a terse description of what we know about the biological system.

CASE STUDY I • *Development and Evolution of Pigment Patterns in Salamanders*

THE NEURAL CREST AS A MODEL SYSTEM

All pigment cells in vertebrates, except for those in the retina, are derived from a transient embryonic structure called the "neural crest" (NC). The NC cells migrate from an original position on top of the neural tube to many different regions in the embryo, giving rise to a variety of tissues in diverse structures, such as the peripheral nervous system, the head, and the adrenal gland, besides giving rise to pigment cells (see Hall, 1999, and Le Douarin and Kalcheim,

A

B

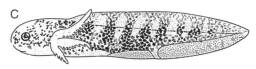

C

FIGURE 10.7 Newly hatched larvae of *Triturus alpestris* (A), *Ambystoma tigrinum tigrinum* (B), and *Ambystoma mexicanum* (C). The drawings are slightly idealized for clarity. From Olsson (1993a).

later do xanthophores leave the aggregates and move ventrally. The melanophores start to form a dorsal and a lateral horizontal stripe. Where xanthophores spread ventrally, melanophores recede to the nearest melanophore region. The xanthophores migrate into the area between the melanophore stripes but do not interfere with the lateral melanophore stripe. In this way a pattern forms with an intact lateral melanophore stripe and a dorsal melanophore stripe interrupted by xanthophore bars. In an elegant study, Parichy (1996a, 1996c) showed that the developing lateral line plays an important role in the formation of this pigment pattern by affecting the melanophores. This also seems to be a trait that is shared between salamandrids and ambystomatids.

THE EVOLUTION OF PATTERNS AND MECHANISMS

The analysis of pigment pattern formation was extended to four more ambystomatids, *Ambystoma talpoideum*, *A. maculatum*, *A. barbouri*, and *A. annulatum* (Olsson, 1993a, 1993b, 1994). Pigment pattern formation in all four species is based on the aggregate formation mechanism described in *A. mexicanum* and *A. tigrinum*. Un-

like *A. mexicanum* and *A. tigrinum*, the aggregates that form in *A. talpoideum* are devoid of melanophores. A major difference compared to those species is that relatively few melanophores migrate on the flank. Only the dorsal horizontal melanophore stripe becomes well developed. Very few melanophores migrate further to reach the area where a lateral melanophore stripe develops in other species. A few remain in the interstripe area. When xanthophores migrate ventrally, they seem to "push" the melanophores in the dorsal melanophore stripe to the sides, creating vertical bars. Gradually, they fill the area below the dorsal melanophore stripe and migrate dorsally into the fin.

A major difference in *A. barbouri* compared to other species is that the number of pigment cells is very large. Most of these cells differentiate into melanophores; i.e., the melanophore-to-xanthophore ratio is high. Melanophores spread out evenly on the whole flank of the embryo, giving it a dark appearance. Xanthophores do not create clear vertical bars as in other species, nor does a distinct interstripe area filled with xanthophores form. In *A. annulatum*, the early NC development closely resembles that in *A. mexicanum*. In an initially homogeneous string of NC cells, aggregates start to develop. At the same time, prospective melanophores spread onto the flank of the embryo. Xanthophores are concentrated to the mixed aggregates and later leave them and migrate ventrally to form vertical xanthophore bars. Xanthophores come to occupy an interstripe area between the dorsal and lateral melanophore stripes and spread up into the dorsal fin. Later, the melanophores change to a rosette form and the end of the tail becomes darker.

In addition to published descriptions of pigment pattern formation, descriptions from the literature of the pigment patterns of larval *A. jeffersonianum*, *A. opacum*, *A. cingulatum*, and *A. texanum* (Bishop, 1941; Brandon, 1961; Orton, 1942) were used. Phylogenetic data on North American ambystomatids reported by Shaffer et al. (1991) and reanalyzed by Jones et al. (1993)

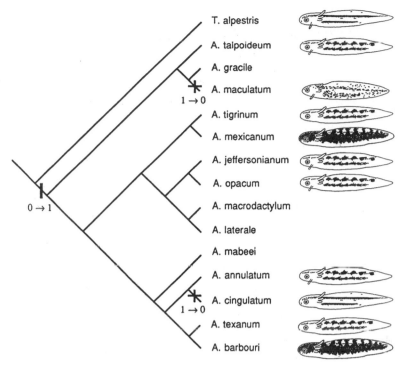

FIGURE 10.8 Evolution of vertical barring in ambystomatid salamanders. Larval pigment patterns are shown for species where they have been investigated. Under this scenario, the differential cell–cell adhesion mechanism that leads to the formation on vertical bars on salamander larvae has evolved once and been lost twice. o = absence of vertical bars, 1 = presence of vertical bars.

provided the phylogeny, and *A. mexicanum* was inserted as a sister taxon to *A. tigrinum* (Shaffer, 1984). The salamandrids were placed as outgroups in accordance with Larson's (1991) rRNA-based family-level phylogeny of salamanders. Using other phylogenies led to similar results.

The pigment patterns were coded as character states: o = horizontal stripes; 1 = vertical bars and horizontal stripes. The character tracing resulted in the pattern presented in Figure 10.8. According to this analysis, the primitive condition is to have horizontal stripes, and vertical bars have evolved only once. Two reversals have taken place, where vertical bars have been lost (indicated by crossed bars in the cladogram). Another equally parsimonious mapping exists, which requires that vertical bars have evolved twice and been lost only once. Both these sce-

narios are rather conservative, indicating that pattern formation mechanisms evolve slowly. The reasons for this can be both internal and external. Early events in development, such as aggregate formation in the NC, are often constrained from changing by the dynamics of the developmental system. Once the aggregate formation mechanism and the vertical bars that it forms have arisen, they can also be constrained from changing by stabilizing selection.

In a more comprehensive study, Parichy (1996b) concluded that the presence of a melanophore-free region is a relatively basal character shared by salamandrids and ambystomatids, whereas the vertical bars are found only in ambystomatids (Figure 10.9). Parichy has also contributed significantly to our understanding of the mechanisms underlying the formation of these patterns (Parichy 1996a, 1996b).

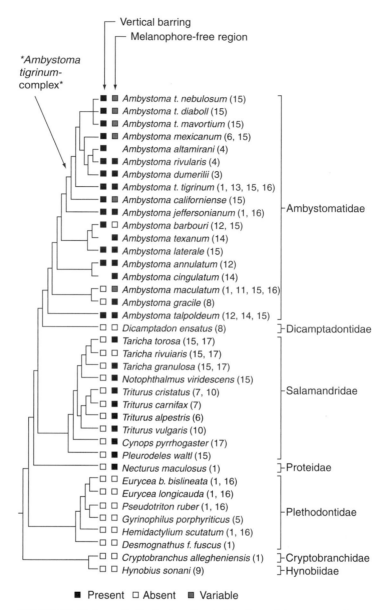

FIGURE 10.9 Phylogenetic distribution of vertical barring and the presence of a melanophore-free region among several urodele families. From Parichy (1996c). Used with the author's permission.

Thus, unlike the examples at the beginning of this chapter, where little is known about the processes actually taking place in the biological system when the pattern forms, here we have a reasonably well-studied pigment pattern formation system that seems promising for a modeling approach. I now turn to a brief description of this work.

A CELLULAR AUTOMATA MODEL FOR SALAMANDER PIGMENT PATTERN FORMATION

Because the empirical work has suggested that salamander pigment patterns form by cell–cell and cell–ECM interactions, models were built that include only those mechanisms. This makes it possible to test how powerful they are as an explanation of the pattern and its formation. An early version was published by Grahn et al. (1994). Andreas Deutsch produced a much more sophisticated model (Deutsch and Dormann, 2005, chap. 9). All models are discrete, based on using cellular automata to model the pigment cells. Using a discrete model to simulate discrete entities like pigment cells is intuitively appealing. Such a model is easy to adapt to what we actually know about the cells and their behavior. This approach is different in principle from the continuous models used in the reaction–diffusion models mentioned earlier but can be used to explain the same type of phenomena. Local interactions are the only "forces" in the model. Cells in the model are equivalent to pigment cells in the animal, and the cells follow simple rules during the simulations. The model was first designed with the Mexican axolotl in mind, the best-studied salamander species.

Based on what we know about the biology of pigment-cell migration in axolotl embryos and larvae, it seems reasonable that a homotypic cell–cell adhesion, stronger between xanthophores than between melanophores, is an important underlying mechanism for several of the processes described empirically. These include chromatophore group formation, melanophores starting to migrate earlier than xanthophores, xanthophores migrating in groups (as opposed to as single cells), and melanophores forced to recede where xanthophores invade.

The model is based on differential cell–cell adhesion, and the strength of adhesion can be set for homotypic (melanophore–melanophore or xanthophore–xanthophore) and heterotypic (melanophore–xanthophore) interactions. Interaction between the cells and the ECM can also be varied. The main result is that differential cell–cell adhesion can in itself explain most aspects of the migration and pattern formation on the flanks on axolotl larvae. The only additional mechanism required to produce a realistic pigment pattern with vertical bars is that the cells must be given a directionality so that they have a higher probability of migrating ventrally than in other directions. Given this, a very realistic pattern and, more importantly, a realistic pattern formation process are seen in the simulations.

A sorting phenomenon in the chromatophore groups before they start to migrate, with melanophores more peripheral than xanthophores, is produced by the differential cell–cell adhesion in the model (stronger between xanthophores than melanophores). Preliminary histological data support this notion. When the xanthophores migrate, a sorting phenomenon has been observed. The melanophores (which have already migrated and spread out on the flank) yield their previous positions to the migrating xanthophores. The model can be used as a link between these kinds of observations and the adhesion hypothesis by giving more detailed predictions than would otherwise be possible. The model is also useful when testing how easy (or difficult) it is to change the pattern in accordance with the evolutionary changes that have occurred during the history of the group under study. This is important for a deeper understanding of how developmental mechanisms constrain and direct evolution.

CASE STUDY II • *Development and Evolution of Zebrafish Pigment Patterns*

Finally, I present a system in which the major drawback of the salamander system, the lack of knowledge about the genetic mechanisms involved in pigment pattern formation, has been overcome. This is the zebrafish, *Danio rerio*, and its close relatives. I refer to *D. rerio* as "the zebrafish" and to the group as "zebrafish" or "danios." The wild-type adult pigment pattern consists of horizontal stripes and is similar to the primitive pattern common to the larvae of salamandrids and ambystomatids. However, in the zebrafish several pigment pattern mutants have been described (Kelsch et al., 1996), and for some of the mutants the gene (or genes) mutated has been identified. Furthermore, experimental embryological studies of pattern formation and regeneration have been performed (e.g., Yamaguchi et al., 2007); and the genetic, embryological, and theoretical approaches are becoming synthesized, rather than just existing in isolation. Here, I briefly describe some aspects of this work but do not aim to give a comprehensive overview. For reviews, see, e.g., Parichy (2003, 2006, 2007) and Rawls et al. (2001).

PIGMENT PATTERN DEVELOPMENT AND DIVERSIFICATION

All zebrafish have the same types of pigment cells as salamanders: yellow xanthophores, black melanophores, and reflecting iridophores (some species also have additional red erythrophores). The larval pigment pattern is formed by the migration of pigment cells from the NC and by their interaction with each other and with surrounding tissues, as in the salamanders. It consists of three horizontal melanophore stripes with xanthophores and iridophores in between. As zebrafish grow and metamorphose into adults, the pattern changes. When metamorphosis starts, melanophores appear outside the larval horizontal stripes, the larval pattern dissolves, and a new pattern, initially made up of two "primary" horizontal melanophore stripes surrounding a light xanthophore–iridophore stripe, forms. With increased size, extra "secondary" melanophore stripes are added until five or six are present. Most melanophores that make up the adult stripes are newly differentiating ones, but a few are larval melanophores that survive metamorphosis and migrate into the adult stripes.

The presence of mutants such as *puma* and *picasso*, which affect only the melanophores that differentiate from precursor cells at metamorphosis, shows that the two melanophore populations are genetically distinct. The larval pigment pattern is unaffected, but later the melanophores that normally differentiate to form the adult pattern fail to do so in these mutants. However, in a species (*D. nigrofasciatus*) with an adult pigment pattern similar to that in *D. rerio*, many more larval pigment cells survive metamorphosis and far fewer melanophores

differentiate at metamorphosis. Thus, the same type of pattern can be made in different ways. A broader comparison indicates that the mode of formation found in *D. rerio* is common among danios, that it is the ancestral mode, and that *D. nigrofasciatus* has a derived mode unique to this species (Quigley et al., 2004).

How are the melanophores organized into adult stripes at metamorphosis? In wild-type zebrafish, the dispersed metamorphic melanophores migrate into the stripe-forming areas; but in the *fms/panther* mutant, they fail to do so and remain disorganized. This mutant also lacks xanthophores completely. Because *fms* is expressed in xanthophores but not in melanophores, it seems clear that xanthophores are required for normal melanophore stripe formation. This is also supported by transplantation experiments (Parichy and Turner, 2003). A change in the *fms* gene may also explain the lack of stripes in the danio species *D. albolineatus*.

Differential cell–cell adhesion is likely to play an important role in pigment pattern formation in zebrafish, as it does in salamanders. What determines the horizontal orientation of the melanophore stripes is less clear. In salamanders, the lateral line plays an important role, and this might be the case also in the zebrafish as indicated by, e.g., the fact that the *puma* mutant has defects in both horizontal melanophore stripe formation and lateral line development.

Evidence that the interaction between xanthophores and melanophores determines the stripe pattern is derived from the *leopard* and *jaguar/obelix* mutants. The *leopard* mutant shows a whole range of pattern defects, from wavy and irregular stripes to a spotted pattern and uniform distribution of pigment cells (Asai et al., 1999). In the *jaguar* mutant, the melanophore stripes are broader and there is an excess of xanthophores, which can also be found in the melanophore stripes. Also, secondary adult stripes do not develop. Data from transplantation experiments and genetic analyses have shown that both *leopard* and *jaguar* are normally expressed in melanophores and promote their

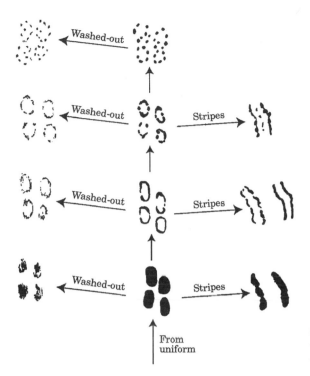

FIGURE 10.10 Weigel's scenario for pigment pattern evolution in felids. From Werdelin and Olsson (1997).

aggregation, also in the absence of xanthophores. *Leopard* also acts in xanthophores and is important for restricting them to areas between the melanophore stripes. These mutants have in common that they destroy the normal sorting out between different pigment cell types. Normal stripe formation is dependent on *jaguar* function for melanophore cell–cell interactions and on the *leopard* gene for both homotypic and heterotypic cell–cell interactions (Maderspacher and Nüsslein-Volhard, 2003).

MODELING PIGMENT PATTERN FORMATION

We have seen earlier in the chapter that the same mathematical model often can account for both striped and spotted patterns. One would predict that transitions between these patterns are easy to achieve in the biological system if they function according to the model. A nice example that supports this view is the *leopard* mutant in the zebrafish (Asai et al., 1999), where different alleles of the *leopard* gene cause different aberrations to the pigment pattern. This could be modeled as an epigenetic phenomenon using a reaction–diffusion model that self-organizes without a prepattern (Asia et al., 1999; Kondo, 2002). By changing the value of a parameter (c) in the mathematical model, pigment patterns could be produced that showed a remarkable similarity to the patterns associated with specific mutant alleles. This suggests that the pattern can be formed by a reaction–diffusion mechanism and that the *leopard* gene product plays a central role in this process.

The development of the late larval and metamorphic patterns has also been modeled using a cellular automata based on differential cell adhesion by Moreira and Deutsch (2005). Their model can also take into account the effects of several mutants, including the *leopard* mutant. Reducing the adhesion between xanthophores in this model mimics the spotted *leopard* phenotype. An advantage with the cellular automata model is that only local interactions between cells need be assumed. There is no need for hypothetical long-range interactions. As we have seen, there is good evidence in zebrafish for a prominent role for cell–cell interactions, such as differential adhesion, being crucial to pigment pattern formation. Evidence for long-range interactions remains elusive in most biological systems, including zebrafish.

CONCLUDING REMARKS

I hope the examples in this chapter have shown convincingly that in order to explain a complex epigenetic phenomenon such as pigment pattern formation, a synthesis of different approaches is necessary. Unfortunately, few model systems have all the desired properties needed for a comprehensive analysis. Instead, different systems have different strengths and weaknesses, and only a comparative approach can be fruitful in producing general explanations for pigment pattern *development*. In order to understand the *evolution* of pigment patterns in different groups of animals, it will be necessary to use approaches such as the one used by zebrafish pigment pattern researchers, i.e., to first use one model species for detailed studies of the genetic and molecular mechanisms underlying pigment pattern formation and to then compare with related species to test, e.g., whether the genes mutated in the model system are the same as those that have changed in the evolution of pigment patterns in the group.

I end with two pleas. The first is that we need more collaboration between experimentalists and theoreticians. In the past, modelers have often taken an interest in systems that have fascinating color patterns, such as the large species of felids (Murray, 1988), but where we do not know anything about how the patterns actually form during embryogenesis. Elegant mathematical models of pigment pattern formation in animals where empirical studies are not possible remain interesting ideas without the possibility of being tested. At the same time, empirical work on pigment pattern development is sometimes conducted without the input from mathematical modelers. Neither situation is ideal.

The second plea is for inclusion of phylogenetic information in studies of the evolution and development of pigment patterns, something that does not come naturally to all theoreticians and developmental biologists. However, for comparative analyses, a phylogenetic perspective is absolutely necessary. An example from the evolution of felid pigment patterns can illustrate this. We have investigated the evolution of pigment patterns in this group by phylogenetic analysis (Werdelin and Olsson, 1997). The hypothesis erected by Weigel (1961), before cladistics became commonly used among systematists, predicts that the ancestral state within felids is a pattern consisting of large spots and that these spots tend to break down during evolution (Figure 10.10). According to this scenario, the large spots first evolve a lighter center and then break down into smaller spots or flecks spaced as rosettes or randomly. At each step in this general decay of the basic pattern, striped patterns may develop (Figure 10.10). However, a phylogenetic perspective shows that this scenario is simply wrong. Mapping the patterns observed in all species of felids onto phylogenies for the group, we could show that the ancestral pattern is not large spots but small spots and the other patterns have evolved from this basic pattern (Werdelin and Olsson, 1997). This demonstrates the importance of incorporating phylogenetic information whenever possible.

ACKNOWLEDGMENTS

I thank Andreas Deutsch (Dresden, Germany) for helping me understand and appreciate the modeling approach and for many interesting discussions over the years. I am grateful to Lars Werdelin (Stockholm, Sweden) for sharing my interest in cats and their color patterns and for his expertise in felid evolution. Dave Parichy, Scott Camazine, and Shigeru Kondo kindly allowed me to use illustrations from their work; and Jan Löfberg and Gunnar Brehm provided drawings. My research on evolutionary developmental biology is supported by the Deutsche Forschungsgemeinschaft.

REFERENCES

Asai, R., E. Taguchi, Y. Kume, M. Saito, and S. Kondo. 1999. Zebrafish Leopard gene as a component of a putative reaction–diffusion system. *Mech Dev* 89:87–92.

Ball, P. 1999. *The Self-Made Tapestry*. Oxford: Oxford University Press.

Bard, J. B. L. 1977. A unity underlying the different zebra striping patterns. *J Zool* 183:527–39.

Bard, J. B. L. 1981. A model for generating aspects of zebra and other mammalian coat patterns. *J Theor Biol* 93:365–85.

Bishop, S. C. 1941. The salamanders of New York. *N Y State Mus Bull* 324:1–365.

Brandon, R. A. 1961. A comparison of the larvae of five northeastern species of *Ambystoma* (Amphibia, Caudata). *Copeia* 1961:377–83.

Camazine, S. 2003. Patterns in nature. *Nat Hist Magazine* 112:34–41.

Camazine, S., J. L. Deneubourg, N. Franks, J. Sneyd, E. Bonabeau, and G. Theraulaz. 2001. *Self-Organization in Biological Systems*. Princeton, NJ: Princeton University Press.

Deutsch, A., and S. Dormann. 2005. *Cellular Automaton Modelling of Biological Pattern Formation*. Boston: Birkhäuser.

Epperlein, H. H., and J. Löfberg. 1984. Xanthophores in chromatophore groups of the premigratory neural crest initiate the pigment pattern of the axolotl larva. *Roux Arch Dev Biol* 193:357–69.

Epperlein, H. H., and J. Löfberg. 1990. The development of the larval pigment patterns in *Triturus alpestris* and *Ambystoma mexicanum*. *Adv Anat Embryol Cell Biol* 118:1–101.

Epperlein, H. H., J. Löfberg, and L. Olsson. 1996. Neural crest cell migration and pigment pattern formation in urodele amphibians. *Int J Dev Biol* 40:229–38.

Gans, C., and G. Northcutt. 1983. Neural crest and the origin of vertebrates: A new head. *Science* 220:268 74.

Gierer, A., and H. Meinhardt. 1972. A theory of biological pattern formation. *Kybernetik* 12:30–9.

Grahn, A., L. Olsson, L. Sundén, A. Deutsch, and J. Löfberg. 1994. Das Wandern ist der Zellen Lust: Pigmentzellen bilden Farbmuster auf Salamanderlarven. In: *Muster des Lebendigen: Faszination ihrer Entstehung und Simulation*, ed. A. Deutsch, 161–82. Braunschweig: Vieweg-Verlag.

Hall, B. K. 1999. *Evolutionary Developmental Biology*. Dordrecht, the Netherlands: Kluwer.

Held, L. I., Jr. 1992. *Models for Embryonic Periodicity*. Basel: Karger.

Jones, T. R., A. G. Kluge, and J. A. Wolf. 1993. When theories and methodologies clash: A phylogenetic reanalysis of the North American ambystomatid salamanders (Caudata: Ambystomatidae). *Syst Biol* 42:92–102.

Kelsh, R. N., M. Brand, Y. Jiang, C. Heisenberg, S. Lin, P. Haffter, J. Odenthal, et al. 1996. Zebrafish pigmentation mutants and the processes of neural crest development. *Development* 123:369–89.

Kondo, S. 2002. The reaction–diffusion system: A mechanism for autonomous pattern formation in the animal skin. *Genes Cells* 7:535–41.

Kondo, S., and R. Asai. 1995. The viable Turing wave on the skin of *P. imperator*. *Nature* 376:765–8.

Larson, A. 1991. A molecular perspective on the evolutionary relationships of the salamander families. *Evol Biol* 25:211–77.

Le Douarin, N. M., and C. Kalcheim. 1999. *The Neural Crest*. Cambridge: Cambridge University Press.

Maderspacher, F., and C. Nüsslein-Volhard. 2003. Formation of the adult pigment pattern in zebrafish requires leopard and obelix dependent cell interactions. *Development* 130:3447–57.

Meinhardt, H. 1982. *Models of Biological Pattern Formation*. London: Academic Press

Meinhardt, H. 2003. *The Algorithmic Beauty of Sea Shells*, 3rd enl. ed. New York: Springer-Verlag.

Moreira, J., and A. Deutsch. 2005. Pigment pattern formation in zebrafish during late larval stages: A model based on local interactions. *Dev Dyn* 232:33–42.

Murray, J. D. 1981. A pre-pattern formation mechanism for animal coat markings. *J Theor Biol* 88:161–99.

Murray, J. D. 1988. How the leopard gets its spots. *Sci Am* 256:62–9.

Northcutt, R. G. 1996. The origin of craniates: Neural crest, neurogenic placodes, and homeobox genes. *Isr J Zool* 42:273–313.

Northcutt, R. G., and C. Gans. 1983. The genesis of neural crest and epidermal placodes. A reinterpretation of vertebrate origins. *Q Rev Biol* 58: 1–28.

Olsson, L. 1993a. Pigment pattern formation in larval ambystomatid salamanders: Evolutionary and mechanistic perspectives. *Acta Univ Upsal* 466:1–48.

Olsson, L. 1993b. Pigment cell migration and pattern formation in the larval salamander *Ambystoma maculatum*. *J Morphol* 215:151–63.

Olsson, L. 1994. Pigment pattern formation in larval ambystomatid salamanders: *Ambystoma talpoideum*, *Ambystoma barbouri* and *Ambystoma annulatum*. *J Morphol* 220:123–38.

Olsson, L., and J. Löfberg. 1992. Pigment pattern formation in larval ambystomatid salamanders: *Ambystoma tigrinum tigrinum*. *J Morphol* 211: 73–85.

Olsson, L., and J. Löfberg. 1993. Pigment cell migration and pattern formation in salamander larvae. In *Oscillations and morphogenesis*, ed. L. Rensing, 453–62. New York: Marcel Dekker.

Orton, G. L. 1942. Notes on the larvae of certain species of *Ambystoma. Copeia* 1942:170–2.

Painter, K. J., P. K. Maini, and H. G. Othmer. 1999. Stripe formation in juvenile *Pomacanthus* explained by a generalized turing mechanism with chemotaxis. *Proc Natl Acad Sci USA* 96:5549–54.

Parichy, D. M. 1996a. Pigment patterns of larval salamanders (Ambystomatidae, Salamandridae): The role of the lateral line sensory system and the evolution of pattern-forming mechanisms. *Dev Biol* 175:265–82.

Parichy, D. M. 1996b. Salamander pigment patterns: How can they be used to study developmental mechanisms and their evolutionary transformation? *Int J Dev Biol* 40:871–84.

Parichy, D. M. 1996c. When neural crest and placodes collide: Interactions between melanophores and the lateral lines that generate stripes in the salamander *Ambystoma tigrinum tigrinum* (Ambystomatidae). *Dev Biol* 175:283–300.

Parichy, D. M. 2003. Pigment patterns: Fish in stripes and spots. *Curr Biol* 13:R947–50.

Parichy, D. M. 2006. Evolution of *Danio* pigment pattern development. *Heredity* 97:200–10.

Parichy, D. M. 2007. Homology and the evolution of novelty during *Danio* adult pigment patterns development. *J Exp Zoolog B Mol Dev Evol* 308:578–90.

Parichy, D. M., and J. M. Turner. 2003. Temporal and cellular requirements for Fms signalling during zebrafish adult pigment pattern development. *Development* 130:817–33.

Quigley, I. K., J. M. Turner, R. J. Nuckels, J. L. Manuel, E. H. Budi, E. L. Macdonald, and D. M. Parichy. 2004. Pigment pattern evolution by differential deployment of neural crest and post-embryonic melanophore lineages in *Danio* fishes. *Development* 131:6053–69.

Rawls, J. F., E. M. Mellgren, and S. L. Johnson. 2001. How the zebrafish gets its stripes. *Dev Biol* 240:301–14.

Shaffer, H. B. 1984. Evolution in a paedomorphic lineage. I. An electrophoretic analysis of the Mexican ambystomatid salamanders. *Evolution* 38:1194–1206.

Shaffer, H. B., J. M. Clark, and F. Kraus. 1991. When molecules and morphology clash: A phylogenetic analysis of the North American ambystomatid salamanders (Caudata: Ambystomatidae). *Syst Zool* 40:284–303.

Suzuki, N., M. Hirata, and S. Kondo. 2003. Traveling stripes on the skin of a mutant mouse. *Proc Natl Acad Sci USA* 100:9680–5.

Turing, A. M. 1952. The chemical basis of morphogenesis. *Philos Trans R Soc Lond* 237:37–72.

Yamaguchi, M., E. Yoshimoto, and S. Kondo. 2007. Pattern regulation in the stripe of zebrafish suggests an underlying dynamic and autonomous mechanism. *Proc Natl Acad Sci USA* 104:4790–3.

Young, D. A. 1984. A local activator–inhibitor model of vertebrate skin patterns. *Math Biosci* 72:51–8.

Werdelin, L., and L. Olsson. 1997. How the leopard got its spots: A phylogenetic view of the evolution of felid coat patterns. *Biol J Linn Soc* 62:383–400.

Weigel, I. 1961. Das Fellmuster der wildlebenden Katzenarten und der Hauskatze in vergleichender und stammesgeschichtlicher Hinsicht. *Säugetierkund Mitteilung* 9:1–120.

Wolpert, L. 2000. *The Unnatural Nature of Science.* Cambridge, MA: Harvard University Press.

11

Epigenetic Interactions of the Cardiac Neural Crest

Martha Alonzo, Kathleen K. Smith, and Margaret L. Kirby

CONTENTS

Epigenetic interactions as envisioned by Waddington involved animal, tissue, or cell responses to environmental cues such as hormones, cell–cell interactions, and mechanical and electrical forces. These mechanisms shape the "landscape" a cell travels through to reach its final differentiated state. In the case of neural crest cells, which migrate from their origin in the dorsal neural tube to distant target sites, there is a signaling landscape in addition to a changing geographic landscape. The signaling landscape includes both received signals and emitted signals. To fully appreciate the epigenetic interactions of the cardiac neural crest, it is first important to understand the geographic landscape through which these cells migrate and the tissues that they either target or influence, or are influenced by, on their journey. This information allows us to understand the "historical landscape" that shapes their final disposition as well as the cells that they influence. In the case of the cardiac crest, the geographical landscape includes the caudal pharynx and the cardiac outflow tract. In the caudal pharynx, cardiac crest cells are necessary for repatterning the aortic arch arteries to the great arteries and for modulating signaling between the pharyngeal endoderm–ectoderm and cardiac outflow progenitors in the caudal pharynx. This modulation of signaling is essential for addition of myocardial cells from the ventral caudal pharynx to the outflow tract. Finally, cardiac crest cells migrate into the outflow tract and orchestrate outflow tract septation. We will first discuss the migratory path of the cardiac crest and then the potential epigenetic interactions in more detail.

CARDIAC NEURAL CREST

Neural crest cells originate from the border between the neural and surface ectoderm along the rostrocaudal extent of vertebrate embryos. Generally, the crest originating cranial to somite 5 is called the "cranial neural crest," while the crest originating caudal to somite 5 is known as the "trunk neural crest" (Le Lievre and Le Douarin 1975). The cranial crest is capable of generating ectomesenchyme in addition to the neural derivatives and melanocytes generated by all crest cells. The subpopulation of cranial neural crest cells that originates between the mid-otic placode and somite 3 contributes to pharyngeal and cardiac structural development and, because of its unique importance in cardiac development, has been called the "cardiac neural crest" (Kirby et al., 1983; Kirby and Stewart, 1983).

In chick and mouse, the cardiac crest migrates as the most caudal of the three cranial streams of the neural crest that originate from the rhombencephalon (Figure 11.1). The post-otic stream extends to somite 5, and the cardiac neural crest is found within this stream. Quail-to-chick chimeras showed that the cardiac neural crest originated between the mid-otic placode and somite 3 and, thus, represents at least part of the post-otic stream. Some of the crest cells originating from the mid-otic placode to somite 3 migrate to the gut and elsewhere, so the term *cardiac crest* refers only to the cells that initially populate pharyngeal arches 3–6. Transplantation and focal cell–labeling studies in mouse showed that the cardiac crest in the mouse originates between the post-otic hindbrain and somite 4 (Chan et al., 2004). The neural tube opposite somite 2 is the largest source of cardiac crest cells.

The three streams of the cranial crest are conserved across all vertebrates, and this caudal stream can be seen in animals as diverse as the lamprey (McCauley and Bronner-Fraser, 2003), Australian lungfish (Ericsson et al., 2008), Mexican axolotl (Ericsson et al., 2004), zebrafish (Eisen and Weston, 1993), mouse (Serbedzija et al., 1992), opossum (Vaglia and Smith, 2003), and chick (Kuratani and Kirby, 1992). It is not known whether the caudal stream in all vertebrates contains a cardiac crest as described in the chick and mouse, although the cardiac crest has been described in both zebrafish and frog embryos (Li et al., 2003; Martinsen et al., 2004; Sato and Yost, 2003).

The cardiac crest in chick and mouse embryos migrates through and populates the caudal pharyngeal region. In the chick, the cells emigrate first at stage 11 or approximately the 12–14-somite stage (Nishibatake et al., 1987), while in the mouse emigration begins at the seven-somite stage (Chan et al., 2004). In the chick, the cells pause in the circumpharyngeal ridge and reinitiate migration into each arch as the arch is formed by the endodermal pouches (Kuratani and Kirby, 1991; Waldo et al., 1996). The cardiac crest cells form a dense sheath around the nascent aortic arch arteries as they develop (Jiang et al., 2000; Kuratani and Kirby, 1992). Crest migration in the mouse appears to be more fluid as the cells seem to migrate without pause into the pharyngeal arches. The timing of cardiac crest migration into the cardiac outflow tract is somewhat different in the chick and mouse. With direct marking of the crest in mouse embryos, Chan et al. (2004) did not see cells in the outflow tract until embryonic day E10.25–E10.5 (32 somites), although cardiac crest cells have been reported in the aortic sac at E9.5 (20–25 somites) using transgenesis and Wnt1-cre recombinase (Jiang et al., 2000; Waldo et al., 1999). In the chick, the outflow tract is not populated by crest cells until later, stage 22 (43–48 somites) (Waldo et al., 1998, 1999).

Quail-chick chimeras show that the cardiac crest also contributes mesenchyme to the endodermal buds that will form the thyroid, parathyroid, and thymus. Indeed, ablation of the premigratory cardiac crest results in aplasia or

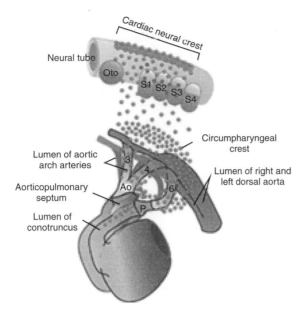

FIGURE 11.1 Schematic of the cardiac neural crest in the chick and mouse. The cardiac crest originates from the dorsal neural tube within the post-otic population. Cells migrate into arches 3, 4, and 6 and contribute to the tunica media of the aortic arch arteries. A subpopulation of cells form the aorticopulmonary septation complex and contribute to cardiac ganglia. Reprinted from Kirby, 2007, with permission.

hypoplasia of these glands, indicating the necessity of these cells for gland development (Bockman and Kirby, 1984). After the developmental period of these glands, the neural crest–derived mesenchyme contributes connective tissue, and it has also been reported that neural crest–derived cells form pericytes and smooth muscle cells in the thymus (Foster et al., 2008). While the cell types adopted by the crest in adult parathyroid and thyroid glands have not been reported, they are likely to be the same.

After populating the caudal arches, the crest cells proliferate (Waldo et al., 1996), and a subpopulation of the cardiac crest cells migrates from the pharynx into the arterial pole of the heart (Kuratani and Kirby, 1991). In the cardiac arterial pole, neural crest–derived ectomesenchymal cells participate in structural development of the septum between the aortic and pulmonary outlets. The neural component of the cardiac crest contributes to the parasympathetic cardiac ganglia, which become the dominant autonomic innervation of the heart and are tonically active to slow the intrinsic beat rate in most species.

Ablation of the premigratory cardiac crest in the chick results in a constellation of defects that reflect the roles of these cells in repatterning the aortic arch arteries to the great arteries, support of pharyngeal gland development (thyroid, thymus, and parathyroid), and septation of the cardiac outflow tract. Most relevant for cardiovascular development is that ablated embryos show abnormal patterning of the great arteries, with stochastic persistence or disappearance of these vessels yielding unpredictable patterns. In addition to this, the outflow tract remains unseptated as a single common outflow vessel rather than separated aorta and pulmonary trunks. As mentioned above, additional defects after ablation include hypoplasia or aplasia of the thyroid, parathyroid, and thymus glands. All of the defects seen with mechanical ablation of the cardiac neural crest in the chick are seen in various neural crest conditional gene deletions in the mouse.

While the cardiac neural crest has been best studied in the chick and mouse, recent studies have addressed the existence of the cardiac crest in other species, such as zebrafish and frogs, that do not have septated outflow tracts (Li et al., 2003; Martinsen et al., 2004; Sato and Yost, 2003). Carefully timed studies have not been done in these species to document cardiac crest cells in the branchial region and heart of these species, although the initial migratory streams appear similar in these and other more primitive vertebrates (McCauley and Bronner-Fraser, 2003; Schilling and Kimmel, 1994). Even so, it is unclear what relationship, if any, the caudal rhombencephalic migratory stream has with the cardiac crest in these species. While fish have an undivided outflow tract, they do have remodeling of the symmetrical aortic arch arteries into bilaterally symmetrical gill arteries (Figure 11.2) and we can only speculate that the cardiac crest plays a similar role in the remodeling and formation of the smooth muscle tunics. Because fish have permanent branchial arches that are remodeled to gills and amphibians go through a gill stage, it is unclear what role the cardiac crest might play in the formation of these structures. The branchial arch arteries in fish are remodeled to afferent and efferent arterioles in the gills, and the cardiac crest could be involved in this remodeling.

The *Xenopus* cardiac neural crest was postulated to exist via expression of the *id* gene found in the heart as well as in the neural tube. The Id genes are negative regulators of the basic helix–loop–helix products, containing the helix–loop–helix domain but lacking the DNA-binding domain (Massari and Murre, 2000). In the chick, the *ID* genes are found in the cardiac neural crest, the secondary heart field, and cardiac structures. Ablation of the cardiac neural crest in the chick results in a downregulation of *ID* in the outflow tract and atrioventricular cushions. *Xenopus* expression of the *xid* gene was also found in heart structures and neural tube, among other sites of expression in the embryo. When the neural crest was ablated in the *Xenopus* embryo, there was an elongated and unlooped heart and lost expression of *id* in the inflow and outflow portions of the heart as well as the splanchnic mesoderm and myocardium (Martinsen et al., 2004). The cardiac neural crest was not directly traced in *Xenopus*, but the ablation model suggests that the neural crest may indeed play a role in *Xenopus* cardiac development.

The zebrafish cardiac neural crest was also discovered by a combination of cell tracing and ablation (Li et al., 2003; Sato and Yost, 2003). Cells traced from the neural tube contributed to regions of the heart, but one unique feature of the cells traced was that they contributed to myocardium. Ablation of the neural crest population led to an elongated unlooped heart, similar to what was found in the *Xenopus* model of ablation (Li et al., 2003; Martinsen et al., 2004). Further studies in the zebrafish have suggested that cardiac neural crest cells migrate directly into the heart fields via Semaphorin signaling, contributing to the myocardial cell fate (Sato et al., 2006). A novel cardiac neural crest cell marker in the zebrafish, Crip2 (cysteine-rich intestinal protein 2), marks a region of the premigratory cardiac neural crest (Sun et al., 2008). Crip2 is a LIM domain–containing protein and has been shown to be under the control of wnt3a, which also regulates the expression of other genes in the neural tube. Knockdown of Crip2 appears to be more specific, affecting the pharyngeal arches and the heart and not the more cranial arches that give rise to the mandible. *crip2* mRNA can also partially rescue some pharyngeal arch and cardiac function in Wnt3a knockdown (Sun et al., 2008). The cardiac neural crest in both models may be acting differently since the neural crest cells do not have to septate the cardiac arterial pole in these two species, but more work needs to be done to verify whether the contribution of the neural crest is a physical interaction or if neural crest cells indirectly modulate signaling that may affect the heart.

Thus, our primary knowledge about the geographic and signaling landscape of the cardiac crest is largely based on mouse and chick studies. From these studies we know that the direct roles of cardiac crest cells in cardiovascular development are (1) to support normal repatterning of the bilaterally paired aortic arch arteries to the asymmetrical great arteries; (2) to form the tunica media of these vessels; (3) to support development of the thymus, parathyroid, and thyroid glands; and (4) to participate in septation of the aortic and pulmonary outflow. In addition to these direct roles in development of

A

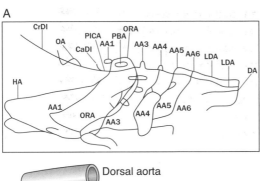

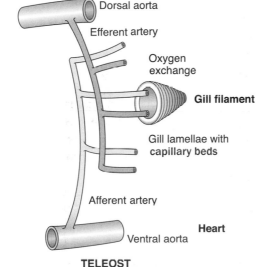

TELEOST

B

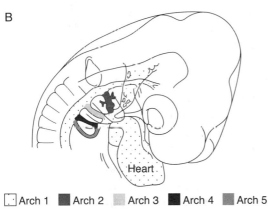

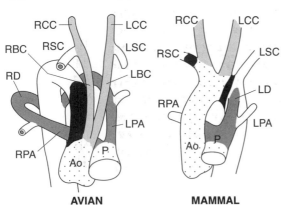

AVIAN   **MAMMAL**

FIGURE 11.2 Aortic arch artery patterns in teleost fish, avians, and mammals. (A) Top panel is a schematic diagram of the zebrafish vasculature at 3–3.5 days postfertilization from Isogai et al. (2001). AA3–AA6 will be remodeled into the afferent and efferent gill arch arteries. Bottom panel is a schematic of the remodeled afferent and efferent gill arch arteries.

(B) Schematic of avian and mammalian aortic arch arteries (top) with the remodeled great arteries (bottom). Both systems begin with bilaterally symmetrical arches. In the avian, the definitive aortic arch is on the right and in the mammalian system the aorta is left-sided. AA1–AA6, aortic arch arteries 1–6; BA, basilar artery; CA, caudal artery; CCV, common cardinal vein; CaDI, caudal division of the internal carotid artery; CrDI, cranial division of the internal carotid artery; CtA, central artery; CV, caudal vein; DA, dorsal aorta; DLAV, dorsal longitudinal anastomotic vessel; DLV, dorsal longitudinal vein; H, heart; HA, hypobranchial artery, IOC 27, inner optic circle; LDA, lateral dorsal aorta; MsV, mesencephalic artery; NCA, nasal ciliary artery; OA, optic artery; ORA, opercular artery; OV, optic vein; PBA, pseudobranchial artery; PCV, posterior (caudal) cardinal vein; PHBC, primordial hindbrain channel; PHS, primary head sinus; PICA, primitive internal carotid artery; PrA, prosencephalic artery; Se, intersegmental vessel; SIA, supraintestinal artery; SIV, subintestinal vein; VA, ventral aorta; Ao, aorta; LBC, left brachiocephalic; LCC, left common carotid; LD, left ductus; LPA, left pulmonary artery; LSC, left subclavian; RBC, right brachiocephalic; RCC, right common carotid; RD, right ductus; RPA, right pulmonary artery; RSC, right subclavian.

the pharynx, a new but indirect role of the cardiac crest in the pharynx has been proposed in the chick, which will be addressed in the next section.

EPIGENETIC INTERACTIONS OF CARDIAC NEURAL CREST CELLS

Three epigenetic interactions of cardiac neural crest cells can be clearly delineated. In the first, cardiac neural crest cells interact with the aortic arch arteries and are necessary for normal repatterning of these bilaterally symmetrical vessels to the asymmetrical great arteries. The second epigenetic role of cardiac crest cells in the caudal pharynx has been recognized only recently, and it is to modulate fibroblast growth factor 8 (FGF8) signaling. This is a significant role because of the extended period of time over which FGF8-sensitive myocardial progenitors are added to the myocardium of the outflow tract of the heart. Without addition of these myocardial cells, significant defects occur in outflow development. Finally, the cardiac crest cells are essential to orchestrate septation of the outflow tract.

CARDIAC NEURAL CREST AS AN EPIGENETIC FACTOR IN AORTIC ARCH ARTERY REMODELING

The aortic arch arteries develop within the pharyngeal arches and are formed as bilaterally symmetrical vascular conduits that connect the outflow (arterial pole) of the heart with the dorsal aorta. In tetrapods, some arch arteries regress and others remodel to form asymmetrical great arteries. In fish, the gill arch arteries remain bilaterally symmetrical and undergo remodeling to afferent and efferent arteries (Figure 11.2) (Goodrich, 1930). It is in the remodeling events of the aortic arch arteries that the most extensive studies have been done concerning the role of the neural crest in the mouse and chick (Figure 11.2).

In both the chick and mouse the cardiac crest cells in arches 3, 4, and 6 form a dense sheath around the endothelium of the forming symmetrical aortic arch arteries (Jiang et al., 2000; Kuratani and Kirby, 1991, 1992). The symmetrical aortic arch arteries are remodeled into the asymmetrical great arteries. After this remodeling is complete, the surrounding crest cells differentiate into the smooth muscle tunica media of the great arteries. The aortic arch arteries are capable of developing without the cardiac neural crest but need the cardiac neural crest in order to maintain and properly remodel the arch arteries into the great arteries (Waldo et al., 1996). The aortic arch artery patterning likely depends on the Hox code expressed in the surrounding neural crest cells because disruption of Hox expression in the cardiac crest cells also causes abnormal patterning of the great arteries even though the crest cells populate the arches normally (Kirby et al., 1997). Despite abnormal patterning in the arches due to Hox code disruption, the cardiac crest can migrate into the cardiac outflow tract and form a normal outflow septum (Kirby et al., 1997), suggesting that normal development of the great arteries is not necessary for normal outflow septation.

Conversely, cardiac neural crest cells are influenced by many signaling pathways that may help to direct them toward the caudal pharynx. The endothelin receptor endothelin-A (ET_A) is one such receptor that is expressed in migrating neural crest cells. One of its ligands, endothelin-1 (ET1), is expressed in the endothelial cells that form the aortic arch arteries (Kurihara et al., 1999; Yanagisawa et al., 1998a). Mice lacking ET_A have aberrant aortic arch artery patterns (Clouthier et al., 1998; Yanagisawa et al., 1998b). In chimeric embryos with mixed populations of wild-type and ET_A null cardiac neural crest cells, the ET_A null cells are excluded from the pharyngeal arches, suggesting that endothelin signaling is needed for neural crest cells to migrate properly toward the aortic arch arteries (Clouthier et al., 2003). A chimera with an increased population of mutant neural crest cells had a ventricular septal defect and a right-sided aortic arch, common in embryos mutant for ET_A. This may suggest a need for endothelin signaling once the cardiac neural crest has populated the arches. If so, then endothelin signaling may provide epigenetic information to crest cells that is important for both migration into the arches and arch repatterning.

Hand2 (dHand) expression by cardiac neural crest–derived structures is induced by ET1 signaling and necessary for normal patterning of the aortic arch arteries (Thomas et al. 1998). HAND2 is a helix–loop–helix transcription factor expressed by pharyngeal neural crest–cell derivatives (Srivastava et al., 1995). Mouse knockouts of Hand2 have shown that the aortic arch arteries fail to pattern correctly (Srivastava et al., 1997). Expression of both Hand2 and Hand1 (eHand) is found in the neural crest–derived mesenchyme just underneath the most exterior portion of the arches (Thomas et al., 1998), but these genes are not expressed in migrating neural crest cells. ET1 cells are adjacent to the Hand-expressing mesenchyme, and in Et1 null mice both Hand1 and Hand2 are significantly reduced (Thomas et al., 1998). Unlike the Et_A null, the neural crest cells in Hand2 null mice still migrate to the pharyngeal arches as evidenced by the expression of Msx1 and -2, homeobox genes expressed in cranial neural crest ectomesenchymal cells (Davidson, 1995). Mice that are mutant for Hand2 do not express Msx1, unlike in ET1 null embryos. HAND2 may therefore be important for maintenance of the neural crest–cell population once it arrives and differentiates in the pharyngeal arches.

The transforming growth factor-β (TGFβ) superfamily has also been implicated in the formation of the aortic arch arteries and could also be a factor in the remodeling of the aortic arch arteries. ALK5, a TGFβ type I receptor, is expressed by cardiac neural crest cells and required for normal patterning of the pharyngeal arch arteries (Wang et al., 2006). Cardiac neural crest cells mutant for Alk5 maintain their migration pattern and populate the aortic arch arteries and aortic sac. However, the aortic arch arteries fail to repattern correctly as the great arteries, suggesting that ALK5 could be involved in the downstream action of remodeling the aortic arch arteries.

Ablation of premigratory cardiac neural crest cells is associated with migration of cells derived from the nodose placode into all of the unoccupied cardiac neural crest pathways and target sites. The nodose placode cells populate all of the target sites of the cardiac crest including the cardiac outflow tract, and they appear to form ectomesenchyme and neural derivatives that are similar to the cardiac neural crest derivatives (Kirby, 1988a, 1988b). However, the ectomesenchyme formed from the nodose placode is not able to orchestrate outflow septation or support normal patterning of the great arteries from the aortic arch arteries, although the neural derivatives of the nodose placode do form a functional cardiac plexus (Kirby et al., 1989). This suggests that cardiac neural crest–derived ectomesenchyme is not replaceable by just any ectomesenchyme but is specialized for aortic arch repatterning and outflow septation. It is not known how the cardiac crest cells influence their migratory pathways to make them inaccessible to migrating nodose placode cells. The neural crest is dependent on particular substrates for migration, and one study shows that substrate-dependent integrin recycling is essential for rapid cranial neural crest migration (Strachan and Condic, 2004). While this shows a potential mechanism for neural crest alteration of its migratory substrates, no evidence currently exists that this happens.

CARDIAC CREST CELLS MODULATE FGF8 SIGNALING IN THE PHARYNX TO REGULATE ARTERIAL POLE DEVELOPMENT

A major function of neural crest cells in the pharyngeal arches is to modulate signaling. It has become apparent recently that this is a critical role for the cardiac crest in supporting normal arterial pole development. Our understanding of early heart development has changed dramatically over the last decade, which has served to both complicate and expand our understanding of the functional role of the cardiac crest in heart development. Previously, it was thought that all of the cells that comprise the myocardium and endocardium of the heart were contained in the initial heart tube (Rosenquist and DeHaan, 1966; Stalsberg and DeHaan, 1969). In this view, the heart tube was thought to be later invaded by extracardiac populations like the cardiac crest and epicardium, without the addition of more myocardium. However, we now know that the myocardium and endocardium that initially form the heart tube are supplemented over an extended period of development by additional myocardial and endocardial cells added to the poles of the heart tube (Abu-Issa and Kirby, 2007). The population of added cells has been given various names but is currently designated "second heart field" (Abu-Issa and Kirby, 2008). A substantial number of cells are added to both the arterial (outflow) pole and the venous (inflow) pole of the heart tube. Cells added to the venous pole contribute to atrial development including the free walls and atrial septation (Galli et al., 2008; Snarr et al., 2007), and there is no current evidence that the cardiac crest influences development of the venous pole.

Cells added to the arterial pole from the second heart field include the right ventricle, proximal and distal outflow myocardium, and smooth muscle at the base of the arterial trunks (Kelly et al., 2001; Waldo et al., 2005a). Cells that will become the right ventricle and proximal part of the outflow myocardium appear to be added from mesoderm located in the position of future pharyngeal arches 1 and 2 before these arches are actually formed, and it is unclear whether neural crest cells in these arches interact with these myocardial progenitors (Kelly et al., 2001; Abu-Issa and Kirby, unpublished observations). The distal outflow myocardium and proximal smooth muscle of the arterial pole are added from second heart field cells that are located in the ventral splanchnic mesoderm of the pharynx underlying pharyngeal arches 3, 4, and 6 (Waldo et al., 2001, 2005a, 2005b). This region of splanchnic mesoderm has also been called "secondary heart field" and represents the final subset of cells added to the arterial pole.

Recent data indicate that these populations of myocardial and endocardial cells are also added to the heart tube over an extended period of time in zebrafish and frogs (Hami et al., 2010; Brade et al., 2007).

Of particular interest is evidence that the cardiac crest in the caudal pharynx decreases FGF8 signaling (Farrell et al., 2001; Hutson et al., 2006). The endoderm is a rich source of growth factors which have a continuing effect on heart development throughout the early embryonic period (Schultheiss et al., 1995). Several factors are produced by the endoderm as cardiac crest cells migrate into the caudal pharynx. These include sonic hedgehog (Shh) expressed in the ventral midline endoderm cells and FGF8, expressed by lateral endoderm and ectoderm (Farrell et al., 2001; Moore-Scott and Manley, 2005). The secondary heart field progenitors of the outflow myocardium located in the caudal ventral pharynx are added to the heart tube during the period that cardiac neural crest cells populate this region. The secondary heart field adds first myocardium and then smooth muscle to the arterial pole, where the outflow myocardium abuts the smooth muscle tunics at the base of the aorta and pulmonary trunk (Waldo et al., 2005a).

During the time that secondary heart field adds myocardium to the arterial pole of the heart (stages 14–18), neural crest cells are beginning their migration from the circumpharyngeal region into the caudal pharyngeal arches; and it is during this period that abnormal cardiac development first appears in neural crest–ablated embryos (Figure 11.3) (Leatherbury et al., 1990; Waldo et al., 1996). When the cardiac neural crest population is ablated, the secondary heart field fails to add myocardium to the outflow tract (Waldo et al., 2005b; Yelbuz et al., 2002). Thus, although the neural crest does not directly interact with the myocardium of the secondary heart field, it is necessary for its normal development.

Because the cardiac crest is not in direct contact with secondary heart field mesoderm when arterial pole development goes awry, it is clear that the interaction of the two populations is through an intermediary, which has been

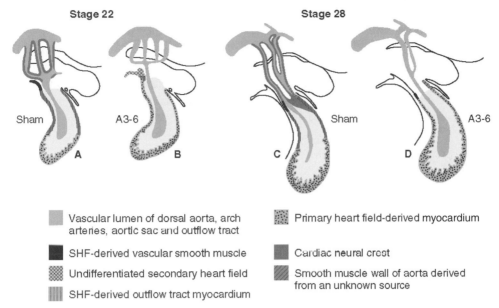

Stage 22

Sham A3-6

Stage 28

Sham A3-6

A B C D

Vascular lumen of dorsal aorta, arch arteries, aortic sac and outflow tract

SHF-derived vascular smooth muscle

Undifferentiated secondary heart field

SHF-derived outflow tract myocardium

Primary heart field-derived myocardium

Cardiac neural crest

Smooth muscle wall of aorta derived from an unknown source

FIGURE 11.3 Addition of secondary heart field to the heart. (A, B, C, D) Sagittal sections of the arterial pole and adjoining secondary heart field in the embryonic chick. Myocardium from this field of progenitors migrates from the secondary heart field into the outflow tract. (A) A sham embryo with normal myocardial migration from the secondary heart field to the outflow. (B) A neural crest–ablated embryo where the myocardial cells fail to migrate. (C) A schematic of the addition of secondary heart field cells to the arterial pole in sham and (D) neural crest–ablated embryos at stages 22 and 26.

identified as FGF8. FGF8 is produced by the pharyngeal endoderm and ectoderm. Cardiac neural crest–ablated chick embryos display depressed myocardial Ca²⁺ transients (Creazzo et al., 1998), which could be rescued by the addition of FGF8b-neutralizing antibody, suggesting an elevation of FGF8 signaling after neural crest ablation (Farrell et al., 2001). Indeed, it was later shown that FGF8 signaling was increased in the caudal pharynx after neural crest ablation because both FGF8 signaling and downstream target genes were elevated (Figure 11.4) (Hutson et al., 2006). The increase in downstream targets of FGF8 coincided temporally with the addition of myocardium by the secondary heart field. The addition of this cell population could be rescued by blocking FGF8 signaling in neural crest–ablated embryos (Hutson et al., 2006). Indeed, blocking FGF8 signaling in neural crest–ablated embryos restored the myocardial calcium transient to normal, as well as restoring proper alignment of the arterial pole (Hutson et al., 2006). These experi-

ments show that cardiac neural crest cells in the pharynx regulate FGF8 signaling, which directly affects the formation of the arterial pole of the heart. How the neural crest cells regulate the FGF8 signal in the pharynx is not yet understood.

CARDIAC NEURAL CREST CELLS AS AN EPIGENETIC FACTOR IN OUTFLOW TRACT SEPTATION

A subpopulation of cardiac neural crest cells located in the caudal pharynx enter the heart through the cardiac outflow cushions (Phillips et al., 1987). The cardiac cushions of the outflow tract are composed of a proximal pair, which is closer to the ventricle, and a distal pair, which is closer to the pharynx. The proximal cushions are populated largely by mesenchyme derived from the endocardium via epithelial–mesenchymal transformation (Park et al., 2008). The distal outflow cushions are populated by neural crest and other mesenchymal cells that migrate into the distal outflow tract

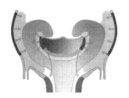
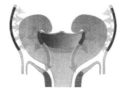

Normal   Cardiac neural
                crest-ablated

FIGURE 11.4 Schematic of a section through the caudal pharynx of a chick showing how the cardiac crest modulates fibroblast growth factor 8 (FGF8) signaling. In the normal panel, FGF8 (dark gray) is produced by the pharyngeal endoderm and ectoderm and neural crest–derived ectomesenchymal cells (light gray) are in close proximity to the areas of FGF8 signaling. When the cardiac neural crest is ablated, right panel, FGF8 signaling increases in the area, as depicted with larger arrows.

from the pharynx (Dodou et al., 2004; Waldo et al., 1998, 1999). The cardiac neural crest cells that populate the distal outflow tract cushions form a condensed mesenchymal structure called the "aorticopulmonary septation complex," which consists of a bridge or shelf that joins two prongs extending into the distal cushions (Waldo et al., 1998). The bridge of neural crest cells crosses the outflow tract between the fourth and sixth arch arteries. Once the septation complex is in place, the shelf of neural crest cells elongates distally into the outflow tract along the length of the prongs. Since the prongs have formed following the spiraling distal outflow cushions, the shelf also spirals as it elongates. Thus, the distal outflow septum is comprised entirely of cells derived from the cardiac neural crest. This spiraling formation is important for the proper alignment of the aorta and pulmonary vessels with the left and right ventricles, respectively (Figure 11.5).

The proximal cushions close in a different manner that does not involve condensed mesenchyme (Waldo et al., 1998). The proximal cushions are invaded by myocardial cells, which cause them to bulge toward each other. As the cushions bulge, the endocardium underlying each cushion touches and subsequently begins to break down, allowing the mixing of myocardial and mesenchymal cells to form a septum.

When the septum is formed, neural crest cells can be seen at the seam where the fusion of the cushions occurred due to their position underneath the endocardium that fused together. While most of the details of outflow septation are known from quail-to-chick chimeras, studies of the aorticopulmonary septation complex in mammals have been made possible due to the creation of Cre recombinant lines such as Cx43-cre, Wnt1-cres, Pax3-cre, Po-cre, and PlexinA2-cre (Brown et al., 2001; Jiang et al., 2000; Lee et al., 1997; Lo et al., 1997); and these studies suggest that the details of mouse outflow tract septation are essentially the same, with some differences in timing.

Although endocardial cushion formation does not require the presence of neural crest cells, the septation of the outflow tract cannot be accomplished without cardiac crest cells. Many of the same signaling pathways that are involved in aortic arch artery development are also important in septation, but often due to defects occurring along the pathway of cardiac neural crest cell migration, outflow septation defects can seem secondary. Yet there are signals that are involved at the time of cardiac neural crest cell migration and its interaction with cardiogenic fields that require specific signaling to occur.

Meltrin β, a member of the ADAM (a disintregin and metalloprotease) family of proteins, has been shown to be important for the proper septation of the outflow tract and may be a factor in neural crest patterning (Kurohara et al., 2004). When mice mutant for Meltrin β were observed, they appeared to have normal epithelial-to-mesenchymal transition in endothelial cells and cardiac neural crest migration (Komatsu et al., 2007); but although these processes appeared normal, ventricular septal defect (VSD) and defective pulmonary and tricuspid valves were found (Kurohara et al., 2004; Zhou et al., 2004). When Meltrin β was expressed in neural crest cells using Cre-Po and Tie2-Cre, respectively, in a Meltrin β mutant background, VSD and to some extent the pulmonary and tricuspid valves were rescued. The expression of

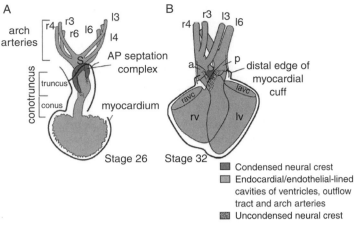

FIGURE 11.5 Schematic of the aorticopulmonary (AP) septation complex (A) and the resulting aorticopulmonary septum (B) by the cardiac neural crest. (A) The cardiac neural crest is condensed mesenchyme that forms a horseshoe-shaped complex at stage 26 located in the cardiac cushions in the distal outflow tract. (B) At stage 32 the aorta and pulmonary vessels are septated by a solid neural crest–derived septum derived from the aorticopulmonary septation complex. Proximal to this septum, neural crest cells (stippled gray) can be seen in the seam where the fusion of the proximal cushions occurred. Reprinted from Waldo et al., 1998, with permission.

Meltrin β in the endocardial epithelial cells did not rescue VSD or outflow valve formation. This can offer some molecular evidence of neural crest importance in septation of the outflow.

Cardiac neural crest cells can have epigenetic actions by having physical contact with the tissues they interact with, thus changing the environment in which they travel. Signaling conducted through the cardiac neural crest can affect the population it targets such as the aortic arch arteries and cardiac cushions. The addition of the secondary heart field to the outflow tract coincides with the cardiac neural crest entering the caudal pharynx, and its presence in the pharynx is needed to create the right amount of signaling for this population to add properly. The cardiac neural crest provides developmental cues and initiates signaling pathways that lead to further remodeling and maintenance of tissues.

REFERENCES

Abu-Issa, R., and M. L. Kirby. 2007. Heart field: From mesoderm to heart tube. *Annu Rev Cell Dev Biol* 23:45–68.

Abu-Issa, R., and M. L. Kirby. 2008. Patterning of the heart field in the chick. *Dev Biol* 319:223–33.

Bockman, D. E., and M. L. Kirby. 1984. Dependence of thymus development on derivatives of the neural crest. *Science* 223:498–500.

Brade, T., S. Gessert, M. Kuhl, and P. Pandur. 2007. The amphibian second heart field: *Xenopus* islet-1 is required for cardiovascular development. *Dev Biol* 311:297–310.

Brown, C. B., L. Feiner, M. M. Lu, J. Li, X. K. Ma, A. L. Webber, L. Jia, J. A. Raper, and J. A. Epstein. 2001. PlexinA2 and semaphorin signaling during cardiac neural crest development. *Development* 128:3071–80.

Chan, W. Y., C. S. Cheung, K. M. Yung, and A. J. Copp. 2004. Cardiac neural crest of the mouse embryo: Axial level of origin, migratory pathway and cell autonomy of the splotch (Sp2H) mutant effect. *Development* 131:3367–79.

Clouthier, D. E., K. Hosoda, J. A. Richardson, S. C. Williams, H. Yanagisawa, T. Kuwaki, M. Kumada, R. E. Hammer, and M. Yanagisawa. 1998. Cranial and cardiac neural crest defects in endothelin-A receptor–deficient mice. *Development* 125:813–24.

Clouthier, D. E., S. C. Williams, R. E. Hammer, J. A. Richardson, and M. Yanagisawa. 2003. Cell-autonomous and nonautonomous actions of endothelin-A receptor signaling in craniofacial and cardiovascular development. *Dev Biol* 261:506–19.

Creazzo, T. L., R. E. Godt, L. Leatherbury, S. J. Conway, and M. L. Kirby. 1998. Role of cardiac neural crest cells in cardiovascular development. *Annu Rev Physiol* 60:267–86.

Davidson, D. 1995. The function and evolution of Msx genes: Pointers and paradoxes. *Trends Genet* 11:405–11.

Dodou, E., M. P. Verzi, J. P. Anderson, S.-M. Xu, and B. L. Black. 2004. Mef2c is a direct transcriptional target of ISL1 and GATA factors in the anterior heart field during mouse embryonic development. *Development* 131:3931–42.

Eisen, J. S., and J. A. Weston. 1993. Development of the neural crest in the zebrafish. *Dev Biol* 159:50–9.

Ericsson, R., R. Cerny, P. Falck, and L. Olsson. 2004. Role of cranial neural crest cells in visceral arch muscle positioning and morphogenesis in the Mexican axolotl, *Ambystoma mexicanum*. *Dev Dyn* 231:237–47.

Ericsson, R., J. Joss, and L. Olsson. 2008. The fate of cranial neural crest cells in the Australian lungfish, *Neoceratodus forsteri*. *J Exp Zoolog B Mol Dev Evol* 310:345–54.

Farrell, M. J., J. L. Burch, K. Wallis, L. Rowley, D. Kumiski, H. Stadt, R. E. Godt, T. L. Creazzo, and M. L. Kirby. 2001. FGF-8 in the ventral pharynx alters development of myocardial calcium transients after neural crest ablation. *J Clin Invest* 107:1509–17.

Foster, K., J. Sheridan, H. Veiga-Fernandes, K. Roderick, V. Pachnis, R. Adams, C. Blackburn, D. Kioussis, and M. Coles. 2008. Contribution of neural crest–derived cells in the embryonic and adult thymus. *J Immunol* 180:3183–9.

Galli, D., J. N. Dominguez, S. Zaffran, A. Munk, N. A. Brown, and M. E. Buckingham. 2008. Atrial myocardium derives from the posterior region of the second heart field, which acquires left–right identity as Pitx2c is expressed. *Development* 135:1157–67.

Goodrich, E. S. 1930. *Studies on the Structure and Development of Vertebrates*. London: McMillan.

Hami, D., H.-J. Grimes, M. L. Tsai, and M. L. Kirby. (2010). Zebrafish cardiac development requires a conserved secondary heart field. Manuscript submitted for publication.

Hutson, M. R., P. Zhang, H. A. Stadt, A. Sato, Y.-X. Li, J. Burch, T. L. Creazzo, and M. L. Kirby. 2006. Cardiac arterial pole alignment is sensitive to FGF8 signaling in the pharynx. *Dev Biol* 295:486–97.

Isogai, S., M. Horiguchi, and B. M. Weinstein. 2001. The vascular anatomy of the developing zebrafish: An atlas of embryonic and early larval development. *Dev Biol* 230:278–301.

Jiang, X., D. H. Rowitch, P. Soriano, A. P. McMahon, and H. M. Sucov. 2000. Fate of the mammalian cardiac neural crest. *Development* 127:1607–16.

Kelly, R. G., N. A. Brown, and M. E. Buckingham. 2001. The arterial pole of the mouse heart forms from Fgf10-expressing cells in pharyngeal mesoderm. *Dev Cell* 1:435–40.

Kirby, M. L. 1988a. Nodose placode contributes autonomic neurons to the heart in the absence of cardiac neural crest. *J Neurosci* 8:1089–95.

Kirby, M. L. 1988b. Nodose placode provides ectomesenchyme to the developing heart in the absence of cardiac neural crest. *Cell Tissue Res* 252:17–22.

Kirby, M. L. 2007. *Cardiac Development*. New York: Oxford University Press.

Kirby, M. L., T. L. Creazzo, and J. L. Christiansen. 1989. Chronotropic responses of chick hearts to field stimulation following various neural crest ablations. *Circ Res* 65:1547–54.

Kirby, M. L., T. F. Gale, and D. E. Stewart. 1983. Neural crest cells contribute to aorticopulmonary septation. *Science* 220:1059–61.

Kirby, M. L., P. Hunt, K. T. Wallis, and P. Thorogood. 1997. Normal development of the cardiac outflow tract is not dependent on normal patterning of the aortic arch arteries. *Dev Dyn* 208:34–47.

Kirby, M. L., and D. E. Stewart. 1983. Neural crest origin of cardiac ganglion cells in the chick embryo: Identification and extirpation. *Dev Biol* 97:433–43.

Komatsu, K., S. Wakatsuki, S. Yamada, K. Yamamura, J. Miyazaki, and A. Sehara-Fujisawa. 2007. Meltrin beta expressed in cardiac neural crest cells is required for ventricular septum formation of the heart. *Dev Biol* 303:82–92.

Kuratani, S. C., and M. L. Kirby. 1991. Initial migration and distribution of the cardiac neural crest in the avian embryo: An introduction to the concept of the circumpharyngeal crest. *Am J Anat* 191:215–27.

Kuratani, S. C., and M. L. Kirby. 1992. Migration and distribution of circumpharyngeal crest cells in the chick embryo. Formation of the circumpharyngeal ridge and E/C8+ crest cells in the vertebrate head region. *Anat Rec* 234:263–80.

Kurihara, H., Y. Kurihara, R. Nagai, and Y. Yazaki. 1999. Endothelin and neural crest development. *Cell Mol Biol* 45:639–51.

Kurohara, K., K. Komatsu, T. Kurisaki, A. Masuda, N. Irie, M. Asano, K. Sudo, Y. Nabeshima, Y. Iwakura, and A. Sehara-Fujisawa. 2004. Essential roles of Meltrin beta (ADAM19) in heart development. *Dev Biol* 267:14–28.

Le Lievre, C.S., and N.M. Le Douarin. 1975. Mesenchymal derivatives of the neural crest. Analysis of chimaeric quail and chick embryos. *J Embryol Exp Morphol* 34:125–54.

Leatherbury, L., H.E. Gauldin, K.L. Waldo, and M.L. Kirby. 1990. Microcinephotography of the developing heart in neural crest–ablated chick embryos. *Circulation* 81:1047–57.

Lee, M.J., A. Brennan, A. Blanchard, G. Zoidl, Z. Dong, A. Tabernero, C. Zoidl, M.A.R. Dent, K.R. Jessen, and R. Mirsky. 1997. P0 is constitutively expressed in the rat neural crest and embryonic nerves and is negatively and positively regulated by axons to generate non-myelin-forming and myelin-forming Schwann cells, respectively. *Mol Cell Neurosci* 8:336–50.

Li, Y.X., M. Zdanowicz, L. Young, D. Kumiski, L. Leatherbury, and M.L. Kirby. 2003. Cardiac neural crest in zebrafish embryos contributes to myocardial cell lineage and early heart function. *Dev Dyn* 226:540–50.

Lo, C.W., M.F. Cohen, G.Y. Huang, B.O. Lazatin, N. Patel, R. Sullivan, C. Pauken, and S.M.J. Park. 1997. Cx43 gap junction gene expression and gap junctional communication in mouse neural crest cells. *Dev Genet* 20:119–32.

Martinsen, B.J., A.J. Frasier, C.V. Baker, J.L. and Lohr. 2004. Cardiac neural crest ablation alters Id2 gene expression in the developing heart. *Dev Biol* 272:176–90.

Massari, M.E., and C. Murre. 2000. Helix–loop–helix proteins: Regulators of transcription in eucaryotic organisms. *Mol Cell Biol* 20:429–40.

McCauley, D.W., and M. Bronner-Fraser. 2003. Neural crest contributions to the lamprey head. *Development* 130.2317–27.

Moore-Scott, B.A., and N.R. Manley. 2005. Differential expression of Sonic hedgehog along the anterior–posterior axis regulates patterning of pharyngeal pouch endoderm and pharyngeal endoderm-derived organs. *Dev Biol* 278:323–35.

Nishibatake, M., M.L. Kirby, and L.H. van Mierop. 1987. Pathogenesis of persistent truncus arteriosus and dextroposed aorta in the chick embryo after neural crest ablation. *Circulation* 75:255–64.

Park, E.J., Y. Watanabe, G. Smyth, S. Miyagawa-Tomita, E. Meyers, J. Klingensmith, T. Camenisch, M. Buckingham, and A.M. Moon. 2008. An FGF autocrine loop initiated in second heart field mesoderm regulates morphogenesis at the arterial pole of the heart. *Development* 135:3599–610.

Phillips, M.T., M.L. Kirby, and G. Forbes. 1987. Analysis of cranial neural crest distribution in the developing heart using quail-chick chimeras. *Circ Res* 60:27–30.

Rosenquist, G.C., and R.L. DeHaan. 1966. Migration of precardiac cells in the chick embryo: A radioautographic study. *Carnegie Inst Wash Publ 625 Contrib Embryol* 38:111–21.

Sato, M., H.J. Tsai, and H.J. Yost. 2006. Semaphorin3D regulates invasion of cardiac neural crest cells into the primary heart field. *Dev Biol* 298:12–21.

Sato, M., and H.J. Yost. 2003. Cardiac neural crest contributes to cardiomyogenesis in zebrafish. *Dev Biol* 257:127–39.

Schilling, T.F., and C.B. Kimmel. 1994. Segment and cell type lineage restrictions during pharyngeal arch development in the zebrafish embryo. *Development* 120:483–94.

Schultheiss, T.M., S. Xydas, and A.B. Lassar. 1995. Induction of avian cardiac myogenesis by anterior endoderm. *Development* 121:4203–14.

Serbedzija, G.N., M. Bronner-Fraser, and S.E. Fraser. 1992. Vital dye analysis of cranial neural crest cell migration in the mouse embryo. *Development* 116:297–307.

Snarr, B.S., J.L. O'Neal, M.R. Chintalapudi, E.E. Wirrig, A.L. Phelps, S.W. Kubalak, and A. Wessels. 2007. Isl1 expression at the venous pole identifies a novel role for the second heart field in cardiac development. *Circ Res* 101:971–4.

Srivastava, D., P. Cserjesi, and E.N. Olson. 1995. A subclass of bHLH proteins required for cardiac morphogenesis. *Science* 270:1995–99.

Srivastava, D., T. Thomas, Q. Lin, M.L. Kirby, D. Brown, and E.N. Olson. 1997. Regulation of cardiac mesodermal and neural crest development by the bHLH transcription factor, dHAND. *Nat Genet* 16:154–60.

Stalsberg, H., and R.L. DeHaan. 1969. The precardiac areas and formation of the tubular heart in the chick embryo. *Dev Biol* 19:128.

Strachan, L.R., and M.L. Condic. 2004. Cranial neural crest recycle surface integrins in a substratum-dependent manner to promote rapid motility. *J Cell Biol* 167:545–54.

Sun, X., R. Zhang, X. Lin, and X. Xu. 2008. Wnt3a regulates the development of cardiac neural crest cells by modulating expression of cysteine-rich intestinal protein 2 in rhombomere 6. *Circ Res* 102:831–9.

Thomas, T., H. Kurihara, H. Yamagishi, Y. Kurihara, Y. Yazaki, E.N. Olson, and D. Srivastava. 1998. A signaling cascade involving endothelin-1, dHAND

and msx1 regulates development of neural crest–derived branchial arch mesenchyme. *Development* 125:3005–14.

Vaglia, J., and K. K. Smith. 2003. Early development of cranial neural crest in the marsupial, *Monodelphis domestica*. *Evol Dev* 5:121–35.

Waldo, K. L., M. R. Hutson, C. C. Ward, M. Zdanowicz, H. A. Stadt, D. Kumiski, R. Abu-Issa, and M. L. Kirby. 2005a. Secondary heart field contributes myocardium and smooth muscle to the arterial pole of the developing heart. *Dev Biol* 281:78–90.

Waldo, K. L., M. R. Hutson, M. Zdanowicz, H. A. Stadt, J. Zdanowicz, and M. L. Kirby. 2005b. Cardiac neural crest is necessary for normal addition of the myocardium to the arterial pole from the secondary heart field. *Dev Biol* 281:66–77.

Waldo, K. L., D. Kumiski, and M. L. Kirby. 1996. Cardiac neural crest is essential for the persistence rather than the formation of an arch artery. *Dev Dyn* 205:281–92.

Waldo, K. L., D. H. Kumiski, K. T. Wallis, H. A. Stadt, M. R. Hutson, D. H. Platt, and M. L. Kirby. 2001. Conotruncal myocardium arises from a secondary heart field. *Development* 128:3179–88.

Waldo, K. L., C. W. Lo. and M. L. Kirby. 1999. Connexin 43 expression reflects neural crest patterns during cardiovascular development. *Dev Biol* 208:307–23.

Waldo, K., S. Miyagawa-Tomita, D. Kumiski, and M. L. Kirby. 1998. Cardiac neural crest cells pro-vide new insight into septation of the cardiac outflow tract: Aortic sac to ventricular septal closure. *Dev Biol* 196:129–44.

Wang, J., A. Nagy, J. Larsson, M. Dudas, H. M. Sucov, and V. Kaartinen. 2006. Defective ALK5 signaling in the neural crest leads to increased post-migratory neural crest cell apoptosis and severe outflow tract defects. *BMC Dev Biol* 6:51.

Yanagisawa, H., R. E. Hammer, J. A. Richardson, S. C. Williams, D. E. Clouthier, and M. Yanagisawa. 1998a. Role of endothelin-1/endothelin-A receptor-mediated signaling pathway in the aortic arch patterning in mice. *J Clin Invest* 102: 22–33.

Yanagisawa, H., M. Yanagisawa, R. P. Kapur, J. A. Richardson, S. C. Williams, D. E. Clouthier, D. De Wit, N. Emoto, and R. E. Hammer. 1998b. Dual genetic pathways of endothelin-mediated intercellular signaling revealed by targeted disruption of endothelin converting enzyme-1 gene. *Development* 125:825–36.

Yelbuz, T. M., K. L. Waldo, D. H. Kumiski, H. A. Stadt, R. R. Wolfe, L. Leatherbury, and M. L. Kirby. 2002. Shortened outflow tract leads to altered cardiac looping after neural crest ablation. *Circulation* 106:504–10.

Zhou, H. M., G. Weskamp, V. Chesneau, U. Sahin, A. Vortkamp, K. Horiuchi, R. Chiusaroli, et al. 2004. Essential role for ADAM19 in cardiovascular morphogenesis. *Mol Cell Biol* 24:96–104.

12

Epigenetics in Bone and Cartilage Development

Tamara A. Franz-Odendaal

CONTENTS

This chapter describes the epigenetic processes involved in bone and cartilage development (osteogenesis and chondrogenesis, respectively). As defined earlier in this book, features, characters, or developmental processes are epigenetic if they can be understood only in terms of interactions that occur above the gene level. While the mapping of genomes is expanding at a great pace, our understanding of how gene networks result in morphological distinction or phenotypic variation is lagging behind. Understanding the epigenetic factors involved in skeletal (bone and cartilage) development includes understanding how epithelial and mesenchymal tissues interact; how positional signals, for example, are interpreted by cells; how cells aggregate

Epigenetics: Linking Genotype and Phenotype in Development and Evolution, ed. Benedikt Hallgrímsson and Brian K. Hall. Copyright © by The Regents of the University of California. All rights of reproduction in any form reserved.

to form condensations; and what cues are important for cell differentiation. An appreciation and understanding of these epigenetic processes lies at the root of understanding how different skeletal phenotypes develop and how they may be perturbed during abnormal development. Without this knowledge it will be impossible for biologists, anatomists, and clinicians to be able to unravel the complexities of skeletal development and their associated disorders.

The skeleton is comprised of all the bones and cartilages of the head, trunk, and tail region. The cells that make up the skeleton have one of two origins: the neural crest or mesoderm. The skull comprises mostly neural crest–derived skeletal elements (e.g., lacrimal, nasal, squamous temporal, zygomatic, maxillary, and mandibular bones), while the trunk skeleton is mesodermally derived (e.g., limb skeleton forms from lateral plate mesoderm). Those parts of the head skeleton that are not neural crest–derived have their origin in the cranial, lateral, and/or paraxial mesoderm (Noden, 1982; Couly et al., 1993). Some parts of the skeleton and even individual bones can, however, have multiple origins. The cranial base of the mouse, for example, has a mixed-cell origin; it develops from 14 pairs of cartilages that are derived from either mesoderm or neural crest cells (McBratney-Owen et al., 2008). Table 12.1 provides some other examples.

To add further complexity, in most taxa skeletal elements ossify during development. Ossification can occur at different stages within an organism's life history and via different modes. Some bones form indirectly from cartilage through the replacement of cartilage matrix (i.e., endochondral, perichondral, periskeletal ossification) (e.g., Thorogood, 1993; Hall, 2005; Franz-Odendaal et al., 2006, for review; Franz-Odendaal et al., 2007), while others form directly without a cartilage template (via intramembranous ossification). Most of the limb skeleton, for example, forms indirectly via endochondral ossification, whereas much of the head skeleton forms directly via intramembra-

nous ossification, which is a cartilage-independent process. Still other skeletal elements persist as cartilage throughout the life of the organism (i.e., persistent or permanent cartilage). Other less well-known modes of ossification include metaplastic ossification and ectopic ossification (see, e.g., Hall, 2005; Vickaryous and Hall, 2008). The same mode of ossification does not occur throughout an organism's skeleton for all of its bone: Flat bones are different from long bones, ribs are different from limb bones, etc.

Bones, therefore, vary in their cellular origin and mode of ossification. The rate of ossification is another variable. The bone vault, for example, develops very rapidly very early during development (in conjunction with the developing brain); its growth rate then slows down and ceases at an earlier stage than the growth of the facial and jaw elements of the skull. Bones can also vary in the type that forms (woven or lamellar). The location of the bone, species, age, and gender of the organism can all influence skeletal development.

The formation of bones and cartilages takes place through a series of steps, each step dependent on the step before; and all steps involve epigenetic processes that are often not fully understood. The steps to skeletal development, summarized in Figure 12.1, are as follows:

1. migration of cells to the site of skeletogenesis

2. tissue interactions involving induction, usually with an epithelium

3. condensation formation to a critical size

4. condensation ceases and differentiation of bone and/or cartilage cell types is initiated

For long bones that form via endochondral ossification, there are several additional steps after the cartilage template has formed. Chondrocytes located at the ends of a condensation deposit extracellular matrix that is cartilage-specific and undergo unidirectional proliferation that results in parallel columns of dividing cells forming growth plates. The cells in the

TABLE 12.1

Tissue Interactions Involved in Embryonic Induction of the Skeleton

SKELETAL PART	INDUCING TISSUE TYPE	INDUCING TISSUE ORIGIN	RESPONDING TISSUE TYPE	RESPONDING TISSUE ORIGIN	INTERACTION TYPE	REFERENCES
Skull	Calvaria[a]	Dura mater: epithelium	Ectoderm	Mesenchyme	Neural crest or cephalic mesoderm[b]	Diffusible factors	Dunlop and Hall, 1995; Mehrara et al., 1999; Rice et al., 2003; Opperman and Rawlins, 2005
Skull	Scleral ossicles	Conjunctival epithelium	Ectoderm	Ectomesenchyme	Neural crest	Diffusible factors	Coulombre et al., 1962; Coulombre and Coulombre, 1973; Pinto and Hall 1991; Franz-Odendaal, 2008
Skull	Scleral cartilage	Retinal pigmented epithelium	Ectoderm	Ectomesenchyme	Neural crest	Unknown	Newsome, 1972

(continued)

TABLE 12.1 (*Continued*)

SKELETAL PART		INDUCING TISSUE TYPE	INDUCING TISSUE ORIGIN	RESPONDING TISSUE TYPE	RESPONDING TISSUE ORIGIN	INTERACTION TYPE	REFERENCES
Skull: jaw	Mandible	Mandibular epithelium	Ectoderm	Ectomesenchyme	Neural crest	Cell–cell contact; + diffusible factors + matrix-mediated factors	Macdonald and Hall, 2001
Trunk	Limb bones	Apical ectodermal ridge	Ectoderm	Limb bud mesenchyme	Lateral plate mesoderm	Diffusible factors	Gilbert, 2003
Trunk: pectoral girdle	Ribs	Neural tube and notochord	Ectoderm	Sclerotome of somites	Paraxial mesoderm	Diffusible factors	Aoyama et al., 2005
Trunk: pectoral girdle	Scapula	Epithelium and mesenchyme	Lateral plate mesoderm and ectoderm	Dermatotome of somites (caudal part) and somatopleure (cranial part)	Paraxial mesoderm	Diffusible factors	Wang et al., 2005; Ehehalt et al., 2004; Huang et al., 2006

NOTE: Other inductive mechanisms may exist, but only those that have been identified in the literature are included. Known factors involved have also been omitted.

[a]Sutures between skull bones are different from the skull bones themselves and are considered a distinct region.

[b]Occipital bones derive from paraxial mesoderm (sclerotome) in some organisms.

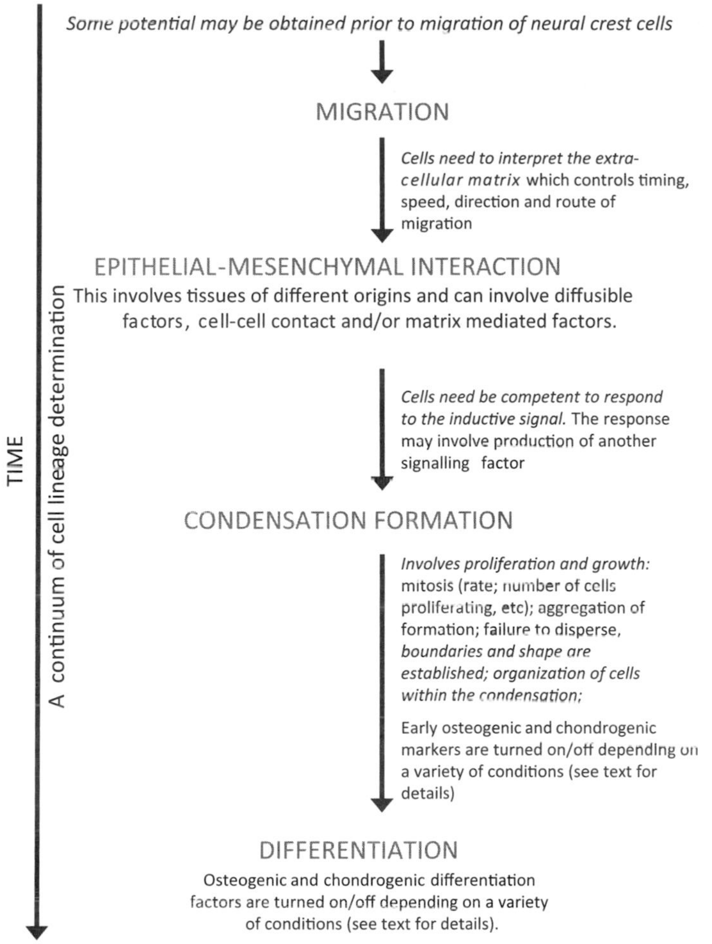

Some potential may be obtained prior to migration of neural crest cells

MIGRATION

Cells need to interpret the extra-cellular matrix which controls timing, speed, direction and route of migration

EPITHELIAL-MESENCHYMAL INTERACTION

This involves tissues of different origins and can involve diffusible factors, cell-cell contact and/or matrix mediated factors.

Cells need be competent to respond to the inductive signal. The response may involve production of another signalling factor

CONDENSATION FORMATION

Involves proliferation and growth: mitosis (rate; number of cells proliferating, etc); aggregation of formation; failure to disperse, *boundaries and shape are established; organization of cells within the condensation;*

Early osteogenic and chondrogenic markers are turned on/off depending on a variety of conditions (see text for details)

DIFFERENTIATION

Osteogenic and chondrogenic differentiation factors are turned on/off depending on a variety of conditions (see text for details).

TIME

A continuum of cell lineage determination

FIGURE 12.1 The four phases of skeletal development (gray) that involve novel epigenetic processes (bold and italic). Emphasis is placed on neural crest-derived elements. Alterations to any step (regardless of the cause) will result in altered bone or cartilage morphology.

center of the cartilage template then leave the cell cycle and become hypertrophic. This process is necessary for the subsequent invasion of blood vessels and replacement of the cartilage by bone matrix. The hypertrophic chondrocytes die, osteoclasts degrade the cartilage matrix, and osteoblasts produce bone-specific matrix using the degraded cartilage matrix as a scaffold (De Crombrugghe et al., 2001). At the same time, mesenchymal cells within the perichondrium surrounding this central region of the cartilage template begin to contain osteoblasts, which secrete a bone collar. There is an interaction between the perichondrium and periosteum on the outside of the cartilage and the central hypertrophic zone that is not fully understood but is likely reciprocal. This interaction is responsible for regulating the growth of the cartilage, ensuring that it achieves the cor-

rect length and size; it involves Indian hedgehog, bone morphogenetic protein (bmp) genes, transforming growth factor β (TGF-β), and their receptors (see Pathi et al., 1999, and more recent work by T. Linsenmeyer's group, such as Crochiere et al., 2008; Bandyopadhyay et al., 2008). Clearly, several epigenetic processes are involved in the above processes; some of these will be discussed.

The characteristics of bone and cartilage outlined above make the study of their development extremely complex. Despite this and quite remarkably, many similarities in bone and cartilage development are shared across type, location, and organism; and some common genes have emerged. The manner in which these gene products interact and their spatial and temporal expression patterns, however, vary among the different skeletal elements. In this chapter I discuss the epigenetic processes that occur during skeletal development in vertebrates. That is, those developmental processes that are important for determining the size and boundaries of skeletogenic condensations, for inducing cells to differentiate into osteocytes or chondrocytes, and ultimately for providing the information needed by cells and tissues to pattern and differentiate correctly. Several authors have discussed epigenetics as it pertains to the skeleton (Hall, 1984, 1990; Young and Badyaev, 2007; Depew and Compagnucci, 2008), and this chapter is therefore limited to some examples; it is not a comprehensive overview of the skeletogenic field.

Most of the following discussion will concentrate on condensations as the morphogenetic unit (or module) central to bone and cartilage formation, the epigenetic processes involved in forming this module, how this module interacts with other tissues, and the genes expressed within condensations. Some of the other processes involved in skeletal development are covered in other chapters in this book, namely, chapters on neural crest cells (Chapter 11), strain/stress on bone (Chapter 21), and muscle–bone interactions (Chapter 13). After the discussion on condensations, I discuss examples of how skeletal variation arises and how mutants can help in unraveling several aspects of skeletal biology. Two classes of mutants are discussed, one with mutations in the key skeletogenic genes (*runx2* and *Sox9*) and the other involving mutations that specifically affect condensations. Finally, I discuss some of the unsolved mysteries of the development of four parts of the skeleton (ribs, scapula, mandible, and sclerotic ring) to highlight the epigenetic processes involved.

With the advance in zebrafish developmental genetics, the potential use of mutant or transgenic zebrafish to understand human bone-related disorders will be significant despite differences between fish skeletal biology and that of mammalian systems. These differences include the presence of acellular bone (osteocytes outside the bone matrix) and resorp-

tion via mononucleated and not multinucleated osteoclasts (Witten et al., 2001). There is, however, much about fish skeletal biology that is not understood; and because of significant differences between the skeletons of fish and terrestrial animals—e.g., the bones of fish are not weight-bearing and calcium metabolism occurs via the gills in fish—the fish skeleton is not discussed in this chapter.

MIGRATION

Successful cell migration is critical for proper skeletal development. Neural crest cells (destined to form the skull, lower jaws, and branchial arch skeletons) migrate out of the closing neural tube during embryonic development to the cephalic regions, where they interact with epithelia to form the head skeleton. Neural crest cells are not the only populations of cells that need to migrate successfully for proper skeletal development; mesodermal cells also contribute to some parts of the skeleton (e.g., ribs, turtle shell, and vertebrae), and these cells also need to migrate successfully to their sites of skeletogenesis (Hall, 2005).

The process of cell migration is highly coordinated intracellularly and is not fully understood (reviewed in Kurosaka and Kashina, 2008). Although much is known about the genes involved in cell migration, the extracellular matrix through which cells migrate defines the timing, direction, speed, and final destination of migrating cells.

For neural crest cells several studies have identified extracellular matrix molecules that impede (e.g., ephrin) or promote (e.g., thrombospondin) neural crest–cell migration. Many genes have been identified that are turned on or off during cell migration. Analyses of cell culture experiments investigating cell migration are not straightforward as researchers have found that several aspects of cell migration in the developing embryo are distinct from those in cell culture. The speed and distances that cells need to migrate are also variable—cranial neural crest cells making the skull, for example, migrate at a rate of 40 μm/hour over a distance of 1,000 μm (Kurosaka and Kashina, 2008, table 1). A human disorder known as Di George syndrome, which is characterized by craniofacial defects, heart defects, and severe mental retardation, has been linked to deletions in chromosome 22 that result in defects specific to neural crest–cell migration (Epstein, 2001; Stoller and Epstein, 2005; Hutson and Kirby, 2007; Kurosaka and Kashina, 2008). This syndrome illustrates how knowing the genetic defect underlying disease does not necessarily mean we understand the resulting phenotype. What is the mechanism by which this particular deletion affects neural crest–cell migration, for example? How does this deletion arise? These and other similar questions are addressed in several chapters of this book.

Crucial epigenetic information resides in migrating neural crest cells prior to them reaching their destination or being induced to differentiate into bone or cartilage. One classic evo–devo example demonstrating the importance of correct neural crest–cell migration on skeletal development is in the development and evolution of the reptilian jaw bones into the mammalian middle ear ossicles. An early hypothesis of how the jaw bones of reptiles moved into the middle ear of mammals was that it was the result of expansion of the brain: The gap between the reduced jaw bones and the middle ear widened and pulled these bones apart from the mandible in mammals (Luo et al., 2001). This hypothesis is now refuted by paleontological evidence (Wang et al., 2001). The more accepted hypothesis is that the development of the middle ear ossicles occurred (over evolutionary time) because of a separation of the postdentary ossicles (Wang et al., 2001). Changes in the size and migration of the postdentary ossicles did not occur randomly. This information resides in the neural crest cells. The middle ear ossicles are derived from the neural crest of the midbrain and hindbrain (Köntges and Lumsden, 1996; Mallo, 2003). In this example, these crest cells receive epigenetic information for migration and the shape and size of these bones, before they start to migrate. This example demonstrates that epigenetic information is critical for the correct patterning of this part of the skeleton.

Other examples include the avian mandible, in which neural crest cells have chondrogenic potential prior to their migration (Hall, 2005). The intrinsic potential of the mesodermal cells that form the avian scapula is acquired prior to migration, when the cells are within the somites (Ehehalt et al., 2004). Scientists have also suggested that migrating neural crest cells that enter a presumptive skeletogenic zone are responding to a signal emanating from the zone itself (Hall and Miyake, 2000; Hall, 2005).

In summary, cells destined to form the skeleton can receive some epigenetic information prior to migration, during migration, or on entering a skeletogenic zone. Unraveling this epigenetic information is critical to understanding skeletal development.

INDUCTION

Once the cells that are destined to make the skeleton have migrated to their location, the next step in cartilage and bone development is induction. Embryonic *induction* is the process whereby a factor from one cell type affects a different responding cell type, such that the responding cell changes its fate or differentiation pathway. The responding cell has to be competent to respond to the inducing factor.

Induction and establishing tissue competence are epigenetic processes. Epithelial–mesenchymal inductions are critically important processes for both chondrogenesis and osteogenesis. They involve interactions between epithelial and mesenchymal tissue; the origins of these tissues may be different, and these interactions may involve cell–cell contact, diffusible factors, and/or matrix-mediated factors located in the basal lamina (Table 12.1). In the mouse mandible, skeletogenesis involves all three of these mechanisms of interaction (MacDonald and Hall, 2001). The four families of proteins that interact during skeletal development are the TGF-β, fibroblast growth factor

(FGF), Wnt, and Hedgehog families. Both positive and negative feedback loops are involved in induction, together with redundant signaling in some cases. These interactions are also often reciprocal. Once complete, the tissues involved become incapable of responding to further inductive signals or producing them. The inducing and responding tissue is not the same for all parts of the skeleton (e.g., skull vault vs. limb bones) or even in the same parts of the skeleton in different organisms (e.g., frontal bones in mammals, reptiles, and amphibians). The factors and mechanisms for inducing the skeleton are therefore variable, and for many skeletal elements these are not yet known.

The first visible morphological characteristic that induction has taken place is the formation of a skeletogenic condensation. Skeletogenic condensations are the central module of skeletal development and will be discussed below in some detail.

SKELETOGENIC CONDENSATIONS

MORPHOLOGY, FORMATION, AND CELLS A *skeletogenic condensation* is a self-organizing dynamic structure that has distinct characteristics. Morphologically, the condensation is defined as a dense aggregation of cells that ultimately will start producing collagen I or II (bone- or cartilage-specific matrix products) (Figure 12.2) in its center at a time when, among other things, it has reached a critical size (i.e., critical number of cells). The manner in which the cell aggregation forms is through either the proliferation of cells or the migration of cells into the site of bone formation. In some cases, failure to disperse from a center can also result in condensation formation (Hall and Miyake, 1995, 2000). Cells of the precondensation have unique properties that distinguish them from neighboring cells that are not condensed. For example, cell surface molecules binding peanut agglutinin lectin (Hall and Miyake, 2000) or alkaline phosphatase, a marker for early osteoblasts, can be used to visualize condensations prior to matrix deposition (Figure 12.2).

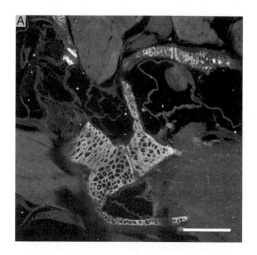

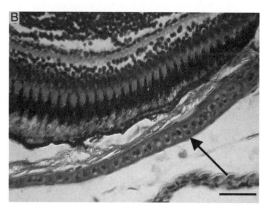

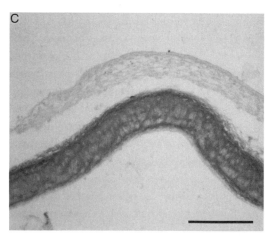

FIGURE 12.2 Zebrafish cartilage stained with an antibody to cartilage matrix, collagen II (A), and with a histological stain, toluidine blue in B (arrow). Scale bars = 50 μm. (C) Tenascin C in the scleral cartilage of a stage 35.5 chicken embryo. Scale bar = 100 μm.

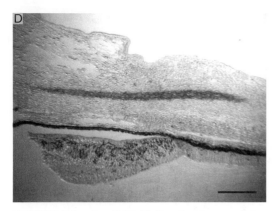

FIGURE 12.2 (continued) (D) Alkaline phosphatase staining of an early scleral ossicle condensation at stage 37 chicken embryo. Scale bar = 20 μm.

Earlier in this chapter, the development of the skeleton was described as a multistep process in which condensation formation is one step. Similarly, the process of forming a condensation consists of a series of ministeps: initiating the condensation, setting condensation boundaries, proliferation of cells, cell adhesion, growth, stopping growth, and overt differentiation.

The cells within a condensation express particular genes (e.g., *Sox9* is a marker for both cartilage and bone condensations). The portfolio of expressed genes, however, changes during development of the condensation from initiation to the laying down of collagen matrix; this occurs concurrently with cell differentiation. For example, some factors initiate the condensation (e.g., TGF-β and neural cell adhesion molecule [N-CAM]), others set boundaries for the condensation (e.g., tenascin C and syndecan), others are involved in cell proliferation to increase the size of the condensation, others are involved in aggregating cells by making them adhere to the extracellular matrix (e.g., N-CAM and N-cadherin), others stop growth of the condensation (e.g., Noggin acting on BMPs, N-CAM), and finally some are involved in differentiation (e.g., *Runx2*). That is, the expression of extracellular matrix molecules, membrane proteins, secreted proteins, and transcription factors varies from cell type to cell type during bone and cartilage development

and for different modes of ossification and different bone locations.

With respect to the differentiation of bone cells (osteocytes) or cartilage cells (chondrocytes), the lineage process is continuous such that many bone- and cartilage-specific cell types have been identified. This identity is confirmed based either on morphological descriptions of the cell types (e.g., electron micrographs or histological features) or on molecular characteristics (e.g., gene profiles) (see Vickaryous and Hall, 2006; Franz-Odendaal et al., 2006, for discussions of these topics). The most recognized cell types are, for bones, the preosteoblast, osteoblast, and osteocyte and, for cartilage, the prechondroblast, chondroblast, and chondrocyte (Figure 12.3). Since these cells represent a lineage of cell differentiation, there are states of differentiation between, for example, the osteoblast and the osteocyte stage. I have previously referred to these as "transitional cells" (Franz-Odendaal et al., 2006). More recently, Abzhanov and colleagues (2007) have included a chondrocyte-like osteoblast between the preosteoblast and osteoblast cell types; this cell type has been recognized by only a few individuals and in a few dermal bones of the mouse.

The difficulty in unraveling the process of cell lineage determination and differentiation is that not only is the process a continuous one but the pathway of differentiation is not conserved for every bone, cartilage, organism, location, etc. (e.g., see asterisk in Figure 12.3). That is, in some regions of the skeleton, one lineage determination pathway is followed and in others, another; both outcomes produce osteoblasts from osteoprogenitor cells (see Abzhanov et al., 2007, and Figure 12.3). In endochondral bones, for example, bone sialoprotein is absent in preosteoblasts but present in osteoblasts and osteocytes, yet the same protein may be present in the preosteoblasts, osteoblasts, and osteocytes of some intramembranous bones (see Franz-Odendaal et al., 2006, table 1). Put differently, the gene-expression profile of, for example, an osteoblast is not the same for all osteoblasts; differences occur based on the type of

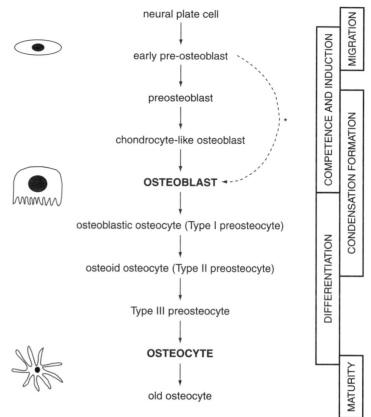

FIGURE 12.3 Cell differentiation of a neural crest-derived osteocyte showing the morphology of the cell types involved. Cell identity is based largely on morphological features rather than on molecular ones. *See Abzhanov et al. (2004) for a situation in which this lineage pathway is followed. See Franz-Odendal et al. (2006) for further discussions.

bone that is laid down, the origin of the cells, the mode of ossification, the location of the bone, the organism, etc. (Franz-Odendaal et al., 2006).

GENE PROFILES   Cells within the condensation change identity as they become surrounded by matrix—osteoblasts become osteocytes, and chondroblasts become chondrocytes. This change in location (outside matrix to within matrix) occurs concurrently with a change in gene expression and cell shape (the latter is due to internal cytoskeletal rearrangements within the cell). Condensations have elevated levels of a wide range of extracellular matrix and cell surface molecules, such as tenascin, syndecan, N-CAM, heparin sulfate, and chondroitin sulfate proteoglycans. Once the condensation has reached a critical size for differentiation to proceed, condensation growth must be stopped so that differentiation can begin. This process requires the downregulation of some genes (e.g.,

N-CAM) and the upregulation of other genes (e.g., *Runx2* for bone). Different condensations take different lengths of time before differentiation within them is initiated and before growth is stopped (Hall and Miyake, 2000, and references therein).

Osteogenic and chondrogenic condensations can be distinguished by their gene-expression portfolio before any overt cellular characteristics distinguish them from one another (Eames et al., 2003). For example, *Sox9* is the dominant factor in chondrogenic condensations, whereas *Runx2* is dominant in osteogenic condensations. The molecular characteristics of the condensations for a replacement versus a permanent cartilage also differ, with the former requiring *Runx2* expression. These two genes are also coexpressed during the early stages of condensation formation, before the cells of the condensation are fate-restricted to a cartilage (chondrocyte) or bone (osteocyte) lineage.

Recently, it was shown that *Sox9* function is dominant over *Runx2* during the condensation stage of skeletal development (Zhou et al., 2006). *Sox9* is involved in specifying osteochondroprogenitors into chondrocytes and/or into osteocytes (in endochondral ossification) (Akiyama et al., 2002; Zhou et al., 2006). This is possible because *Sox9* is a positive regulator for chondrocyte differentiation (Bi et al., 1999; Zhou et al., 2006) and a negative regulator of osteoblast differentiation via *Runx2* and *Osterix* (Akiyama et al., 2002). Inactivation of *Sox9* in cranial neural crest cells results in a complete defect of cranial neural crest–derived endochondral bone formation. These cranial neural crest cells express osteoblast markers instead (Mori-Akiyama et al., 2003). This example demonstrates the complex interactions of one single gene (*Sox9*) during skeletal development.

Despite these differences between bone types, origins, and mode of ossification (to name a few), Eames and Helms (2004) succeeded in identifying some common markers:

1. All histological markers of replacement cartilage, bone, and permanent cartilage are conserved among limb and head skeletal elements.

2. Most molecular markers (including some transcription factors) are the same between the trunk and head skeleton.

3. All condensations (regardless of skeletal fate: permanent cartilage, replacement cartilage, bone) have the same association with the developing vasculature.

4. As condensations initiate formation, they are distinguishable by their gene-expression profiles.

Chondrogenic condensations fated to form replacement cartilage and those fated to form permanent cartilage can be distinguished from one another by the expression of *Runx2* (Eames and Helms, 2004). These authors suggest that a fundamental set of genes directs skeletal cell differentiation throughout the body regardless of cell origin or location; however, the epigene-

tic processes involved in progressing from this gene profile to a particular skeletal fate are not clearly understood or straightforward. It involves understanding all the processes outlined in this chapter (e.g., condensation interactions, organization). The cautionary note here is that the above study was conducted on the chicken and may not apply to all vertebrates.

Even in terms of genetics and cell identity, the story is not clear-cut. Abzhanov et al. (2007) showed that the intramembranous dentary bone expresses only some chondrogenic markers (e.g., collagen II and collagen IX) and not others (*Sox9*, aggrecan). They showed that four major cell types can be distinguished based on skeletogenic markers during the formation of the dentary: (1) early preosteoblast expressing *Runx2*; (2) preosteoblast expressing Runx2, collagen II, and collagen IX; (3) chondrocyte-like osteoblast expressing osteopontin, collagen II, and collagen IX; and (4) mature osteoblast expressing osteopontin and bone sialoprotein II. In addition, they suggest that some cells can differentiate directly from (1) to (4) without passing through the other stages. How this occurs at the cell level and why this mechanism has evolved remain to be determined.

Several studies throw other interesting genes and expression patterns into the skeletogenic pot. In the avian scapula, for example, the anlagen express *Pax1* and *Sox 9* in a cranial-to-caudal direction (Huang et al., 2000a). The process of chondrogenesis follows a similar directional development within the scapula.

CONDENSATION SIZE AND BOUNDARIES Several advances in our understanding of the epigenetic processes involved in skeletal development revolve around the condensation. One very interesting aspect that has been uncovered is that skeletogenic and chondrogenic condensations have to be the right size in order to initiate skeletogenesis or chondrogenesis within. If the condensation is too small, then osteogenesis is not initiated; if it is too large, then abnormally large bones develop (Hall, 1978; Atchley and Hall, 1991; Thorogood, 1993; Hall and Miyake, 2000; reviewed in Willmore et al., 2007).

Condensation size is monitored and can be adjusted even if too few cells migrate from the neural tube. In these cases, other processes are initiated to result in a normal skeletal element: (1) increased rate of migration, (2) increasing proportion of cells undergoing cell division, (3) increasing rate of cell division, (4) decreasing rate and proportion of cell death, or (5) a combination of the above (Willmore et al., 2007). Although it is known that condensation size is regulated through signaling pathways involving BMP2 and BMP4, how this is done is not clear. The experiments that suggest size regulation made use of beads soaked in a recombinant form of BMP to overexpress these growth factors; the result is to dramatically increase the size and alter the shape of the condensation, which ultimately changes the size and shape of the resulting skeletal element.

Interestingly, regardless of whether the condensation is neural crest– or mesoderm-derived, similar molecules (e.g., tenascin C and syndecan) are involved in setting condensation boundaries. For example, Meckel's cartilage and the membrane bones surrounding it, which together form the mandibular skeleton, have different origins but similar condensation boundary markers (Hall and Miyake, 2000; Gluhak et al., 1996).

However, other condensations for similarly derived elements are not identical in morphology or gene-expression profiles. For example, wing condensations are broad and flat, whereas leg condensations are compact and spherical. This cell density is the result of different levels of fibronectin. Fibronectin is present at higher levels in leg condensations than in wing condensations. Wing condensations are more sensitive to TGF-β, which can increase condensation size via elevating fibronectin levels. In addition, treating leg mesenchyme with fibronectin antibody in culture has no effect on the condensations, whereas doing the same experiment with wing tissue completely inhibits condensation formation. The molecules involved in setting condensation boundaries include matrix molecules such as tenascin and syndecan, growth factors (such as FGFs, BMPs), and transcription factors (such as *Msx*) (Hall and Miyake, 2000, and references therein). These experiments (Downie and Newman, 1995) and others demonstrate that epigenetic phenomena act within the condensation to set condensation size. The condensation size can be influenced by both intrinsic and extrinsic factors (discussed below).

Recent studies have investigated the significance of neural crest boundaries in the development of the skull vault (reviewed in Chai and Maxson, 2006; Gross and Hanken, 2008). For example, the extent of the neural crest–cell contribution to the calvarial bones in birds is controversial. Several studies using quail–chicken chimeras or retroviral infection have produced fate maps suggesting that the boundary between neural crest cells and nonneural crest cells lies within the frontal bone (Noden and Trainor, 2005; Evans and Noden, 2006), while others show that the entire cranial vault is of neural crest origin (Couly et al., 1993). Thus, although fate-mapping studies can help to identify the origin of cell populations, they are not always conclusive. Gross and Hanken (2008) make a point that part of the controversy lies in the "cleanness" of graft tissue and the use of slightly different anatomical nomenclature by different research groups. Without a clear understanding of the epigenetic processes involved in the tissue interactions that are active, interpreting these data is tricky.

CELLULAR DYNAMICS WITHIN CONDENSATIONS There is one additional yet fundamental aspect of bone and cartilage development that I would like to point out and which we do not fully understand: that is, how osteoblasts are buried in matrix to become osteocytes (see Franz-Odendaal et al., 2006, for detailed discussion). The proportion of osteoblasts that are buried is not always the same and can range from 10% to 65% in deer antlers compared to human cancellous bone, respectively (Banks, 1974; Parfitt, 1990). Recently, we presented four schemes by which osteoblasts lay down the first collagen strands in a condensation and how they become entrapped in bone matrix (Franz-Odendaal

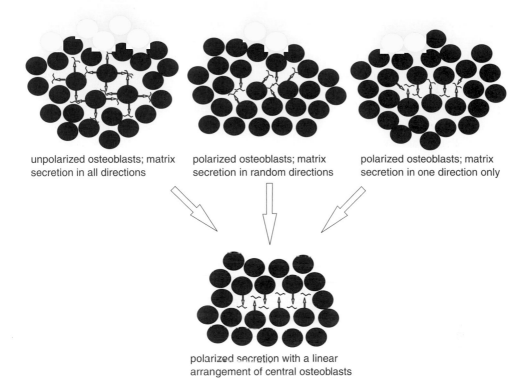

unpolarized osteoblasts; matrix secretion in all directions

polarized osteoblasts; matrix secretion in random directions

polarized osteoblasts; matrix secretion in one direction only

polarized secretion with a linear arrangement of central osteoblasts

FIGURE 12.4 Schematic showing the different ways in which osteoblasts may deposit collagen fibers within the center of a condensation. Either osteoblasts are unpolarized and deposit matrix in all directions or they are polarized with either a completely random secretion of fibers or an organized secretion toward the center of the condensation. For full details of these scenarios see Franz-Odendaal et al. (2006).

et al., 2006) (Figure 12.4). These schemes vary because we do not yet understand the cellular dynamics of osteoblasts. Are osteoblasts polarized in their secretion of bone matrix? How do the osteoblasts in the center of the condensation know in which direction to deposit collagen fibers? What is the mechanism by which osteoblasts are organized to lay down bone? Is one generation burying the next? Are all osteoblasts within the condensation secreting bone matrix at the same rate? How is bone laid down synchronously by osteoblasts within a condensation? This fundamental aspect of osteocyte differentiation has been overlooked by many skeletal biologists.

PARTITIONING There is some evidence that (at least some) condensations can split. That is, more than one bone or cartilage can arise from a single condensation. Two bones of the middle ear (the malleus and incus) arise from a single condensation, the pair of frontal bones arise from a single condensation, seven bones of the lower beak of chicken embryos arise from a single condensation, three bones in the zebrafish head arise from a single condensation, the tibia and fibula as well as the ulna and radius in the limbs arise from single condensations— these condensations are later separated by apoptosis (Hall and Miyake, 2000; Hall, 2005). Other condensations have never been reported to split (e.g., scleral ossicle condensations— here, a series of almost identical bones, 6–16, are induced independently within the sclera during eye development). (Scleral ossicles are discussed further below.)

Not only is it possible for condensations to split to form several bones but a single condensation can be partitioned into a part that forms permanent cartilage (articulates with adjacent elements) and another region that differentiates into replacement cartilage. This partitioning (at least in the mouse limb) is under the action of *Gdf5* and the BMP family of proteins (Storm and Kingsley, 1996). The replacement cartilage later induces bone formation in the perichondrium, transforming it into a perichondrium (St. Jacques et al., 1999; Ueta et al., 2001; Eames and Helms, 2004). Another example is the occipital bone, in which a portion ossifies intramembranously from two ossification centers, while the rest of this bone ossifies endochondrally. The mechanism by which cells within the calvarial tissue are restricted into the osteogenic versus the chondrogenic lineage has been examined (e.g., Toma et al., 1997). These authors show that this restriction is a result of negative selection against the growth of chondrogenic cells in the absence of an inductive environment (i.e., extrinsic environmental factors can modulate lineage determination) and that there is a progressive reduction in the number of cells with chondrogenic potential as development progresses. Clearly, several epigenetic processes are involved, each one the source of skeletal variation.

DEPENDENCE ON OTHER TISSUES Condensations do not exist in isolation but interact with other tissues and cell types (e.g., muscle, nervous tissue) to produce a functioning organ or appendage. A few are mentioned below; some are more fully discussed elsewhere in this book.

1. *Vasculature* The regression of blood vessels is observed around the boundary of presumptive skeletogenic condensations. This regression begins before any molecular, cellular, or histological evidence for the development of a condensation is detected; the factors governing this process and the mechanism of vascular clearing are not understood and are often overlooked. It is, however, essential for proper skeletogenesis. Inhibiting vascular regression by application of vascular endothelium–derived growth factor (VEGF) can prevent skeletal elements from forming (Yin and Pacifici, 2001).

2. *Musculature* Bones or parts of bones vary in their dependence on muscle interaction during normal growth. Development of the anterior portion of the mammalian dentary is influenced by the developing dentition, while the posterior portion is influenced by muscle contraction and not tooth development. Animals fed different diets have variable cranial morphologies due to the effects of contraction of the masticatory muscle. The strength and orientation of muscle fibers can affect cranial morphology (Willmore et al., 2007). Strains imposed on the cranium also affect skull shape by creating differential growth at cranial sutures and by adding bone at sites of muscle attachment (Herring and Tend, 2000; Rafferty et al., 2000; reviewed in Willmore et al., 2007). The expanding growing brain imposes tensile forces on the developing skull vault and sutures, which directly influence its growth. Not surprisingly, the interaction with musculature is dependent on the type of bone formed—the furcula is more dependent and the mandible less dependent on muscle activity than the long bones.

3. *Nervous tissue* The calvariae (skull vault) form directly by intramembranous ossification in a sandwich between the developing brain and overlying skin. Signaling from both of these tissues (epithelial–mesenchymal interactions and vice versa) coordinates calvarial development (Dunlop and Hall, 1995; Opperman et al., 1995). Rice et al. (2003) show that the expression patterns of *Twist*; *Msx2*; *Fgfr1*, *-2*, and *-3*; and *Fgf2* are overlapping and spatiotemporally regulated during mouse calvarial development. The exact mechanisms governing these inductive events are unknown. Ossification of the intramembranous calvarial bones depends on the presence of the brain; in its absence (*anencephaly*), no bony calvariae form. In

contrast, zebrafish mutants which are eyeless have normal-sized skull orbits, despite having no eye.

The coordination of the delicate balance between bone resorption (by osteoclasts) and bone deposition (by osteoblasts) takes place via osteocytes (bone cells), but how osteocytes actually detect and interpret strain and stress is not fully understood. It is known that bone cells form a functional network which includes all the cell types from the preosteoblast to the mature osteocyte (Franz-Odendaal et al., 2006) (Figure 12.3), but the details of how the individual cells within a skeletal element communicate with one another to respond to (outside) stressors in a meaningful manner are not clear.

One unique and often overlooked function of osteocytes is to resorb and deposit bone around the osteocyte lacuna (*osteocytic osteolysis*). This process is not found in humans but is present in some vertebrates (e.g., bats, hamsters, squirrels, rats, rabbits, snakes, carp, salmon) (reviewed in Franz-Odendaal et al., 2006). How and under what circumstances osteocytes obtain essentially osteoblastic and osteoclastic abilities is not clear.

SKELETAL VARIATION

Recent studies exploring phenotypic variation of the skeleton have implicated variations in the expression of one particular group of proteins, the bone morphogenetic proteins, as primary players inducing adaptive and innovative changes in bone and cartilage development (Young and Badyaev, 2007).

Two examples of their role in innovation in the trunk skeleton are (1) the formation of the turtle carapace, which is induced by BMP and/or BMP regulators (e.g., *Indian hedgehog*) possibly secreted by the developing ribs (Cebra-Thomas et al., 2005), and (2) the formation of the bat wing, in which *BMP2* expression is increased, resulting in elongation of wing digits (Sears et al., 2006). More recently, SHH–FGF signaling has also been shown to increase the length of digits as well as to ensure survival of the interdigital (web) tissue in two bat species (Hockman et al. 2008).

In the mandible, adaptive mechanisms that contribute to the diversity of cichlid jaw morphologies (Terai et al., 2002; Albertson et al., 2005) and bird bills (Abzhanov et al., 2004; Wu et al., 2004, 2006) have been identified. Both are associated with levels of *BMP4* expression and with shifts in epithelial–mesenchymal signaling.

Skeletogenic condensations can also be induced to form in areas outside the skeleton by implanting beads soaked in BMPs. When these beads are implanted outside the skeleton, this elicits a cascade of events involving condensation formation, vascular invasion, cartilage differentiation, and cartilage replacement into bone (Hall and Miyake, 2000; Hall, 2005, and references therein).

For a detailed discussion of how this variation in BMP expression during skeletal development comes about, the reader is referred to Young and Badyaev (2007). They highlight three primary mechanisms:

1. External stresses on developing tissues initiate changes in gene expression via modification of the cellular and intracellular environments (Skerry, 2000; Rauch and Schoenau, 2001; Moore, 2003).

2. Mutations in regulatory regions alter the timing, location, and level of gene expression (Terai et al., 2002).

3. Neutral genetic variation facilitates the development of phenotypic traits that may be adaptive in novel environments (Rodríguez-Trelles et al., 2005).

The above epigenetic mechanisms are covered elsewhere in this book. As noted in the preceding text, there are many steps in skeletal development that involve epigenetic processes (cell movements, cell signaling, etc.), which if altered can result in skeletal variation and account for the phenotypic variation observed among vertebrate skeletons.

DECIPHERING MUTANTS

The study of mutants can help to decipher the epigenetic processes involved in skeletal development. Here, I discuss mutations of two key transcription factors involved in bone and cartilage development (*Sox9* and *Runx2*) as well as mutations affecting the condensation stage of skeletal development.

MUTATIONS IN CRITICAL SKELETOGENIC GENES Two transcription factors that play essential roles in osteogenesis and, to a lesser extent, in chondrogenesis are *Runx2* and *Sox9*. Using model organisms, one can study the phenotype of mutants with homozygous null deletions of a gene or heterozygotes with a single null allele in an attempt to understand the function of a gene. For example, *Sox 9*$^{-/-}$ mice lack cartilage, whereas *Runx2*$^{-/-}$ mice lack all membrane and endochondral bones.

Heterozygous mutations in *Runx2* results in mice with cleidocranial dysplasia, a dominantly inherited skeletal dysplasia characterized by hypoplastic clavicles, large fontanels, dental anomalies, and delayed skeletal development (Otto et al., 1997). *Sox9* mutations result in campomelic dysplasia, a disorder characterized by generalized hypoplasia of endochondral bones. Both *Sox9* and *Runx2* are expressed in the mesenchymal condensations prior to overt differentiation of osteoblasts or chondroblasts. For cells to commit to the chondrogenic lineage, *Runx2*, which is a strong osteoblast-specific gene, needs to be inhibited. Zhou et al. (2006) showed that *Sox9* plays a key role as a transcriptional repressor for osteoblast differentiation, in part via inhibition of *Runx2* transactivation of its target genes.

Looking more closely at Runx2$^{-/-}$ mutant mice, several authors have noted that the replacement cartilages in these mice fail to express collagen 10, Indian hedgehog, *bmp6*, and bone sialoprotein (Ducy et al., 1997; Hoshi et al., 1999; Inada et al., 1999; Kim et al., 1999; Eames and Helms, 2004). Collagen 10 is a specific marker of replacement cartilage (Eames and Helms, 2004). By treating embryos with

Runx2, Eames and Helms (2004) succeeded in transforming a permanent cartilage into a replacement cartilage. They further showed that *Runx2* (or a downstream target) can drive *Col10* transcription in the presence of chondrogenic factors. In addition, they showed that *Sox9* (or a downstream target) can drive *Col10* transcription in the presence of osteogenic factors. The array of defects outlined above demonstrate that epigenetic factors play a key role in phenotypic outcome and that we do not fully understand the interactions of some of the key skeletogenic genes, despite their discovery many years ago.

Deciphering mutants is, however, not always easy since other genes can compensate for the knocked-out or silenced gene. This is because the knocked-out gene could have a redundant function such that another gene can "step in" and perform its function when needed. For example, the single-copy knockout of tenascin C in mice results in normal mice, suggesting that either tenascin is not essential for condensation boundaries or another gene product compensates for its absence. This ability to compensate has severely hampered original hopes of deciphering the role of particular genes via mutants in developmental biology.

Other knockouts, which we expect to have profound effects based on known gene expression, surprisingly do not have the predictable effects. For example, the important role of N-CAM in condensations was described earlier, yet with N-CAM knocked out, mice are normal. This, again, indicates that other factors and epigenetic processes are acting and can compensate for the knocked-out N-CAM gene (Hall and Miyake, 2000; Hall, 2005).

Why are some genes and gene products seemingly redundant and others not? In order to fully understand the mutant phenotypes described above, a thorough understanding of the epigenetic processes involved in skeletal development (at the cell and tissue levels) as well as the interactions of genes and gene networks is required. Questions (e.g., in *Runx2*$^{+/-}$ mice) such as why clavicles are affected but not the scapula to which they connect or why the ribs are affected but not the similarly shaped long bones remain unanswered and require knowledge of the bone type, mode of ossification, cell origin, etc.—factors that we now know influence skeletal development.

MUTATIONS AFFECTING CONDENSATIONS Mutations affecting each stage of condensation formation are known. Once again, these mutant phenotypes can be difficult to decipher, but all provide some insight into the epigenetic processes involved in condensation formation.

1. *Alx-4* knockout: A paired mouse homeobox gene, *Alx-4* (aristaless), is expressed in condensations of the skull, hair, teeth, and mammary glands of murine embryos. Null mutants, however, have only preaxial polydactyly, indicating that either *Alx-4* plays a minor role in skeletogenesis despite being expressed in distinct condensations or its functions overlap with those of other genes (Hall and Miyake, 2000).

2. *Talpid* (chicken) and *Short Ear* (mouse): Several mutants involving different genetic pathways have missing, small, or abnormally large bones; *Talpid* involves the overexpression of N-CAM, while *Short Ear* is a mutation in BMP5. These defects can be traced back to the condensation stages (Hall and Miyake, 2000).

3. *Hoxa-2* knockout: *Hoxa-2* knockout mice have duplications in some elements of the mandible but not all elements (Rijli et al., 1994), despite all of them arising from a single condensation. This single condensation divides into a series of separate condensations (for Meckel cartilage, incus, malleus, etc.), each persisting for varying lengths of time and interacting with different tissues. In *Hoxa-2* knockout mice, however, only Meckel cartilage is unaffected. Some bones are duplicated (e.g., tympanum, malleus, incus, and squamosal); some are missing (e.g., stapes) and extra cartilages form (e.g., between the skull and the incus) (Hall and

Miyake, 2000). Some of these effects have been attributed to an excess of mesenchymal cells in the condensations (Smith and Schneider, 1998). This is a clear example where gene products are important in differentiation and morphogenetic pathways rather than specifying a particular phenotype.

In summary, although studying mutants can certainly be extremely helpful in pinpointing which genes are critical, which are redundant, and which are interlinked, it often cannot provide complete answers and the results need to be interpreted in conjunction with knowledge of the other important factors in skeletal development described above.

SOME CASE STUDIES

A few skeletal elements were selected to highlight particular aspects of their development that clearly involve epigenetic processes beyond the gene level and because they serve to highlight some of the more traditional methods of understanding tissue interactions. Only a limited set of data for each element is discussed.

THE RIBS

The ribs are derived from *somites*, which are blocks of mesoderm that form alongside the neural tube. Several transplantation studies have shown that whether a somite forms ribs depends on its location along the vertebral column. If a somite is transplanted to the cervical region, it will form ribs; if cervical somatic mesoderm is transplanted to the thoracic region, it does not form ribs (Chevallier, 1975). This early regionalization appears to occur in the presumptive streak stage of chick development (Sakamoto et al., 2001; Aoyama et al., 2005).

The somitic mesoderm comprises three parts (dermotome, myotome, sclerotome), and there has been some controversy as to which parts contribute to the ribs. Parts of the somatic mesoderm have been transplanted from quail to chick embryos in order to determine which parts are involved in rib development (Huang et al., 2000b); these authors and others (Evans, 2003) show that the ribs have a sclerotome origin. More recently, other researchers have considered a single rib as having three developmentally different parts (proximal, vertebrodistal, and sternodistal parts) (Aoyama et al., 2005). It appears as if only the sternal (or sternodistal) rib depends on the lateral plate mesoderm for development but the proximal and vertebrodistal ribs do not. This example demonstrates that even parts of the same bone element can be dependent on and induced by different over- or underlying tissues. Another example is that of the turtle or tortoise shell. Here, the ribs and parts of the vertebrae fuse with dermal elements to make the carapace, resulting in a carapace of dual origin (mesoderm and neural crest) (Hall, 2005).

THE SCAPULA

Although much is known about the molecular regulation of the scapula by studying mouse mutants in particular (see Huang et al., 2006), the origin of the scapula remains controversial. Recently, it has been demonstrated through transplantation studies and cell labeling that the scapula is derived from the dermomyotome of the somite (Wang et al., 2005; Huang et al., 2006) and not from the sclerotome as previously thought, that lateral plate mesoderm is essential for scapula blade formation, that several somites contribute cells to the scapula, and that these cells do not mix but rather maintain their segmental organization. In addition, the derivation of the scapula may not be the same in all vertebrates; in reptiles the scapula is derived from both the somite and the lateral plate mesoderm (see Vickaryous and Hall, 2006, for a full discussion of the pectoral apparatus). Why cell mixing does not occur within the scapula condensation is not understood. This example demonstrates that condensations can have mixed origins and that elucidating something as simple as a bone's origin is not straightforward and cannot be assumed to be similar across all organisms.

THE MANDIBLE

The mandible develops from the first branchial arch, also known as the "mandibular arch," which is first apparent at six- to eight-somite stage (in chicken embryos). Its induction involves three of the methods of interaction outlined in Table 12.1 (Atchley and Hall, 1991; MacDonald and Hall, 2001).

In 1983, Drew Noden showed that morphogenesis of the mandibular skeleton is independent of the environment in which the cells find themselves—that is, that it is perhaps not under epigenetic control. When he transplanted first-arch mandibular neural crest cells into the future second-arch region of the neural tube, these first-arch cells migrated into the second arch and developed into first-arch skeletal tissues, ultimately forming a duplicated mandibular skeleton (Noden, 1983). We now know that mandibular development depends on interactions between the pharyngeal endoderm, oral epithelium, and cranial neural crest–derived mesenchyme within the first branchial arch (reviewed in Chai and Maxson, 2006). The pharyngeal endoderm is thought to prepattern the oral epithelium, which in turn provides instructive signals to the underlying mesenchyme (Haworth et al. 2004). Signaling molecules within the oral epithelium are expressed in a region-specific manner. They include BMP, TGF-ß, and FGF and are thought to regulate homeobox-containing genes (such as *Dlx, Lhx, Msx, Gsc*) within the underlying neural crest–derived mesenchyme to generate early polarity and to pattern the first arch (Chai et al., 1994, 2000; Trumpp et al., 1999; Ito et al., 2002; Depew et al., 2002; Santagati and Rijli, 2003; Tucker and Sharpe, 2004; Liu et al., 2005). Thus, cranial neural crest cells act on positional information to form the jaw skeleton. In Noden's study, the transplanted first-arch neural crest cells were already patterned and determined to form first-arch derivatives before transplantation to the neural tube. His remarkable study therefore showed that once fate is determined and the neural crest cells are patterned, they cannot be altered by surrounding factors.

Loss-of-function studies have been carried out to try to unravel this signaling. Loss of BMP signaling in the oral ectoderm and pharyngeal endoderm results in extreme phenotypes ranging from almost completely missing mandible to severe defects in only the distal region (Liu et al., 2005). Mutations in the *Msx1* and *Msx2* genes result in a midline cleft of the first arch and severe mandibular defects (Satokata et al., 2000; Ishii et al., 2005; Chai and Maxson, 2006). These types of experiments (and others) indicate that mandibular morphogenesis requires tightly controlled BMP and FGF signaling in the ectoderm (Chai and Maxson, 2006). More recently, Oka et al. (2008) have shown that the cranial neural crest–derived cells in the proximal region of the mandible have a cell-intrinsic requirement for TGF-β signaling. These cells have the potential to become either osteoblasts or chondrocytes, with TGF-β signaling regulating the expression of transcription factors that either promote chondrogenesis or inhibit osteoblasts. Thus, TGF-β signaling plays a critical role in cell lineage determination during endochondral ossification in the mandible.

Exactly how this regulation occurs is still not fully understood. For example, TGF-β signaling can promote chondrogenesis by controlling *Sox9* expression. However, in *Tgfbr2^{fl/fl}, Wnt-Cre* mice, which have small mandibles with defects in several regions (Oka et al., 2007), cartilage still forms in some tissues, suggesting that other factors in addition to TGF-β must regulate *Sox9* expression. Teasing out these interactions at the gene level will set the stage for deciphering the complexity of cartilage and bone regulation at the cell and tissue levels within this region of the embryo.

THE SCLEROTIC RING

The development of a series of bones that form the sclerotic ring within the eye of many vertebrates will be discussed in light of the epigenetic processes described above. The sclerotic ring is comprised of a series of bone plates (or scleral

ossicles) that overlap one another to form a complete ring encircling the eye. The variation in the morphology of this ring within Reptilia is show in Figure 12.5; for extant reptiles, the typical number of scleral ossicles is 6–16 per eye.

The sequence of events leading to ossicle formation as it is currently understood is as follows:

1. development of epithelial (conjunctival) papillae (the inducing tissue)

2. development of responding neural crest–derived mesenchymal tissue below the papillae that is competent to respond to the inductive signal

3. production of a diffusible signal in the epithelium

4. diffusion of the inducing signal

5. mesenchymal cells respond by condensing

6. the inducing conjunctival papillae disappear

7. critical size of the condensation is reached

8. osteoblasts differentiate within the condensation and bone is laid down

9. bone elements increase in size

Numbers 1–5 are fairly typical of epithelial–mesenchymal interactions. However, note that diffusible factors mediating epithelial–mesenchymal interactions are less common than matrix-mediated interactions. Numbers 7–9 are relevant to typical bone development. Each step involves epigenetic processes beyond the level of the gene that are, for the most part, not understood. Our understanding of these processes can be briefly summarized as follows:

- Each bone plate (or ossicle) is induced separately and independently from its neighboring plate. This was shown by experiments in the 1960s by Coulombre and colleagues, who removed the inductive tissue of one ossicle and prevented only this ossicle from forming (Coulombre et al., 1962; Coulombre and Coulombre, 1973).

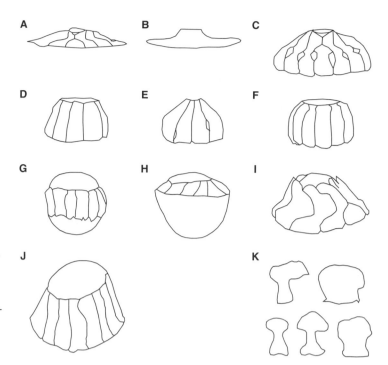

FIGURE 12.5 Variation in the sclerotic ring in some reptilia, scaled to similar size. (A) *Varanus bengalensis*. (B) *Calotes jubatus*. (C) *Lacerta viridis*. (D) *Heloderma suspectum*. (E) *Anguis fragilis*. (F) *Gehyra mutilata*. (G) *Gallus gallus*. (I) *Chameleon oustaleti*. (K) individual ossicles from *Lacerta bedrigae*, *Sphenodon*, and *Lacerta viridis*.

- Induction is by an unknown factor that diffuses from an epithelial (conjunctival) papilla about 200–300 μm to the site of the future ossicle.

- Induction takes 2–3 days (Hall, 1981).

- Experimentally manipulating the ring results in other ossicles compensating for the alteration (Coulombre et al., 1962; Coulombre and Coulombre, 1973; Franz-Odendaal, 2008).

- *Hedgehog* signaling is involved in ossicle induction (Franz-Odendaal, 2008).

- Papillae do not develop simultaneously but rather in groups (see Franz-Odendaal, 2008); the result of this is that developing ossicles within the eye are at different stages of development.

It is known that the interossicle regions (regions between the ossicle anlagen) do not develop ossicles, presumably because papillae do not form; but it is unclear why papillae develop in only some parts of the epithelium, in 6–16 different places around the eye (Franz-Odendaal and Hall, 2006; Franz-Odendaal, 2008). These papillae are evenly spaced around the eye, regardless of the number of papillae that form.

The sequence in which papillae develop around the eye is not simply clockwise or counterclockwise from a given point to form a circle but is, nevertheless, predictable (to some degree). Papillae on one side of the eye develop first, then the papillae on the opposite side, followed by a dorsal and ventral group of papillae (reviewed in Franz-Odendaal, 2008). What processes and molecules underlie this patterning is not known but is being investigated in my laboratory.

The size of the sclerotic ring (compared to total eye size), the size of the aperture, the number of elements, the thickness of the elements, and the shape of the scleral ossicles can vary among vertebrates (Figure 12.5). How this variation in the sclerotic ring is achieved can be understood only when we unravel the processes and interactions involved in their development. Some aspects of sclerotic ring morphology and development are discussed below.

NUMBERS OF PLATES Among vertebrates there is a wide variation in the number of scleral ossicles per eye (Franz-Odendaal and Hall, 2006). This variation is present not only among vertebrates but within species (Franz-Odendaal and Hall, 2006) as well as within a single embryo (Franz-Odendaal, 2008). For example, the variation in the number of ossicles in the domestic quail is 12–18 elements per eye (Canavese et al., 1987), and the variation within the domestic chicken *Gallus gallus* is 12–16 (Curtis and Miller, 1938; Coloumbre et al., 1962; Coloumbre and Coloumbre, 1973; Franz-Odendaal, 2008). What is the developmental mechanism that determines the number of plates? Since there is a one-to-one relationship between the number of conjunctival papillae that develop and the number of ossicles that are present and removing one papilla results in one less ossicle in the sclerotic ring, the number of plates is most likely altered by changing the number of papillae that develop. In this system there is no evidence for condensations splitting or partitioning as described previously. At present, we do not yet understand how papillae are induced and, therefore, cannot answer the above question. We do know that not only are the papillae morphologically distinct structures but they can also be distinguished from interpapillary epithelium by the genes that are expressed. The first gene identified with this unique expression pattern is *sonic hedgehog*, which is expressed only in mid-stage papillae in HH stage 35–36 embryos (Franz-Odendaal, 2008). I have recently shown that inhibiting *Hedgehog* signaling within the chicken embryonic eye can disrupt the normal patterning of scleral ossicles, reducing both the size and number of ossicles within the ring (Franz-Odendaal, 2008).

Surprisingly, a variation of up to two ossicles per eye within the same wild-type chicken embryo can occur. Asymmetry by one ossicle is, however, more common (44% of sclerotic rings were asymmetrical, $n = 68$) (Franz-Odendaal, 2008). What is even more curious is that this asymmetry is random (with regard to side, left or right) but that the direction of asymmetry is not random—a reduction from the modal number is more common than an increase from the modal number (Franz-Odendaal, 2008). This suggests that inhibiting a papilla is perhaps easier than inducing an extra one; perhaps inhibition is the more parsimonious process.

One part of the ocular skeleton not yet mentioned, because it develops earlier and independently of scleral ossicles, is the cartilage within the sclera (known as "scleral cartilage"). Both the scleral cartilage and the scleral ossicles can have their form and pattern disrupted by perturbing eye growth. Early work by Coulombre et al. (1962) showed that after draining fluid from the developing optic cup, the normal differentiation of cartilage and bone was not affected. However, the form and pattern were affected. The scleral cartilage developed on schedule but was reduced in size, corresponding to the smaller eye which it encapsulated; this cartilage was also thicker than normal. The scleral ossicles were also affected—fewer ossicles formed—correlating to the reduced eye size. In both elements, the necessary tissue interactions took place and normal skeletal differentiation resulted; however, the pattern and form were altered by local influences (in this case, eye growth).

PLATE SHAPE What is similarly unclear is how the variety of shapes for the individual scleral ossicles and the sclerotic ring is established (Figure 12.5). Some lizards, for example, have hourglass-shaped ossicles, and some owls have very elongated ossicles. The elongated ossicles of the owl develop probably as a result of the tubular eyes that are present in this group. Exactly how this is accomplished is not clear. Altering the shape of the ossicle condensation must come about through interactions between the developing soft tissues of the eye and developing bone within the sclera; how this is accomplished and what epigenetic processes are involved are not known.

INDUCTION Interestingly, scleral cartilage is induced by the retinal pigmented epithelium, unlike scleral ossicles which are induced by conjunctival epithelium. The scleral cartilage induction is mediated by elements of epithelial extracellular matrices localized within the basement membrane, similar to the induction of mandibular membrane bone in response to the mandibular epithelium. Pinto and Hall (1991) showed that mandibular epithelium can elicit osteogenesis from scleral mesenchyme and conjunctival (scleral) epithelium can elicit osteogenesis from mandibular mesenchyme, suggesting that the epigenetic mechanisms are fundamentally similar. It is possible that both epithelia produce a similar factor (or epigenetic agent), which is trapped in the mandibular basement membrane, creating matrix-mediated interaction, but allowed to pass through the basement membrane of the scleral or conjunctival epithelium, creating a diffusion-mediated interaction (Hall, 1990). The differences between the two epigenetic inductive events (for mandibular bone and for scleral ossicles) may therefore lie in the structural organization of the epithelial basement membrane.

We are only beginning to identify the key factors involved in the scleral ossicle system. Once identified, we can begin to understand how these factors are interpreted by cells and tissues, how the cells and tissues respond and interact, and how the ultimate variation in scleral ossicles arises. Through embryonic manipulation and an epigenetic approach my lab hopes to unravel this intriguing skeletal system.

SUMMARY

Throughout this chapter, I have described the epigenetic factors involved in skeletal development and tried to demonstrate how phenotypic variability is influenced by the cumulative effect of these factors. These epigenetic phenomena include tissue, cell, and protein interactions, with the critical step of condensation formation. In many cases, the epigenetic processes involved are not clearly identified, let alone understood. In order to understand how skeletal variation arises during evolution, we need to understand how the molecular and cellular mechanisms underlying the vertebrate body

plan develop and are maintained, coordinated, and modified. Ultimately, the more traditional methods of direct embryonic tissue manipulation in conjunction with studies at the gene level will be very useful in trying to fully understand the epigenetic processes involved in bone and cartilage development.

ACKNOWLEDGMENTS

I would first and foremost like to thank Brian K. Hall (Dalhousie University) for his mentorship over the last 5 years. I am also very grateful to Natural Sciences and Engineering Research Council of Canada for funding my research on vertebrate skeletal biology. I appreciate the comments made by B. K. Hall and B. Hallgrímsson on earlier versions of this chapter. My students and previous lab colleagues are thanked for their enthusiastic conversations revolving around the dynamic vertebrate skeleton.

REFERENCES

Abzhanov, A., W. P. Kuo, C. Hartmann, R. Grant, R. P. Grant, and C. J. Tabin. 2004. The calmodulin pathway and the evolution of elongated beak morphology in Darwin's finches. *Nature* 442:563–7.

Abzhanov, A., S. J. Robba, A. P. McMahon, and C. J. Tabin. 2007. Regulation of skeletogenic differentiation in cranial dermal bone. *Development* 134:3133–44.

Akiyama, H., M.-C. Chaboissier, J. F. Martin, A. Schedl, and B. de Crombrugghe. 2002. The transcription factor Sox9 has essential roles in successive steps of the chondrocyte differentiation pathway and is required for expression of *Sox5* and *Sox*. *Genes Dev* 16:2813–28.

Albertson, R. C., J. T. Streelman, T. D. Kocher, and P. C. Yelick. 2005. Integration and evolution of the cichlid mandible: The molecular basis of alternate feeding strategies. *Proc Natl Acad Sci USA* 102:16287–92.

Aoyama, H., Y. Moitzutani-Koseki, and H. Koseki. 2005. Three developmental compartments involved in rib formation. *Int J Dev Biol* 49:325–33.

Atchley, W. R., and B. K. Hall. 1991. A model for development and evolution of complex morphological structures. *Biol Rev Camb Philos Soc* 66:101–57.

Bandyopadhyay, A., J. K. Kubilus, M. L. Crochiere, and T. F. Linsenmayer. 2008. Identification of unique molecular subdomains in the perichondrium and periosteum and their role in regulating gene expression in the underlying chondrocytes. *Dev Biol* 321:162–74.

Banks, W. J. 1974. The ossification process of the developing antler in the white-tailed deer (*Odocoileus virginianus*). *Calcif Tissue Res* 14:257–74.

Bi, W., J. M. Deng, Z. Zhang, R. R. Behringer, and B. de Crombrugghe. 1999. *Sox9* is required for cartilage formation. *Nat Genet* 22:85–9.

Canavese, B., C. Vignolini, U. Fazzini, and S. Bellardi. 1987. Pardcolarica morfologiche dell'anello osseo sclerae della quaglia domestica. *Ann Fac Med Vet Di Torino* 32:3–10.

Cebra-Thomas, J., F. Tan, S. Sistla, E. Estes, G. Bender, C. Kim, P. Riccio, and S. F. Gilbert. 2005. How the turtle forms its shell: A paracrine hypothesis of carapace formation. *J Exp Zool* 304:558–69.

Chai, Y., X. Jiang, Y. Ito, P. Bringas, Jr., J. Han, D. H. Rowitch, P. Soriano, A. P. McMahon, and H. M. Sucov. 2000. Fate of the mammalian cranial neural crest during tooth and mandibular morphogenesis. *Development* 127:1671–9.

Chai, Y., A. Mah, C. Crohin, S. Groff, P. Bringas, Jr., T. Le, V. Santos, and H. C. Slavkin. 1994. Specific transforming growth factor-beta subtypes regulate embryonic mouse Meckel's cartilage and tooth development. *Dev Biol* 162:85–103

Chai, Y., and R. E. Maxson, Jr. 2006. Recent advances in craniofacial morphogenesis. *Dev Dyn* 235: 2353–75.

Chevallier, A. 1975. Rôle du mésoderme somitique dans le développement de la cage thoracique de l'embryon d'oiseau. I. Origine du segment sternal et mécanismes de la différenciation des côtes. *J Embryol Exp Morphol* 33:291–311.

Coulombre, A. J., and J. L. Coulombre. 1973. The skeleton of the eye. II. Overlap of the scleral ossicles of the domestic fowl. *Dev Biol* 33:257–67.

Coulombre, A. J., J. L. Coulombre, and H. Mehta. 1962. The skeleton of the eye. I. Conjunctival papillae and scleral ossicles. *Dev Biol* 5:382–401.

Couly, G. F., P. M. Coltey, and N. M. Le Douarin. 1993. The triple origin of skull in higher vertebrates: A study in quail-chick chimeras. *Development* 117:409–29.

Crochiere, M. L., J. K. Kubilus, and T. F. Linsenmayer. 2008. Perichondral-mediated TGF-beta regulation of cartilage growth in avian long bone development. *Int J Dev Biol* 52:63–70.

Curtis, E. L., and R. C. Miller. 1938. The sclerotic ring in North American birds. *Auk* 55:225–43.

De Crombrugghe, B., V. Lefebvre, and K. Nakashima. 2001. Regulatory mechanisms of cartilage and bone formation. *Curr Opin Cell Biol* 13:721–7.

Depew, M. J., and C. Compagnucci. 2008. Tweaking the hinge and caps: Testing a model of the organization of jaws. *J Exp Zoolog B Mol Dev Evol* 310:315–35.

Depew, M. J., T. Lufkin, and J. L. R. Rubenstein. 2002. Specification of jaw subdivisions by *Dlx* genes. *Science* 298:381–5.

Downie, S. A., and S. A. Newman. 1995. Different roles for fibronectin in the generation of fore and hind limb precartilage condensations. *Dev Biol* 172:519–30.

Ducy, P., R. Zhang, V. Geoffroy, A. L. Ridall, and G. Karsenty. 1997. Osf2/Cbfa1: A transcriptional activator of osteoblast differentiation. *Cell* 89:747–54.

Dunlop, L. T., and B. K. Hall. 1995. Relationships between cellular condensation, preosteoblast formation and epithelial–mesenchymal interactions in initiation of osteogenesis. *Int J Dev Biol* 39:357–71.

Eames, B. F., L. de la Fuente, and J. A. Helms. 2003. Molecular ontogeny of the skeleton. *Birth Defects Res C Embryo Today* 69:93–101.

Eames, B. F., and J. A. Helms. 2004. Conserved molecular program regulating cranial and appendicular skeletogenesis. *Dev Dyn* 231:4–13.

Ehehalt, F., B. Wang, B. Christ, K. Patel, and R. Huang. 2004. Intrinsic cartilage-forming potential of dermomyotome cells requires ectodermal signals for the development of the scapula blade. *Anat Embryol (Berl)* 208:431–7.

Epstein, J. A. 2001. Developing models of DiGeorge syndrome. *Trends Genet* 17:S13–17.

Evans, D. J. R. 2003. Contribution of somitic cells to the avian ribs. *Dev Biol* 256:115–27.

Evans, D. J. R., and D. M. Noden. 2006. Spatial relations between avian craniofacial neural crest and paraxial mesoderm cells. *Dev Dyn* 235: 1310–25.

Franz-Odendaal, T. A. 2008. Towards understanding the development of scleral ossicles in the chicken, *Gallus gallus*. *Dev Dyn* 237:3240–51.

Franz-Odendaal, T. A., and B. K. Hall. 2006. Skeletal elements within teleost eyes and a discussion on their homology. *J Morphol* 267:1326–37.

Franz-Odendaal, T., B. K. Hall, and P. E. Witten. 2006. Buried alive: How osteoblasts become osteocytes. *Dev Dyn* 235:176–90.

Franz-Odendaal, T., K. Ryan, and B. K. Hall. 2007. Developmental and morphological variation in the teleost craniofacial skeleton reveals an unusual mode of ossification. *J Exp Zool B Mol Dev Evol* 308:709–21.

Gilbert, S. F. 2003. *Developmental Biology*, 7th ed. Sunderland, MA: Sinauer Associates.

Gluhak, J., A. Mais, and M. Mina. 1996. Tenascin-C is associated with early stages of chondrogenesis by chick mandibular ectomesenchymal cells in vivo and in vitro. *Dev Dyn* 205:24–40.

Gross, J.B., and J. Hanken. 2008. Review of fate-mapping studies of osteogenic cranial neural crest in vertebrates. *Dev Biol* 317:389–400.

Hall, B.K. 1978. *Developmental and Cellular Skeletal Biology*. New York: Academic Press.

Hall, B.K. 1981. Specificity in the differentiation and morphogenesis of neural-crest-derived scleral ossicles and of epithelial scleral papillae in the eye of the embryonic chick. *J Embryol Exp Morphol* 66:175–90.

Hall, B.K. 1984. Genetic and epigenetic control of connective tissues in the craniofacial structures. *Birth Defects Res* 20:1–17.

Hall, B.K. 1990. Genetic and epigenetic control of vertebrate embryonic development. *Neth J Zool* 40:352–61.

Hall, B.K. 2005. *Bones and Cartilage: Developmental and Evolutionary Skeletal Biology*. London: Elsevier.

Hall, B.K., and T. Miyake. 1995. Divide, accumulate, differentiate: Cell condensation in skeletal development revisited. *Int J Dev Biol* 39:881–93.

Hall, B.K., and T. Miyake. 2000. All for one and one for all: Condensations and the initiation of skeletal development. *BioEssays* 22:138–47.

Haworth, K.E., C. Healy, P. Morgan, and P.T. Sharpe. 2004. Regionalisation of early head ectoderm is regulated by endoderm and prepatterns the orofacial epithelium. *Development* 131:4797–4806.

Herring, S.W., and S. Tend. 2000. Strain in the braincase and its sutures during function. *Am J Phys Anthropol* 112:575–93.

Hockman, D., C.J. Cretekos, M.K. Mason, R.R. Behringer, D.S. Jacobs, and N. Illing. 2008. A second wave of Sonic hedgehog expression during the development of the bat limb. *Proc Natl Acad Sci USA* 105:16982–987.

Hoshi, K., T. Komori, and H. Ozawa. 1999. Morphological characterization of skeletal cells in Cbfa1-deficient mice. *Bone* 25:639–51.

Huang, R., B. Christ, and K. Patel. 2006. Regulation of scapula development. *Anat Embryol* 211 (Suppl 1): 65–71.

Huang, R., Q. Zhi, K. Patel, J. Wilting, and B. Christ. 2000a. Dual origin and segmental organization of the avian scapula. *Development* 127:3789–94.

Huang, R., Q. Zhi, C. Schmidt, J. Wilting, B. Brand-Saberi, and B. Christ. 2000b. Sclerotomal origin of the ribs. *Development* 127:527–32.

Hutson, M.R., and M.L. Kirby. 2007. Model systems for the study of heart development and disease. Cardiac neural crest and conotrunc malformations. *Semin Cell Dev Biol* 18:101–10.

Inada, M., T. Yasui, S. Nomura, S. Miyake, K. Deguchi, M. Himeno, M. Sato, et al. 1999. Maturational disturbance of chondrocytes in Cbfa1-deficient mice. *Dev Dyn* 214:279–90.

Ishii, M., J. Han, H.Y. Yen, H.M. Sucov, Y. Chai, and R.E. Maxson, Jr. 2005. Combined deficiencies of Msx1 and Msx2 cause impaired patterning and survival of the cranial neural crest. *Development* 132:4937–50.

Ito, Y., P. Bringas, Jr., A. Mogharei, J. Zhao, C. Deng, and Y. Chai. 2002. Receptor-regulated and inhibitory Smads are critical in regulating transforming growth factor beta-mediated Meckel's cartilage development. *Dev Dyn* 224:69–78.

Kim, I.S., F. Otto, B. Zabel, and S. Mundlos. 1999. Regulation of chondrocyte differentiation by Cbfa1. *Mech Dev* 80:159–70.

Köntges, G., and A. Lumsden. 1996. Rhombencephalic neural crest segmentation is preserved throughout craniofacial ontogeny. *Development* 122:3229–42.

Kurosaka, S., and A. Kashina. 2008. Cell biology of embryonic migration. *Birth Defects Res C Embryo Today* 84:102–22.

Liu, W., J. Selever, D. Murali, X. Sun, S.M. Brugger, L. Ma, R.J. Schwartz, R. Maxson, Y. Furuta, and J.F. Martin. 2005. Threshold-specific requirements for *Bmp4* in mandibular development. *Dev Biol* 283:282–93.

Luo, Z.-X., A.W. Crompton, and A.-L. Sun. 2001. A new mammaliaform from the early Jurassic and evolution of mammalian characteristics. *Science* 292:1535–9.

MacDonald, M.E., and B.K. Hall. 2001. Altered timing of the extracellular-matrix mediated epithelial–mesenchymal interaction that initiates mandibular skeletogenesis in three inbred strains of mice: Development, heterochrony, and evolutionary change in morphology. *J Exp Zool* 291:258–73.

Mallo, M. 2003. Formation of the outer and middle ear, molecular mechanisms. *Curr Top Dev Biol* 57:85–103.

McBratney-Owen, B., S. Iseki, S.D. Bamford, B.R. Olsen, and G.M. Morris-Kay. 2008. Development and tissue origins of the mammalian cranial base. *Dev Biol* 322:121–132.

Mehrara, B.J., D. Most, J. Chang, S. Bresnick, A. Turk, S.A. Schendel, G.K. Gittes, and M.T. Longaker. 1999. Basic fibroblast growth factor and transforming growth factor β1 expression in the developing dura mater correlates with calvarial bone formation. *J Am Soc Plast Surg* 104:435–44.

Moore, S. W. 2003. Scrambled eggs: Mechanical forces as ecological factors in early development. *Evol Dev* 5:61–6.

Mori-Akiyama, Y., H. Akiyama, D. H. Rowitch, and B. de Crombrugghe. 2003. Sox9 is required for determination of the chondrogenic cell lineage in the cranial neural crest. *Proc Natl Acad Sci USA* 100:9360–5.

Newsome, D. A. 1972. Cartilage induction by retinal pigmented epithelium of chick embryos. *Dev Biol* 27:575–9.

Noden, D. M. 1982. Patterns and organization of craniofacial skeletogenic and myogenic mesenchyme: A perspective. *Prog Clin Biol Res* 101:167–203.

Noden, D. M. 1983. The role of the neural crest in patterning avian cranial skeletal, connective and muscle tissues. *Dev Biol* 96:144–65.

Noden, D. M., and P. A. Trainor. 2005. Relations and interactions between cranial mesoderm and neural crest populations. *J Anat* 207:575–601.

Oka, K., S. Oka, R. Hosokawa, P. Bringas, H. C. Brockhoff II, K. Nonaka, and Y. Chai. 2008. TGF-β mediated Dlx5 signaling plays a crucial role in osteo-chondroprogenitor cell lineage determination during mandible development. *Dev Biol* 321:303–9.

Oka, K., S. Oka, T. Sasaki, Y. Ito, P. Bringas, Jr., K. Nonaka, and Y. Chai. 2007. The role of TGF-β signaling in regulating chondrogenesis and osteogenesis during mandibular development. *Dev Biol* 303:391–404.

Opperman, L. A., R. W. Passerelli, E. P. Morgan, M. Reintjies, and R. C. Ogle. 1995. Cranial sutures require tissue interactions with dura mater to resist osseous obliteration in vitro. *Bone Miner Res* 10:1978–87.

Opperman, L. A., and J. T. Rawlins. 2005 The extracellular matrix environment in suture morphogenesis and growth. *Cells Tissues Organs* 181: 127–35.

Otto, F., A. P. Thornell, T. Crompton, A. Denzel, K. C. Gilmour, I. R. Rosewell, G. W. H. Stamp, et al. 1997. Cbfa1, a candidate gene for cleidocranial dysplasia syndrome, is essential for osteoblast differentiation and bone development. *Cell* 89:765–71.

Parfitt, A. M. 1990. Bone-forming cells in clinical conditions. In: *Bone*, ed. B. K. Hall, 351–429. Boca Raton, FL: Telford Press and CRC Press.

Pathi, S., J. B. Rutenberg, R. L. Johnson, and A. Vortkamp. 1999. Interaction of Ihh and BMP/Noggin signaling during cartilage differentiation. *Dev Biol* 209:239–53.

Pinto, C. B., and B. K. Hall. 1991. Toward an understanding of the epithelial requirement for osteo-

genesis in scleral mesenchyme of the embryonic chick. *J Exp Zool* 259:92–108.

Rafferty, K. L., S. W. Herring, and F. Artese. 2000. Three dimensional loading and growth of the zygomatic arch. *J Exp Biol* 203:2093–3004.

Rauch, F., and E. Schoenau. 2001. The developing bone: Slave or master of its cells and molecules? *Pediatr Res* 50:309–14.

Rice, D. P. C., R. Rice, and I. Thesleff. 2003. Molecular mechanisms in calvarial bone and suture development, and their relation to craniosynostosis. *Eur J Orthod* 25:139–48.

Rijli, F. M., M. Mark, S. Lakkaraju, A. Dierich, P. Dolle, and P. Chambon. 1994. A homeotic transformation is generated in the rostral branchial region of the head by disruption of *Hoxa-2*, which acts as a selector gene. *Cell* 75:1333–49.

Rodríguez-Trelles, F., R. Tarrío, and F. J. Ayala. 2005. Is ectopic expression caused by deregulatory mutations or due to gene-regulation leaks with evolutionary potential? *BioEssays* 27:592–601.

Sakamoto, N., N. Kanatani, and H. Aoyama. 2001. The rostrocaudal patterning of paraxial mesoderm in the primitive streak. *Dev Growth Differ* 43(Suppl): S31.

Santagati, F., and F. M. Rijli. 2003. Cranial neural crest and the building of the vertebrate head. *Nat Rev Neurosci* 4:806–18.

Satokata, I., L. Ma, H. Ohshima, M. Bei, I. Woo, K. Nishizawa, T. Maeda, et al. 2000. *Msx2* deficiency in mice causes pleiotropic defects in bone growth and ectodermal organ formation. *Nat Genet* 24:391–5.

Sears, K. E., R. R. Behringer, J. J. Rasweiler IV, L. A. Niswander. 2006. Development of bat flight: Morphological and molecular evolution of bat wing digits. *Proc Natl Acad Sci USA* 103:6581–6.

Skerry, T. 2000. Biomechanical influences on skeletal growth and development. In *Development, Growth and Evolution: Implications for the Study of the Hominid Skeleton*, ed. P. O'Higgins and M. J. Cohn, 29–39. New York: Academic Press.

Smith, K. K., and R. A. Schneider. 1998. Have gene knockouts caused evolutionary reversals in the mammalian first arch? *BioEssays* 20:245–55.

St. Jacques, B., M. Hammerschmidt, and A. McMahon. 1999. Indian hedgehog signaling regulates proliferation and differentiation of chondrocytes and is essential for bone formation. *Genes Dev* 13:2072–86.

Stoller, J. Z., and J. A. Epstein. 2005. Cardiac neural crest. *Semin Cell Dev Biol* 16:704–15.

Storm, E., and D. Kingsley. 1996. Joint patterning defects caused by single and double mutations in members of the bone morphogenetic protein BMP family. *Development* 122:3969–79.

Terai, Y., N. Morikawa, and N. Okada. 2002. The evolution of the pro-domain of bone morphogenetic protein 4 (*Bmp4*) in an explosively speciated lineage of East African cichlid fishes. *Mol Biol Evol* 19:1628–32.

Thorogood, P. 1993. Differentiation and morphogenesis of cranial skeletal tissues. In *The Skull*, ed. J. Hanken and B. K. Hall. Chicago: University of Chicago Press.

Toma, C. D., J. L. Schaffer, M. C. Meazzini, D. Zurakowski, H.-D. Nah, and L. C. Gerstenfeld. 1997. Developmental restriction of embryonic calvarial cell populations as characterized by their in vitro potential for chondrogenic differentiation. *J Bone Miner Res* 12:2024–39.

Trumpp, A., M. J. Depew, J. L. R. Rubenstein, J. M. Bishop, and G. R. Martin. 1999. Cre-mediated gene inactivation demonstrates that FGF8 is required for cell survival and patterning of the first branchial arch. *Genes Dev* 13:3136–48.

Tucker, A. S., and P. Sharpe. 2004. The cutting-edge of mammalian development: How the embryo makes teeth. *Nat Rev Genet* 5:499–508.

Ueta, C., M. Iwamoto, N. Kanatani, C. Yoshida, Y. Liu, M. Enomoto-Iwamoto, T. Ohmori, et al. 2001. Skeletal malformations caused by overexpression of Cfba1 or its dominant negative form in chondrocytes. *J Cell Biol* 153:87–100.

Vickaryous, M. K., and B. K. Hall. 2006. Homology of the reptilian coracoid and a reappraisal of the evolution and development of the amniote pectoral apparatus. *J Anat* 208(3):263–85.

Vickaryous, M. K., and B. K. Hall. 2008. Development of the dermal skeleton in *Alligator mississipiensis* (Archosaura, Crocodylia) with comments on the homology of osterderms. *J Morphol* 269: 398–422.

Wang, B., L. He, F. Ehehalt, P. Geetha-Loganathan, S. Nimmagadda, B. Christ, M. Scaal, and R. Huang. 2005. The formation of the avian scapula blade takes place in the hypaxial domain of the somites and requires somatopleure-derived BMP signals. *Dev Biol* 287:11–18.

Wang, Y., Y. Hu, J. Meng, and C. Li. 2001. An ossified Meckel's cartilage in two Cretaceous mammals and origin of the mammalian middle ear. *Science* 294:357–61.

Willmore, K. E., N. M. Young, and J. T. Richtsmeier. 2007. Phenotypic variability: Its components, measurement and underlying developmental processes. *Evol Biol* 43:99–120.

Witten, P. E., A. Hansen, and B. K. Hall. 2001. Features of mono- and multinucleated bone resorbing cells of the zebrafish *Danio rerio* and their contribution to skeletal development, remodelling and growth. *J Morphol* 250:197–207.

Wu, P., T. X. Jiang, S. Suksaweang, R. B. Widelitz, and C. M. Chuong. 2004. Molecular shaping of the beak. *Science* 305:1465–6.

Wu, S., L. Page, and N. M. Sherwood. 2006. Brain regionalization and eye development: A role for GnRH in zebrafish embryo. *Dev Biol* 295:371.

Yin, M., and M. Pacifici. 2001. Vascular regression is required for mesenchymal condensation and chondrogenesis in the developing limb. *Dev Dyn* 222:522–33.

Young, R., and A. Badyaev. 2007. Evolution of ontogeny: Linking epigenetics remodeling and genetic adaptation in skeletal structures. *Integr Comp Biol* 47:234–44.

Zhou, G., Q. Zheng, F. Engin, E. Munivez, Y. Chen, E. Sebald, D. Krakow, and B. Lee. 2006. Dominance of SOX9 function over RUNX2 during skeletogenesis. *Proc Natl Acad Sci USA* 103: 19004–9.

13

Muscle–Bone Interactions and the Development of Skeletal Phenotype

JAW MUSCLES AND THE SKULL

Susan W. Herring

CONTENTS

The association between the musculature and the skeleton is both physiological and physical. Cell origins for the two tissues are not identical. For example, the cranial neural crest forms much of the skull but contributes only connective tissue to the muscles (Noden and Trainor, 2005). Nevertheless, muscular and skeletal cells go through a common mesenchymal stage of differentiation. Muscles and bones are also linked by genes and hormones which affect both tissues. Muscles and bones are attached to each other and jointly share the responsibility for support and movement of body parts.

The most fundamental aspect of the interaction between muscles and bones is mechanical. Muscles load bones—indeed, are the major source of bone loading—probably even before birth or hatching and certainly after, especially in flying and aquatic vertebrates, whose skeletons are partially relieved of weight bearing. As discussed elsewhere in this volume and in numerous reviews (Pearson and Lieberman, 2004; Allori et al., 2008), bones respond to mechanical loading at many levels, including proliferation of progenitor cells, skeletogenic differentiation, and modeling and remodeling of existing skeletal elements. This responsiveness is epigenetic in the Waddingtonian sense of noninherited (but controllable) changes in gene activity (Pearson, 2008) that lead to a particular phenotype. Understanding the influence of muscles on bones at this level may be as straightforward (although experimentally difficult) as understanding the static and dynamic biomechanics of muscle-contraction patterns.

Mechanical forces arising from muscle are particularly interesting epigenetic stimuli from an evolutionary point of view because muscle-contraction patterns are an intrinsic part of the way animals interact with their environment, both uterine and external. While gross aspects of muscle-contraction pattern are intrinsic ("hardwired"), such as the alternation of jaw openers and jaw closers, many finer aspects of behavior develop gradually and involve learning, such as the acquisition of a transverse chewing stroke (Herring, 1985a). In addition to considerable intra- as well as interindividual variability (German et al., 2008), muscle-contraction patterns necessarily vary with conditions. Ecological conditions that alter the landscape and sources of food can be expected to cause changes in the activity of locomotory and feeding muscles, and these in turn may affect skeletal morphology.

The extent of the skeletal response to muscular stimuli, like any aspect of physiological adaptation, depends not only on the strength of the stimulus but also on the general reactivity of the system, which varies with metabolic rate in different taxa and with different stages of life within individuals. In addition, individual skeletal elements vary in the relative importance of muscular stimuli as opposed to other mechanical or nonmechanical influences.

In this chapter I will review some of the suggestions made for coregulation (and possibly coevolution) of the muscular and skeletal systems, discuss general issues of skeletal responsiveness to muscle activity, and then offer more detail in examples drawn from the mammalian skull, which, because it does not bear body weight and yet has extensive muscle attachments, provides opportunities for experimental investigation.

HOW DO MUSCLES AFFECT THE SKELETON?

MECHANISMS OF INTERACTION

Muscle and bone cells have much in common, including their passage through a mesenchymal stage. Myogenic and skeletogenic differentiation are among the fates available for both embryonic (Koh et al., 2009) and adult (Crisan et al., 2008) mesenchymal stem cells. Some attachment processes of bones arise within muscle anlagen, possibly from myogenic cells (Rot-Nikcevic et al., 2007). Interestingly, one of the most significant influences for directing stem cells toward myogenic vs. osteogenic differentiation is the stiffness of the matrix on which cells are grown (Engler et al., 2006; Rowlands et al., 2008), a clearly mechanical, as well as epigenetic, effect. Given their common cellular history, it might be anticipated that these tissue types might also have common regulatory elements.

A recent review has tabulated candidate genes and pathways for pleiotropic action on both muscle and bone (Karasik and Kiel, 2008). Chief among these pleiotropic pathways are the steroid hormones. Both muscle and bone cells have receptors for sex steroids, and musculoskeletal growth is one of the most obvious features of puberty. Estrogens are chiefly known for their (mostly) antiresorptive and promechanical adaptation actions on bone (Westerlind et al., 1997; Lee et al., 2003; Saxon and Turner, 2006; Zaman et al., 2006), in both males and females (Orwoll, 2003); but their effects on muscle are negligible (Zofková, 2008). Androgens, on the contrary, are well-known (indeed, notorious) promoters of muscle growth (Graham et al., 2008) while also being strongly anabolic for bone in both males and females, by direct action on osteocytes and by conversion to estrogen via aromatase (Kasra and Grynpas, 1995; Notelovitz, 2002). Larger muscles, of course, also increase the mechanical loading of bones, and this action probably supplements the direct effect of androgens on bone cells (Notelovitz, 2002).

The growth hormone–insulin-like growth factor 1 (IGF-1) axis also promotes growth in both tissue types. Growth hormone (somatotropin) promotes protein synthesis and, hence, an increase in muscle mass (again contributing to skeletal loading). Independently, it activates osteogenesis on bone surfaces, at least in part by making the bones more sensitive to mechanical stimuli (Forwood et al., 2001; Zofková, 2008). Growth hormone further leads to the local production of IGF-1 (somatomedin), a potent stimulator of cartilage growth (Laron, 2001). Leptin, a hormone which regulates body and bone mass through the central nervous system (Takeda et al., 2002), stimulates this axis and is closely correlated with serum IGF-1 levels (Hamrick and Ferrari, 2008; Hamrick et al., 2008). IGF-1 can also be elicited and/or up-regulated by mechanical strain in muscular (Cheema et al., 2005) and skeletal (Hirukawa et al., 2005) tissues, with anabolic effects on both. This chain of events appears to work in reverse as well. Evidence reviewed by Bikle (2008) suggests that the rapid loss of bone mass following skeletal unloading can be ascribed to blocked IGF-1 signaling. Bikle (2008) goes on to suggest a possible mechanism, specifically that absence of mechanical stimulation reduces the expression of integrins, which are necessary for formation of an IGF–receptor complex in osteoblasts.

It is notable that both the steroid hormone and growth hormone–IGF pathways are intertwined with mechanics; loading is a stimulus for IGF-1 production, and estrogen is a facilitator of mechanical adaptation. Other candidate pathways for pleiotropic action on muscle and bone are less well studied, but muscle activity itself is often implicated as the common element influencing both tissues via mechanics. For example, polymorphisms in LRP5, a Wnt modulator, are associated with bone mineral density and with physical activity, at least in human males (Kiel et al., 2007). Genetic changes that enlarge muscles (such as mutations in myostatin/GDF-8 [Hamrick et al., 2000]) or inactivate them (such as defects in calcium channels [Herring and Lakars, 1981; Powell et al., 1996]) also affect bone morphology, but the skeletal changes are secondary to those of the muscle and, in most cases, can be ascribed to increased or decreased mechanical action of the muscle.

The remainder of this chapter will therefore be devoted to a consideration of muscle activity as the basis of communication between muscles and bones. Developmental coordination between force production (the muscles) and force reception (the skeleton) is an example of "ontogenetic integration of form and function" (Herring, 2003a) and is presumably the result of selection for a mechanism for matching body systems that work in concert. Muscle–bone interactions that depend on muscle activity begin relatively late in the chronology of development (Hallgrímsson and Lieberman, 2008) but are long-lasting. Muscle activity begins when myoblasts and myotubes become contractile during organogenesis and continues throughout life. The mechanisms by which muscle activity communicates with the skeleton include (1) overall physical activity and accompanying effects on metabolic rate; (2) straightforward mechanical effects in which the muscle forces could theoretically be replaced by a nonmuscular source of load with the same response from the skeleton; and (3) events relating to the interface of the two tissues at the attachment sites, for example, due to a shared blood or nerve supply. Although overall physical activity probably affects all skeletal elements similarly, events at the attachment site and mechanical effects are clearly specific to particular elements or even parts of elements, such as the greater trochanter of the femur or the coronoid process of the mandible.

OVERALL EFFECTS: PHYSICAL ACTIVITY, METABOLIC RATE, SET POINTS, ENDOTHERMY

Physical activity in birds and mammals is typically associated with increased skeletal mass (Plochocki et al., 2008), even though total body mass may decrease (Morgan et al., 2003). This response, which takes the form of increased bone density and/or diameter, is in large part due to mechanical loading (including load bearing due to the increased muscle mass itself). Mechanical loading is usually considered to be a local effect and is dealt with in detail below. However, some aspects of the response to physical activity appear to be related to more general phenomena such as metabolic rate and are dependent on factors such as age and gender.

The response to physical activity is mediated by the constant bony remodeling that characterizes birds and mammals and is probably related to their elevated metabolic rate. Bony remodeling, which subserves the functions of calcium homeostasis and hematopoiesis (Parfitt, 2002), is faster than that needed to maintain structural integrity (Heaney, 2003). Even though bony remodeling is not always targeted to mechanically "needy" areas (Parfitt, 2002), it provides an opportunity for the skeleton to rebuild itself in accordance with current mechanical conditions. Even when specifically triggered by localized loading, there is evidence for a neurally regulated, multiple-bone remodeling reaction (Sample et al., 2008).

The skeletal response to physical activity is most noticeable during the growth period (Suominen, 2006). The older skeleton continues or even accelerates remodeling but loses some of its capacity for mounting an osteogenic response (Pearson and Lieberman, 2004; Seeman, 2008). Thus, although adaptations continue to occur in maturity, they differ in both quality and quantity from those in juveniles (Bouvier and Hylander, 1996; Nara-Ashizawa

et al., 2002). Physical activity and muscle mass decrease with age, and this provides an additional explanation of why bone mass is lost in older humans (Melton et al., 2006). Interestingly, a drop in physical activity, at least in mice, precedes age decreases in leptin level and musculoskeletal mass (Hamrick et al., 2006). Age and gender interact in the skeleton. Possibly because of differences in body composition, muscle cross section is a better predictor of bone strength in males than in females (Fricke et al., 2008). Further, because estrogen apparently makes bone more sensitive to mechanical stimulation, its loss at menopause changes the set point for skeletal adaptation in females (Turner, 1999).

If metabolic rate is an important determinant of the skeletal response to physical activity, it is reasonable to ask whether ectothermic vertebrates show the same phenomena as endothermic birds and mammals. In terms of total body mass, ectotherms may be more affected by overall physical activity than endotherms for the simple reason that their energy budgets are more limited. General physical activity requires frequent, although not necessarily powerful, contraction from many muscle groups and, therefore, uses energy. Physical activity can be elevated in ectotherms by a number of different environmental conditions, of which one entertaining example is the hyperactivity of *Xenopus* tadpoles in a mirrored aquarium, examining and attempting to school with their own image, which results in a decrease in growth rate (Gouchie et al., 2008). However, evidence for a specific skeletal reaction to physical activity is scanty at best, despite claims for teleosts as promising animal models for studying bone disorders related to unloading (Renn et al., 2006). The best examples, such as increased head size of eastern tiger snakes (*Notechis scutatus*) that consume larger prey (Aubret et al., 2004), concern overall growth patterns rather than a specific mechanical adaptation. Further, the process of remodeling is probably highly energy-dependent and, thus, unlikely to occur to any great extent in ectotherms. Indeed, little turnover occurs in the bones of ectotherms, as illustrated by the scarcity of histological features that facilitate remodeling, such as cortical vascular canals (de Buffrénil et al., 2008) and osteogenic endosteal surfaces (Cubo et al., 2008). Rather, the details of skeletal organization in ectotherms are probably better explained as a result of growth dynamics than of muscle–bone interactions (Lee, 2004).

LOCAL EFFECTS: DIFFERENT SKELETAL ELEMENTS AND LOADING CONDITIONS

Local stress concentrations from muscle pull may be one factor involved in the initiation of ossification. Differences in postcranial ossification patterns between marsupials and placentals suggest an influence of muscular stress (Weisbecker et al., 2008), and correlations between centers of ossification and local muscle attachment have been noted for cranial bones and scleral ossicles in a number of vertebrate groups (Spyropoulos, 1977; Wake and Hanken, 1982; Rieppel, 1987; Franz-Odendaal et al., 2007). These effects, however, may be minor relative to the total phylogenetic and ontogenetic signaling apparatus for ossification (Smith, 1994).

Experimental paralysis of embryos or knockouts of genes necessary for myogenesis also suggest that muscle contraction plays a role in skeletal morphogenesis that varies in importance in various bones. For example, chicks paralyzed early in ontogeny show minor shape and size changes in the mandible but much more serious defects in the furcula and sternum, including failure of the sternum to fuse (Figure 13.1) (Hall and Herring, 1990). The severity of changes in the sternum mostly likely involves the direct or indirect dependence of movement of the sternal plates on muscle activity. Other than the sternum and furcula, many skeletal elements can form relatively normally and grow somewhat normally even in the complete absence of muscle tissue, although they lack muscle processes and the nuances of shape that typify normal bones (Rot-Nikcevic et al., 2006). One reason for the differential effect of

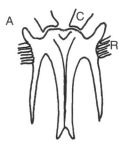

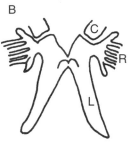

FIGURE 13.1 Ventral views of the sternum of 12-day chick embryos (cranial is toward the top). A is a control, and B was paralyzed on day 7 using injected decamethonium iodide. The sternum of the control chick is fused except for the caudal tip, but that of the paralyzed chick is fused only at the cranial margin, and the right and left (L) sternal processes are widely separated. Unlike the articulations in the control, the sternal ribs (R) of the paralyzed embryo are fused with the sternal processes. The articulations with the coracoid bones (C) are normal however. Modified from Hall and Herring (1990).

muscle on various bones is secondary cartilage, which is poorly maintained in the absence of muscle; thus, bones that depend on secondary cartilage for shape or growth, such as the mammalian clavicle and mandible, are severely affected (Rot-Nikcevic et al., 2007).

Once ossification has taken place, loading, whether by body weight or by muscle contraction, promotes osteogenesis and thus increases bone density and often diameter, although this may vary with genotype (Middleton et al., 2008). That this is a local, rather than a global, effect can be shown by treatments that alter loading of a restricted set of bones. Loss of loading from either body weight or muscle contraction results in loss of bone density. For example, under the weightless conditions of space travel, astronauts lose bone even if they exercise regularly (Sibonga et al., 2007). In this case some muscle loading is present but body weight is absent. The reciprocal example is paralysis of hindlimb muscles using botulinum toxin, which causes limb bones to lose density even though weight bearing has been resumed (Warner et al., 2006).

The skull has a relatively minor weight-bearing role, and therefore, the importance of muscle loading is relatively greater. Craniofacial bones, like limb bones, respond to decreased muscle loading with local changes. For example, removal of tissue from the anterior tongue in growing pigs resulted in decreased density and decreased dimensions but only in the adjacent premaxilla and anterior mandible (Liu et al., 2008). A convenient and noninvasive method of examining the influence of muscle contraction on the skull is to manipulate dietary consistency, altering muscle usage without any surgical or chemical interference. Animals fed a water-softened diet typically show changes in cranial and mandibular growth patterns from their standard-diet controls, and their craniofacial bones are thinner and less dense (Engström et al., 1986; Lieberman et al., 2004; Mavropoulos et al., 2004; Katsaros et al., 2006; Burn, 2007). If a hard diet is restored after the growth period is over, bone density may recover but there is little, if any, "catch-up" growth to restore normal dimensions (Ödman et al., 2008). An example from a trial using pigs is shown in Figure 13.2. Notably, soft diets still require some processing of food by muscle activity; and this treatment does not reduce muscle usage for behaviors other than mastication (agonistic encounters, for example), yet the effect on the skull is often quite marked. Conversely, greater jaw muscle–generated biting force in humans is correlated with more robust alveolar bone (after controlling for body size) (Thongudomporn et al., 2009).

In examples such as the soft-diet experiments, the osteopenic effect of low muscle loading in growing animals is mainly due to decreased apposition, whereas that in mature animals involves remodeling in which apposition does not keep up with resorption, which is

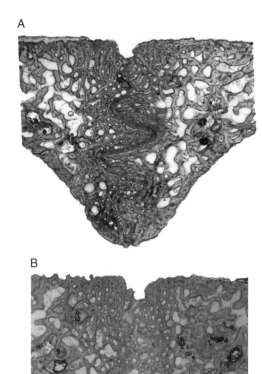

FIGURE 13.2 Coronal sections (undecalcified, toluidine blue stain) through the internasal suture of pigs fed regular pig chow plus dried corn (A) vs. water-softened chow with corn flour (B) from 6 to 18 weeks of age (specimens courtesy of D. Lieberman and K. Duncan). The hard-diet animals had thicker skulls overall (original vertical dimension 6.0 mm in A compared to 5.2 mm in B and denser cortical bone both ectocranially (top) and endocranially (bottom). Modified from Burn (2007).

presumed to occur at normal rates. The osteogenic effect of high loading primarily involves increased apposition, but microdamage to the bone can cause targeted remodeling to repair the defect (Lee et al., 2002).

THE INTERFACE AT ATTACHMENT SITES

The locus at which loads are transmitted from muscle to skeleton is attachment of muscle fibers or tendon to the periosteum. Muscle cells and the cells of their skeletal attachments are evolutionarily stable even in the face of extreme morphological modification, as shown by cell-tracing studies of neural crest– and mesodermally derived neck and shoulder attachments (Matsuoka et al., 2005). The fidelity between muscle and periosteal attachment is also seen during the growth period as the periosteum stretches and slides along growing skeletal elements, always carrying with it the attaching muscle (for review, see Herring, 1994), even severing the Sharpey fibers that anchor the tendons into the osseous matrix (Dörfl, 1980).

The attachment is the point where the load due to muscle contraction is most concentrated, and a local osteogenic effect is often seen, usually in the form of a bony process or tuberosity. Good examples are the deltoid tuberosity of the humerus, the trochanters of the femur, and the coronoid process of the mandible. These processes are thought to result from tensile forces pulling the periosteum away from the bony surface, eliciting osteogenesis (Chierici and Miller, 1984; Schmidt et al., 2002). Enlarging muscles by mutating myostatin (GDF-8) produces notable extensions of these attachment tuberosities (Hamrick et al., 2000). It has been more difficult to rationalize why some muscle-attachment areas (with or without Sharpey fibers) are resorptive, leading some researchers to deny a relationship between muscle loading and osteogenesis (Hoyte and Enlow, 1966). At least part of the difficulty arises from the fact that muscle loading is not a simple tensile force perpendicular to the surface of the bone. Muscles attach at an acute angle to bony surfaces, exerting in most cases only a minor tensile component. Furthermore, as muscles shorten, their cross sections increase. This transverse swelling, especially if it occurs in a restricted space, could produce compressive forces on the bones adjacent to the muscles (Hoyte, 1971). Because pressures against the periosteum typically cause resorption, their absence and the consequent absence of resorption could explain why concave surfaces become convex and curved bones become straight after paralysis (Lanyon, 1980). Using flat pressure transducers, we were able to verify the existence of remarkably high contrac-

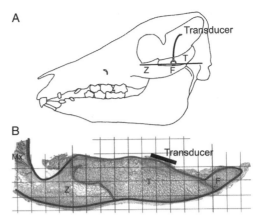

FIGURE 13.3 The medial side of the zygomatic arch in pigs is the origin of the deep part of the masseter muscle, yet this location is consistently resorptive during growth. An experiment in which a small pressure transducer was placed on the medial surface of the arch confirmed that muscle contraction places pressure (37 ± 20 kPa, about 10 times higher than systolic blood pressure), rather than tension, on the bone, consistent with the observed resorption. (A) The position of the transducer relative to the arch, which is composed of the temporal bone (T) and the zygomatic bone (Z), which has a caudal flange-like extension (F). Modified from Teng and Herring (1998). (B) A horizontal section through the arch at the plane indicated in A (methyl green), showing the profile of the transducer lying against the temporal bone. The overlying grid facilitated the localization of cellular activity; the grid squares were 2.3 mm/side. The dotted gray lines are the sutures between bones (Mx, maxillary bone). Heavy lines along the periosteal margins indicate areas undergoing active apposition (dark gray, lateral surface of temporal bone and most surfaces of zygomatic bone) or resorption (light gray, medial surfaces of temporal bone and zygomatic flange). Overlapping dark and light gray lines (zygomatic bone near the suture with the maxillary bone) indicate remodeling. Modified from Ochareon (2004).

tion pressures on the bones of attachment of the jaw muscles of pigs (Teng and Herring, 1998). These compressed areas included some that are partially (temporal fossa) or fully (medial surface of the zygomatic arch) resorptive, as illustrated in Figure 13.3. Therefore, the fact that muscle-attachment sites are sometimes resorptive cannot be taken as evidence that muscle loading is not involved.

The mechanism by which muscle attachments elicit either apposition or resorption could well involve targeted remodeling; the concentrated loads at these locations could produce

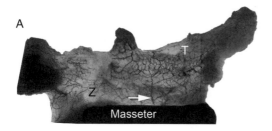

A

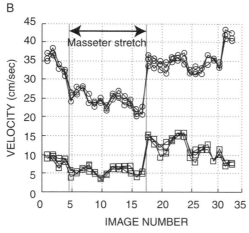

B

FIGURE 13.4 (A) The cleared whole-mount periosteum from the lateral surface of the zygomatic arch of a pig that had been injected with a vascular fill. Z and T overlie the zygomatic and temporal bones, respectively. The masseter muscle appears as a black mass attaching to the lower border of the arch. The white arrow indicates a small artery that leaves the masseter to ramify within the temporal periosteum, which is appositional in this area. In anesthetized animals, the equivalent vessel was recorded with Doppler ultrasound while the masseter muscle was stretched or tetanized. Modified from Ochareon (2004). (B) A trial in which the masseter was stretched and released. The superimposed symbols and lines are from three different cardiac cycles and demonstrate repeatability of the measurements; the upper set (circles) are systolic blood flow, and the lower set (squares) are diastolic. The total recording was 6.5 minutes in duration; images were taken at irregular intervals. Blood flow to the periosteum decreased while the muscle was stretched but after release exceeded the control levels for several minutes.

microdamage. Microdamage at attachments, as elsewhere, would be expected to cause local turnover. Indeed, muscle-attachment sites on the rabbit mandible do have lower bone mineral density (that is, the bone is younger and, hence, less mineralized) than nonattachment areas, suggesting higher turnover (Langenbach et al., 2008).

Loads are not the only thing transmitted between muscles and bones at their attachments. Attachments are also sites where vessels pass, in many cases from muscles to osteogenic surfaces of bones. Historically, this fact was one of the chief arguments in the debate about why attachment areas such as the coronoid process regressed when the attaching muscles (temporalis in this case) were removed because blood supply as well as load was eliminated (Boyd et al., 1967). Similarly, it was argued that the growth of skeletal elements in response to muscle contraction was a consequence of increased vascularity rather than mechanics (Humphrey, 1971). The debate was mainly settled in favor of load on the basis of experiments that inactivated muscles without physically removing them (Moss and Meehan, 1970) and by the examination of species with varying ontogenetic patterns (Lakars and Herring, 1980). However, it remains possible that blood supply to an osteogenic surface might be influenced by muscular-contraction patterns. In an effort to ascertain whether such a phenomenon is possible, we measured blood flow in one small artery which runs from the masseter muscle to supply the appositional lateral surface of the zygomatic arch in pigs. Stretching the masseter by opening the jaw usually decreased blood flow in the vessel (Figure 13.4), as did tetanizing the muscle, although occasionally the opposite effect was seen. These observations suggest that muscle-activity patterns could modify periosteal vascularity and, hence, osteogenesis.

AN EXAMPLE • *Jaw Muscles and the Mandibular Condyle*

CONDYLAR LOADING IS A CONSEQUENCE OF MUSCLE CONTRACTION AND JAW MECHANICS

The best way to see how muscle loading influences skeletal morphology is to investigate specific examples. This section highlights the mandibular condyle, the moving element of the mammalian jaw joint (Figure 13.5). The condyle

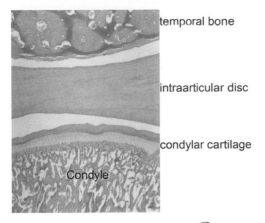

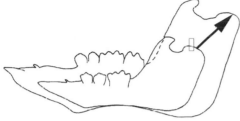

FIGURE 13.5 The drawing shows a superimposition of pig mandibles and illustrates the contribution of condylar cartilage growth to the enlargement of the mandible (arrow). The small rectangle on the younger mandible shows the approximate location of the inset, which is a sagittal section through the jaw joint of a 10-month-old pig (hematoxylin and eosin). The condylar cartilage is still actively growing and is separated from the lower joint cavity only by a layer of fibrous tissue. Fibrous tissue also forms the articular surface of the temporal bone. The mammalian jaw joint is divided into upper and lower cavities by a fibrous intraarticular disc.

derives from one of several secondary cartilages that arise in or near the periosteum of the intramembranously ossifying dentary bone (reviewed by Hall, 2005). Unlike the other secondary cartilages, the condylar cartilage is an important contributor to the length of the mandible both prenatally (Baume, 1962) and postnatally (Luder, 1996). Thus, the condylar cartilage functions simultaneously as a growth plate and as an articular cartilage. In addition to this dual role, the condyle is interesting and useful because (1) secondary cartilages are thought to be particularly liable to mechanical influences (Rot-Nikcevic et al., 2007); (2) articulations are always the sites of load transmission between bones; and (3) like the skull as a whole, the condyle is minimally affected by body weight, and loading arises from contraction of the jaw muscles. In the case of the condyle, the load is an indirect one, a reaction force that results from the fact that most mammalian jaws act as third-class levers (Hylander, 1975; Herring, 2003b). Because the muscles are closer to the joint than the teeth, at rotational equilibrium the muscle force is greater than the bite force. For translational equilibrium, the difference between the upwardly directed muscle force and the downwardly directed bite force must be balanced by an additional downward force, specifically a compressive load at the jaw joint, on the condyle.

The existence of a strong component of compressive loading at the lateral condylar neck during natural jaw muscle contraction has been demonstrated using strain gauges in anthropoid primates (*Macaca*) (Hylander, 1979; Hylander and Bays, 1979) and pigs (*Sus scrofa*) (Marks et al., 1997; Liu and Herring, 2000; Rafferty et al., 2007). Our preliminary data from pigs indicate that the medial condylar neck is also compressed, leading to the conclusion that the condylar head and its cartilaginous cap are also under compressive loading. Furthermore, in pigs these loads can be traced directly to the contraction of the same-side masseter and medial pterygoid muscles, with a lesser contribution from the third jaw-closing muscle, the temporalis (Marks et al., 1997; Rafferty et al., 2006).

That muscle forces do affect the osseous makeup of the condylar process is shown by greater bone mineral density in the condyles of myostatin-deficient mice compared to normal controls and hard-diet rabbits in comparison to their soft-diet siblings (Ravosa et al., 2008). In adult rabbits, just 1 month after one masseter muscle was paralyzed by an injection of botulinum toxin, dramatic differences in bone content were seen between the paralyzed-side and opposite-side condyles (Herring, 2007), indicating either rapid resorption of the former or osteogenesis in the latter.

The question then arises as to whether, and how, changes in the muscle-derived loading of the condylar cartilage might influence its growth and eventual morphology. Cartilage as a tissue flourishes in a pressurized environment (Carter and Beaupré, 2001), to which it is adapted by avascularity. Cartilage cross section, like that of bone, tends to increase when loading increases, an effect observed in cross-species comparisons of lightly and heavily loaded mandibular condyles (Herring, 1985b). Nevertheless, even though some level of loading and movement is required for cartilage health, increased physiological loading decreases the height of articular cartilage (Wong and Carter, 2003) and increases chondrocytic apoptosis (Ravosa et al., 2007). These observations support experimental work on cartilaginous growth plates showing that growth in length is retarded by imposed compressive loads but enhanced by distraction, with effects on both proliferation and enlargement of cells (Stokes et al., 2007).

As a rough, heuristic approximation, then, increasing the axial compressive load on a cartilage would be expected to lead to a growth pattern that rendered the cartilage short but fat, while reduced axial loading would produce a long, skinny structure. With regard to the mandibular condyle, this general idea has been applied to the practice of orthodontics, although efficacy is debatable. Specifically, appliances that compress the condyle against the back wall of its fossa (chin cup or cap) are used to retard mandibular growth, while appliances that position the condyle forward, out of contact with the fossa wall (functional appliances), are used in the hope of accelerating condylar elongation (Proffit, 1986). Because jaw-muscle contraction is a natural source of compressive axial loading on the condylar cartilage, we can hypothesize that increasing the force or duration of muscle activity should inhibit the growth in length (height) of the condylar cartilage while increasing its cross-sectional dimensions. Decreases in muscle force should be associated with longer and more slender condylar cartilages.

EVICENCE FOR AN INVERSE RELATIONSHIP OF LOADING AND CONDYLAR CARTILAGE LENGTH

Most reduced-function models support the idea that the condyle becomes more gracile with less muscle activity, but increases in condylar axial growth have rarely been demonstrated. Botulinum paralysis of the masseter reduces the transverse width of the condyle in growing rabbits but does not alter mandibular length or height (Matic et al., 2007). A similar study using rats showed decreases in mandibular length, rather than the elongation that would be expected from unloading; but conclusions could not be drawn because of problems in controlling for body weight (Kim et al., 2008). In mouse models lacking muscle tissue (Rot-Nikcevic et al., 2007), lacking muscle contraction (Herring and Lakars, 1981), or with muscular dystrophy (Vilmann et al., 1985), the dimensions of the condylar head (which represent the cross section of the condylar cartilage) are markedly reduced but the length of the mandible in these studies showed little change relative to normal mice. Studies on humans with Duchenne muscular dystrophy confirm that condylar length grows at a normal rate, although not in the normal direction (Matsuyuki et al., 2006). Rats and rabbits fed a soft diet develop smaller condyles than do their hard-diet controls (Kiliaridis et al., 1999; Ravosa et al., 2007), although a study in pigs failed to find a difference (Lindsten et al., 2004); none of these noted length changes. Significantly, however, the one diet-altering study that specifically examined condylar elongation documented increased length in soft-diet rats (Tuominen et al., 1994).

These studies are convincing in showing that reducing muscle activity leads to a smaller condyle but offer little support for the notion that the unloaded condyle should become longer. Measurement difficulties may play a role; because the condylar cartilage is progressively ossified from its base within the mandible, its length is not an accurate reflection of its growth rate. Mandibular length is also not a good surrogate for condylar elongation because bony apposition anteriorly also contributes to the total (Figure 13.5). Particularly if increases in condylar elongation are short-term accelerations, they would be hard to detect. Another problem with all the studies mentioned is that the loading environment of the condyle is inferred rather than measured. Although loading is surely decreased, we do not know how much or whether parameters in addition to magnitude are changed. This information can be provided only by empirical studies in which some aspect of loading, such as bone strain, is measured directly. Our own studies using pigs seek to assess both condylar loading and condylar growth, and these have provided stronger support for the hypothesized inverse relationship.

Perhaps the most convincing evidence that reduction in functional muscle activity can increase axial growth rate in the condyle comes from a surgical procedure, distraction osteogenesis. This example also is a case study showing why it is essential to measure the mechanical environment experimentally. Distraction osteogenesis of the mandible is a method to lengthen the lower jaw by creating an osteotomy (Figure 13.6) and gradually separating the fragments to elicit osteogenesis in the gap. Clinicians performing this procedure have been concerned that it would compress the condyle against the posterior wall of its fossa and therefore retard condylar growth, an undesirable event (Stucki-McCormick, 1998). Surprisingly, pigs undergoing unilateral mandibular distraction had thicker (i.e., longer) condylar cartilage on the distracted side than on the intact side, implying more rather than less growth (Thurmüller et al., 2006). This finding is not unique to one species. Enhanced growth of the distraction-side condyle has also been reported for sheep (Karaharju-Suvanto et al., 1996) and rabbits (Kim et al., 2004). These authors suggested that increased compressive loading on the instrumented side caused a reactive chondrogenesis, exactly the opposite of the general hypothesis

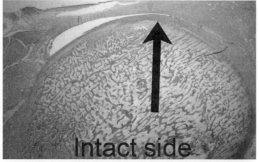

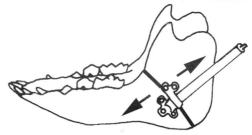

FIGURE 13.6 Distraction osteogenesis of the mandible in pigs. The drawing illustrates the procedure, which is thought to propel the posterior fragment, bearing the mandibular condyle, backward into the fossa in addition to lengthening the jaw at the osteotomy site (large black arrows). Sagittal sections through the condyles of one individual show that whereas the intact side was growing postero-superiorly (normal, see Figure 13.5), the distracted side was growing vertically. Elongation of the condyle was faster on the distracted side, as indicated by cartilage thickness and mineralization rate. The small arrows show the average compressive strain during mastication as measured by strain gauges on the condylar necks. The distracted side was unloaded relative to the intact side, and compressive strain was oriented more posteriorly. Modified from Rafferty et al. (2007).

of an inverse relationship between loading and elongation. Further examination, however, has shown that the assumption of greater loading on the distracted side was in error, at least for pigs. While confirming a higher growth rate on the distracted side, Rafferty et al. established, first, that activating the appliance did not load the condyle (Rafferty et al., 2004), probably because in pigs the fossa for the condyle lacks a posterior wall (Herring et al., 2002), and, second, that during mastication the condyle on the distracted side was strikingly underloaded rather than overloaded (Rafferty et al., 2007). Furthermore, the orientation as well as the magnitude of the compressive strain was altered, and this corresponded to a change in the direction of growth so that maximal growth continued to occur roughly perpendicular to the loading axis, i.e., in an unloaded direction (Figure 13.6). In this example, muscle activity itself was not altered by the experiment, but the same muscle activity loaded the condyle differently because the procedure changed the mechanics of the jaw. Nevertheless, muscle activity was far more important than the appliance as an influence on condylar growth, and the example supports the inverse relationship between condylar loading and elongation.

Perhaps the most interesting question is whether individual variation in muscle activity among normal subjects is sufficient to affect skeletal growth and morphology. If so, then this would constitute a literal reading of "function determining form." Using pigs, we assessed jaw joint loading by measuring the masticatory strain of both bony elements, the condylar neck and the temporal bone. On the same day, we labeled replicating cells using the thymidine analogue 5-bromo-2'-deoxyuridine (Herring et al., 2002). The labeled (proliferating) cells in the condylar cartilage and the cartilage of the articular eminence were counted and correlated with the bone strains observed in the same animals. All correlations between cell replication and strain parameters were negative (Figure 13.7), once again supporting an inverse relationship between loading and cartilage elongation, even on a time scale as short as 1 day and in normal animals. Clearly, individual differences in muscle usage do have the potential to change skeletal morphology, although to what extent is not yet known.

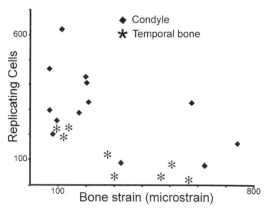

FIGURE 13.7 Scatterplots showing the inverse relationship between the number of replicating cells in the cartilages covering the mandibular condyle (diamonds) and the articular eminence of the temporal bone (asterisks) and bone strain recorded on the same day from the condylar neck and temporal bone lateral to the eminence. Replication was assessed by labeling with bromodeoxyuridine. Compressive strain is plotted for the condyle, and shear strain is plotted for the temporal bone. Modified from Herring et al. (2002).

DO MUSCLE–BONE INTERACTIONS CONFUSE EVOLUTIONARY INFERENCES?

Mineralized tissues occupy a place of unique importance in evolutionary biology simply because these are the structures that are preserved in the fossil record. In the absence of DNA from most extinct species, phylogenetic trees are constructed based on these relics. The underlying assumption is that the details of preserved morphology reflect true genetic change that has evolved and can be followed through geological strata.

To a considerable degree, this assumption is correct and the method has been successful in reconstructing the phylogenies of vertebrate groups. Modern molecular methods have confirmed most of the general patterns that were outlined by paleontologists in the last century using mineralized tissue morphology alone, for example, the close relationship of cetaceans (whales) to artiodactyls (even-toed ungulates) and the early branching of edentates from the placental lineage (Novacek, 1992). Osseous features, such as the trio of middle ear ossicles that defines mammals and the characteristic astragalus of artiodactyls, are as critical as teeth in outlining the major features of vertebrate phylogeny. As successful as the strategy has been overall, it is nevertheless clear that the assumption of a skeleton which is completely determined by "bone genes" is not only simplistic but, for the most part, false. The details of skeletal morphology are not predetermined but result from interactions of cells with their environment, including that provided by the musculature.

The ultimate phenotype of a skeletal element is the product of its growth and modeling. Not uncommonly, the constituents present at birth are completely replaced by adulthood. A good example is the mammalian mandible, in which the cells and matrix that form the muscle-bearing ramus end up in the tooth-bearing body, while the adult's ramus is entirely new, formed during growth (Figure 13.5). The important factor here is not that the mature skeletal phenotype is formed postnatally but, rather, that postnatal growth and ultimate phenotype reflect environmentally manipulated variables such as exercise and muscle contraction as well as genotype. DNA does not encode the size or shape of mature bones but only a set of structural molecules and a complex, probably stochastic, network of rules for responding to external stimuli. The skeleton must be considered as a product of the interactions between the osteogenic tissues and their functional surroundings, of which the muscular activity pattern is one of the most important. It remains an open question as to whether trends of apparently evolutionary change in the fossil record are examples of epigenetic (*sensu lato*) plasticity based on ecological variation and muscle–bone interactions rather than actual genetic change in the population.

ACKNOWLEDGMENTS

Work from my laboratory reported here was supported by the U.S. Public Health Service through National Institute of Dental and Craniofacial Research awards DE08513, DE11962, and DE14336. The procedures were performed by Drs. Amrit Burn, Jay Decker, Daniel Leotta, Zi-Jun Liu, Pannee Ochareon, Katherine Rafferty, Zongyang Sun, and Shengyi Teng, to all of whom I am grateful. The soft-diet pig experiment illustrated in Figure 13.2 was conducted by Dr. Daniel Lieberman and Katherine Duncan, Harvard University, and supported by the American School of Prehistoric Research.

REFERENCES

Allori, A.C., A.M. Sailon, J.H. Pan, and S.M. Warren. 2008. Biological basis of bone formation, remodeling, and repair. Part III: Biomechanical forces. *Tissue Eng* 14:285–93.

Aubret, F., R. Shine, and X. Bonnet. 2004. Adaptive developmental plasticity in snakes. *Nature* 431:261–2.

Baume, L.J. 1962. Ontogenesis of the human temporomandibular joint: 1. Development of the condyles. *J Dent Res* 41:1327–39.

Bikle, D.D. 2008. Integrins, insulin like growth factors, and the skeletal response to load. *Osteoporos Int* 19:1237–46.

Bouvier, M., and W.L. Hylander. 1996. The mechanical or metabolic function of secondary osteonal bone in the monkey *Macaca fascicularis*. *Arch Oral Biol* 41:941–50.

Boyd, T.G., W.A. Castelli, and D.F. Huelke. 1967. Removal of the temporalis muscle from its origin: Effects on the size and shape of the coronoid process. *J Dent Res* 46:997–1001.

Burn, A.K. 2007. Dietary consistency and the midline sutures: A study in growing pigs. MSD thesis, University of Washington, Seattle.

Carter, D.R., and G.S. Beaupré. 2001. *Skeletal Function and Form*. Cambridge: Cambridge University Press.

Cheema, U., R. Brown, V. Mudera, S.Y. Yang, G. McGrouther, and G. Goldspink. 2005. Mechanical signals and IGF-1 gene splicing in vitro in relation to development of skeletal muscle. *J Cell Physiol* 202:67–75.

Chierici, G., and A.J. Miller. 1984. Experimental study of muscle reattachment following surgical detachment. *J Oral Maxillofac Surg* 42:485–90.

Crisan, M., S. Yap, L. Casteilla, C.-W. Chen, M. Corselli, T.S. Park, G. Andriolo, et al. 2008. A perivascular origin for mesenchymal stem cells in multiple human organs. *Cell Stem Cell* 3:301–13.

Cubo, J., P. Legendre, A. de Ricqles, L. Montes, E. de Margerie, J. Castanet, and Y. Desdevises. 2008. Phylogenetic, functional, and structural components of variation in bone growth rate of amniotes. *Evol Dev* 10:217–27.

de Buffrénil, V., A. Houssaye, and W. Bohme. 2008. Bone vascular supply in monitor lizards (Squamata: Varanidae): Influence of size, growth, and phylogeny. *J Morphol* 269:533–43.

Dörfl, J. 1980. Migration of tendinous insertions. I. Cause and mechanism. *J Anat* 131:179–95.

Engler, A. J., S. Sen, H. L. Sweeney, and D. E. Discher. 2006. Matrix elasticity directs stem cell lineage specification. *Cell* 126:677–89.

Engström, C., S. Kiliaridis, and B. Thilander. 1986. The relationship between masticatory function and craniofacial morphology. II. A histological study in the growing rat fed a soft diet. *Eur J Orthod* 8:271–9.

Forwood, M. R., L. Li, W. L. Kelly, and M. B. Bennett. 2001. Growth hormone is permissive for skeletal adaptation to mechanical loading. *J Bone Miner Res* 16:2284–90.

Franz-Odendaal, T. A., K. Ryan, and B. K. Hall. 2007. Developmental and morphological variation in the teleost craniofacial skeleton reveals an unusual mode of ossification. *J Exp Zoolog B Mol Dev Evol* 308:709–21.

Fricke, O., Z. Sumnik, B. Tutlewski, A. Stabrey, T. Remer, and E. Schoenau. 2008. Local body composition is associated with gender differences of bone development at the forearm in puberty. *Horm Res* 70:105–11.

German, R. Z., A. W. Crompton, and A. J. Thexton. 2008. Variation in EMG activity: A hierarchical approach. *Integr Comp Biol* 48:283–93.

Gouchie, G. M., L. F. Roberts, and R. J. Wassersug. 2008. The effect of mirrors on African clawed frog (*Xenopus laevis*) larval growth, development, and behavior. *Behav Ecol Sociobiol* 62:1821–9.

Graham, M. R., B. Davies, F. M. Grace, A. Kicman, and J. S. Baker. 2008. Anabolic steroid use: Patterns of use and detection of doping. *Sports Med* 38:505–25.

Hall, B. K. 2005. *Bones and Cartilage: Developmental and Evolutionary Skeletal Biology*. London: Elsevier Academic.

Hall, B. K., and S. W. Herring. 1990. Paralysis and growth of the musculoskeletal system in the embryonic chick. *J Morphol* 206:45–56.

Hallgrímsson, B., and D. E. Lieberman. 2008. Mouse models and the evolutionary developmental biology of the skull. *Integr Comp Biol* 48:373–84.

Hamrick, M. W., and S. L. Ferrari. 2008. Leptin and the sympathetic connection of fat to bone. *Osteoporos Int* 19:905–12.

Hamrick, M. W., A. C. McPherron, C. O. Lovejoy, and J. Hudson. 2000. Femoral morphology and cross-sectional geometry of adult myostatin-deficient mice. *Bone* 27:343–9.

Hamrick, M. W., K. H. Ding, C. Pennington, Y. J. Chao, Y. D. Wu, B. Howard, D. Immel, et al. 2006. Age-related loss of muscle mass and bone strength in mice is associated with a decline in physical activity and serum leptin. *Bone* 39:845–53.

Hamrick, M. W., K.-H. Ding, S. Ponnala, S. L. Ferrari, and C. M. Isales. 2008. Caloric restriction decreases cortical bone mass but spares trabecular bone in the mouse skeleton: Implications for the regulation of bone mass by body weight. *J Bone Miner Res* 23:870–8.

Heaney, R. P. 2003. Is the paradigm shifting? *Bone* 33:457–65.

Herring, S. W. 1985a. The ontogeny of mammalian mastication. *Am Zool* 25:339–49.

Herring, S. W. 1985b. Morphological correlates of masticatory patterns in peccaries and pigs. *J Mammal* 66:603–17.

Herring, S. W. 1994. Development of functional interactions between skeletal and muscular systems. In: *Bone*, vol 9, ed. B. K. Hall, 165–91. Boca Raton, FL: CRC Press.

Herring, S. W. 2003a. Ontogenetic integration of form and function. In: *Keywords and Concepts in Evolutionary Developmental Biology*, ed. B. K. Hall and W. M. Olson, 275–9. Cambridge, MA: Harvard University Press.

Herring, S. W. 2003b. TMJ anatomy and animal models. *J Musculoskelet Neuronal Interact* 3:391–4.

Herring, S. W. 2007. Masticatory muscles and the skull: A comparative perspective. *Arch Oral Biol* 52:296–9.

Herring, S. W., and T. C. Lakars. 1981. Craniofacial development in the absence of muscle contraction. *J Craniofac Genet Dev Biol* 1:341–57.

Herring, S. W., J. D. Decker, Z. J. Liu, and T. Ma. 2002. The temporomandibular joint in miniature pigs: Anatomy, cell replication, and relation to loading. *Anat Rec* 266:152–66.

Hirukawa, K., K. Miyazawa, H. Maeda, Y. Kameyama, S. Goto, and A. Togari. 2005. Effect of tensile force on the expression of IGF-I and IGF-I receptor in the organ-cultured rat cranial suture. *Arch Oral Biol* 50:367–72.

Hoyte, D. A. N. 1971. Mechanisms of growth in the cranial vault and base. *J Dent Res* 50:1447–59.

Hoyte, D. A. N., and D. H. Enlow. 1966. Wolff's law and the problem of muscle attachment on resorptive surfaces of bone. *Am J Phys Anthropol* 24:205–14.

Humphrey, T. 1971. Development of oral and facial motor mechanisms in human fetuses and their relation to craniofacial growth. *J Dent Res* 50:1428–41.

Hylander, W. L. 1975. The human mandible: Lever or link? *Am J Phys Anthropol* 43:227–42.

Hylander, W. L. 1979. An experimental analysis of temporomandibular joint reaction force in macaques. *Am J Phys Anthropol* 51:433–56.

Hylander, W. L., and R. Bays. 1979. An in vivo strain-gauge analysis of the squamosal-dentary joint reaction force during mastication and incisal biting in *Macaca mulatta* and *Macaca fascicularis*. *Arch Oral Biol* 24:689–97.

Karaharju-Suvanto, T., J. Peltonen, R. Ranta, O. Laitinen, and A. Kahri. 1996. The effect of gradual distraction of the mandible on the sheep temporomandibular joint. *Int J Oral Maxillofac Surg* 25:152–6.

Karasik, D., and D. P. Kiel. 2008. Genetics of the musculoskeletal system: A pleiotropic approach. *J Bone Miner Res* 23:788–802.

Kasra, M., and M. D. Grynpas. 1995. The effects of androgens on the mechanical properties of primate bone. *Bone* 17:265–70.

Katsaros, C., A. Zissis, A. Bresin, and S. Kiliaridis. 2006. Functional influence on sutural bone apposition in the growing rat. *Am J Orthod Dentofacial Orthop* 129:352–7.

Kiel, D. P., S. L. Ferrari, L. A. Cupples, D. Karasik, D. Manen, A. Imamovic, A. G. Herbert, and J. Dupuis. 2007. Genetic variation at the low-density lipoprotein receptor–related protein 5 (LRP5) locus modulates Wnt signaling and the relationship of physical activity with bone mineral density in men. *Bone* 40:587–96.

Kiliaridis, S., B. Thilander, H. Kjellberg, N. Topouzelis, and A. Zafiriadis. 1999. Effect of low masticatory function on condylar growth: A morphometric study in the rat. *Am J Orthod Dentofacial Orthop* 116:121–5.

Kim, J.-Y., S.-T. Kim, S.-W. Cho, H.-S. Jung, K.-T. Park, and H.-K. Son. 2008. Growth effects of botulinum toxin type A injected into masseter muscle on a developing rat mandible. *Oral Dis* 14:626–32.

Kim, S. G., J. W. Ha, and J. C. Park. 2004. Histological changes in the temporomandibular joint in rabbits depending on the extent of mandibular lengthening by osteodistraction. *Br J Oral Maxillofac Surg* 42:559–65.

Koh, C. J., D. M. Delo, J. W. Lee, M. M. Siddiqui, R. P. Lanza, S. Soker, J. J. Yoo, and A. Atala. 2009. Parthenogenesis-derived multipotent stem cells adapted for tissue engineering applications. *Methods* 47:90–7.

Lakars, T. C., and S. W. Herring. 1980. Ontogeny of oral function in hamsters (*Mesocricetus auratus*). *J Morphol* 165:237–54.

Langenbach, G. E. J., B. van der Zwan, L. Mulder, and T. van Eijden. 2008. Influence of muscle activity on the degree of bone mineralisation. Special Issue B, *J Dent Res* 87:Abstract 3571, www.dentalresearch.org.

Lanyon, L. E. 1980. The influence of function on the development of bone curvature. An experimental study on the rat tibia. *J Zool* 192:457–66.

Laron, Z. 2001. Insulin-like growth factor 1 (IGF-1): A growth hormone. *Mol Pathol* 54:311–6.

Lee, A. H. 2004. Histological organization and its relationship to function in the femur of *Alligator mississippiensis*. *J Anat* 204:197–207.

Lee, K., H. Jessop, R. Suswillo, G. Zaman, and L. Lanyon. 2003. Bone adaptation requires oestrogen receptor-α. *Nature* 424:389.

Lee, T. C., A. Staines, and D. Taylor. 2002. Bone adaptation to load: Microdamage as a stimulus for bone remodelling. *J Anat* 201:437–46.

Lieberman, D. E., G. E. Krovitz, F. W. Yates, M. Devlin, and M. St. Claire. 2004. Effects of food processing on masticatory strain and craniofacial growth in a retrognathic face. *J Hum Evol* 46:655–77.

Lindsten, R., T. Magnusson, B. Ogaard, and E. Larsson. 2004. Effect of food consistency on temporomandibular joint morphology: An experimental study in pigs. *J Orofac Pain* 18:56–61.

Liu, Z. J., and S. W. Herring. 2000. Masticatory strains on osseous and ligamentous components of the jaw joint in miniature pigs. *J Orofac Pain* 14:265–78.

Liu, Z. J., V. Shcherbatyy, G. Gu, and J. A. Perkins. 2008. Effects of tongue volume reduction on craniofacial growth: A longitudinal study on orofacial skeletons and dental arches. *Arch Oral Biol* 53:991–1001.

Luder, H.-U. 1996. *Postnatal Development, Aging, and Degeneration of the Temporomandibular Joint in Humans, Monkeys, and Rats*. Ann Arbor: Center for Human Growth and Development, University of Michigan.

Marks, L., S. Teng, J. Årtun, and S. Herring. 1997. Reaction strains on the condylar neck during mastication and maximum muscle stimulation in different condylar positions: An experimental study in the miniature pig. *J Dent Res* 76:1412–20.

Matic, D. B., A. Yazdani, R. G. Wells, T. Y. Lee, and B. S. Gan. 2007. The effects of masseter muscle paralysis on facial bone growth. *J Surg Res* 139:243–52.

Matsuoka, T., P. E. Ahlberg, N. Kessaris, P. Iannarelli, U. Dennehy, W. D. Richardson, A. P. McMahon, and G. Koentges. 2005. Neural crest origins of the neck and shoulder. *Nature* 436:347–55.

Matsuyuki, T., T. Kitahara, and A. Nakashima. 2006. Developmental changes in craniofacial morphology in subjects with Duchenne muscular dystrophy. *Eur J Orthod* 28:42–50.

Mavropoulos, A., S. Kiliaridis, A. Bresin, and P. Ammann. 2004. Effect of different masticatory functional and mechanical demands on the structural adaptation of the mandibular alveolar bone in young growing rats. *Bone* 35:191–7.

Melton, L.J., III, B.L. Riggs, S.J. Achenbach, S. Amin, J.J. Camp, P.A. Rouleau, R.A. Robb, A.L. Oberg, and S. Khosla. 2006. Does reduced skeletal loading account for age-related bone loss? *J Bone Miner Res* 21:1847–55.

Middleton, K.M., C.E. Shubin, D.C. Moore, P.A. Carter, T. Garland, Jr., and S.M. Swartz. 2008. The relative importance of genetics and phenotypic plasticity in dictating bone morphology and mechanics in aged mice: Evidence from an artificial selection experiment. *Zoology (Jena)* 111:135–47.

Morgan, T.J., T.J. Garland, and P. Carter. 2003. Ontogenies in mice selected for high voluntary wheel-running activity. I. Mean ontogenies. *Evolution* 57:646–57.

Moss, M.L., and M.A. Meehan. 1970. Functional cranial analysis of the coronoid process in the rat. *Acta Anat* 77:11–24.

Nara-Ashizawa, N., L.J. Liu, T. Higuchi, K. Tokuyama, K. Hayashi, Y. Shirasaki, H. Amagai, and S. Saitoh. 2002. Paradoxical adaptation of mature radius to unilateral use in tennis playing. *Bone* 30:619–23.

Noden, D.M., and P.A. Trainor. 2005. Relations and interactions between cranial mesoderm and neural crest populations. *J Anat* 207:575–601.

Notelovitz, M. 2002. Androgen effects on bone and muscle. *Fertil Steril* 77 (Suppl 4): S34–41.

Novacek, M.J. 1992. Mammalian phylogeny: Shaking the tree. *Nature* 356:121–5.

Ochareon, P. 2004. Craniofacial periosteal cell capacities. PhD diss., University of Washington, Seattle.

Ödman, A., A. Mavropoulos, and S. Kiliaridis. 2008. Do masticatory functional changes influence the mandibular morphology in adult rats. *Arch Oral Biol* 53:1149–54.

Orwoll, E.S. 2003. Men, bone and estrogen: Unresolved issues. *Osteoporos Int* 14:93–8.

Parfitt, A.M. 2002. Targeted and nontargeted bone remodeling: Relationship to basic multicellular unit origination and progression. *Bone* 30:5–7.

Pearson, H. 2008. Disputed definitions. *Nature* 455:1023–24.

Pearson, O.M., and D.E. Lieberman. 2004. The aging of Wolff's "law": Ontogeny and responses to mechanical loading in cortical bone. *Ybk Phys Anthropol* 47:63–99.

Plochocki, J.H., J.P. Rivera, C. Zhang, and S.A. Ebba. 2008. Bone modeling response to voluntary exercise in the hindlimb of mice. *J Morphol* 269:313–8.

Powell, J.A., L. Petherbridge, and B.E. Flucher. 1996. Formation of triads without the dihydropyridine receptor α subunits in cell lines from dysgenic skeletal muscle. *J Cell Biol* 134:375–87.

Proffit, W.R. 1986. *Contemporary Orthodontics*. St. Louis: Mosby.

Rafferty, K.L., Z. Sun, M.A. Egbert, and S.W. Herring. 2004. Bony strains in response to distractor activation. In: *Biological Mechanisms of Tooth Movement and Craniofacial Adaptation*, ed. Z. Davidovitch and J. Mah, 59–62. Bangkok: Harvard Society for the Advancement of Orthodontics.

Rafferty, K.L., Z. Sun, M.A. Egbert, E.E. Baird, and S.W. Herring. 2006. Mandibular mechanics following osteotomy and appliance placement. II. Bone strain on the body and condylar neck. *J Oral Maxillofac Surg* 64:620–7.

Rafferty, K.L., Z. Sun, M.A. Egbert, D.W. Bakko, and S.W. Herring. 2007. Changes in growth and morphology of the condyle following mandibular distraction in minipigs: Overloading or underloading? *Arch Oral Biol* 52:967–76.

Ravosa, M.J., R. Kunwar, S.R. Stock, and M.S. Stack. 2007. Pushing the limit: Masticatory stress and adaptive plasticity in mammalian craniomandibular joints. *J Exp Biol* 210:628–41.

Ravosa, M.J., E.K. López, R.A. Menegaz, S.R. Stock, M.S. Stack, and M.W. Hamrick. 2008. Using "mighty mouse" to understand masticatory plasticity: Mostatin-deficient mice and musculoskeletal function. *Integr Comp Biol* 48:345–59.

Renn, J., C. Winkler, M. Schartl, R. Fischer, and R. Goerlich. 2006. Zebrafish and medaka as models for bone research including implications regarding space-related issues. *Protoplasma* 229:209–14.

Rieppel, O. 1987. The development of the trigeminal jaw adductor musculature and associated skull elements in the lizard *Podarcis sicula*. *J Zool Lond* 212:131–50.

Rot-Nikcevic, I., T. Reddy, K.J. Downing, A.C. Belliveau, B. Hallgrimsson, B.K. Hall, and B. Kablar. 2006. Myf5⁻/⁻:MyoD⁻/⁻ amyogenic fetuses reveal the importance of early contraction and static loading by striated muscle in mouse skeletogenesis. *Dev Genes Evol* 216:1–9.

Rot-Nikcevic, I., K.J. Downing, B.K. Hall, and B. Kablar. 2007. Development of the mouse mandibles and clavicles in the absence of skeletal myogenesis. *Histol Histopathol* 22:51–60.

Rowlands, A.S., P.A. George, and J.J. Cooper-White. 2008. Directing osteogenic and myogenic differentiation of MSCs: Interplay of stiffness and adhesive ligand presentation. *Am J Physiol Cell Physiol* 295:C1037–C1044.

Sample, S.J., M. Behan, L.D. Smith, W.E. Oldenhoff, M.D. Markel, V.L. Kalscheur, Z. Hao, V. Miletic, and P. Muir. 2008. Functional adaptation to loading of a single bone is neuronally regulated and involves multiple bones. *J Bone Miner Res* 23: 1372–81.

Saxon, L. K., and C. H. Turner. 2006. Low-dose estrogen treatment suppresses periosteal bone formation in response to mechanical loading. *Bone* 39:1261–7.

Schmidt, B. L., L. Kung, C. Jones, and N. Casap. 2002. Induced osteogenesis by periosteal distraction. *J Oral Maxillofac Surg* 60:1170–5.

Seeman, E. 2008. Bone quality: The material and structural basis of bone strength. *J Bone Miner Metab* 26:1–8.

Sibonga, J. D., H. J. Evans, H. G. Sung, E. R. Spector, T. F. Lang, V. S. Oganov, A. V. Bakulin, L. C. Shackelford, and A. D. LeBlanc. 2007. Recovery of spaceflight-induced bone loss: Bone mineral density after long-duration missions as fitted with an exponential function. *Bone* 41:973–8.

Smith, K. K. 1994. Development of craniofacial musculature in *Monodelphis domestica* (Marsupialia, Didelphidae). *J Morphol* 222:149–73.

Spyropoulos, M. N. 1977. The morphogenetic relationship of the temporal muscle to the coronoid process in human embryos and fetuses. *Am J Anat* 150:395–409.

Stokes, I. A. F., K. C. Clark, C. E. Farnum, and D. D. Aronsson. 2007. Alterations in the growth plate associated with growth modulation by sustained compression or distraction. *Bone* 41:197–205.

Stucki-McCormick, S. U. 1998. The effect of distraction osteogenesis on the temporomandibular joint. In: *Distraction Osteogenesis and Tissue Engineering*, ed. J. A. J. McNamara and C.-A. Trotman, 75–103. Ann Arbor: Center for Human Growth and Development, University of Michigan.

Suominen, H. 2006. Muscle training for bone strength. *Aging Clin Exp Res* 18:85–93.

Takeda, S., F. Elefteriou, R. Levasseur, X. Liu, L. Zhao, K. L. Parker, D. Armstrong, P. Ducy, and G. Karsenty. 2002. Leptin regulates bone formation via the sympathetic nervous system. *Cell* 111:305–17.

Teng, S., and S. W. Herring. 1998. Compressive loading on bone surfaces from muscular contraction: An in vivo study in the miniature pig, *Sus scrofa*. *J Morphol* 238:71–80.

Thongudomporn, U., V. Chongsuvivatwong, and A. F. Geater. 2009. The effect of maximum bite force on alveolar bone morphology. *Orthod Craniofac Res* 12:1–8.

Thurmüller, P., M. J. Troulis, A. Rosenberg, S.-K. Chuang, and L. B. Kaban. 2006. Microscopic changes in the condyle and disc in response to distraction osteogenesis of the minipig mandible. *J Oral Maxillofac Surg* 64:249–58.

Tuominen, M., T. Kantomaa, and P. Pirtiniemi. 1994. Effect of altered loading on condylar growth in the rat. *Acta Odontol Scand* 52:129–34.

Turner, R. T. 1999. Mechanical signaling in the development of postmenopausal osteoporosis. *Lupus* 8:388–92.

Vilmann, H., M. Juhl, and S. Kirkeby. 1985. Bone–muscle interactions in the muscular dystrophic mouse. *Eur J Orthod* 7:185–92.

Wake, M. H., and J. Hanken. 1982. Development of the skull of *Dermophis mexicanus* (Amphibia: Gymnophiona), with comments on skull kinesis and amphibian relationships. *J Morphol* 173: 203–23.

Warner, S. E., D. A. Sanford, B. A. Becker, S. D. Bain, S. Srinivasan, and T. S. Gross. 2006. Botox induced muscle paralysis rapidly degrades bone. *Bone* 38:257–64.

Weisbecker, V., A. Goswami, S. Wroe, and M. R. Sánchez-Villagra. 2008. Ossification heterochrony in the therian postcranial skeleton and the marsupial–placental dichotomy. *Evolution* 62:2027–41.

Westerlind, K. C., T. J. Wronski, E. L. Ritman, Z.-P. Luo, K.-N. An, N. H. Bell, and R. T. Turner. 1997. Estrogen regulates the rate of bone turnover but bone balance in ovariectomized rats is modulated by prevailing mechanical strain. *Proc Natl Acad Sci USA* 94:4199–204.

Wong, M., and D. R. Carter. 2003. Articular cartilage functional histomorphology and mechanobiology: A research perspective. *Bone* 33:1–13.

Zaman, G., H. L. Jessop, M. Muzylak, R. L. De Souza, A. A. Pitsilides, J. S. Price, and L. L. Lanyon. 2006. Osteocytes use estrogen receptor α to respond to strain but their ERα content is regulated by estrogen. *J Bone Miner Res* 21:1297–306.

Zofková, I. 2008. Hormonal aspects of the muscle-bone unit. *Physiol Res* 57 (Suppl 1): S159–69.

14

Evolution of the Apical Ectoderm in the Developing Vertebrate Limb

Lisa Noelle Cooper, Brooke Autumn Armfield,
and J. G. M. Thewissen

CONTENTS

Vertebrate limbs display a diverse array of morphologies, including the fins of teleosts and lungfish, wings of birds and bats, arms of humans, and flippers of cetaceans. Due to this morphological diversity, limbs are a topic of intense study in paleontology, phylogenetic systematics, descriptive embryology, and functional morphology. Evolutionary developmental biology (evo–devo) in particular has focused on understanding the developmental pathways that establish diverse limb phenotypes by integrating data from gene-expression and protein-signaling with transplantation and ablation experiments. As a result of the increase in the

Epigenetics: Linking Genotype and Phenotype in Development and Evolution, ed. Benedikt Hallgrímsson and Brian K. Hall.

number of evo–devo studies on diverse taxa, additional variants in the limb developmental pathway have been discovered.

Limb evo–devo research attempts to explain how the signaling centers within the developing vertebrate limb control patterning and how phenotypic and expression variation within these signaling centers shape vertebrate limb morphology. Limb development is controlled by two main signaling centers: (1) the apical ectoderm of the limb, which is a specialized region of cells at the limb tip that controls outgrowth and patterning along the proximodistal axis, and (2) the zone of polarizing activity, which regulates patterning along the anteroposterior axis.

This chapter compares limb apical ectodermal morphologies and the associated signaling patterns across vertebrates. We initially review the morphology and function of the apical ectoderm. We then describe expression of fibroblast growth factors (FGF) in the apical ectoderm as the primary signaling molecules. This chapter also provides a taxonomically broad comparison of ectodermal morphologies and associated FGF-expression patterns across vertebrates (bony fish, lungfish, amphibians, squamates, birds, and mammals). Lastly, we present new morphological and protein-signaling data regarding the apical ectoderm of the pantropical spotted dolphin (*Stenella attenuata*) and the domestic pig (*Sus scrofa*) developing limbs.

GENERAL DESCRIPTION OF THE APICAL ECTODERMAL RIDGE

The apical ectodermal ridge (AER) is aptly named as it typically takes on a ridge-like morphology in vertebrates, is nipple-shaped in cross section (Saunders, 1948), and is usually composed of stratified or pseudostratified columnar epithelial tissue (Richardson et al., 1998). Ridge morphology is most frequently studied in chicks (e.g., Saunders, 1948; Jurand, 1965; Rubin and Saunders, 1972; Pizette and Niswander, 1999; Talamillo et al., 2005) and mice (e.g., Jurand, 1965; Lee and Chan, 1991; Talamillo

et al., 2005). The AER is a specialized thickened epithelium at the distal apex of a developing limb bud (and in lungfish and teleosts, a fin bud), along the dorsoventral boundary, that secretes morphogens necessary for limb outgrowth and patterning. It originates from the ectodermal tissues associated with the medial somatopleure (Michaud et al., 1997).

A detailed description of the morphology of the AER and speculations about its function (based on transplantation and ablation experiments) were first reported in the chick (*Gallus gallus*; for review, see Saunders 1948, 1998), but the earliest mention of different apical thickenings occurred as early as 1879 (for review, see footnote in Saunders, 1998). Early embryological descriptions reported the AER as an *ektodermkappe* (an "ectodermal cap"; Köllicker, 1879; Braus, 1906; Fischel, 1929), *epithelfalte* (an "epithelial fold"), *randfalte* (an "edge fold"), *epithelverdickung* (an "epithelial thickening"; Fischel, 1929), *extremitätenscheitelleiste* (an "extremity crest"; Peter, 1903), and a ring (Steiner, 1928; Schmidt, 1898; O'Rahilly and Müller, 1985).

This chapter presents a comparison of developing vertebrate limb morphologies and documents three apical ectodermal morphologies (Figure 14.1): AE-1, a thick, or prominent, ridge-shaped ectoderm; AE-2, a slightly thickened apical ectoderm; and AE-3, an apical ectoderm that is not thick compared to the adjacent ectoderm (Figure 14.1A–F). Lastly, the regenerating limb blastema of amphibians develops an apical "epithelial" cap (AEC; Figure 14.1G,H), which functions to direct regrowth of a severed limb. Within hours of limb amputation, epithelial cells migrate to the wound surface and proliferate to form a multilayered AEC (Christensen and Tassava, 2000; Han et al., 2005). The AEC is necessary for limb regeneration and functionally homologous to the apical ectoderm in the tetrapod limb (Christensen and Tassava, 2000; Han et al., 2005) but displays several morphological differences (Table 14.1). The AEC covers the entire end of a limb stump (Figure 14.1D), whereas the apical ectoderm in

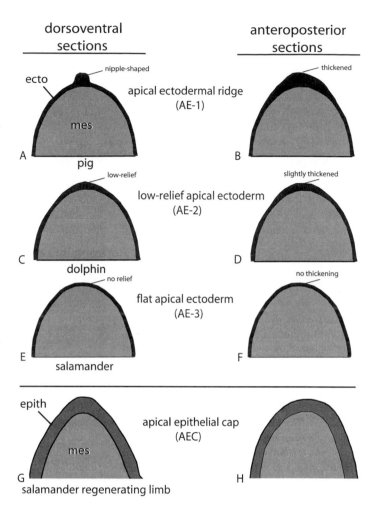

dorsoventral sections

anteroposterior sections

nipple-shaped

ecto

mes

apical ectodermal ridge
(AE-1)

thickened

A

B

pig

low-relief

low-relief apical ectoderm
(AE-2)

slightly thickened

C

D

dolphin

no relief

flat apical ectoderm
(AE-3)

no thickening

E

F

salamander

epith

mes

apical epithelial cap
(AEC)

G

H

salamander regenerating limb

FIGURE 14.1 Schematics of apical ectodermal (AE) morphologies. (A, B) The typical ridge-shaped AE ridge (AE-1) of most vertebrates as seen in the domestic pig (*Sus scrofa*), in dorsoventral (A) and anteroposterior (B) sections. (C, D) The less prevalent low-relief AE (AE-2) that displays slight relief in dorsoventral (C) and anteroposterior (D) sections as seen in the pantropical spotted dolphin (*Stenella attenuata*). (E, F) A flattened AE as reported in the normal developing limb of salamanders in dorsoventral (E) and anteroposterior (F) sections. (G, H) The regenerating salamander limb with an elongating blastema made of epithelial and mesenchymal tissues in dorsoventral (G) and anteroposterior (H) sections (modified from Christensen and Tassava, 2000). AEC, apical ectodermal cap; ecto, ectoderm; epith, epithelium; mes, mesenchyme.

normal developing vertebrate limbs is localized to the dorsoventral boundary (Figure 14.1A; Christensen and Tassava, 2000). Furthermore, the AEC is made of 4–15 stratified layers of cells, whereas the apical ectoderm of vertebrates is composed of only a single layer of pseudostratified cells (Christensen and Tassava, 2000) that may be overlain with nonstratified cells.

ORGANIZATION OF THE APICAL ECTODERM IN THE DEVELOPING LIMB

While several experimental studies on the chick have revealed the function of the apical ectoderm in the developing limb, a specific organization of cells in the ectoderm is not essential for proper limb development and patterning

(for review, see Saunders, 1998). For example, AER removal causes a cessation of limb outgrowth for distal skeletal elements, resulting in limb truncation (Saunders, 1948). Alternatively, the addition of AER tissue to the terminus of a developing limb causes distal outgrowth to resume. Similarly, two limbs can be produced by the transplantation of an isolated AER that lacks associated dorsal and ventral ectodermal tissues onto a limb bud that already possesses a normal AER (Saunders and Gasseling, 1968). Removal and replacement of ectodermal cells with an inverted ectodermal jacket also produces a normal limb (Errick and Saunders, 1974). If AER cells are removed, disassociated, mixed, and then placed on the distal limb bud, a normal limb de-

TABLE 14.1

Comparison Between the Apical Ectoderm of Vertebrates and the Apical Epithelial Cap of Regenerating Salamander (Urodele) Limbs

	APICAL ECTODERM	APICAL EPITHELIAL CAP	REFERENCES
Origin of tissue	Ectoderm associated with medial somatopleure	Limb epidermis	Michaud et al., 1997; Christensen and Tassava, 2000
Function	Limb outgrowth	Regenerating limb outgrowth	Christensen and Tassava, 2000
Gross morphology	Low to ridgelike, localized to dorso-ventral boundary	Uniformly smooth, broadly covers wound stump	Christensen and Tassava, 2000
Number of stratified cell layers	Single pseudostratified layer of ectodermal cells topped with additional nonstratified cells	Several stratified layers (4–15 layers)	Christensen and Tassava, 2000
Basement membrane	Present	Absent	Christensen and Tassava, 2000
FGF expression	Throughout cell layers	Basal-most layer of cells, underlying mesenchyme	Han et al., 2001; Sun et al., 2002

velops (Errick and Saunders, 1974). Taken together, these experimental manipulations show that the apical ectoderm is required for limb development and proximodistal outgrowth but that the organization of ectodermal cells within individuals appears to be inconsequential for proper limb development.

FGFs IN THE APICAL ECTODERM CONTROL PROXIMODISTAL OUTGROWTH AND LIMB PATTERNING

FGFs are expressed in apical ectodermal cells of developing limbs (Mariani et al., 2008) and are members of the heparin-binding growth factor family. They function in promoting cell survival and proliferation of undifferentiated mesenchymal cells (Niswander et al., 1994a, 1994b; Hara et al., 1998; Ngo-Muller and Muneoka, 2000; Han et al., 2001; Niswander, 2002; Weatherbee et al., 2006) as well as specifying cell fate during

digit formation (Mariani et al., 2008; Lu et al., 2008). *Fgf4* (genes indicated by italicized text, whereas proteins are indicated by Roman font), *Fgf8*, *Fgf9*, and *Fgf17* are some of the many genes expressed in the apical ectoderm of developing limbs; but *Fgf8* is the most important for normal limb outgrowth (Niswander et al., 1994a, 1994b; Mahmood et al., 1995; Vogel et al., 1996; Hara et al., 1998; Moon and Capecchi, 2000; Ngo-Muller and Muneoka, 2000; Sun et al., 2002; Talamillo et al., 2005; Verheyden and Sun, 2008). It is expressed earlier and at higher concentrations compared to other FGFs (Fernandez-Teran and Ros, 2008; Mariani et al., 2008). Ancillary FGFs (*Fgfs 4, 9,* and *17*) are functionally redundant and can rescue limb outgrowth and patterning in the absence of *Fgf8* (Hara et al., 1998; Moon and Capecchi, 2000; Niswander, 2002; Mariani et al., 2008; Verheyden and Sun, 2008).

Fgf8 expression is normally localized to apical ectodermal cells in order to signal to the underlying mesenchymal cells during limb outgrowth. However, a few notable exceptions have documented *Fgf8* expression within limb mesenchyme (Vogel et al., 1996; Moon and Capecchi, 2000; Pizette et al., 2001; Weatherbee et al., 2006). The developing forelimbs of bats express *Fgf8* both in the apical ectoderm and in the interdigital mesenchyme, presumably to direct outgrowth of the digits and promote cell survival and proliferation of interdigital mesenchymal cells for generation of a wing membrane (Weatherbee et al., 2006). Furthermore, in the regenerating limbs of salamanders, *Fgf8* expression was found in the basal-most layer of the apical ectoderm and the underlying mesenchymal tissues, also presumably to promote cell survival and proliferation (Han et al., 2001; Christensen et al., 2002).

FGFs produced by the limb apical ectoderm are necessary for the initiation and maintenance of sonic hedgehog (*Shh*) expression, and together FGFs and *Shh* create a positive feedback loop that is essential for limb outgrowth, polarizing, and patterning (Niswander et al., 1994a; Vogel et al., 1996; Niswander, 2002; Boulet et al., 2004; Panman et al., 2006; Tickle, 2006; Mariani et al., 2008; Tabin and McMahon, 2008). This positive feedback loop is also connected to an *Fgf/Gremlin1* inhibitory feedback loop that terminates limb bud outgrowth (Verheyden and Sun, 2008). If either FGF or *Shh* experiences a cessation in expression, a normal limb will not form. For instance, in primitive snakes the hindlimbs fail to form an AER with associated *Fgf8* expression, preventing *Shh* expression and causing a cessation of limb development (Cohn and Tickle, 1999). Adult snakes lack visible hindlimbs as they are vestigial and contained in the body wall (Cohn and Tickle, 1999). In pantropical spotted dolphin embryos, both Fgf8 and Shh protein signals were present during incipient limb bud stages, but a later cessation in Fgf8 signaling (presumably concomitant with the hiatus of all ectodermally derived

FGF expression) arrested limb development. Adult dolphins lack external hindlimbs, but incomplete hindlimb and pelvic girdle vestiges are encased in the body wall near the vertebral column (Figure 14.2) (Thewissen et al., 2006) to various degrees and in different cetacean species. By interrupting *Fgf8* expression at different times during limb development, both snakes and dolphins convergently evolved a streamlined body with hindlimbs encased in the body wall.

Duration of *Fgf8* expression appears to be essential for normal limb development and is correlated with both phalangeal number and digit length. Increased duration of *Fgf8* expression results in polydactyly (Vogel et al., 1996; Talamillo et al., 2005), inhibits terminal phalanx formation, and directs the development of supernumerary phalanges (Sanz-Ezquerro and Tickle, 2003; Richardson et al., 2004). Conversely, experimentally induced decreases in *Fgf8* expression result in the generation of deformed limbs (Sun et al., 2002) with fewer skeletal elements (Vogel et al., 1996; Sun et al., 2002; Mariani et al., 2008) and increased apoptotic activity (Moon and Capecchi, 2000; Boulet et al., 2004; Talamillo et al., 2005). If an *Fgf8* inhibitor is present, premature formation of the terminal phalanx will occur, in some cases creating fewer numbers of phalanges (Sanz-Ezquerro and Tickle, 2003).

MOLECULAR PATHWAYS DETERMINING APICAL ECTODERMAL MORPHOLOGY

Through study of modern taxa, such as chicks and mice, some of the developmental pathways creating a ridge-shaped ectoderm are well-known. These findings may offer insight into the mechanisms that may inhibit formation of a ridge-like AER and allow for development of a low-relief or flattened apical ectoderm (AE-2, -3).

The AER lies between the dorsal and ventral ectodermal surfaces of the limb bud (Kimmel et al., 2000). Cells of the adjacent dorsal ectoderm display *Wnt7a* and *Lmx1*, while the homeobox transcription factor *EN1* is expressed in the ven-

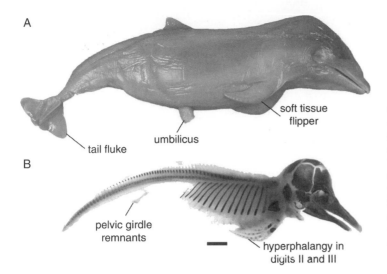

FIGURE 14.2 Morphology of an approximately 110-day-old (Carnegie stage 23) pantropical spotted dolphin (*Stenella attenuata*, LACM 94285). (A) Whole fetus. (B) The same fetus clear and stained, revealing pelvic girdle and hindlimb remnants as well as hyperphalangy in the principal digits of the flipper. Clearing and staining completed by Dr. Sirpa Nummela. Scale bar = 1 cm.

A

soft tissue
flipper

umbilicus

tail fluke

B

pelvic girdle
remnants

hyperphalangy in
digits II and III

tral ectoderm (Talamillo et al., 2005). If *EN1* is misexpressed or the dorsoventral border is lost, the distinctive ridge shape of the AER is lost, resulting in a flattened apical ectoderm (Kimmel et al., 2000). Furthermore, interruption of *Fgf8* expression along the AER when *EN1* is misexpressed leads to missing and in some cases ectopic digits (Kimmel et al., 2000).

Bone morphogenic proteins (BMPs) also play a key role in regulating the height of the apical ectoderm (Ahn et al., 2001; Pizette et al., 2001). Inhibition of BMP signaling, through application of the BMP antagonist *Noggin*, at early stages of chick limb development resulted in an increase in AER height (Pizette and Niswander, 1999). *Gremlin1* also regulates AER height by inhibiting BMP (for review, see Fernandez-Teran and Ros, 2008).

The *Wnt/β-catenin* pathway lies upstream of BMP signaling and associated patterning and is essential to establishing mouse AER morphology (Barrow et al., 2003; Narita et al., 2005; Lu et al., 2008). A *Wnt/β-catenin/Fgf* regulatory loop was found to be essential to the establishment and survival of a morphological AER in mice, and *Wnt3* was a key signal regulating AER thickness in this organism (Barrow et al., 2003). Mouse mutants with disrupted *Wnt3* expression displayed a 50% reduction in dorsoventral

thickness of the AER and variably displayed fewer limb skeletal elements (Barrow et al., 2003). However, *Fgf8* expression was only mildly affected and was localized to only those ectodermal cells that were slightly thickened (Barrow et al., 2003). *Wnt3a* carried out a similar role in chicks (Kengaku et al., 1998; Niswander, 2002).

COMPARATIVE MORPHOLOGY OF THE APICAL ECTODERM AMONG VERTEBRATES

The AER (AE-1; Figure 14.1A,B) was fully developed in most vertebrates studied to date (Hanken et al., 2001), although rare exceptions in apical ectoderm shape have been documented in tetrapods (Table 14.2). We conducted a broad literature review, focusing primarily on morphological variation within the distal limb ectoderm of nonmodel vertebrates (fish, lungfish, amphibians, squamates, birds, and mammals; see Table 14.2). Additionally, morphology of the apical ectoderm of the limb and, if possible, patterns of *Fgf4* and *Fgf8* gene expression or protein-signal localization were also investigated (Table 14.2). We first describe a typical vertebrate AER (AE-1) using a pig model and then discuss variations, such as the low-relief apical

TABLE 14.2

Patterns of Morphological Variation and Fibroblast Growth Factor Expression in the Limb Ectoderm of Developing Vertebrates During Normal Development

ORDER	TAXON	COMMON NAME	APICAL ECTODERMAL MORPHOLOGY	APICAL ECTODERMAL EXPRESSION	REFERENCES
Cypriniformes	*Danio rerio*	Zebrafish	AE-1	*Fgf8*	Grandel and Schulte-Merker, 1998; Reifers et al., 1998; Mercader, 2007
Cyprinodontiformes	*Aphyosemion scheeli*	Killifish	AE-1		Wood, 1982
Ceratodontiformes	*Neoceratodus forsteri*	Australian lungfish	AE-1	Fgf8 protein	Hodgkinson et al., 2007
Urodeles	*Ambystoma mexicanum*	Mexican axolotl	AE-3	*Fgfs 4, 8*	Han et al., 2001; Christensen et al., 2002
			AEC[rg]	*Fgf8*[rg], lacks *Fgf4*[rg]	
Anura	*Eleutherodactylus coqui*	Tree frog	AE-2	DLX	Fang and Elinson, 1996; Richardson et al., 1998
Anura	*Xenopus laevis*	African clawed frog	AE-2	*Fgf8*	Tarin and Sturdee, 1971; Fang and Elinson, 1996; Christen and Slack, 1997
			AEC[rg]	*Fgf8*[rg]	
Chelonia	*Chelonia mydas*	Green turtle	AE-1		Milaire, 1957; Vasse, 1972; Miller, 1985
	Chelonia depressa	Flatback turtle	AE-1		
	Caretta caretta	Loggerhead turtle	AE-1		
	Eretmochelys imbricata	Hawksbill turtle	AE-1		
	Lepidochelys olivacea	Pacific ridley turtle	AE-1		
	Dermochelys coriacea	Leatherback turtle	AE-1		
	Emys orbicularis	European pond turtle	AE-1		
	Testudo graeca	Greek tortoise	AE-1		
	Pseudemys	Pond turtle	AE-1		

Order	Species	Common name	AE	Fgf	References
Squamata	*Lacerta vivipara*	Common lizard	AE-1		Milaire, 1957; Dufaure and Hubert, 1961; Goel and Mathur, 1977
	Calotes versicolor	Garden lizard	AE-1		
	Chamelaeo	Chameleon	AE-1		
	Mabuya	Long-tailed skink	AE-1		
Crocodilia	*Alligator mississippiensis*	American alligator	AE-1		Honig, 1984; Ferguson, 1985
	Crocodylus porosus	Saltwater crocodile	AE-1		
	Crocodylus johnsoni	Freshwater crocodile	AE-1		
Marsupialia	*Monodelphis domestica*	Short-tailed opossum	AE-2	*Fgf8*	Smith, 2003; Sears, pers. comm.
Rodentia	*Mus musculus*	Mouse	AE-1	*Fgfs 4, 8*	e.g, Sun et al., 2002; Boulet et al., 2004
Galliformes	*Gallus gallus*	Chick	AE-1	*Fgfs 4, 8*	e.g., Niswander et al., 1994a; Kengaku et al., 1998; Narita et al., 2005
Chiroptera	*Carollia perspicillata*	Short-tailed fruit bat	AE-1	*Fgf8*	Weatherbee et al., 2006; Cretekos, et al., 2007; Sears, 2008
Primata	*Homo sapiens*	Human	AE-1		Bardeen and Lewis, 1901; Steiner, 1929; O'Rahilly et al., 1956; Kelley, 1973; O'Rahilly and Müller, 1985; Hallgrímsson et al., 2002
Cetartiodactyla	*Stenella attenuata*	Pantropical spotted dolphin	AE-2	Fgf4, Fgf8 proteins	This study
	Sus scrofa	Domestic pig	AE-1	Fgf8 protein	This study

NOTE: AE, apical ectodermal; AEC, apical epithelial cap; [rg], regenerating limb.

ectoderm of dolphins (AE-2). For some vertebrates, such as amphibians and teleosts, limb development is quite different from that in model organisms (chicks, mice), which are subsequently discussed in detail. The apical ectodermal morphologies and expression patterns documented in model taxa (i.e., chicks, mice, humans) are listed in Table 14.2.

AN AER (AE-1) IS THE TYPICAL ECTODERMAL MORPHOLOGY OF VERTEBRATES

DOMESTIC PIG (SUS SCROFA)

Limb morphogenesis in the domestic pig (*Sus scrofa*) progresses from a typical mammalian handplate to a four-digit limb with two elongated central digits (Hamrick, 2002), characteristic of artiodactyls (even-toed ungulates). At approximately 17 days' gestation the forelimb projects from the body wall and forms a handplate on the following day. By approximately 24 days digital condensations are visible (Patten, 1943). The digit I anlage usually does not form in the pig, as reported by Hamrick (2002), although Patten (1943) observed pentadactyly. Regardless, the limb develops such that, as in almost all artiodactyls, digits III and IV are the longest and most robust and become the main load-bearing elements, while digits II and V are reduced. However, in a number of cases digit I developed in some artiodactyls (Prentiss, 1903). The metacarpals are much longer than the phalanges in both the manus and pes (Hamrick, 2001). Apoptosis of interdigital webbing then creates four separate digits, and eventually small hooves will develop along the superficial aspects of the ungual phalanges of all four digits. We describe the morphology of the distal forelimb ectoderm of the developing domestic pig (*Sus scrofa*) and document the presence of Fgf protein signals within that ectoderm.

After day 20 of embryogenesis, a morphological AER (AE-1) is present in the developing pig forelimb. By day 21, the limb bud has passed the handplate stage and is instead blunted and rectangular, due to extensive proximodistal lengthening (Figure 14.3A). In anterior view the limb is conical and the raised apical ectoderm

(AE-1) is variably discernable (Figure 14.3A). Histological sections of the limb at this stage reveal a prominent, classically shaped AER with more rounded, rather than columnar, basal cells (Figure 14.3B). This porcine AER (AE-1) is clearly stratified. Extensive Fgf8 protein signals were found in the apical ectoderm and adjacent ectodermal tissues; however, no protein signals were found in the underlying mesodermal tissues.

At day 24 of embryogenesis, the limb is paddle-shaped with a wide diameter, in lateral view, and Fgf8 protein signaling is localized to the AER (Figure 14.3C). Slight digital condensations are also apparent. Embryos harvested near day 28 display distinct condensations in digits III and IV, with the AER present only along the distal ends of those digits.

TELEOSTS

The fins of some teleost fish display an AER (AE-1) and typical patterns of *Fgf8* expression (e.g., zebrafish, *Danio rerio* [Reifers et al., 1998; for review, see Mercader, 2007]), even though they possess fins with rays (lepitotrichia) that lack the appendicular elements seen in the vertebrate autopod and evolved the ability to regenerate parts of pectoral fins (Poss et al., 2000; Galis et al., 2003). The teleost AER (AE-1) does not undergo apoptosis, as in most tetrapods (Wood, 1982), but instead folds and elongates distally to form a fin fold (Figure 14.4). The fin fold then grows distally until a semicircular swimming paddle develops (Wood, 1982). Fin fold cells form dorsal and ventral layers, express similar markers, and perform similar functions as the tetrapod AER (Mercader, 2007).

Detailed descriptions of the apical ectoderms of some teleosts have documented how they differ from those of most tetrapods. For instance, killifish (*Aphyosemion scheeli*) have a morphologically distinct apical ectoderm (AE-1) that, relative to the fin bud, is larger than the AER (AE-1) of most tetrapods (Wood, 1982). The killifish apical ectoderm spans the entire distal margin of the fin bud along the anteroposterior

AE-1

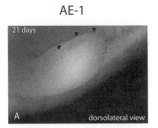

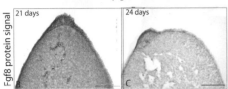

domestic pig (*Sus scrofa*)

AE-2

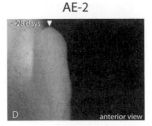

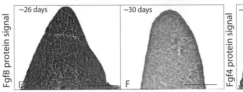

pantropical spotted dolphin (*Stenella attenuata*)

FIGURE 14.3 Apical ectodermal morphologies and associated protein signaling of (A–C) the domestic pig (*Sus scrofa*) fore-limb and (D–G) the pantropical spotted dolphin (*Stenella attenuata*). Pigs (NEOUCOM-P6012 [A], NEOUCOM-P6014 [B], and NEOUCOM-P111 [C]) display a characteristic vertebrate apical ectodermal ridge (AER) (A–C) and associated Fgf8 protein signaling (brown color) (B, C). Dolphins (LACM 94613, Carnegie stage 15 [D]; LACM 94594, Carnegie stage 15 [E]; LACM 94770, Carnegie stage 16 [F]; LACM 94817, Carnegie stage 19 [F]), however, are unique among most tetrapods in that they display a flattened apical ectoderm (D–G) but localize Fgf8 (E, F) and Fgf4 (G) protein signals to the distal limb ectoderm. Scale bars = 100 μm.

plane (Wood, 1982; Grandel and Schulte-Merker, 1998). Compared to non–apical ecto-dermal cells, those cells within the basal layer of the apical ectoderm are elongated and pseu-dostratified. The AE-1 of trout taxa *Salmo trutta fario* and *S. gairdneri* exhibits an area of elevated ectoderm (Bouvet, 1968) made of pseudostrati-fied columnar cells (Géraudie and François, 1973; Géraudie, 1978; Wood, 1982).

LUNGFISH

Lungfishes possess an AER and Fgf8 protein signaling similar to most vertebrates. However, lungfishes also are able to regenerate both the soft tissues and skeletal elements of their fins (Galis et al., 2003). The distal fin bud ectoderm of the lungfish *Neoceratodus* is initially a strati-fied bilayer of cuboidal cells covered by a squa-mous periderm, and after a period of out growth, morphological AER emerges (Hodg-kinson et al., 2007). This AER is composed of pseudostratified columnar cells in the basal membrane of the distal fin bud epithelium (Hodgkinson et al., 2007). Fgf8 protein signals have also been documented in the *Neoceratodus* AER (Hodgkinson et al., 2007), indicating pro-

tein signals consistent with most vertebrates during limb development.

SQUAMATA

The garden lizard (*Calotes versicolor*) has an AER that is nipple-shaped in cross section, much like that of most tetrapods (Goel and Mathur, 1977).

CHIROPTERA

Like most vertebrates, the developing limb of bats displays both an AER and associated *Fgf8* expression. The bat AER is initially present over the entire anterior–posterior distal aspect of the developing handplate, but as digits of the fore-limb elongate, the AER and associated gene ex-pression become localized over digits II and III (Weatherbee et al., 2006). Compared to that of similar-aged mice, the Fgf8 AER expression domain of bats is approximately three times wider than that of model taxa (Cretekos et al., 2007).

Bats also display a novel domain of *Fgf8* ex-pression in the interdigital tissues between dig-its II–V, presumably to aid in the proliferation and survival of these tissues during wing mem-brane formation (Weatherbee et al., 2006). By

FIGURE 14.4 Schematic of the transition from an apical ecto-dermal ridge (AE-1) to a fin fold in the killifish (*Aphyosemion scheeli*, modified from Wood, 1982). (A) Apical ectodermal ridge (AE-1) is present at ~128 hours. (B) At 135 hours, the apical ectoderm is folded, caused by differential mitosis in the dorsal and ventral ectoderms. (C) At 144 hours, a fin fold has formed and is made of a tightly appressed ectodermal bilayer. Not to scale.

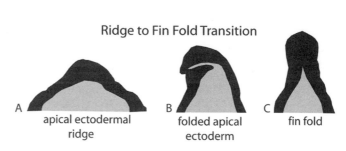

Ridge to Fin Fold Transition

A   apical ectodermal ridge   B   folded apical ectoderm   C   fin fold

altering the domain of *Fgf8* expression in both the AER and interdigital tissues, bats display relatively elongated metacarpals and phalanges (Sears et al., 2006) that are connected by a thin wing membrane (Cretekos et al., 2001; Weatherbee et al., 2006; Sears, 2008). Mesenchymal expression of *Fgf8* is similar to that of axolotls but is unique compared to most amniotes (Weatherbee et al., 2006).

UNUSUAL VERTEBRATES LACKING AN AER (AE-1)

PANTROPICAL SPOTTED DOLPHIN (STENELLA ATTENUATA)

Descriptive embryological studies document that the forelimb of the pantropical spotted dolphin (*Stenella attenuata*) begins as a typical mammalian handplate that is as long as it wide until about 28 days' gestation (Richardson and Oelschläger, 2002). Five digital condensations form (Sedmera et al., 1997), and in some cases, a low-relief epithelial thickening appears along the distal aspect of the limb bud (Richardson and Oelschläger, 2002). After 30 days' gestation, a weakly organized and thickened epithelium is present at the ends of the central digits II and III (Richardson and Oelschläger, 2002). Toward the end of the embryonic period, near 48 days' gestation, digits II and III have an increased number of phalanges relative to the other digits, creating a chisel-shaped flipper in lateral view, and the thickened ectoderm is localized to the ends of these developing digits (Richardson and Oelschläger, 2002).

In the developing *Stenella* forelimb, neither gross anatomical nor sectioned fore- or hindlimbs display an AER (AE-1) at any examined ontogenetic stage. Instead, the distal apex of the limb buds is encapsulated by a thickened ectoderm that has a smoothed contour (Figure 14.1, AE-2). This ectodermal morphology is consistent with the morphology of the apical ecto-dermal cap reported in the tree frog *Eleuthrodac-tylus coqui*, as well as the modest limb ectoderm of *Xenopus* (see above, Figures 14.1, 14.3D–G). At approximately 26 days' gestation (Carnegie stage 15), the dolphin limb is cone-shaped in lateral view and the entire distal aspect is lined with a transparent ectoderm. In anterior view, a low-relief apical rise is apparent between the dorsal and ventral surfaces of the limb bud. This apical ectoderm is two to four cell layers thick. Cells of the basal layer are elongated and columnar, while cells of the second layer from the bottom are rounded and approximately half the height of the basal cells. The one or two apical layers consist of slightly flattened cells. Fgf8 protein signals are localized along the superficial apical layers as well as the dorsal surface of the apical ectoderm (Figure 14.3E).

By ~35 days' gestation (Carnegie stage 17), both digits II and III have elongated considerably, creating a blunt-ended limb bud when viewed laterally. In cross section, a distinct apical ridge is absent; instead, only a low-relief rise of epithelium occurs at the border between the dorsal and ventral surfaces. Cross sections through the distal aspect of the limb bud revealed a dorsoventrally narrowed and thickened ectoderm, relative to that of previous ontogenetic stages. The apical ectoderm at this stage consists of tightly packed cells that are slightly elongated and elliptical in shape; however, only a dorsoventral narrow portion of this ectoderm displays a tiny additional cell layer.

At approximately 48 days' gestation, at the end of the embryonic period (Carnegie stage 19), the flipper is almost entirely formed, with digit II being the longest, followed closely by digit III. The other digits are considerably shorter. Five digital condensations are clearly visible, and each digit displays obvious interphalangeal joints. Although ectodermal tissues are thickened along the ends of digit II, no distinct apical ectodermal ridge is present. In anterior view, this region of ectoderm appears as a slight thickening with little relief compared to the adjacent dorsal and ventral ectodermal tissues. Cross sections through the distal end of digit II showed no apical thickening or change in cell shape compared to the nonapical ectoderm. Fgf4 protein signals were localized to the ends of this digit, indicating an expression pattern consistent with that of other tetrapods (Figure 14.3G).

MARSUPIALA

Morphology and patterns of gene expression in the short-tailed opossum (*Monodelphis domestica*) fore- and hindlimb AER are currently under study (Sears, personal communication). Preliminary results indicate that the developing forelimb of these taxa lacks a typical AER characteristic of vertebrates and that instead they possess a low-relief (AE-2) and disassociated apical ectoderm with few cell layers (Sear, personal communication). Forelimb apical ectoderm expresses *Fgf8* much earlier (stage 25) than that of the hindlimb (stage 31) (Smith, 2003).

AMPHIBIA

URODELES The urodele limb is exceptional by undergoing direct growth of the digits as independent buds off the limb bud (Von Dassow

and Munro, 1999; Franssen et al., 2005). Morphogenic differences in limb development among amphibians suggest polyphyly (Hanken, 1986; Von Dassow and Munro, 1999; Franssen et al., 2005) because the pattern of urodele limb development could have evolved separately from that of other amphibians, including Anura (Holmgren, 1933; Jarvik, 1965; Franssen et al., 2005).

The normal urodele limb buds lack an apical ectodermal thickening of the developing limb (AE-3; Galis et al., 2003; Franssen, et al., 2005; Han et al., 2005), but regenerating limbs possess an AEC. Both normal and regenerating limb buds express *Fgf8* in the apical ectoderm (Christensen et al., 2002). The basal layer of the AEC functions much like the amniote AER during normal limb development (Onda and Tassava, 1991; Christensen and Tassava, 2000; Galis et al., 2003; Franssen et al., 2005).

In the Mexican axolotl (*Ambystoma mexicanum*), *Fgf8* expression was detected in both the developing limb bud and the AEC of a regenerating limb blastema (Han et al., 2001; Christensen et al., 2002). Before digit formation, *Fgf8* expression in *Ambystoma* is localized in the epithelium; however, a gradual translocation of *Fgf8* expression to the underlying mesenchymal tissue occurs. This expression is unlike that of *Xenopus*, chicks, and mice, where *Fgf8* expression is isolated to the AER (Han et al., 2001). Similarly, in the AEC of regenerating limbs of *Ambystoma*, *Fgf8* is expressed in the basal-most layer of the AEC and the underlying mesenchymal tissues, further suggesting that the basal-most layer of the AEC is functionally equivalent to the amniote AER (Han et al., 2001; Christensen et al., 2002). *Fgf4* was expressed only slightly in the developing limb and was absent from the AEC (Christensen et al., 2002).

ANURA During limb development, the metamorphosing African clawed frog (*Xenopus laevis*) displays a low-relief apical ectoderm (AE-2) that consists of three layers of ectodermal cells (Tarin and Sturdee, 1971). *Fgf8* is expressed in the distal tip of the developing *Xenopus* hindlimb but in only the epithelium, whereas in a regenerating limb $Fgf8$ expression is localized to both the mesenchyme and basal-most layer of the AEC (Han et al., 2001).

The large neotropical tree frog (*Eleutherodactylus coqui*) directly develops as a froglet as it does not proceed through a tadpole stage (Richardson, 1995). Its limb buds appear at an earlier ontogenetic stage compared to metamorphosing species (e.g., indirect developers like *Xenopus*) (Richardson, 1995; Richardson et al., 1998; Hanken et al., 2001; Bininda-Emonds et al., 2007). *E. coqui* has been the subject of study because it can form a normal vertebrate limb in the absence of a morphological AER (Richardson et al., 1998; Hanken et al., 2001). The ectoderm of *E. coqui* is a low-relief thickened ectodermal cap along the limb apex (Richardson et al., 1998; Hanken et al., 2001).

Excision of the hindlimb apical ectoderm of *E. coqui* resulted in loss and/or fusion of the distal limb elements (Richardson et al., 1998), suggesting the apical ectoderm played a role in controlling limb outgrowth and functioned much like the AER of amniotes. However, truncation of the distal limb elements was not observed (Richardson et al., 1998), possibly because the ectoderm was partially regenerated.

DISCUSSION

MORPHOLOGY OF THE VERTEBRATE LIMB APICAL ECTODERM

This chapter documents that a ridge-like apical ectoderm (AER, AE-1) along the apex of a developing limb is the most common morphology for vertebrates as it is present in model taxa (e.g., mice and chicks) and several other lineages of vertebrates (Table 14.1). Presence of a ridge-like apical ectoderm in teleosts and lungfishes suggests that the ridge morphology is the primitive condition among vertebrates and evolved before the transition from a fin to a limb in the earliest tetrapods. During fin development in teleosts, the AER remains active and morphs from a ridge to a layered fin fold (Figure 14.4) and finally into a swimming paddle (Wood, 1982). In contrast, the AER of most tetrapods is only transitory during embryogenesis. The tetrapod AER will undergo apoptosis first along the interdigital spaces, then at the ends of developing digits (Fernandez-Teran and Ros, 2008). Therefore, the AER is most common among vertebrates, but its function has reduced in the development of tetrapods.

Unrelated lineages of vertebrates (i.e., amphibians, cetaceans, and marsupials) have convergently evolved a low-relief apical ectoderm of the developing limb. In these groups, the limb apical ectoderm either lacks a thickening (AE-3, salamanders) or displays only a slight thickening (AE-2, dolphins, anurans) (Figure 14.1, Table 14.2). Evolution of a low-relief apical ectoderm in dolphins is autapomorphic as an AER (AE-1) is present in their terrestrial artiodactyl relative, the pig (Figure 14.3, Table 14.2). A low-relief apical ectoderm is also present in the marsupial developing forelimb (Table 14.2, Sears personal communication), but unlike the apical ectoderm of most vertebrates, this apical ectoderm is discontinuous, suggesting that this morphology is also an autapomorphy.

Regardless of the gross morphology of the apical ectoderm (high- vs. low-relief), all taxa included in this analysis displayed normal fin or limb development, indicating that gross morphology of the apical ectoderm does not affect its function. Furthermore, correlations between proper apical ectoderm function and its cellular organization are uninformative as experimental manipulations of cellular distribution show no effect on function (Saunders, 1948, 1998; Saunders and Gasseling, 1968; Errick and Saunders, 1974). Only removal of apical ectodermal cells negatively altered function (Saunders, 1948, 1998; Saunders and Gasseling, 1968; Errick and Saunders, 1974). A morphological definition of the apical ectoderm is useful for comparative studies and tracing evolutionary transformations, but correlations between function and morphology are dubious. A normal limb apical ectoderm is probably best described by a molecular (signaling) criterion.

The apical ectoderm of a properly developing limb, regardless of its morphology, secretes morphogens (e.g., FGFs) that control limb outgrowth and digital patterning. We chose the presence of FGFs as a molecular indicator of an active limb apical ectoderm as they are the foundation of several pathways involved in limb outgrowth, patterning, and arrest of growth and their expression is consistent among taxa with varying limb apical ectodermal morphologies (Table 14.2) (e.g., Barrow et al., 2003; Verheyden and Sun, 2008). This chapter documented that all taxa expressed FGFs within the apical ectoderm of normal developing limbs (Table 14.2). Our results therefore indicate that a molecular definition of an active limb apical ectoderm is a conservative and reliable alternative to a morphological definition.

POTENTIAL CORRELATES OF LIMB APICAL ECTODERM HEIGHT

Thickness of the limb apical ectoderm is directly related to the number of cells populating that region of tissue. Constituent cells of the normal limb apical ectoderm include those signaling for limb development as well as apoptotic cells. These apoptotic cells are distributed throughout the limb apical ectoderm and are absent from adjacent ectodermal tissues (Fernandez-Teran and Ros, 2008). Their presence has been documented in the limb apical ectoderm of the chick and mouse throughout its life span (Jurand, 1965; Todt and Fallon, 1984; Fernandez-Teran and Ros, 2008); however, little is known of the abundance of these cells in nonmodel taxa, including those presented here. It could be that the ratio of cells signaling for limb growth and patterning versus apoptotic cells is different between taxa with high-relief (i.e., fishes, lungfishes, chelonians, squamates, crocodilians, chicks, mice, chiropterans, primates, and pigs) and low-relief (i.e., marsupials, cetaceans, and amphibians) limb apical ectoderms. Alternatively, the ratio of different cell types may be equivalent across these taxa but regulated differently via those genes directly affecting height of the limb ectoderm (e.g., BMP, *Noggin*, *Gremlin*). The activity level of apoptotic cells has been shown to be a chief determinant of the rate of

epithelial morphogenesis and could directly affect the rate of limb outgrowth and digital development (Davidson, 2008; Toyama et al., 2008). Indeed, taxa with low-relief apical ectoderms (i.e., marsupials, cetaceans, and amphibians) have become the topics of intense study as their limb development is either significantly delayed or precocial relative to most vertebrates (McCrady, 1938; Richardson, 1995; Richardson and Oelschläger, 2002; Galis et al., 2003; Smith, 2003; Sears, 2004; Keyte et al., 2006; Bininda-Emonds et al., 2007). It may be that the abundance and/or activity of apoptotic cells in the apical ectoderm plays a significant role not only in shaping the limb apical ectoderm but also in directing the rate of limb development.

IMMUNOHISTOCHEMICAL METHODS

Embryonic specimens of the pantropical spotted dolphin (*Stenella attenuata*) were supplied by the Los Angeles Museum of Natural History. Embryos were immersion-fixed, preserved in 70% ethanol, and stored without refrigeration for time periods ranging from 15 to 32 years. The embryos were staged according to a modified version of the Carnegie system (Thewissen and Heyning, 2007). The immunohistochemical data are based on six dolphin embryos (Los Angeles County Museum [LACM]), ranging from Carnegie stage 13 to Carnegie stage 19. Each embryo was embedded in paraffin and sectioned at 6 µm. Protocols were optimized with immersion-fixed, ethanol-preserved mouse embryos. Nonlimb embryonic dolphin tissue was then tested and optimized. Because of the variance in fixation and storage times, slightly different procedures were used for different specimens to obtain optimal results. In addition, negative control samples (minus primary antibody) were used to determine the level of background staining for all experiments.

The *Sus scrofa* embryos were obtained from sows with timed pregnancies supplied by Tank Farms (Fremont, OH). The forelimb AERs were viewed at approximately 20 days' gestation. For our purposes, two stages of limb development were reviewed (~21 and 24 days after gestation) to illustrate AER morphology and Fgf8 protein expression. The embryos were prepared and stained in a similar method as the dolphin embryos but were fixed in 4% paraformaldehyde for 24 hours, followed by storage in 1× PBS.

The following antibodies were used in this study: anti-Fgf8 (Santa Cruz Biotechnology, Santa Cruz, CA; sc-6958); anti-Fgf4 (Santa Cruz Biotechnology, sc-1361).

ACKNOWLEDGMENTS

We thank Dr. Benedikt Hallgrímsson and Dr. Brian K. Hall for the invitation to submit this work for inclusion in their book. We thank Dave Janiger and the late John Heyning for experimental use of dolphin embryos; Verity Hodgkinson, Mike Jorgensen, and Verne Simmons for discussions; and Tobin L. Hieronymus, Mike Selby, Burt Rosenman, Christopher J. Vinyard, Karen E. Sears, and Amy L. Mork for comments on this manuscript. Funding for this study came from grants to L. N. C. from the Lerner-Gray Fund for Marine Research, a Sigma Xi Grant in Aid of Research, and the Skeletal Biology Fund of the Northeastern Ohio Universities College of Medicine. Funding for portions of this study also came from National Science Foundation grants to B. A. A. (BCS-0725951) and J. G. M. T. (EAR 0207370).

REFERENCES

Ahn, K., Y. Mishina, M. C. Hanks, R. R. Behringer, and E. B. Crenshaw III. 2001. BMPR-IA signaling is required for the formation of the apical ectodermal ridge and dorsal-ventral patterning of the limb. *Development* 128:4449–61.

Bardeen, C. R., and W. H. Lewis. 1901. The development of the limbs, body-wall and back in man. *Am J Anat* 1:1–36.

Barrow, J. R., K. R. Thomas, O. Boussadia-Zahui, R. Moore, R. Kemler, M. R. Capecchi, and A. P. McMahon. 2003. Ectodermal *Wnt3/β-catenin* signaling is required for the establishment and maintenance of the apical ectodermal ridge. *Genes Dev* 17:394–409.

Bininda-Emonds, O. R. P., J. E. Jeffery, M. R. Sánchez-Villagra, J. Hanken, M. Colbert, C. Pieau, L. Selwood, et al. 2007. Forelimb–hindlimb developmental timing changes across tetrapod phylogeny. *BMC Evol Biol* 7:182.

Boulet, A. M., A. M. Moon, B. R. Arenkiel, and M. R. Capecchi. 2004. The roles of *Fgf4* and *Fgf8* in limb bud initiation and outgrowth. *Dev Biol* 273 (2): 361–72.

Bouvet, J. 1968. Histogenèse précoce et morphogenèse du squelette cartilagineux des ceintures primaires et des nageoires paires chez la truite (*Salmo trutta fario* L.). *Arch Anat Microsc* 57: 35–52.

Braus, H. 1906. Die Entwickelulng der form der Extremitäten und des Extremitätenskeletts. *Hertwig Hbh Entwicklung Wirbelt* 3:167–338.

Christen, B., and J. M. W. Slack. 1997. FGF-8 is associated with anteroposterior patterning and limb regeneration in *Xenopus*. *Dev Biol* 192:455–66.

Christensen, R. N., and R. A. Tassava. 2000. Apical epithelial cap morphology and fibronectin gene expression in regenerating axolotl limbs. *Dev Dyn* 217:216–24.

Christensen, R. N., M. Weinstein, and R. A. Tassava. 2002. Expression of fibroblast growth factors 4, 8, and 10 in limbs, flanks, and blastemas of *Ambystoma*. *Dev Dyn* 223:193–203.

Cohn, M. J., and C. Tickle. 1999. Developmental basis of limblessness and axial patterning in snakes. *Nature* 399:474–9.

Cretekos, C. J., J. M. Deng, E. D. Green, J. J. Rasweiler, and R. R. Behringer. 2007. Isolation, genomic structure and developmental expression of Fgf8 in the short-tailed fruit bat, *Carollia perspicillata*. *Int J Dev Biol* 51:333–8.

Cretekos, C. J., J. J. Rasweiler IV, and R. R. Behringer. 2001. Comparative limb morphogenesis in mice and bats: A functional genetic approach towards a molecular understanding of diversity in organ formation. *Reprod Fertil Dev* 13:691–5.

Davidson, L. A. 2008. Apoptosis turbocharges epithelial morphogenesis. *Science* 321:1641–2.

Dufaure, J. P., and J. Hubert. 1961. Table de développment du lézard vivipare: *Lacerta (Zootoca) vivipara* Jacquin. *Arch Anat Micr Morph Exp* 50:309–27.

Errick, J., and J. W. Saunders. 1974. Effects of an "inside–out" limb bud ectoderm on development of the avian limb. *Dev Biol* 41:338–51.

Fang, H., and R. P. Elinson. 1996. Patterns of distalless gene expression and inductive interactions in the head of the direct developing frog *Eleutherodactylus coqui*. *Dev Biol* 179:160–72.

Ferguson, M. W. J. 1985. Reproductive biology and embryology of the crocodilians. In *Biology of the Reptilia*, ed. C. Gans, F. Billett, and P. F. A. Maderson, 329–491. New York: John Wiley & Sons.

Fernandez-Teran, M., and M. A. Ros. 2008. The apical ectodermal ridge: Morphological aspects and signaling pathways. *Int J Dev Biol* 52:857–71.

Fischel, A. 1929. *Lerbuch der Entwicklung des Menschen*. Berlin: Springer.

Franssen, R. A., S. Marks, D. Wake, and N. Shubin. 2005. Limb chondrogenesis of the seepage salamander, *Desmognathus aeneus* (Amphibia: Plethodontidae). *J Morphol* 265:87–101.

Galis, F., G. P. Wagner, and E. L. Jockusch. 2003. Why is limb regeneration possible in amphibians but not in reptiles, birds, and mammals? *Evol Dev* 5 (2): 208–20.

Géraudie, J. 1978. Scanning electron microscope study of the developing trout pelvic fin bud. *Anat Rec* 191:391–6.

Géraudie, J., and Y. François. 1973. Les premiers stades de la formation de l'ébauche de nageoire pelvienne de truite (*Salmo fario* et *Salmo gairdneri*). I. Etude anatomique. *J Embryol Exp Morphol* 29:221–37.

Goel, S. C., and J. K. Mathur. 1977. Morphogenesis in reptilian limbs. In *Vertebrate Limb and Somite Morphogenesis*, ed. D. A. Ede, J. R. Hinchliffe, and M. Balls, 387–404. Cambridge: Cambridge University Press.

Grandel, H., and S. Schulte-Merker. 1998. The development of paired fins in the zebrafish (*Danio rerio*). *Mech Dev* 79:99–120.

Hallgrímsson, B., K. Willmore, and B. K. Hall. 2002. Canalization, developmental stability, and morphological integration in primate limbs. *Ybk Phys Anthropol* 45:131–58.

Hamrick, M. W. 2001. Primate origins: Evolutionary change in digital ray patterning and segmentation. *J Hum Evol* 40:339–51.

Hamrick, M. W. 2002. Developmental mechanisms of digit reduction. *Evol Dev* 4 (4): 247–8.

Han, M.-J., J.-Y. An, and W.-S. Kim. 2001. Expression patterns of *Fgf-8* during development and limb regeneration of the axolotl. *Dev Dyn* 220:40–8.

Han, M., X. Yang, G. Taylor, C. A. Burdsal, R. A. Anderson, and K. Muneoka. 2005. Limb regeneration in higher vertebrates: Developing a roadmap. *Anat Rec B New Anat* 287:14–24.

Hanken, J. 1986. Developmental evidence for amphibian origins. *Evol Biol* 20:389–417.

Hanken, J., T. F. Carl, M. K. Richardson, L. Olsson, G. Schlosser, C. K. Osabutey, and M. W. Klymkowsky. 2001. Limb development in a "nonmodel" vertebrate, the direct-developing frog *Eleutherodactylus coqui*. *J Exp Zoolog B Mol Dev Evol* 291:375–88.

Hara, K., J. Kimura, and H. Ide. 1998. Effects of FGFs on the morphogenetic potency and AER-maintenance activity of cultured progress zone cells of chick limb bud. *Int J Dev Biol* 42:591–9.

Hodgkinson, V. S., Z. Johanson, R. Ericsson, and J. M. P. Joss. 2007. Apical ectodermal ridge (AER) development in the pectoral fin of the Australian lungfish (*Neoceratodus forsteri*). *J Morphol* 268 (12): 1085.

Holmgren, N. 1933. On the origin of the tetrapod limb. *Acta Zool* 14:185–295.

Honig, L. S. 1984. Pattern formation during development of the amniote limb. In *The Structure, Development, and Evolution of Reptiles*, ed. M. J. W. Ferguson, 197–221. London: Academic Press.

Jarvik, E. 1965. On the origin of girdles and paired fins. *Isr J Zool* 14:141–72.

Jurand, A. 1965. Ultrastructural aspects of early development of the fore-limb buds in the chick and the mouse. *Proc R Soc Lond B Biol Sci* 162 (988): 387–405.

Kelley, R. O. 1973. Fine structure of the apical rim—mesenchyme complex during limb morphogenesis in man. *J Embryol Exp Morphol* 29:117–31.

Kengaku, M., J. Capdevila, C. Rodriguez-Esteban, J. de la Peña, R. L. Johnson, J. C. I. Belmonte, and C. J. Tabin. 1998. Distinct WNT pathways regulating AER formation and dorsoventral polarity in the chick limb bud. *Science* 280:1274–7.

Keyte, A. L., T. Imam, and K. K. Smith. 2006. Limb heterochrony in a marsupial, *M. domestica*. *Dev Biol* 295:414–22.

Kimmel, R. A., D. H. Turnbull, V. Blanquet, W. Wurst, C. A. Loomis, and A. L. Joyner. 2000. Two lineage boundaries coordinate vertebrate apical ectodermal ridge formation. *Genes Dev* 14 (11): 1377–89.

Köllicker, A. 1879. *Entwicklungsgeschichte des Menschen und der höhren Thiere, Zweite Auflage*. Leipzig: W. Englemann.

Lee, K. K. H., and W. Y. Chan. 1991. A study of the regenerative potential of partially excised mouse embryonic fore-limb bud. *Anat Embryol* 184:153–7.

Lu, P., Y. Yu, Y. Perdue, and Z. Werb. 2008. The apical ectodermal ridge is a timer for generating distal limb progenitors. *Development* 135:1395–1405.

Mahmood, R., J. Bresnick, A. Hornbruch, C. Mahony, N. Morton, K. Colquhoun, P. Martin, A. Lumsden, C. Dickson, and I. Mason. 1995. A role for FGF-8 in the initiation and maintenance of vertebrate limb outgrowth. *Curr Biol* 5 (7): 797–806.

Mariani, F. V., C. P. Ahn, and G. R. Martin. 2008. Genetic evidence that FGFs have an instructive role in limb proximal–distal patterning. *Nature* 453:401–5.

McCrady, E. 1938. Embryology of the opossum. *Am Anat Mem* 16:1–233.

Mercader, N. 2007. Early steps of paired fin development in zebrafish compared with tetrapod limb development. *Dev Growth Differ* 49:421–37.

Michaud, J.L., F. Lapointe, and N.M. Le Douarin. 1997. The dorsoventral polarity of the presumptive limb is determined by signals produced by the somites and by the lateral somatopleure. *Development* 124:1443–52.

Milaire, J. 1957. Contribution à la connaissance morphologique et cytochimique des bourgeons de membres chez quelques reptiles. *Arch Biol* 68:429–512.

Miller, J.D. 1985. Embryology of marine turtles. In *Biology of the Reptilia*, ed. C. Gans, F. Billett, and P.F.A. Maderson, 269–328. New York: John Wiley & Sons.

Moon, A.M., and M.R. Capecchi. 2000. Fgf8 is required for outgrowth and patterning of limbs. *Nat Genet* 26:455–9.

Narita, T., S. Sasaoka, K. Udagawa, T. Ohyama, N. Wada, S.I. Nishimatsu, S. Takada, and T. Nohno. 2005. Wnt10a is involved in AER formation during chick limb development. *Dev Dyn* 233:282–7.

Ngo-Muller, V., and K. Muneoka. 2000. Influence of FGF4 on digit morphogenesis during limb development in the mouse. *Dev Biol* 219:224–36.

Niswander, L. 2002. Interplay between the molecular signals that control vertebrate limb development. *Int J Dev Biol* 46:877–81.

Niswander, L., S. Jeffrey, G.R. Martin, and C. Tickle. 1994a. A positive feedback loop coordinates growth and patterning in the vertebrate limb. *Nature* 371:609–12.

Niswander, L., C. Tickle, A. Vogel, and G. Martin. 1994b. Function of FGF-4 in limb development. *Mol Reprod Dev* 39 (1): 83–9.

Onda, H., and R.A. Tassava. 1991. Expression of the 9G1 antigen in the apical cap of axolotl regenerates requires nerves and mesenchyme. *J Exp Zool* 257:336–49.

O'Rahilly, R., E. Gardner, and D.J. Gray. 1956. The ectodermal thickening and ridge in the limbs of staged human embryos. *J Embryol Exp Morphol* 4:254–64.

O'Rahilly, R., and F. Müller. 1985. The origin of the ectodermal ring in staged human embryos of the first 5 weeks. *Acta Anat* 122:145–57.

Panman, L., A. Galli, N. Lagarde, O. Michos, G. Soete, A. Zuniga, and R. Zeller. 2006. Differential regulation of gene expression in the digit forming area of the mouse limb bud by SHH and gremlin 1/FGF-mediated epithelial–mesenchymal signaling. *Development* 133:3419–28.

Patten, B.M. 1943. *The Embryology of the Pig*, 2nd ed. Philadelphia: Blakiston.

Peter, K. 1903. Mitteilungen zur Entwicklungsgeschichte der Eidechse. IV. Die Extremitätenscheitelleiste der Amnioten. *Arch Mikr Anat* 61:509–21.

Pizette, S., C. Abate-Shen, and L. Niswander. 2001. BMP controls proximodistal outgrowth, via induction of the apical ectodermal ridge, and dorsoventral patterning in the vertebrate limb. *Development* 128:4463–74.

Pizette, S., and L. Niswander. 1999. BMPs negatively regulate structure and function of the limb apical ectodermal ridge. *Development* 126:883–94.

Poss, K.D., J. Shen, A. Nechiporuk, G. McMahon, B. Thisse, C. Thisse, and M.T. Keating. 2000. Roles for Fgf signaling during zebrafish fin regeneration. *Dev Biol* 222:347–58.

Prentiss, C.W. 1903. Polydactylism in man and the domestic animals, with especial reference to digital variations in swine. *Bull Mus Comp Zool Harv Coll* 40:1–341.

Reifers, F., H. Böhli, E.C. Walsh, P.H. Crossley, D.Y. R. Stainier, and M. Brand. 1998. *Fgf8* is mutated in zebrafish *acerebellar* (*ace*) mutants and is required for maintenance of midbrain–hindbrain boundary development and somitogenesis. *Development* 125:2381–95.

Richardson, M.K. 1995. Heterochrony and the phylotypic period. *Dev Biol* 172:412–21.

Richardson, M.K., T.F. Carl, J. Hanken, R.P. Elinson, C. Cope, and P. Bagley. 1998. Limb development and evolution: A frog embryo with no apical ectodermal ridge (AER). *J Anat* 192:379–90.

Richardson, M.K., J.E. Jeffery, and C.J. Tabin. 2004. Proximodistal patterning of the limb: Insights from evolutionary morphology. *Evol Dev* 6 (1): 1–5.

Richardson, M.K., and H.A. Oelschläger. 2002. Time, pattern, and heterochrony: A study of hyperphalangy in the dolphin embryo flipper. *Evol Dev* 4:435–44.

Rubin, L., and J.W. Saunders. 1972. Ectodermal–mesodermal interactions in the growth of limb buds in the chick embryo: Constancy and temporal limits of the ectodermal induction. *Dev Biol* 28:94–112.

Sanz-Ezquerro, J.J., and C. Tickle. 2003. Fgf signaling controls the number of phalanges and tip formation in developing digits. *Curr Biol* 13 (20): 1830–6.

Saunders, J.W., Jr. 1948. The proximodistal sequence of origin of the parts of the chick wing and the role of the ectoderm. *J Exp Zool* 108:363–403.

Saunders, J.W., Jr. 1998. Apical ectodermal ridge in retrospect. *J Exp Zool* 282:669–76.

Saunders, J.W., and M.T. Gasseling. 1968. Ectodermal–mesodermal interactions in the origin of limb symmetry. In *Epithelial–Mesenchymal Interactions*, ed. R. E. Fleischmajer and R. Billingham, 78–97. Baltimore: Williams & Wilkins.

Schmidt, H. 1898. Über die Entwicklung der Milchdruse und die Hyperthelie menschlicher Embryogen. *Morph Arb Jena* 8:157–93.

Sears, K. E. 2004. Constraints on the morphological evolution of marsupial shoulder girdles. *Evolution* 58 (10): 2353–70.

Sears, K. E. 2008. Molecular determinants of bat wing development. *Cells Tissues Organs* 187: 6–12.

Sears, K. E., R. R. Behringer, J. J. Rasweiler IV, and L. A. Niswander. 2006. Development of bat flight: Morphologic and molecular evidence of bat wing digits. *Proc Natl Acad Sci USA* 103 (17): 6581–6.

Sedmera, D., I. Míšek, and M. Klima. 1997. On the development of cetacean extremities: II. Morphogenesis and histogenesis of the flippers in the spotted dolphin (*Stenella attenuata*). *Eur J Morphol* 35:117–23.

Smith, K. K. 2003. Time's arrow: Heterochrony and the evolution of development. *Int J Dev Biol* 47:613–21.

Steiner, K. 1928. Entwicklungsmechanische Untersuchungen über die Bedeutung des ektodermalen Epithels der Extremitätenknospe von Amphibienlarven. *Roux Arch Entwm Org* 113:1–11.

Steiner, K. 1929. Über die Entwicklung und Differenzierungsweise der menschlichen Haut. I. Über die fruhembryonale Entwicklung der menschlichen Haut. *Z Zellforsch Mikrosk Anat* 8: 691–720.

Sun, X., F. V. Mariani, and G. R. Martin. 2002. Functions of FGF signaling from the apical ectodermal ridge in limb development. *Nature* 418:501–8.

Tabin, C. J., and A. P. McMahon. 2008. Grasping limb patterning. *Science* 321:350–2.

Talamillo, A., M. F. Bastida, M. Fernandez-Teran, and M. A. Ros. 2005. The developing limb and the control of the number of digits. *Clin Genet* 67:143–53.

Tarin, D., and A. P. Sturdee. 1971. Early limb development of *Xenopus laevis*. *J Embryol Exp Morphol* 26 (2): 169–79.

Thewissen, J. G. M., M. J. Cohn, L. S. Stevens, S. Bajpai, J. Heyning, and W. E. Horton, Jr. 2006. Developmental basis for hind-limb loss in dolphins and origin of the cetacean body plan. *Proc Natl Acad Sci USA* 103:8414–18.

Thewissen, J. G. M., and J. Heyning. 2007. Embryogenesis and development in *Stenella attenuata* and other cetaceans. In *Reproductive Biology and Phylogeny of Cetacea: Whales, Dolphins, and Porpoises*, vol 7. ed. D. L. Miller, 307–29. Enfield, NH: Science Publishers.

Tickle, C. 2006. Making digit patterns in the vertebrate limb. *Nat Rev Mol Cell Biol* 7:45–53.

Todt, W. L., and J. F. Fallon. 1984. Development of the apical ectodermal ridge in the chick wing bud. *J Embryol Exp Morphol* 80:24–41.

Toyama, Y., X. G. Peralta, A. R. Wells, D. P. Kiehart, and G. S. Edwards. 2008. Apoptotic force and tissue dynamics during *Drosophila* embryogenesis. *Science* 321:1683–6.

Vasse, J. 1972. Sur les activités de synthèse dans la crête épiblastique apicale de l'ébauche du membrane antérieur chez les embryons de tortue (*Testudo graeca* L. et *Emys orbicularis* L.); étude histologique et autoradiographique. *C R Hebd Séanc Acad Sci Paris D* 274:284–7.

Verheyden, J. M., and X. Sun. 2008. An Fgf/Gremlin inhibitory feedback loop triggers termination of limb bud outgrowth. *Nature* 454:638–41.

Vogel, A., C. Rodriguez, and J.-C. Izpisúa-Belmonte. 1996. Involvement of FGF-8 in initiation, outgrowth and patterning of the vertebrate limb. *Development* 122:1737–50.

Von Dassow, G., and E. Munro. 1999. Modularity in animal development and evolution: Elements of a conceptual framework for evodevo. *J Exp Zoolog B Mol Dev Evol* 285:307–25.

Weatherbee, S. D., R. R. Behringer, J. J. Rasweiler IV, and L. A. Niswander. 2006. Interdigital webbing retention in bat wings illustrates genetic changes underlying amniote diversification. *Proc Natl Acad Sci USA* 103 (41): 15103–7.

Wood, A. 1982. Early pectoral fin development and morphogenesis of the apical ectodermal ridge in the killfish, *Aphyosemion scheeli*. *Anat Rec* 204: 349–56.

15

Role of Skeletal Muscle in the Epigenetic Shaping of Organs, Tissues, and Cell Fate Choices

Boris Kablar

CONTENTS

Since the inception of my independent laboratory in July 2000, I have been able to study the role of muscle in the shaping of developing tissues, which is an important example of Waddington epigenetics. This has been the focus of

my research program. Muscle tissue is one of the four basic tissue types of which the body consists. There are three types of muscle tissue, and we are interested in one of them, the skeletal or striated muscle. We can study the developmental role of muscle in the whole mouse embryo or fetus because it is enough to knock out two myogenic regulatory factors (MRFs), Myf5 and MyoD, to obtain an embryo without any skeletal musculature. Obviously, such a fetus cannot survive after birth, but it is viable as long as it is in the womb. This experiment was performed for the first time in 1993 by my postdoctoral supervisor, Dr. Michael A. Rudnicki, while he was a postdoctoral fellow in Dr. Rudolph Jaenisch's laboratory (Rudnicki et al., 1993). Even though it is understandable that the muscle may have numerous functions during development, we think of muscle as either an executor of various movements or a provider of neurotrophic factors. Therefore, I will concentrate on the description of two major research programs performed in this laboratory. The first one, also known as "developmental morphodynamics," deals with studies that examine the ability of muscle to provide mechanical cues for organogenesis. In this program, we are trying to understand mechanical control of tissue morphogenesis during development (Ingber, 2006). In fact, the analysis of Myf5:MyoD compound nulls reveals that several organs have difficulties in fully developing in the absence of the musculature. Organs that depend on continuity between pre- and postnatal motility are the lungs, retina, inner ear, and some parts of the skeleton (e.g., mandible, clavicle, sternum, and palate). The second research program is composed of experiments that test the neurotrophic hypothesis. In this program, we are trying to find out if there is a muscle-provided trigger of motor neuron death ultimately relevant to the motor neuron diseases such as amyotrophic lateral sclerosis (ALS). The reason for this kind of thinking is the fact that a complete absence of lower and upper motor neurons, which is the pathological definition of ALS, is achieved only in the complete absence of the muscle (Kablar and Rudnicki, 1999).

Myf5:MyoD COMPOUND NULLS

Due to the inability of Myf5 nulls to survive after birth, it is necessary to perform a two-step breeding scheme in order to obtain muscleless embryos. In the first generation of breeding, Myf5 heterozygous mice are crossed with MyoD null mice to obtain Myf5:MyoD double heterozygous mice. In the second generation of breeding, these double heterozygous mice are interbred, resulting in a 1:16 probability of finding a muscleless embryo. Embryos and fetuses without any skeletal myoblasts and musculature are characterized by the existence of unspecified myogenic precursor cells (MPCs) and the coexistence of Myf5- and MyoD-dependent MPCs that erroneously contribute to the bone tissue, undergo apoptosis, become adipocytes, or simply remain as mesenchyme (Kablar et al., 2003, and references therein).

In order to be able to make conclusions pertinent to the absence of the muscle and not to the absence of Myf5 and/or MyoD, a careful analysis of Myf5 and MyoD expression and distribution patterns was necessary. To that end, $Myf5^{nlacZ}$ knockin mice, $258/-2.5lacZ$ transgenic mice (for the −20 kb enhancer of MyoD promoter), $MD6.0-lacZ$ transgenic mice (for the −5 kb enhancer of MyoD promoter), in situ hybridization, and immunohistochemistry were employed. We concluded that neither Myf5 nor MyoD was expressed or contained in any of the tissues of interest. In other words, lungs, bones, and the inner ear neuroepithelial fields did not show any expression or distribution of Myf5 or MyoD, but the central nervous system (CNS) and the neural retina contained ectopic expression of Myf5 and MyoD. However, the function of Myf5 and MyoD was found to be inhibited in the neural tissues, leaving no functional consequences of Myf5 and MyoD in neurogenesis (Tajbakhsh and Buckingham, 1995; Kablar, 2002, 2004).

LUNG DEVELOPMENT IN THE ABSENCE OF FETAL BREATHING-LIKE MOVEMENTS

A large body of literature confirms that in the absence of fetal breathing-like movements (i.e., possibly spontaneous neuronal firings in the brain stem, transmitted via peripheral nerves and executed by the striated musculature), the developing lung fails to grow appropriately, resulting in pulmonary (lung) hypoplasia (Inanlou and Kablar, 2005; Inanlou et al., 2005, and references therein). In mice, the histopathological appearance of the hypoplastic lung at term corresponds to the earlier canalicular stage of normal lung development. Moreover, the ratio between body weight and lung weight is less than 4%, due to downregulated cell proliferation and upregulated cell death in the alveolar epithelium and the lung mesenchyme. Several cell cycle molecular players are also downregulated, such as platelet-derived growth factor (PDGF)-β, its receptor (PDGFR-β), and insulin-like growth factor (IGF)-I. Finally, thyroid transcription factor (TTF)-1 loses its proximal-to-distal expression and distribution gradient.

In addition to the failure in growth, the lungs exhibit failures in cell differentiation. Whereas the conductive system of the lung (i.e., trachea, bronchi, and bronchioli) is unaffected, the respiratory portion, and in particular the alveolar epithelium, fails to differentiate properly (Inanlou and Kablar, 2005). Specifically, type II pneumocytes, responsible for the synthesis of surfactant, in spite of normal distribution of surfactant proteins (SPs) A, B, C, and D, fail to assemble (i.e., do not utilize glycogen adequately), store (i.e., have irregular lamellar bodies), and secrete (i.e., have irregular myelin figures) the surfactant. At the same time, type I pneumocytes, responsible for gas exchange, fail to differentiate from a cuboidal cell type into the squamous cell type (i.e., fail to flatten) in order to become a part of the blood–air barrier, the site of oxygen and carbon dioxide exchange.

The striking normality of various components of the lung tissue in the muscleless fetuses, coupled with the very specific differentiation failures of type II and type I pneumocytes, led us to believe that the systemic subtractive microarray analysis approach (SSMAA) (Figure 15.1) would reveal a profile of genes involved in type I and type II pneumocyte differentiation. In other words, we hypothesized that the difference in gene-expression patterns between the control and the mutant lung would be related to the specific differentiation failures of the alveolar epithelium. Indeed, our Affymetrix Gene Chip cDNA microarray analysis revealed nine upregulated and 54 downregulated genes (Baguma-Nibasheka et al., 2007). Out of 54 downregulated genes, the literature and database search detected 24 viable and fertile knockout mice that did not have a lung phenotype as single knockouts. Furthermore, two knockout mice died too early during development to be useful for studies relevant to lung organogenesis. Finally, our analysis revealed four molecules whose knockouts die at birth due to respiratory failure. The knockout mice that die at birth due to pulmonary hypoplasia are connective tissue growth factor (*Ctgf*), special AT-rich sequence binding protein 1 (*Satb1*), myeloblastosis oncogene (*Myb*), and T-cell receptor β, variable 13 (*Tcrb-V13*). We are currently performing experiments to verify the features of lung hypoplasia in each of the aforementioned mouse knockouts. Our analysis of *Ctgf* nulls revealed all the criteria for mouse pulmonary hypoplasia and specifically a failure of type II pneumocytes to properly assemble and store the surfactant (Baguma-Nibasheka and Kablar, 2008). In addition, four more mouse knockouts are in the pipeline of the European Conditional Mouse Mutagenesis (EUCOMM) Consortium. Our ultimate goal is to identify new molecular players with precisely attributed functions in lung development and disease, while defining which features of the phenotype are the result of the absence of the gene and which are due to the absence of mechanical forces. We intend to use human fetal hypoplastic lung tissues from fetal akinesia and oligohydramnios to verify if any of the molecules detected via mouse mutagenesis and pathology apply to the human conditions in order to eventually identify a marker for predicting pulmonary hypoplasia. We also intend to apply mechanical factors, such as stretch, to the single alveolar epithelial cell or a layer of epithelial cells to study the role of the identified molecules in the mechanochemical signal-transduction pathways in order to ultimately propose improvements to the protocols for tissue engineering.

RETINAL DIFFERENTIATION IN THE ABSENCE OF FETAL OCULAR MOVEMENTS

Another project carried out in the laboratory deals with the fact that the neural retina of muscleless term fetuses does not contain any cholinergic amacrine cells (CACs) (Kablar, 2003, and references therein). CACs are apparently responsible for motion vision and directional selectivity (Moran and Schwartz, 1999; Yoshida et al., 2001, and references therein). Whether this is an example of cell differentiation being

MUSCLE
↓
LUNG

GROWTH

Histopathological appearance (arrested in canalicular stage)
Body weight to lung weight ratio (less than 4%)
Cell proliferation and death (up-regulated and down-regulated, respectively)
Platelet-derived growth factor (PDGF)-β (down-regulated)
PDGF-β receptor (down-regulated)
Insulin growth factor (IGF)-I (down-regulated)
Thyroid transcription factor (TTF)-1 (proximal-to-distal gradient of distribution not maintained)

DIFFERENTIATION

Type II pneumocytes (assembly, storage and secretion of surfactant affected)
Type I pneumocytes (flattening into squamous epithelium affected)

SSMAA
↓

CTGF (7 criteria for pulmonary hypoplasia met, assembly and storage of surfactant affected)

FIGURE 15.1 Lung model for the systematic subtractive microarray analysis approach (SSMAA). In the absence of the respiratory musculature, and therefore in the absence of the executors of the fetal breathing-like movements, the growth of the lung is severely affected, to the point that the seven criteria for pulmonary hypoplasia in mice can be met. Importantly, in spite of the correct expression of molecular markers such as surfactant proteins for type II pneumocytes and Gp38, a controversial marker for type I pneumocytes, transmission electron microscopy elucidated that further differentiation steps of the alveolar epithelium were affected in the hypoplastic lung. The cuboidal type II pneumocytes, currently thought to be the source of type I pneumocytes, had failures in assembly, storage, and secretion of surfactant. Meanwhile, the type I pneumocytes could not flatten to become squamous epithelial cells and function in the blood–air barrier for gas exchange. To discover new molecular players with precisely attributed functions in lung development and disease, we performed Affymetrix Gene Chip cDNA microarray analysis, followed by the analysis of mouse mutants. In fact, *Ctgf* is an example of the SSMAA that worked because out of approximately 25,000 genes we did identify one whose knockout mouse had specific differentiation failures as "predicted" by the original phenotype (i.e., the phenotype described in *Myf5:MyoD* nulls). Moreover, defining which features of the phenotype were the result of the absence of the individual genes (e.g., *Ctgf*) and which were due to the general absence of mechanical forces (i.e., *Myf5:MyoD* nulls) was another advantage of this approach. This approach could be successful for the analysis of the molecular basis of hair-cell differentiation in the crista a mpullaris since it has some analogies with the lung example. Importantly, this approach systematically addressed various tissues and organs of the fetal or embryonic body that were affected by the absence of the skeletal musculature (e.g., retina, inner ear, skeleton) or Myf5 and MyoD (e.g., limb and back myogenic precursor cells). Finally, the approach benefited from the fact that the mutant tissues and cells had particular differentiation failures (e.g., alveolar epithelium, hair cells in the crista, myogenic precursor cells) or were completely absent (e.g., cholinergic amacrine cells, palate, secondary cartilage), and therefore, the analysis potentially provided a profile of genes involved in that particular differentiation failure by subtraction from the normal control.

dependent on its function and vice versa (i.e., form–function mutual interaction) or just a co-incidence we do not know at this point, and we cannot address it more directly than we already have. In fact, another way of addressing the mechanisms of this phenotype would be to iden-tify another mouse mutant without extraocular musculature and examine its neural retina. For instance, *Pitx2* nulls apparently specifically lack the extraocular muscle and are viable at term (Diehl et al., 2006), therefore representing an ideal occasion for testing our hypothesis. In the meantime, we used the opportunity provided by the retinas from muscleless fetuses to assess the differences in gene-expression pattern be-tween the control and the mutant retinas, em-ploying 15K mouse cDNA microarray slides from Ontario Cancer Institute (Baguma-Nibasheka et al., 2006). In this way, otherwise inaccessible CACs have been assessed for dif-ferences in gene expression, and the profile of genes obtained after SSMAA revealed some molecular players relevant to the differentiation of CACs. Our analysis discovered two molecules whose mouse mutants lack CACs: adapter-related protein complex 3, δ 1 subunit (*Ap3δ1*) and β-transducin repeat containing (*Btrc*) (Ba-guma-Nibasheka and Kablar, 2009a, 2009b). While *Ap3δ1* mutants are not going to be very useful for further studies of CAC function, be-cause they become blind soon after birth (Qiao et al., 2003), the *Btrc* nulls are viable and not blind and, therefore, represent a model for fur-ther motion vision and directional selectivity studies. Additionally, two more genes have been identified, and their knockout mice are in the EUCOMM pipeline (protein sorting nexin 17, or *Snx17*, and WD repeat domain 5, or *Wdr5*). Finally, the identified molecules could also serve as specific markers for CACs (Figure 15.2).

DIFFERENTIATION OF THE CRISTA AMPULLARIS IN THE ABSENCE OF FETAL ANGULAR ACCELERATION

As previously mentioned, extrinsic mechanical stimuli have been shown to affect differentiation

MUSCLE
↓
RETINA
Cholinergic amacrine cells (complete absence)
↓
SSMAA
↓
AP3D1 (absence of cholinergic amacrine cells)
BTRC (absence of cholinergic amacrine cells)

FIGURE 15.2 Retina model for the systematic subtractive microarray analysis approach (SSMAA). In the absence of the extraocular musculature, and therefore in the ab-sence of the executors of the fetal ocular movements, the differentiation of the retina is affected in a very peculiar way. The precise mechanisms of this relationship are un-known, but the complete absence of cholinergic ama-crine cells, normally involved in motion vision and direc-tional selectivity, was found. To discover new molecular players with precisely attributed functions to retinal de-velopment, we performed cDNA microarray analysis, fol-lowed by the analysis of mouse mutants. We discovered two mutants whose retinas do not contain cholinergic amacrine cells. Closing the circle seemed to be more suc-cessful in the case of the complete absence of a cell type, as seen in the retina example. This approach could prob-ably be as successful for the other two examples where the tissue is completely absent, as occurs in the case of the palate (maxilla) and secondary cartilage (mandible or clavicle).

of specific cell subpopulations in the lungs and retina of mouse fetuses when devoid of mechan-ical stimulation from skeletal musculature dur-ing embryonic development. In this section, we focus on the link between mechanical stimuli, both acoustic and static, on the embryonic de-velopment of inner ear neuroepithelial fields. As usual, we employed double mutant $Myf5^{-/-}$: $MyoD^{-/-}$ mouse fetuses that completely lacked skeletal musculature and analyzed the develop-ment of sensory fields in both the vestibular and the auditory components of the inner ear. Embryos and fetuses that lack skeletal muscles are deficient in the ability to move the chain of three middle ear ossicles and consequently are unable to properly transfer sound vibrations into the inner ear. They also cannot tilt their head due to the fusion of cervical vertebrae and the lack of musculature (Rot-Nikcevic et al., 2006), which prevents the perception of angular accel-eration. Our results show that the development of cristae ampullaris, vestibular sensory fields sensitive to the angular acceleration, was the most affected (e.g., the size of the mutant hair-cell area was significantly smaller in comparison to the control by 43%). A somewhat lesser effect was observed in the vestibular macula sacculi, responsible for the perception of lateral acceleration and gravity (i.e., the supporting cells were affected but not the hair cells). Cochlear sensory field, the spiral organ of Corti, which serves as a receptor for acoustic signals, displayed normal development in mutant embryos (Rot and Kablar, 2010). This initial analysis of the complex relationship between the inner ear neuroepithelial fields' cell differentiation and the mechanical stimuli they perceive during the last stages of development will be followed by the next step of our SSMAA in order to compare gene expression in cristae ampullaris of $Myf5^{-/-}:MyoD^{-/-}$ term fetuses to the analogous normal neuroepithelial field of wild-type term fetuses. By employing this approach, it will be possible to identify genes and to measure the expression of genes associated with the detected phenotype and specific cell subpopulations that may reside within the inner ear's neuroepithelial fields.

DEVELOPMENT OF THE SKELETON IN THE COMPLETE ABSENCE OF THE MUSCULATURE

Both the vertebrate skeleton and the musculature function as one system in which the muscles execute the movements and the skeleton serves as the support. Importantly, the connecting feature between the two systems appears to be their ability to react to mechanical stimuli (Herring, 1994, and references therein). Mutant mice that completely lack skeletal myogenesis and the musculature, as obtained in $Myf5^{-/-}:MyoD^{-/-}$ embryos and fetuses, are providing us with an opportunity to explore the mechanical aspects of a complex developmental relationship between the muscles and the bones. In this section, I will concentrate on the role of mechanical cues provided by the skeletal muscle

(i.e., the static loading from the skeletal muscle) in determining the timing and the character of morphogenetic events during skeletogenesis (e.g., the formation of the secondary cartilage) and comment on the role of skeletal muscle in bone mergence, fusion, and the determination of size and shape of bones and joints.

Secondary cartilage is the tissue that provides growth sites and articulations during the subsequent steps of bone and joint morphogenesis and, therefore, is critical for the later growth and shaping of bones and joints. In mice, secondary cartilage is seen in the mandible (Fang and Hall, 1997, and references therein) and in the clavicles (Hall, 2001, and references therein). Formation of the secondary cartilage is apparently dependent on mechanical stimulation from the skeletal muscle (Herring, 1994). Indeed, we investigated the development of the mandibles and the clavicles with special attention to the formation of secondary cartilage in muscleless fetuses in comparison to normal controls. We also paid particular attention to the initiation versus maintenance of the secondary cartilage in the two bones of interest in the absence of mechanical stimulation from the muscles. Our findings conclude that in most cases secondary cartilage formation can be initiated in the absence of the striated muscle but the amount of tissue is reduced (with the exception of the angular process of the mandible that cannot be initiated and the condylar process that is unaffected). On the other hand, maintenance of the secondary cartilage is impossible in the absence of the musculature since all of the processes are either reduced or completely absent (Rot-Nikcevic et al., 2007).

An important example of skeletal muscle requirement for a morphogenetic event is the process of secondary palate fusion in mammals (Herring, 1994, and references therein). Earlier in development, the palatal shelves extending from the maxilla into the oral cavity are widely separated by the tongue. At a later stage, the tongue is moved out, clearing the way for the palatal shelves to merge and fuse in the midline. The current belief is that the tongue is moved superoxide dismutase 1 or *SOD1*[-/-] mice (Park and Vincent, personal communication), and (3) *Mdx:MyoD*[-/-9th] fetuses have normal-appearing limb and back musculature (Inanlou et al., 2003) and normal MMC neurons but their LMC neurons are 50% reduced (Kablar, unpublished data), indicating that a muscle-expressed factor (or a group of factors), potentially related to MyoD and situated within the MyoD-dependent MPCs (and not Myf5-dependent MPCs), triggers this phenotype. In addition, a series of in utero treatments with neurotrophic factors indicated that the ability of a particular neurotrophic factor to support the survival of motor neurons is dependent on and modified by the presence of the skeletal musculature (Geddes et al., 2006; Angka and Kablar, 2007; Angka et al., 2008; Angka and Kablar, 2009).

More than 15 factors from different gene families are now known to enhance motor neuron survival during development and to be expressed in a manner consistent with a role in regulating motor neuron numbers. Different factors may act on different subpopulations of motor neurons, and these factors may act in synergy on a given motor neuron (reviewed by Henderson et al., 1998). In our experiment, we are comparing gene expression, first, between control and MyoD null limbs and, second, between control and Myf5 null back musculature. In the first generation of subtraction, we will obtain lists of down- and upregulated genes for each of the two experiments, in comparison to the normal control. Subsequently, we will be able to perform another subtraction, comparing the MyoD to the Myf5 lists of genes (e.g., a downregulated gene from the MyoD experiment may be upregulated in the Myf5 experiment). Using various databases, we are currently gaining insight into the information so far known about the identified genes, and we are making priority lists for further actions. In other words, either we will find the information we are looking for in the literature (e.g., a mouse mutant of the gene of interest has a decreased number of LMC neurons, the expression pattern of the gene is consistent with its role in regulation of the motor neuron numbers, the protein is retrogradely transported to motor neuron cell bodies), we will have to analyze the mouse mutant and the expression pattern ourselves (e.g., a mouse mutant of the gene of interest has a lung phenotype, but its neuronal phenotype has not been examined; the gene is expressed in the spinal cord, but it is not known if its protein is contained within the motor neurons), a mouse mutant will have to be generated (e.g., a mouse mutant of the gene of interest is in the EUCOMM or another pipeline), or we will perform in utero treatments with the protein of interest. In conclusion, I believe that by this systematic approach we will generate a large amount of very useful comprehensive information that will be testable by us and others in various ways.

Fortunately, we can already share some of our findings. For example, a well-known candidate BDNF is normally present in the developing limb musculature at E13.5 (Kablar and Belliveau, 2005), it is retrogradely transported by the motor neurons (Koliatsos et al., 1993), and it is known to be a survival factor for motor neurons (Oppenheim et al., 1992); but it is completely absent in *MyoD*[-/-] E13.5 limb muscle, indicating that it may not be essential for the survival of LMC neurons because LMC is completely normal in *MyoD*[-/-] nulls (Kablar and Belliveau, 2005). Consistently, *BDNF*[-/-] mice also have normal LMC neurons (Conover et al., 1995), and BDNF preferentially rescues MMC neurons (Geddes et al., 2006). Together, these data suggest that BDNF is a preferred survival factor for MMC, and not LMC, neurons, potentially excluding BDNF as a candidate for the etiology of ALS with limb symptoms (see Table 15.1 for more examples).

Another interesting finding from our current muscle microarrays (Baguma-Nibasheka and Kablar, unpublished data) is that E13.5 *MyoD*[-/-] limb muscle has downregulated the amyotrophic lateral sclerosis 2 (juvenile) chromosome region, candidate 13 (human) gene (*Als2cr13*); but unfortunately the molecular and biological functions of this gene have not been

TABLE 15.1

The Lack of MMC Data Incorrectly Suggests That LMC and MMC Neurons Share the Same Neurotrophic Requirements

NEUROTROPHIC FACTOR KOs	MMC	LMC	REFERENCES
$BDNF^{-/-}$	?	N.S.	Liebl et al. (1997)
$NT\text{-}3^{-/-}$	?	28%*	Woolley et al. (1999)
$NT\text{-}4/5^{-/-}$	?	N.S.	Conover et al. (1995)
$GDNF^{-/-}$	?	22%*	Moore et al. (1996)

NOTE: Data from muscle developmental biology only recently indicated the existence of important differences between the limb and the back musculature (Kablar et al., 1997), while our work, based on myogenic diversities, has been furthering our knowledge on differences in neurotrophic requirements that exist between the respective muscle innervating neurons. For instance, according to our data (Kablar and Belliveau, 2005) and the published knockout (KO) data shown in the table, BDNF and NT-4/5 seem not to have an effect on LMC neurons, while NT-3 and GDNF have an effect on LMC neurons. We find it essential to reexamine the existing KOs to determine the MMC neuronal numbers and, with them, their neurotrophic requirements. In fact, the first contribution so far has been a recent report in which it seems that the MMC neurons react to the absence of GDNF in a fashion dissimilar to that of LMC neurons (MMC vs. LMC = 22% vs. 37%) (Oppenheim et al., 2000).

?, not studied before 2000; N.S., no significant difference in neuron numbers when compared to the control; *, reported as a significant decrease in the literature.

specified yet. Additionally, two other downregulated genes with existing knockouts have been identified to have decreased motor neuron numbers: kinesin family member 5C (*Kif5c*), whose molecular function is structural and cytoskeletal (Kanai et al., 2000), and syntaxin binding protein 1 (*Stxbp1*), whose molecular function is in transport and carrier activity and whose knockouts have massive motor neuronal apoptosis, which causes death at birth due to respiratory failure (Verhage et al., 2000). Meanwhile, among upregulated genes, we detected the polymerase (DNA-directed), beta gene (*Polβ*), whose molecular function is metabolic and housekeeping but whose knockout also has excessive neuronal apoptosis (Sugo et al., 2000).

On the other hand, our current muscle microarrays (Baguma-Nibasheka and Kablar, unpublished data) with E13.5 *Myf5⁻/⁻* back muscles have identified downregulated genes whose knockout mice also have motor abnormalities: peroxisome proliferator–activated receptor, gamma, coactivator 1 alpha (*Ppargc1α*), with molecular function in receptor and signal-transduction activity (Cui et al., 2006), and glycine receptor, beta subunit regulator of G-protein signaling 7 binding (*Glrb*), with molecular function in receptor and signal-transduction activity, whose knockouts have extensive motor neuron loss (Karaplis et al., 1994). Meanwhile, among upregulated genes, we detected homeobox D10 (*Hoxd10*) transcription factor, whose knockout mouse spinal cord had motor neuron column shifts (Carpenter et al., 1997).

CONCLUSION

Mutual embryonic inductive interactions between different tissue types and organs, between individual cell types belonging to the same or different lineages, and between various kinds of molecular players are only some examples of the complex machinery that operates to connect genotype and phenotype. Our studies so far indicate that some aspects of this interplay can indeed be studied as proposed, confirming the role of skeletal muscle contractile and secretory activity in the epigenetic shaping of organs, tissues, and cell fate choices. We will therefore continue this analysis as outlined to gain insight into the nature of the epigenetic events that lead to the emergent properties of a phenotype.

ACKNOWLEDGMENTS

This chapter represents a summary of the work performed since the inception of my independent laboratory in July 2000. I would like to thank some exceptional individuals who performed various parts of this

Woolley, A., P. Sheard, K. Dodds, and M. Duxson. 1999. Alpha motoneurons are present in normal numbers but with reduced soma size in neurotrophin-3 knockout mice. *Neurosci Lett* 272: 107–10.

Yoshida, K., D. Watanabe, H, Ishikane, M. Tachibana, I. Pastan, and S. Nakanishi. 2001. A key role of starburst amacrine cells in originating retinal directional selectivity and optokinetic eye movement. *Neuron* 30:771–80.

PART FOUR

Epigenetics in Evolution and Disease

16

Epigenetic Integration, Complexity, and Evolvability of the Head

RETHINKING THE FUNCTIONAL MATRIX HYPOTHESIS

Daniel E. Lieberman

CONTENTS

As I get older, I find myself increasingly hesitant to use the word *epigenetics* because I worry about employing a term that is so liable to engender confusion and disagreement. Many biologists define *epigenetics* in a narrow sense solely as heritable changes in the phenotype that derive from molecular mechanisms other than sequence changes in the genotype (the classic example being methylation). However,

Waddington, who coined the word in 1942, and other early users of the term had a broader concept in mind, one that captured the variable effects of interactions between genes, embryonic development, and the environment. According to this original and more encompassing definition, epigenetics refers to the vast set of processes by which alternative, variable phenotypes—cellular, anatomical, physiological, even behavioral—derive from a given genotype (see Haig, 2004). Not surprisingly, there is a rich literature on epigenetics in complex organisms because many layers and types of epigenetic interactions are essential to initiate the development and integration of diverse units so that they grow and function together appropriately (Wagner, 2001; Kirschner and Gerhart, 2005; West-Eberhard, 2003). Epigenetic interactions occur at each hierarchical level of development including within the genome; among cells, tissues, and organs; and between an organism and its environment.

Epigenetics: Linking Genotype and Phenotype in Development and Evolution, ed. Benedikt Hallgrímsson and Brian K. Hall.
Copyright © by The Regents of the University of California. All rights of reproduction in any form reserved.

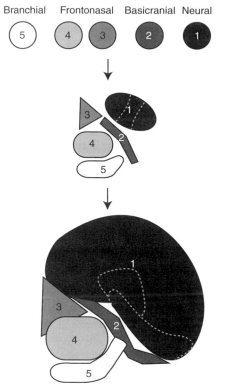

Branchial Frontonasal Basicranial Neural

5 4 3 2 1

FIGURE 16.1 Schematic and highly simplified view of the head's early morphogenesis (in lateral and anterior views), showing movements of the frontonasal prominences (black arrows) and branchial arches (gray arrows) over the top and around the head. Note that the prominences and branchial arches interact with and envelop various placodes and organizing centers as they stream into final position. Thus, the eyes, nose, pharynx, and other organs and spaces develop surrounded by a shared structural framework.

tonasal prominence (Figure 16.1). As they migrate, cells in the prominence interact with neural placodes (especially optic and olfactory) and other tissues to differentiate into many of the units that eventually form the upper and middle facial skeleton around the eyes, nose, and the upper part of the pharynx. The other set of neural crest cells, the branchial arches, migrate like collars around the side of the head to create the middle ear, the lower jaw, and the throat around the pharynx (Figure 16.1). These cells, too, are patterned by inductive interactions with cells they encounter as they stream into position (see below).

To reiterate, complex epigenetic interactions are involved in patterning every region of the body, but a key point about the head is that its initial units become intricately integrated with each other in an architecturally convoluted arrangement from the start so that skeletal units grow around and between organs and functional spaces. For example, the cells of the frontonasal prominence that migrate over the forebrain to differentiate into the skeletal framework around the eyes and nose depend on numerous interactions with tissues they encounter along the way in order to bifurcate and diverge around these primordial organs and then re-fuse below them. To accomplish these tricky maneuvers, the frontonasal prominence relies on inductive signals from the nasal and olfactory placodes and vice versa so that the eyes and nose form *surrounded by* regions of mesenchyme that form a skeletal framework (see Marcucio et al., 2005). Not surprisingly, the most common type of craniofacial disorders is palatal clefting, which results from improper coordination of these events (see Marazita and Mooney, 2004). In addition, derivatives of the frontonasal prominence must integrate with the cranial base along with the pharynx and branchial arches to grow around the structures that form the lower face. Similarly convoluted migrations and interactions characterize most other craniofacial structures, setting up an embryonic head in which it quickly becomes difficult to discern straightforward boundaries between units. For example, the four bones of the cranial base derive from 13 pairs of condensations—some mesodermal in origin, others neuroectodermal in origin—whose formation appears to be dependent on interactions with the notochord, brain, sensory placodes, nerves, and other tissues (most of these inductive interactions have yet to be worked out). As a consequence, many of the traditional divisions of the skull (e.g., basicranium, neurocranium, and face; prechordal and postchordal) do not map on directly to any bones. The sphenoid, for instance, is often considered an endochondral, basicranial bone of mesodermal origin, but it actually contains some elements that are neural crest and others that are paraxial mesoderm (Jiang et al., 2002); in addition, some of its components are basicranial and grow via endochondral ossification, while others are neurocranial and grow via intramembranous ossification. As we shall see below, there is reason to believe that this byzantine arrangement, made possible by innumerable epigenetic interactions, underlies some of the head's ability to be successfully complex.

MORPHOGENESIS

A second level of epigenetic interaction important to craniofacial development occurs during morphogenesis, when precursor cells in particular units differentiate into the cell types that form distinct tissues and organs. In the head, as in the rest of the body, differentiation typically occurs because of inductive interactions with neighboring tissues. As an example, the basioccipital derives from paraxial mesodermal cells above the notochord and below the brain stem that differentiate into sclerotome, then into mesenchyme, and then into chondrocytes and osteoblasts which form the bone through endochondral ossification (see McBratney-Owen et al., 2008). Transplant experiments provide dramatic evidence of the importance of local inductive interactions between the brain and basicranium that drive the fate of these cells. Reorienting an embryonic chick brain by 180° causes a concomitant reversal in the shape of the underlying cranial base (Schowing, 1968). Teeth provide another classic example of induction elsewhere in the head. Epithelial cells transplanted from the incisor to the molar regions of the jaw cause local mesenchymal cells to develop into incisors rather than molars (see Peters and Balling, 1999; Jernvall and Thesleff, 2000).

Although there is nothing necessarily unique about the nature of the inductive interactions involved in morphogenetic differentiation in the vertebrate head, the complex ways that the head's units become arranged with respect to

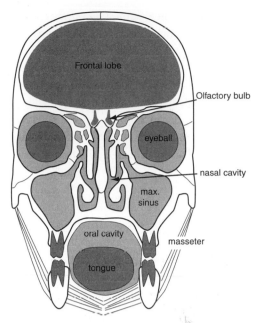

FIGURE 16.2 Coronal section through a head, highlighting selected key functional matrices that are either organs (e.g., the brain, olfactory bulbs, eyeballs, tongue, teeth), spaces (e.g., the pharynx and sinuses), or muscles (e.g., the masseter muscle). Importantly, all these functional matrices share a common skeletal framework. Almost every wall of bone is part of the cranial component of more than one functional matrix.

each other do lead to some special properties. For one, the head has a greater density of organs (such as the eyes, vestibular system, and pituitary gland) and functional spaces (such as the pharynx, oral cavity, and sinuses) compared to other parts of the body. More importantly, because the branchial arches and the derivatives of the frontonasal prominence stream *around* the organs of sense and the pharynx and then differentiate into skeletal condensations, these organs and spaces share a common, architecturally complex skeletal framework. As the coronal section shown in Figure 16.2 illustrates, few walls of bone or cartilage are unique to any single organ or space. For example, the floor of the anterior cranial fossa is the roof of the orbit, the medial wall of the orbit is the lateral wall of the nasal cavity, the roof of the nasopharynx is the floor of the brain stem, and so on. Because the head's organs and spaces grow in and around each other with a common skeletal framework, many different factors can affect the size and shape of each morphogenetic unit. One key factor is the size and shape of a given initial condensation, which Atchley and Hall (1991) modeled as a function of five major parameters: (1) the number of initial mesenchymal stem cells, largely a function of the number of cells that migrated there in the first place; (2) the time of initiation; (3) the rate of cell division; (4) the percentage of mitotically active cells; and (5) the rate of cell death. These initial parameters have important effects on the size of the resulting adult phenotype (Cottrill et al., 1987; Cohen, 2000). In addition, the special way that the head's skeletal units are arranged around organs and spaces both requires and ensures a high degree of integration because the shape and size of each unit necessarily has effects on the shape and size of its neighbors. As shown by numerous studies, covariation among units of the skull is extremely high even in disparate regions, with especially strong levels of covariation between the widths of adjoining units (e.g., Zelditch and Carmichael, 1989; Cheverud, 1995; Lieberman et al., 2000a; Marroig and Cheverud, 2001; Strait, 2001; Polanski and Franciscus, 2006; Hallgrimsson et al., 2007). We will return later to these reciprocal interactions and their effects on integration and evolvability.

GROWTH

The final type of epigenetic integration to consider is how particular units interact during growth to attain their final size and shape. This aspect of ontogeny, which can generate significant levels of phenotypic variation via mechanisms such as heterochrony (see Gould, 1977; Atchley and Hall, 1991; McKinney and McNamara, 1991), involves a broad range of epigenetic interactions that contribute to substantial morphological integration and promote or maintain functional integration (Olson and Miller, 1958; Chernoff and Magwene, 1999; Young and Badyaev, 2006). Many of these interactions occur at the systemic level through shared responses to circulating hormones such as growth hormone and insulin-like growth factor I, permitting disparate regions to grow in a common trajectory and with appropriate scaling (see Shea and Gomez, 1988). For instance, most components of the face grow in a skeletal growth trajectory along with the rest of the body but not the brain and basicranium, thus maintaining constant scaling between the skeletal units that surround the pharynx, the oral cavity, the nasal cavity, and body size (for details, see Lieberman, 2011). At a more regional level, a variety of epigenetic interactions (e.g., cell–cell signaling and responses to mechanical stimuli) enable neighboring organs and spaces and their surrounding skeletal frameworks to accommodate each other's growth. For example, as the brain grows, it generates tension in the dura mater, which secretes fibroblast growth factor 2 (Fgf2) that binds to receptors in the sutures, stimulating bone growth (Opperman, 2000; Morriss-Kay and Wilkie, 2005). Thus, during normal growth, differing amounts of growth in the brain trigger appropriate rates of vault growth; disruptions of this epigenetic pathway, such as mutations to the Fgf receptors in sutures, cause synostoses that lead to abnormal compensatory growth in other sutures (Richtsmeier, 2002; Marie et al., 2005).

An important, general point (critical but not unique to the head) is that mechanisms of epigenetic integration during growth are essential because it would be impossible to preprogram how the organism's many units grow to accommodate each other without compromising function. The masticatory system provides an excellent example of how morphological integration maintains functional integration during growth in response to mechanical and other stimuli (Olson and Miller, 1958; Herring, 1993). In order to have proper occlusion, the teeth from the upper and lower jaws must fit each other precisely. An engineer might build the two jaws as mirrors of each other, but the evolutionary origins of the head require much more complex tinkering for several reasons. First, the mandible and maxilla have different embryonic origins (only the former derives from the first branchial arch) and are patterned somewhat differently (see Mina, 2001; Cerny et al., 2004; Lee et al., 2004; Depew et al., 2002, 2005). Second, the mandible and maxilla grow through a different set of processes: The mandible grows to a large extent in the condyles as a secondary ossification center against the base of the middle cranial fossa, whereas the maxilla is displaced downward and forward from the rest of the face (for reviews, see Enlow, 1990; Sperber, 2001). These differences set up many integrative challenges, including the need for the mandible to grow in a way relative to the maxilla (and vice versa) that maintains proper occlusion of the lower and upper teeth, even though the mandible articulates with the cranial base while the maxilla mostly grows downward from the nasal cavity. If the cranial base were a stable platform, this coordination would not be so complex; but the cranial base changes its angle and length in ways that convolute this arrangement in several ways. Among other factors, expansion of brain volume relative to cranial base length during ontogeny causes the cranial base to flex, whereas facial elongation causes the cranial base to extend, thereby shortening or lengthening, respectively, the distance between the temporomandibular joint and the teeth (Biegert, 1963; Ross and Ravosa, 1993; Lieberman and McCarthy, 1999; Lieberman et al., 2008). In addition, most vectors of facial growth alter the position of the maxillary arch, and hence the upper dentition, relative to the mandibular arch, and hence the lower dentition. Since these and other changes occur at different rates, in different ways, and with different effects, a high degree of epigenetic integration must occur throughout postnatal ontogeny to coordinate growth of the upper and lower jaws to maintain effective occlusion. These integrative mechanisms are poorly known but include alterations in the rate of growth in the mandibular condyles, repositioning the teeth within the alveolar crests from differing amounts of tension and compression, variation in growth rates within the alveolar crest, and differential elongation and widening of both the mandibular

and maxillary arches (Moyers, 1988; Enlow, 1990; Herring, 1993).

THE FUNCTIONAL MATRIX HYPOTHESIS REVISITED

As described above, the basic epigenetic processes necessary for craniofacial growth and development do not differ substantially from those in other parts of the body. However, the way the head's units become arranged and then interact with each other may be special in a certain respect. Namely, the processes by which the head initially forms cause almost every organ and functional space to be partially encapsulated in a skeletal framework that is shared to some extent with other organs and functional spaces. As Figure 16.2 makes clear, frontal lobe growth will affect orbital growth and vice versa, eyeball growth will affect nasal cavity growth and vice versa, and so on. From a functional standpoint, this seemingly byzantine arrangement requires a high degree of epigenetic integration so that units can accommodate each other and still perform their functions effectively in spite of variations in each other's size and shape. For example, if brain growth repositions the orbits, then the orbits need to reposition the nasal and oral cavities to function properly. If so, then it is reasonable to hypothesize that the same mechanisms that accommodate variations among units during growth also accommodate considerable variations among units between individuals, leading to the potential for evolutionary change over many generations.

The idea that the head comprises a number of mutually accommodating functional units— albeit not in an evolutionary context—is well known as the functional matrix hypothesis (FMH), a concept first proposed by Van der Klaauw (1948–1952) and then elaborated by Moss (see Moss and Young, 1960; Moss, 1968, 1997a, 1997b, 1997c, 1997d). According to the FMH, the head is comprised of a series of *functional matrices*, defined as "genetically determined and functionally maintained" soft tissues and the spaces they occupy (Moss, 1968, 69).

Functional matrices include organs such as the olfactory bulbs, spaces such as the inner ear and nasopharynx, and muscles such as the temporalis. Each functional matrix is enclosed to some extent by skeletal tissue, a *functional cranial component*. According to Moss's hypothesis, each functional cranial component derives its shape principally from the shape and/or functions of the soft tissue and spaces it encloses. The functional cranial components of the above functional matrices are the cribriform plate, the petrous temporal, the nasal capsule, and the temporal fossa. Importantly, the FMH posits that the major determinants of head shape are not the functional cranial components but the functional matrices (organs and spaces) they contain. As functional matrices grow and function, they strongly influence the shape of their capsules via epigenesis. As an example, brain growth stimulates growth in the surrounding cranial vault so that the vault fits perfectly around the brain (see Richtsmeier et al., 2006). Similarly, as the eyeball grows, it stimulates growth in the surrounding orbit (Sarnat, 1982). Therefore, according to a strict interpretation of the FMH, the overall shape of the skull is basically an emergent property of the shape of its constituent functional matrices.

The FMH as formulated (perhaps overformulated) by Moss has additional details not reviewed here (e.g., the difference between capsular and periosteal matrices). Although influential, the FMH is not as widely accepted as it deserves because the original hypothesis was oversimplified. One problem is that it is untrue that skeletal units have no intrinsic genetic regulation (for review, see Hall, 2005) and that organ shapes are entirely heritable (e.g., Alsbirk, 1977; Rogers et al. 2007). In addition, it should be evident that functional matrices are not independent of each other, particularly because many of them share parts of the same cranial components (e.g., the lateral wall of the nasal cavity is the medial wall of the orbit). From the perspective of integration, a better way to think about functional matrices is that the head forms an amalgamated complex of units in which any given functional matrix exerts a strong morphogenetic influence on its skeletal capsule (functional cranial component) but in which there are also reciprocal epigenetic interactions, hence integration, among neighboring functional matrices and their cranial components. For example the size and shape of the brain have a strong influence on the shape of the cranial vault, which accounts for why individuals with hydrocephaly or microcephaly grow appropriately sized and shaped vaults around their, respectively, large or small brains (Babineau and Kronman, 1969; Rönning, 1995). However, it is also clear that the shape of the cranial vault itself can strongly influence the shape of the brain, as is evident from changes in brain shape that derive from craniosynostoses or head binding (Antón, 1989; Cheverud et al., 1992).

A more integrated FMH (hereafter, IFMH) that considers reciprocal epigenetic interactions among skeletal and nonskeletal components and other modules of the head may also be a useful way to think about how heads grow and function because it accounts for how the head can manage to have so many different units with complex functions and yet allow these different interdependent units to accommodate each other. As noted above, the floor of the cranial base is part of the skeletal capsule of the brain, but the cranial base floor is also part of the skeletal capsule of the eyeball. In addition, the medial wall of the orbit is the lateral wall of the nasal capsule, the floor of the nasal capsule is the roof of the oral cavity, the roof of the oropharynx is part of the posterior cranial fossa floor, and so on. To reiterate, each of these shared walls necessitates integration via epigenetic interactions. For instance, as the brain grows, it causes growth (via flexion, drift, and/or elongation) in the floor of the anterior cranial fossa, which in turn affects the growth of the orbits and face; in turn, the growth of the eyeball and that of the face also have effects on the growth of the roof of the orbit via the same wall of bone. These mutual, sometimes reciprocal, interactions among units ensure high levels of integration, which helps to explain why analyses of the skull routinely display such high levels of correlation and covariance (see Hallgrímsson et al., 2007). For example, growth of the occipital lobe of the brain affects its skeletal capsule, the posterior cranial fossa, which in turn affects the middle cranial fossa, which in turn affects the midface, and so on. Also, to a lesser extent, the same pathway appears to work in reverse so that changes in facial shape affect the shape of the middle cranial fossa and even the posterior cranial fossa. These reciprocal pathways can lead to some surprising patterns of correlation. For instance, individuals with schizophrenia tend to have significantly wider temporal lobes than controls, which leads to significantly wider middle cranial fossae and, hence, significantly wider midfaces than normal controls (McGrath et al., 2002); similarly, individuals with cleft palates develop asymmetries and other morphological changes not only in the midface but also in the neurocranium (Young et al., 2007; Parsons et al., 2008).

In short, the way in which the head develops lends itself to a more epigenetic, integrated model of functional matrices that takes into account how reciprocal interactions between and among organs, spaces, and their skeletal components creates an integrated whole. This IFMH model, which incorporates more functional feedback among modules, not only helps to account for how the head's many bones, organs, and other tissues manage to perform dozens of disparate functions effectively during growth but also may explain how the head manages to permit so many successful variants. In other words, one can hypothesize that the intensity of epigenetic integration in the head accounts for its paradoxically high degree of evolvability despite its many vital functions.

APPLYING THE MODEL (BRAIN SIZE IN HUMAN EVOLUTION)

It is beyond the scope of this chapter to test the model described above. For one, the IFMH is a general model of craniofacial development that will be difficult to test. One prediction, for example, might be that levels of integration are higher in the skull than the postcranium. Although one can quantitatively compare levels of integration among regions using methods such as those described by Wagner (1984), interpreting such data is complicated by multiple processes that cause integration such as pleiotropy, linkage, and shared mechanical responses to loading. In addition, mechanisms and patterns of integration may be qualitatively different in the skull versus the postcranium, where organs are not encapsulated by a skeletal framework in the same way. Muscles do attach to postcranial bones, creating periosteal matrices; but interpreting these muscle–bone interactions as functional matrices can be difficult to comprehend when many muscles attach to a single bone (more than 20 muscles insert on the femur) and by other functional roles, notably weight bearing during locomotion. Additional models, such as beam models, may be more useful for considering how epigenetic interactions influence postcranial morphology (see Carter and Beaupré, 2001; Currey, 2002).

Another problem with rigorously testing the IMFH model is that it does not lend itself easily to simple, falsifiable predictions (for reviews, see Watson, 1982; Herring, 1993; Moss 1997a, 1997b, 1997c, 1997d; Radlanski and Renz, 2006). The model essentially posits that skulls accommodate considerable variation because the head's structure and the nature of its many epigenetic interactions allow many disparate craniofacial modules to influence each other in multiple ways. One basic prediction, that levels of integration in the skull are high but extremely variable within and between species, has been demonstrated by numerous studies (e.g., Zelditch and Carmichael, 1989; Cheverud, 1995; Marroig and Cheverud, 2001; Strait, 2001; Hallgrimsson et al., 2007). In a more focused test of the hypothesis among primate skulls, McCarthy (2004) found that covariation between matrices decreases with distance (e.g., that anterior cranial fossa shape covaries more strongly with orbit shape than with oral cavity shape) and that these relationships tend to scale with negative allometry as they become more distant. Another prediction of the model is that the narrow sense heritabilities (h^2) of most structures in the skull—both organs and bones—should be low, reflecting their many variable epigenetic influences. Measuring, comparing, and interpreting heritabilities are fraught with complications; but most meta-analyses of the craniofacial complex yield h^2 values between 0.1 and 0.4 for most measurements, with a mean of about 0.3 (see Hunter, 1990). Even tooth crown sizes, typically considered very genetic traits, have h^2 estimates of about 0.6 (Harris and Johnson, 1991). These values, of course, refute Moss's original formulation of the FMH, that the forms of functional matrices such as tooth crowns are entirely genetic and that the forms of functional cranial components such as the mandibular corpus have no intrinsic genetic basis. However, as noted above, the original FMH was oversimplified in this respect, and the evidence suggests that both types of units have some combination of intrinsic genetic and extrinsic regulation.

There is much to gain from examining the details of the IFMH model in greater detail and testing it at different levels, but it is also worthwhile to develop a model from an evolutionary perspective to ask how well the model explains observed patterns of evolvability. In this context, human evolution is a useful test case because we know much about human craniofacial growth and development and because the fossil record of human evolution is so well documented and studied that we have a very good idea of the many craniofacial transformations that occurred since the divergence of the human and chimpanzee lineages about 6–8 million years ago (see Figure 16.3). Of these transformations, one of the most important occurred in the genus *Homo* in terms of brain size (accompanied by reductions in tooth size and facial prognathism). Chimpanzees have endocranial volumes (ECVs) of about 400 cc, and early hominin australopiths have slightly larger ECVs (400–580 cc); but a major transition occurred about 2 million years ago with the origins of the

genus *Homo*, when ECVs started to increase substantially in terms of both absolute and relative size (Figure 16.4). Although modest increases in brain size in the very earliest species of the genus *Homo* may have been driven largely by increases in body size, later increases also involved relatively larger brains (Ruff et al., 1997; Holloway et al., 2004; Rightmire, 2004; Lordkipanidze et al., 2007; Lieberman, 2007). Clearly, these increases had enormous effects on craniofacial growth that were potentially a challenge for mechanisms of integration.

A highly genetic view of skull growth would suggest that encephalization in human evolution required substantial selection in order for the various parts of the skull to accommodate the increases in brain size. In fact, one might argue that constraints on growing and accommodating a big brain within the skull may help to explain why so few animals have very large brains. The IFMH, however, suggests that increases in brain size were not highly constrained from a developmental perspective because the skull in *Homo* was easily able to accommodate bigger brains by taking advantage of epigenetic mechanisms that already operate in mammals to accommodate increases in brain size during ontogeny. Further, these mechanisms account for the wide range of configurations evident among different species of the genus *Homo* in terms of brain size and overall craniofacial shape. Based on our current knowledge of craniofacial growth, three sets of epigenetic mechanisms of integration were probably the most important.

1. The first of these sets of mechanisms are the effects of intracranial pressure (ICP) on neurocranial growth that cause components of the braincase to grow superiorly, laterally, and posteriorly as the brain grows in volume (see Figure 16.5). As noted above, increases in brain mass and/ or fluids within the cranial cavity (e.g., cerebrospinal fluid) generate tension in the dura mater, which expresses signaling factors such as Fgf2 that activate osteo-

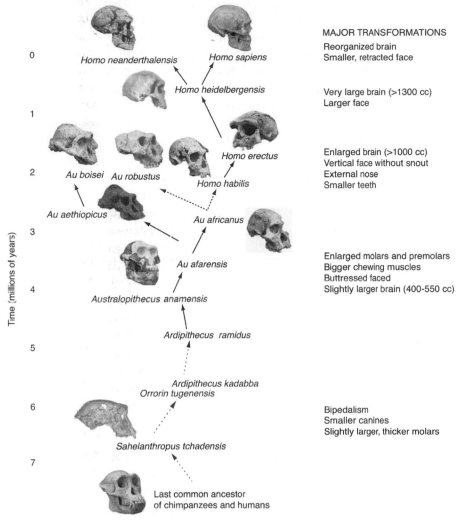

FIGURE 16.3 Major craniofacial transformations in human evolution in the context of a hypothetical hominin phylogeny (dashed lines indicate less secure evolutionary relationships).

The figure includes the following labels and annotations:

- Time (millions of years): axis from 0 to 7

Species (with approximate time positions):
- Homo neanderthalensis (0)
- Homo sapiens (0)
- Homo heidelbergensis (~0.7)
- Homo erectus (~1.7)
- Au boisei, Au robustus (~2)
- Homo habilis (~2)
- Au aethiopicus (~2.7)
- Au africanus (~2.8)
- Au afarensis (~3.5)
- Australopithecus anamensis (~4)
- Ardipithecus ramidus (~4.5)
- Ardipithecus kadabba / Orrorin tugenensis (~5.8)
- Sahelanthropus tchadensis (~6.5)
- Last common ancestor of chimpanzees and humans (~7)

MAJOR TRANSFORMATIONS

Reorganized brain
Smaller, retracted face

Very large brain (>1300 cc)
Larger face

Enlarged brain (>1000 cc)
Vertical face without snout
External nose
Smaller teeth

Enlarged molars and premolars
Bigger chewing muscles
Buttressed faced
Slightly larger brain (400–550 cc)

Bipedalism
Smaller canines
Slightly larger, thicker molars

blasts in sutures or their precursors, the fontanelles (see Cohen, 2000; Wilkie and Morriss-Kay, 2001). Brain expansion also causes drift in the floor of the anterior, middle, and posterior cranial fossae (Duterloo and Enlow, 1970) and some degree of mediolateral and anteroposterior expansion of the synchondroses of the cranial base (reviewed in Enlow, 1990; Lieberman, 2011). There is abundant evidence that these growth mechanisms and constraints accommodated increased brain size in human evolution. As hom-inin brains get bigger, so does the neuro-cranium become predictably longer, taller, wider, and more rounded (see Lieberman et al., 2000a; Bookstein et al., 2003). Moreover, in the one hominin species that appears to have undergone a reduction in brain size from dwarfism, *H. flore-siensis*, the braincase shrank in just the manner predicted by the scaling relationship between shape and size in the genus *Homo* (Baab and McNulty, 2009). Brains, and hence vaults, however, do not grow as perfect spheres because of several

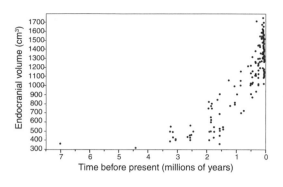

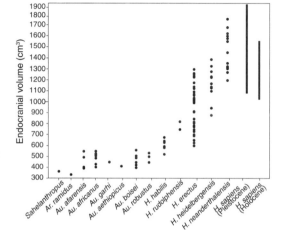

FIGURE 16.4 Endocranial volume versus time (top) and by species (bottom). Note that major increases in volume did not really start until about 2 million years ago in *Homo erectus*. Data from Ruff et al. (1997) and Holloway et al. (2004).

constraints. First, the dural bands—reflections of the dura mater between the two hemispheres (the falx cerebri) and below the occipital lobe above the cerebellum (the tentorium cerebelli)—insert at several locations within the cranial base. These bands appear to act as anchors that constrain the length and width of the basicranium and neurocranium (Moss and Young, 1960; Friede, 1981; Jeffery, 2002). In addition, the area of the cranial base, especially the width, constrains growth of the brain and neurocranium, with relatively wider cranial bases accommodating wider neurocrania (Lieberman et al., 2000a).

2. A second important mechanism of epigenetic integration is angulation of the cranial base. As initially proposed by Bolk (1926), Weidenreich (1941), Biegert (1963), and Gould (1977), the basicranial

platform on which the brain sits can accommodate a larger brain relative to the length of the cranial base by being more flexed (Figure 16.6a). This spatial packing hypothesis has been tested most rigorously in primates (Figure 16.6b), which display a strong correlation across species between cranial base flexion and an index of brain size relative to cranial base length (Ross and Ravosa, 1993; McCarthy, 2001; Lieberman et al., 2000b). There is also some correlation among hominins between ECV and cranial base flexion (Ross and Henneberg, 1995; Spoor, 1997). Cranial base flexion is measured most commonly as cranial base angle 1 (CBA1), the angle in the sagittal plane between the line connecting basion and sella (the center of the hypophyseal fossa) and the line connecting sella to the foramen cecum (the most anteroinferior point on the cra-

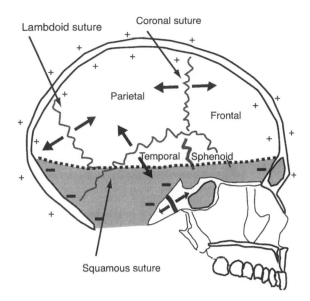

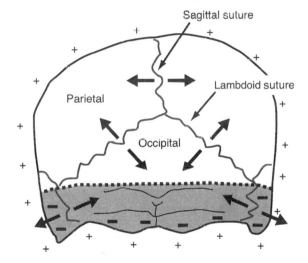

FIGURE 16.5 Vectors and processes of neurocranial and basicranial growth (modified from Lieberman et al., 2000a). Tension in the dura mater generated by the expansion of the brain induces one deposition with sutures and synchondroses. Brain expansion also induces drift (deposition on the outside combined with resorption on the inside) in the basicranium and lower portion of the neurocranium. Note that the relative amount of growth in different vectors is constrained by the dural bands and by other factors that influence the shape and size of the basicranium.

nial base). In chimpanzees, CBA1 averages 159°; in australopiths such as *Australopithecus africanus*, CBA1 is approximately 147°; and in *Homo sapiens*, it averages 137° (Lieberman, 2011). However, the shift from less flexed to more flexed cranial bases does not correlate solely with brain volume because, as emphasized above, the cranial base is sandwiched between the brain and face and, thus, has to accommodate variations in facial size. As shown by Lieberman and McCarthy (1999) for chimpanzees and by Lieberman et al. (2008) for mice, longer and

more prognathic faces are also associated with a kind of reverse spatial packing, which causes the cranial base to extend (although possibly at different synchondroses than where cranial base flexion occurs). The extent to which cranial base flexion and extension accommodate and integrate variations in both facial size and brain size aspect of the cranial base is also evident from variations in CBA1 among hominin species. Some australopiths such as *Au. aethiopicus* have very extended cranial bases (156°) because they have small brains (410 cc) combined with very

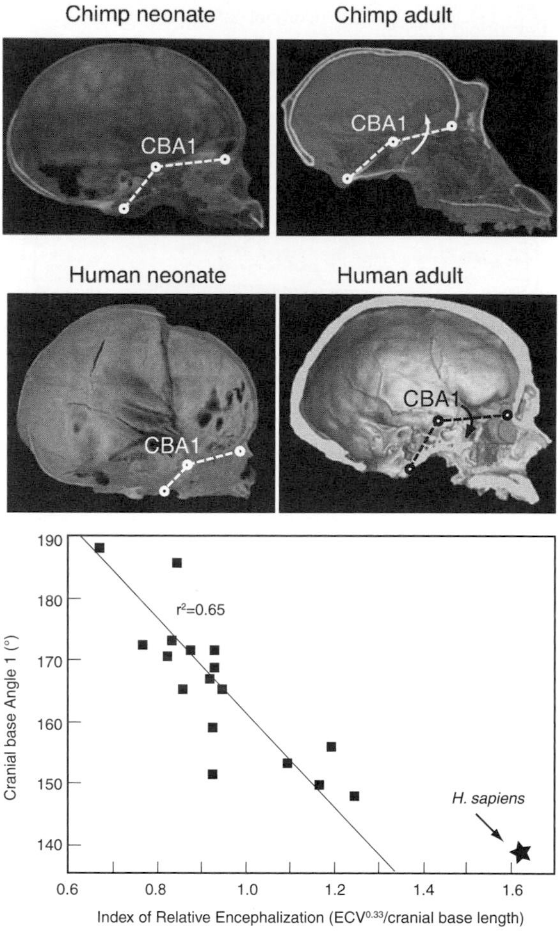

FIGURE 16.6 Top: Changes in the cranial base angle (CBA1) between a neonatal chimpanzee and a human (modified from Lieberman, 2011). Note that CBA1 extends in the chimpanzee but flexes in the human because of the combination of a relatively larger brain and a relatively shorter face. Bottom: The relationship between CBA1 and brain size relative to cranial base length (quantified as the index of relative encephalization) in primates (from McCarthy, 2001). Primates with relatively bigger brains have more flexed cranial bases, but humans are more flexed than predicted by the primate regression. ECV, endocranial volume.

prognathic faces, but other australopiths such as *Au. boisei* have more flexed cranial bases (ca. 135°) because their brains are slightly larger (average 492 cc) and because they have more orthognathic, ventrally rotated faces designed to generate and withstand large masticatory forces (see Rak, 1983; Hylander, 1988; Kimbel et al., 2004). Similarly, there is a considerable variation in CBA1 within the genus *Homo*, with some species such as *H. heidelbergensis* and *H. neanderthalensis* having fairly extended cranial bases (published values for CBA1 range 139°–150°) and some *H. erectus* crania with smaller faces having more flexed cranial bases (Baba et al., 2003; Lieberman, 2011).

3. A final aspect of epigenetic integration especially relevant to increased brain size in human evolution is skull width. Most mammals, including apes, have relatively long and narrow skulls; but increases in brain size in human evolution have been accommodated to a large extent by wider posterior, middle, and anterior cranial fossae (Weidenreich, 1941). Because the face grows downward/ventrally from the anterior cranial fossa and forward from the middle cranial fossa (see Enlow, 1990), a wider basicranium and neurocranium also leads to a wider face, particularly in the upper portions around the orbits but also in the middle and lower portions of the face (Lieberman et al., 2000a, 2004;

Bastir et al., 2008). These width increases and their effects on overall craniofacial integration are probably accommodated by a number of other epigenetic mechanisms, which probably explains why width dimensions are among the strongest sources of covariation in the mammalian skull (Hallgrimsson et al., 2007), including in humans (Polanski and Franciscus, 2006). For example, lateral expansion of the brain and posterior cranial fossa appears to cause lateral (coronal) rotation of the adjoining petrous portions of the temporal (Dean, 1988; Spoor, 1997; but see Jeffery and Spoor, 2002). Note also that rotation of the petrous portion of the temporal also rotates the temporomandibular joints, helping to align the condyle and ramus of the mandible with the orientation of the maxillary tooth row (bigger brains set up wider faces with wider maxillary arches that necessitate a more coronally oriented temporomandibular joint) (Lieberman, 2011).

Few mammals have evolved very big brains, probably because brain tissue is extremely metabolically expensive (Elia, 1992). However, highly encephalized mammals, such as hominins that have been able to pay for larger brains and benefit in terms of fitness, have also been able to accommodate larger brains without substantial reengineering by using existing epigenetic mechanisms that arise from how heads grow in the first place. Notably, the same mechanisms that permit mammals to grow bigger brains during ontogeny without compromising functions such as mastication, vision, and olfaction have also been able to accommodate variations in adult brain size among individuals and, hence, evolutionary shifts over time. Evidence for this accommodation is abundant in the performance of individuals with developmental problems that affect brain growth such as microcephalus and hydrocephalus. Many humans with ECVs more than three standard deviations above and below the species mean can still hear, chew, smile, and locomote effectively. Similar kinds of epigenetic mechanisms also help humans of varying brain volume to function well in spite of wide variations in tooth size, eyeball size, face size, and so on. These, of course, are extreme cases, but they illustrate how learning more about the many mechanisms by which individuals with these disorders do (and do not) adapt to variations, we will surely learn more about the mechanisms that permit skulls to be so variable and evolvable. In turn, unraveling the epigenetic mechanisms that make skulls so variable and evolvable has much potential for yielding generalizable insights into how epigenetic interactions are structured during development to permit mutual accommodation of organs, tissues, spaces, and skeletal structures.

In short, it has long been appreciated by developmental biologists that epigenetic interactions are involved in many aspects of embryogenesis, but I hope that the above examples illuminate how these sorts of general interactions, which involve various kinds of feedback mechanisms, are also vitally important throughout an organism's ontogeny. In the case of the head, there is good reason to believe that the initial (apparently convoluted) way the head grows sets up the potential for many future interactions, thus permitting a wide range of structures that can function in many different ways. Also, as long as skeletal units grow in and around functional units, the many parts of the head can push and pull on each other to permit the head to function well in these different configurations, thus permitting selection to operate. This general model, of course, needs much further testing, and many key details remain poorly studied and barely understood. In addition, functional matrix-type models need to be applied to other complex aspects of phenotype elsewhere in the body. Yet, I am confident that this sort of research on the relationship between complexity and evolvability will continue to provide excellent examples of why Waddington's original, broad definition of *epigenetics* is both necessary and useful. Although some biologists are tempted to impose a simplifying, reductionist definition on the concept, without complex epigenetic processes neither biologists nor their heads would be here.

REFERENCES

Alsbirk, P. H. 1977. Variation and heritability of ocular dimensions. A population study among adult Greenland Eskimos. *Acta Ophthalmol* 55:443–56.

Antón, S. C. 1989. Intentional cranial vault deformation and induced changes in the cranial base and face. *Am J Phys Anthropol* 79:253–67.

Atchley, W. R., and B. K. Hall. 1991. A model for development and evolution of complex morphological structures. *Biol Rev Camb Philos Soc* 66: 101–57.

Baab, K. L., and K. P. McNulty. 2009. Size, shape, and asymmetry in fossil hominins: The status of the LB1 cranium based on 3D morphometric analyses. *J Hum Evol* doi:10.1016/j.jhevol.2008. 08.011.

Baba, H., F. Aziz, Y. Kaifu, G. Suwa, R. T. Kono, and T. Jacob. 2003. *Homo erectus* calvarium from the Pleistocene of Java. *Science* 299:1384–8.

Babineau, T. A., and J. H. Kronman. 1969. A cephalometric evaluation of the cranial base in microcephaly. *Angle Orthod* 39:57–63.

Bastir, M., A. Rosas, D. E. Lieberman, and P. O'Higgins. 2008. Middle cranial fossa anatomy and the origin of modern humans. *Anat Rec* 291:130–40.

Biegert, J. 1963. The evaluation of characters of the skull, hands and feet for primate taxonomy. In *Classification and Human Evolution*, ed. S. L. Washburn, 116–45. Chicago: Aldine de Gruyter.

Bolk, L. 1926. *Das Problem der Menschwerdung*. Jena, Germany: Gustav Fischer.

Bookstein, F. L., P. Gunz, P. Mitteroecker, H. Prossinger, K. Schaefer, and H. Seidler. 2003. Cranial integration in *Homo*: Singular warps analysis of the midsagittal plane in ontogeny and evolution. *J Hum Evol* 44:167–87.

Carter, D. R., and G. S. Beaupré. 2001. *Skeletal Function and Form: Mechanobiology of Skeletal Development, Aging, and Regeneration*. New York: Cambridge University Press.

Cerny, R., P. Lwigale, R. Ericsson, D. Meulemans, H. H. Epperlein, and M. Bronner-Fraser. 2004. Developmental origins and evolution of jaws: New interpretation of "maxillary" and "mandibular." *Dev Biol* 276:225–36.

Chernoff, B., and P. M. Magwene. 1999. Morphological integration: Forty years later. In *Morphological Integration*, 319–53. Chicago: University of Chicago Press.

Cheverud, J. M. 1995. Morphological integration in the saddle-back tamarin (*Saguinus fuscicollis*) cranium. *Am Nat* 145:63–89.

Cheverud, J. M., L. A. Kohn, L. W. Konigsberg, and S. R. Leigh. 1992. Effects of fronto-occipital artificial cranial vault modification on the cranial base and face. *Am J Phys Anthropol* 88:323–45.

Cohen, M. M., Jr. 2000. Sutural biology. In *Craniosynostosis: Diagnosis, Evaluation and Management*, ed. M. M. Cohen and R. E. MacLean, 11–23. New York: Oxford University Press.

Cottrill, C. P., C. W. Archer, and L. Wolpert. 1987. Cell sorting and aggregate formation in micromass culture. *Dev Biol* 122:503–15.

Currey, J. D. 2002. *Bones: Structure and Mechanics*. Princeton, NJ: Princeton University Press.

Dean, M. C. 1988. Growth processes in the cranial base of hominoids and their bearing on morphological similarities that exist in the cranial base of *Homo* and *Paranthropus*. In *Evolutionary History of the "Robust" Australopithecines*, ed. F. E. Grine, 107–12. New York: Aldine de Gruyter.

Depew, M. J., T. Lufkin, and J. R. L. Rubenstein. 2002. Specification of jaw subdivision by *Dlx* genes. *Science* 298:381–85.

Depew, M. J., C. A. Simpson, M. Morasso, and J. L. Rubenstein. 2005. Reassessing the Dlx code: The genetic regulation of branchial arch skeletal pattern and development. *J. Anat* 207:501–61.

Deschamps, J., and J. van Nes. 2005. Developmental regulation of the Hox genes during axial morphogenesis in the mouse. *Development* 132:2931–42

Duterloo, H. S., and D. H. Enlow. 1970. A comparative study of cranial growth in *Homo* and *Macaca*. *Am J Anat* 127:357–68.

Elia, M. 1992. Organ and tissue contribution to metabolic weight. In *Energy Metabolism: Tissue Determinants and Cellular Corollaries*, ed. J. M. Kinney and H. N. Tucker, 61–79. New York: Raven Press.

Enlow, D. H. 1990. *Facial Growth*, 3rd ed. Philadelphia: Saunders.

Friede, H. 1981. Normal development and growth of the human neurocranium and cranial base. *Scand J Plast Reconstr Surg* 115:163–9.

Gould, S. J. 1977. *Ontogeny and Phylogeny*. Cambridge, MA: Belknap Press.

Haig, D. 2004. The (dual) origin of epigenetics. *Cold Spring Harb Symp Quant Biol* 69:1–4.

Hall, B. K. 2005. *Bones and Cartilage: Developmental and Evolutionary Skeletal Biology*. Amsterdam: Elsevier.

Hallgrímsson, B., D. E. Lieberman, W. Lie, A. F. Ford-Hutchinson, and F. R. Jirik. 2007. Epigenetic interactions and the structure of phenotypic variation in the skull. *Evol Dev* 9:76–91.

Harris, E. F., and M. G. Johnson. 1991. Heritability of craniometric and occlusal variables: A longitudinal sib analysis. *Am J Orthod Dentofacial Orthop* 99:258–68.

Herring, S. W. 1993. Epigenetic and functional influences on skull growth. In *The Skull*, ed. J. Hanken and B. K. Hall, 237–71. Chicago: University of Chicago Press.

Holloway, R. L., D. C. Broadfield, M. S. Yuan, J. H. Schwartz, and I. Tattersall. 2004. *The Human Fossil Record*. Vol. 3 of *Brain Endocasts—The Paleoneurological Evidence*. New York: Wiley.

Hunter, W. S. 1990. Heredity in the craniofacial complex. In *Facial Growth*, 3rd ed., ed. D. H. Enlow, 249–66. Philadelphia: Saunders.

Hylander, W. L. 1988. Implications of in vivo experiments for interpreting the functional significance of "robust" australopithecine jaws. In *Evolutionary History of the "Robust" Australopithecines*, ed. F. Grine, 55–83. New York: Aldine de Gruyter.

Jeffery, N. 2002. Differential regional brain growth and rotation of the prenatal human tentorium cerebelli. *J Anat* 200:135–44.

Jeffery, N., and F. Spoor. 2002. Brain size and the human cranial base: A prenatal perspective. *Am J Phys Anthropol* 118:324–40.

Jernvall, J., and I. Thesleff. 2000. Reiterative signaling and patterning during mammalian tooth morphogenesis. *Mech Dev* 92:19–29.

Jiang, X., S. Iseki, R. E. Maxson, H. M. Sucov, and G. M. Morriss-Kay. 2002. Tissue origins and interactions in the mammalian skull vault. *Dev Biol* 241:106–16.

Kimbel, W. H., Y. Rak, and D. C. Johanson. 2004. *The Skull of Australopithecus afarensis*. Oxford: Oxford University Press.

Kirschner, M. W., and J. C. Gerhart. 2005. *The Plausibility of Life*. New Haven, CT: Yale University Press.

Klingenberg, C. P. 2008. Morphological integration and developmental modularity. *Annu Rev Ecol Evol Syst* 39:115–32.

Lee, S. H., O. Bédard, M. Buchtová, K. Fu, and J. M. Richman. 2004. A new origin for the maxillary jaw. *Dev Biol* 276:207–24.

Lieberman, D. E. 2007. Homing in on early *Homo*. *Nature* 449:291–2.

Lieberman, D. E. 2011. *The Evolution of the Human Head*. Cambridge, MA: Harvard University Press.

Lieberman, D. E., B. Hallgrímsson, W. Liu, T. A. Parsons, and H. A. Jamniczky. 2008. Spatial packing, cranial base angulation, and craniofacial shape variation in the mammalian skull: Testing a new model using mice. *J Anat* 212:720–35.

Lieberman, D. E., G. E. Krovitz, and B. McBratney-Owen. 2004. Testing hypotheses about tinkering in the fossil record: The case of the human skull. *J Exp Zoolog B Mol Dev Evol* 302:302–21.

Lieberman, D. E., and R. C. McCarthy. 1999. The ontogeny of cranial base angulation in humans and chimpanzees and its implications for reconstructing pharyngeal dimensions. *J Hum Evol* 36:487–517.

Lieberman, D. E., K. M. Mowbray, and O. M. Pearson. 2000a. Basicranial influences on overall cranial shape. *J Hum Evol* 38:291–315.

Lieberman, D. E., C. F. Ross, and M. J. Ravosa. 2000b. The primate cranial base: Ontogeny, function and integration. *Ybk J Phys Anthropol* 43:117–69.

Lordkipanidze, D., T. Jashashvili, A. Vekua, M. A. Ponce de León, C. P. Zollikofer, G. P. Rightmire, H. Pontzer, et al. 2007. Postcranial evidence from early *Homo* from Dmanisi, Georgia. *Nature* 449:305–10.

Marazita, M. L., and M. Mooney. 2004. Current concepts in the embryology and genetics of cleft lip and cleft palate. *Clin Plast Surg* 31:125–40.

Marcucio, R. S., D. R. Cordero, D. Hu, and J. A. Helms. 2005. Molecular interactions coordinating the development of the forebrain and face. *Dev Biol* 284:48–61.

Marie, P. J., J. D. Coffin, and M. M. Hurley. 2005. FGF and FGFR signaling in chondrodysplasias and craniosynostosis. *J Cell Biochem* 96:888–96.

Marroig, G., and J. Cheverud. 2001. A comparison of phenotypic variation and covariation patterns and the role of phylogeny, ecology and ontogeny during cranial evolution of New World monkeys. *Evolution* 55:2576–600.

McBratney-Owen, B., S. Iseki, S. D. Bamforth, B. R. Olsen, and G. M. Morriss-Kay. 2008. Development and tissue origins of the mammalian cranial base. *Dev Biol* 322:121–32.

McCarthy, R. C. 2001. Anthropoid cranial base architecture and scaling relationships. *J Hum Evol* 40:41–66.

McCarthy, R. C. 2004. Constraints and primate craniofacial growth and form. PhD diss., George Washington University.

McGrath, J., O. El-Saadi, V. Grim, S. Cardy, B. Chaple, D. Chant, D. E. Lieberman, and B. Mowry. 2002. Minor physical anomalies and quantitative measures of the head in psychosis. *Arch Gen Psychiatry* 59:458–64.

McKinney, M. L., and K. J. McNamara. 1991. *Heterochrony: The Evolution of Ontogeny*. New York: Plenum Press.

Mina, M. 2001. Regulation of mandibular growth and morphogenesis. *Crit Rev Oral Biol Med* 12:276–300.

Morriss-Kay, G. M., and A. O. Wilkie. 2005. Growth of the normal skull vault and its alteration in craniosynostosis: Insights from human genetics and experimental studies. *J Anat* 207:637–53.

Moss, M. L. 1968. The primacy of functional matrices in orofacial growth. *Dent Practitioner* 19:63–73.

Moss, M. L. 1997a. The functional matrix hypothesis revisited. 1. The role of mechanotransduction. *Am J Orthod Dentofacial Orthop* 112:8–11.

Moss, M. L. 1997b. The functional matrix hypothesis revisited. 2. The role of an osseous connected cellular network. *Am J Orthod Dentofacial Orthop* 112:221–6.

Moss, M. L. 1997c. The functional matrix hypothesis revisited. 3. The genomic thesis. *Am J Orthod Dentofacial Orthop* 112:338-342.

Moss, M. L. 1997d. The functional matrix hypothesis revisited. 4. The epigenetic antithesis and the resolving synthesis. *Am J Orthod Dentofacial Orthop* 112:410–14.

Moss, M. L., and R. W. Young. 1960. A functional approach to craniology. *Am J Phys Anthropol* 18:281–92.

Moyers, R. L. 1988. *Handbook of Orthodontics*, 4th ed. Chicago: Yearbook Medical Publishers.

Olson, E. R., and R. L. Miller. 1958. *Morphological Integration*. Chicago: University of Chicago Press.

Opperman, L. A. 2000. Cranial sutures as intramembranous bone growth sites. *Dev Dyn* 219:472–85.

Parsons, T. E., E. Kristensen, L. Hornung, V. M. Diewert, S. K. Boyd, R. Z. German, and B. Hallgrímsson. 2008. Phenotypic variability and craniofacial dysmorphology: increased shape variance in a mouse model for cleft lip. *J Ana.* 135–43.

Peters, H., and R. Balling. 1999. Teeth. Where and how to make them. *Trends Genet* 15:59–65.

Polanski, J. M., and R. G. Franciscus. 2006. Patterns of craniofacial integration in extant *Homo*, *Pan*, and *Gorilla*. *Am J Phys Anthropol* 75:195–96.

Radlanski, R. J., and H. Renz. 2006. Genes and forces and forms. Mechanical aspects during prenatal craniofacial development. *Dev Dyn* 235:1219–29.

Rak, Y. 1983. *The Australopithecine Face*. New York: Academic Press.

Richtsmeier, J. T. 2002. Cranial vault morphology and growth in craniosynostoses. In *Understanding Craniofacial Anomalies: The Etiopathogenesis of Craniosynostoses and Facial Clefting*, ed. M. P. Mooney and M. I. Siegel, 321–41. New York: Wiley-Liss.

Richtsmeier, J. T., K. Aldridge, V. B. DeLeon, J. Panchal, A. A. Kane, J. L. Marsh, P. Yan, and T. M. Cole III 2006. Phenotypic integration of neurocranium and brain. *J Exp Zoolog B Mol Dev Evol* 306:360–78.

Rightmire, G. P. 2004. Brain size and encephalization in early to mid-Pleistocene *Homo*. *Am J Phys Anthropol* 124:109–23.

Rogers, J., P. Kochunov, J. Lancaster, W. Shelledy, D. Glahn, J. Blangero, and P. Fox. 2007. Heritability of brain volume, surface area and shape: An MRI study in an extended pedigree of baboons. *Hum Brain Mapp* 28:576–83.

Rönning, H. J. 1995. Growth of the cranial vault: Influence of intracranial and extracranial pressures. *Acta Odontol Scand* 53:192–5.

Ross, C. F., and M. Henneberg. 1995. Basicranial flexion, relative brain size, and facial kyphosis in *Homo sapiens* and some fossil hominids. *Am J Phys Anthropol* 98:575–93.

Ross, C. F., and M. J. Ravosa. 1993. Basicranial flexion, relative brain size, and facial kyphosis in nonhuman primates. *Am J Phys Anthropol* 91:305–24.

Ruff, C. B., E. Trinkaus, and T. W. Holliday. 1997. Body mass and encephalization in Pleistocene *Homo*. *Nature* 387:173–6.

Sarnat, B. G. 1982. Eye and orbital size in the young and adult. Some postnatal experimental and clinical relationships. *Ophthalmologica* 185:74–8.

Schowing, J. 1968. Demonstration of the inductive role of the brain in osteogenesis of the embryonic skull of the chicken. *J Embryol Exp Morphol* 19:83–94.

Shea, B. T., and A. M. Gomez. 1988. Tooth scaling and evolutionary dwarfism: An investigation of allometry in human pygmies. *Am J Phys Anthropol* 77:117–32.

Sperber, G. 2001. *Craniofacial Development*. Hamilton, Canada: B. C. Decker.

Spoor, C. F. 1997. Basicranial architecture and relative brain size of Sts 5 (*Australopithecus africanus*) and other Plio-Pleistocene hominids. *S Afr J Sci* 93:182–6.

Strait, D. S. 2001. Integration, phylogeny, and the hominid cranial base. *Am J Phys Anthropol* 114:273–97.

van der Klaauw, C. 1948–1952. Size and position of the functional components of the skull. *Arch Neerland Zool* 9:1–559.

Wagner, G. P. 1984. On the eigenvalue distribution of genetic and phenotypic dispersion matrices—evidence for a nonrandom organization of quantitative character variation. *J Math Biol* 21:77–95.

Wagner, G. P. 2001. *The Character Concept in Evolutionary Biology*. New York: Academic Press.

Watson, W. G. 1982. The functional matrix revisited. *Am J Orthod* 81:71–3.

Weidenreich, F. 1941. The brain and is rôle in the phylogenetic transformation of the human skull. *Trans Am Philos Soc* 31:328–442.

West-Eberhard, M. J. 2003. *Developmental Plasticity and Evolution*. Oxford: Oxford University Press.

Wilkie, A. O., and G. M. Morriss-Kay. 2001. Genetics of craniofacial development and malformation. *Nat Rev Genet* 2:458–68.

Wilkins, A. S. 2002. *The Evolution of Developmental Pathways*. Sunderland, MA: Sinauer Associates.

Young, N. M., S. Wat, V. M. Diewert, L. W. Browder, and B. Hallgrímsson. 2007. Comparative morphometrics of embryonic facial morphogenesis: implications for cleft-lip etiology. *Anat Rec* 290:123–39.

Young, R. L., and A. V. Badyaev. 2006. Evolutionary persistence of phenotypic integration: Influence of developmental and functional relationships on complex trait evolution. *Evolution* 60:1291–9.

Zelditch, M. L., and A. C. Carmichael. 1989. Ontogenetic variation in patterns of developmental and functional integration in skulls of *Sigmodon fulviventer*. *J Mammal* 70:477–84.

17

Epigenetic Interactions

THE DEVELOPMENTAL ROUTE TO FUNCTIONAL INTEGRATION

Miriam Leah Zelditch and Donald L. Swiderski

CONTENTS

Epigenetic interactions are obviously necessary for normal development—without them there would be no primary embryonic induction, no epithelial–mesenchymal interactions, and no interactions between differentiated tissues such as muscles and bones. These interactions are necessary not only for normal development but also for normal function, if only because they produce the structures that carry out function. The ability of jawed vertebrates to eat typically requires having a jaw, and without epithelial–mesenchymal interactions there would be no jaw. However, eating requires more than just having a jaw, and epigenetic interactions do more than just produce it. Epigenetic interactions also integrate developmentally heterogeneous tissues into a coherent functional whole, coordinating the development of bones with that of the skeletal muscles that move the bones and with that of the teeth, not to mention the tongue, nerves, and blood vessels. All of these, taken together, comprise a single integrated whole—the feeding system. In the more general case, bones and muscles can be regarded as a single functional system because bones provide skeletal struts and levers that are moved by

Epigenetics: Linking Genotype and Phenotype in Development and Evolution, ed. Benedikt Hallgrímsson and Brian K. Hall.
Copyright © by The Regents of the University of California. All rights of reproduction in any form reserved.

the forces supplied by muscles (Herring, 1994). Obviously, the system does not work in the absence of its parts; but even if all the parts are there, the system does not work very well when they are disproportionate relative to each other. Excessively strong muscles coupled to a small, slight mandible could generate forces capable of yanking the mandible out of its joint; conversely, excessively weak muscles inserting on a massive jaw could generate forces too weak to open the jaw at all. Even modest disproportions can imperil function, such as when they lead to a misalignment of the jaws and teeth; in the case of a rodent's ever-growing incisors, malocclusion means that the incisors can grow through the roof of the mouth, which is usually lethal.

That epigenetic interactions benefit individuals by integrating functionally interacting parts seems obvious, but that benefit does not mean that epigenetic interactions evolved to serve that particular biological role. Whether epigenetic interactions, or morphological integration more generally, are adaptations is a topic addressed by Hansen in this volume (see Chapter 20); and as is clear from his chapter, this adaptive scenario can be challenged on several grounds. One alternative hypothesis is that integration is an intrinsic feature of developmentally modular systems, a hypothesis that has an important and novel implication because it means that the developmental basis of integration may be critical for understanding its evolutionary origin. Another alternative theory is that integration is not an adaptation in its own right but, rather, a correlated effect of some other adaptation, another hypothesis that emphasizes the theoretical significance of the developmental basis of integration. As we argue herein, the developmental basis of integration may be just as crucial to the evolution of integration as it is to the development of an individual.

We begin by briefly reviewing the concept of morphological integration and then focus more specifically on the distinction between developmental and functional integration to provide the context for analyzing the developmental basis of functional integration. In that section,

we highlight the distinctive features of epigenetic pleiotropy and why those features might make them distinctive for the evolution of integration in general and for functional integration in particular. Finally, we turn to one classic model system for studies of integration, the mammalian mandible, one that has been regarded as a paradigm for the theory that integration is an adaptation but one which also may exemplify the two alternative theories and that clearly reveals some of the complications of testing theories about integration.

AN OVERVIEW OF MORPHOLOGICAL INTEGRATION

ORGANISMS AS INTEGRATED SYSTEMS

The idea of morphological integration is grounded in the perception that organisms are not collections of isolated parts but, rather, coherent systems. Consequently, fitness depends on the *relationships* among traits rather than on their individual values. Taking the simplest possible case of two traits, fitness depends on their relationship when the adaptive value of one trait is conditional on the phenotype of the other. In the case of a lever, for example, the force that it transmits depends on the ratio between the in- and out-lever arms (mechanical advantage). This ratio gives the ratio of output to input force. It is this ratio, not the lengths of the individual lever arms, that matters to force transmission, so unless the length of one arm is fixed, there is no optimal value for either of them. Rather, the optimal value for one depends on the length of the other (and on the optimal value for the output force generated by the system). Most biological levers are more complex than this simple case, if only because something has to move the lever and that something else is part of the system for generating the force. Additionally, most biological levers confront more than one functional challenge. For the mandible, one of those other challenges is to open the jaw wide enough to engulf the food; that matters for optimizing the form of structure of phenotypic integration (Zelditch et al., 2008, 2009). Comparisons between species also suggest that rodents differ in structure of integration in a way compatible with their different functional demands due to differences in food consistency. Whether differences in consistency of food actually eaten by the animals are consequential for the structure of integration is not yet clear, but that obviously bears on the issue of whether function itself (e.g., gnawing, biting) can induce integration between functionally integrated parts. If so, then epigenetic pleiotropy and epigenetic plasticity may cooperate in building developmental integration into functional systems.

CONCLUSION

Epigenetic and intrinsic genetic pleiotropy are distinct developmental mechanisms of integration, and we have argued that this warrants treating them as theoretically distinct, whether a theory aims to explain developmental or evolutionary origins of integration. Distinguishing these two forms of pleiotropy will no doubt complicate evolutionary theories for integration because, in some respects, epigenetic pleiotropy is more akin to adaptive plasticity than it is to intrinsic genetic pleiotropy; but epigenetic pleiotropy is also distinct from adaptive plasticity in one crucial respect—the internal environment of the organism contains its genes. It may thus seem that epigenetic pleiotropy ought to be viewed as an interaction between genes in that the impact of variation in alleles at one locus is contingent on those at another, but that would miss what makes epigenetic pleiotropy like adaptive plasticity: The tissue doing the signaling not only conditions the expression of genes within the responding tissue but also determines the optimal phenotype for it. That is why epigenetic pleiotropy matters most for functionally interdependent traits that express few genes in common and that interact primarily via signaling interactions, such as muscles and bones. It is through their signaling interactions that they maintain the internal coherence of a functional module.

ACKNOWLEDGMENTS

We thank Benedikt Hallgrímsson and Brian Hall for the invitation to contribute to this volume, and we thank them and Chuck Crumly for their patience.

REFERENCES

Abe, M., L. B. Ruest, and D. E. Clouthier. 2007. Fate of cranial neural crest cells during craniofacial development in endothelin-A receptor-deficient mice. *Int J Dev Biol* 51:97–105.

Ackermann, R. R. 2005. Ontogenetic integration of the hominoid face. *J Hum Evol* 48:175–97.

Alfaro, M. E., D. I. Bolnick, and P. C. Wainwright. 2004. Evolutionary dynamics of complex biomechanical systems: An example using the four-bar mechanism. *Evolution* 58:495–503.

Alfaro, M. E., D. I. Bolnick, and P. C. Wainwright. 2005. Evolutionary consequences of many-to-one mapping of jaw morphology to mechanics in labrid fishes. *Am Nat* 165:E140–E154.

Ancel, L. W., and W. Fontana. 2000. Plasticity, evolvability, and modularity in RNA. *J Exp Zool* 288:242–83.

Anthwal, N., Y. Chai, and A. S. Tucker. 2008. The role of transforming growth factor-beta signalling in the patterning of the proximal processes of the murine dentary. *Dev Dyn* 237:1604–13.

Arnold, S. J. 1983. Morphology, performance and fitness. *Am Zool* 23:347–61.

Atchley, W. R. 1983. A genetic analysis of the mandible and maxilla in the rat. *J Craniofac Genet Dev Biol* 3:409–22.

Atchley, W. R. 1984. Ontogeny, timing of development, and genetic variance–covariance structure. *Am Nat* 123:519–40.

Atchley, W. R. 1987. Developmental quantitative genetics and the evolution of ontogenies. *Evolution* 41:316–30.

Atchley, W. R., D. E. Cowley, C. Vogl, and T. McLellan. 1992. Evolutionary divergence, shape change, and genetic correlation structure in the rodent mandible. *Syst Biol* 41:196–221.

Atchley, W. R., and B. K. Hall. 1991. A model for development and evolution of complex morphological structures. *Biol Rev Camb Philos Soc* 66:101–57.

Atchley, W. R., T. Logsdon, D. E. Cowley, and E. J. Eisen. 1991. Uterine effects, epigenetics, and postnatal skeletal development in the mouse. *Evolution* 45:891–909.

Atchley, W. R., A. A. Plummer, and B. Riska. 1985. Genetics of mandible form in the mouse. *Genetics* 111:555–77.

Atchley, W. R., S. Z. Xu, and C. Vogl. 1994. Developmental quantitative genetic models of evolutionary change. *Dev Genet* 15:92–103.

Atchley, W. R., and J. Zhu. 1997. Developmental quantitative genetics, conditional epigenetic variability and growth in mice. *Genetics* 147:765–76.

Badyaev, A. V., and K. R. Foresman. 2004. Evolution of morphological integration. I. Functional units channel stress-induced variation in shrew mandibles. *Am Nat* 163:868–79.

Badyaev, A. V., K. R. Foresman, and R. L. Young. 2005. Evolution of morphological integration: Developmental accommodation of stress-induced variation. *Am Nat* 166:382–95.

Bailey, D. W. 1985. Genes that affect the shape of the murine mandible: Congenic strain analysis. *J Hered* 76:107–14.

Bailey, D. W. 1986. Genes that affect morphogenesis of the murine mandible: Recombinant-inbred strain analysis. *J Hered* 77:17–25.

Berg, R. L. 1960. The ecological significance of correlation pleiades. *Evolution* 14:171–80.

Beverdam, A., G. R. Merlo, L. Paleari, S. Mantero, F. Genova, O. Barbieri, P. Janvier, and G. Levi. 2002. Jaw transformation with gain of symmetry after *Dlx5/Dlx6* inactivation: Mirror of the past? *Genesis* 34:221–7.

Bock, W. J. 1960. A critique of morphological integration. *Evolution* 14:130–2.

Buettner-Janusch, J. 1959. Morphological integration—Olson and Miller. *Am Anthropol* 61:918–19.

Busser, B. W., M. L. Bulyk, and A. M. Michelson. 2008. Toward a systems-level understanding of developmental regulatory networks. *Curr Opin Genet Dev* 18:521–9.

Carroll, S. B., J. K. Grenier, and S. D. Weatherbee. 2001. *From DNA to Diversity*. Malden, MA: Blackwell Science.

Carter, D. R., and T. E. Orr. 1992. Skeletal development and bone functional adaptation. *J Bone Miner Res* 7:S389–95.

Caton, J., and A. S. Tucker. 2009. Current knowledge of tooth development: Patterning and mineralization of the murine dentition. *J Anat* 214:502–15.

Chai, Y., Y. Ito, and J. Han. 2003. TGF-beta signaling and its functional significance in regulating the fate of cranial neural crest cells. *Crit Rev Oral Biol Med* 14:78–88.

Cheverud, J. M. 1982. Phenotypic, genetic, and environmental morphological integration in the cranium. *Evolution* 36:499–516.

Cheverud, J. M. 1984. Quantitative genetics and developmental constraints on evolution by selection. *J Theor Biol* 110:155–71.

Cheverud, J. M. 1988. A comparison of genetic and phenotypic correlations. *Evolution* 42:958–68.

Cheverud, J. M. 1995. Morphological integration in the saddle-back tamarin (*Saguinus fuscicollis*) cranium. *Am Nat* 145:63–89.

Cheverud, J. M. 1996a. Developmental integration and the evolution of pleiotropy. *Am Zool* 36:44–50.

Cheverud, J. M. 1996b. Quantitative genetic analysis of cranial morphology in the cotton-top (*Saguinus oedipus*) and saddle-back (*S. fuscicollis*) tamarins. *J Evol Biol* 9:5–42.

Cheverud, J. M. 2004. Modular pleiotropic effects of quantitative trait loci on morphological traits. In *Modularity in Development and Evolution*, ed. G. Schlosser and G. P. Wagner, 132–53. Chicago: University of Chicago Press.

Cheverud, J. M., T. H. Ehrich, T. T. Vaughn, S. F. Koreishi, R. B. Linsey, and L. S. Pletscher. 2004. Pleiotropic effects on mandibular morphology II: Differential epistasis and genetic variation in morphological integration. *J Exp Zool B Mol Dev Evol* 302:424–35.

Cheverud, J. M., S. E. Hartman, J. T. Richtsmeier, and W. R. Atchley. 1991. A quantitative genetic analysis of localized morphology in mandibles of inbred mice using finite-element scaling analysis. *J Craniofac Genet Dev Biol* 11:122–37.

Cheverud, J. M., E. J. Routman, and D. J. Irschick. 1997. Pleiotropic effects of individual gene loci on mandibular morphology. *Evolution* 51:2006–16.

Clouthier, D. E., K. Hosoda, J. A. Richardson, S. C. Williams, H. Yanagisawa, T. Kuwaki, M. Kumada, R. E. Hammer, and M. Yanagisawa. 1998. Cranial and cardiac neural crest defects in endothelin-A receptor-deficient mice. *Development* 125:813–24.

Cobourne, M. T., and P. T. Sharpe. 2003. Tooth and jaw: Molecular mechanisms of patterning in the first branchial arch. *Arch Oral Biol* 48:1–14.

Collar, D. C., and P. C. Wainwright. 2006. Discordance between morphological and mechanical diversity in the feeding mechanism of centrarchid fishes. *Evolution* 60:2575–84.

Cowley, D. E., and W. R. Atchley. 1990. Development and quantitative genetics of correlation structure among body parts of *Drosophila melanogaster*. *Am Nat* 135:242–68.

Cowley, D. E., and W. R. Atchley. 1992. Quantitative genetic models for development, epigenetic selection, and phenotypic evolution. *Evolution* 46:495–518.

novel environments is not new. The essential arguments were first articulated more than a century ago (the "Baldwin effect") (Baldwin, 1896, 1902; Morgan, 1896; Osborn, 1896) and have subsequently been refined, most recently by West-Eberhard (2003). The argument runs as follows.

When members of a population encounter novel environments, the ability to produce variant phenotypes (phenotypic plasticity) can expose previously cryptic genetic variation that can enhance or hinder persistence. In those environments in which persistence is possible, the course of subsequent evolution can be influenced by the pattern of trait expression, which can direct the course of future evolution by determining where on the new adaptive landscape a population begins its journey (e.g., Kirkpatrick, 1982; Hinton and Nowlan, 1987; Fear and Price, 1998; Pal and Miklos, 1999; Price et al., 2003). A logical extension of this argument is that the pattern of phenotypic plasticity in populations ancestral to adaptive radiations can influence the overall pattern of diversification in the radiation (West-Eberhard, 2003; Shaw et al., 2007; Wund et al., 2008).

As is so often the case, theoretical treatments outpace (and encourage) empirical research. Although a number of reviews (e.g., West-Eberhard, 1989, 1998, 2003, 2005; Pigliucci and Murren, 2003; Pigliucci, 2005, 2007) have presented examples compatible with the idea that phenotypic plasticity preceded genetic evolution, they offered few unequivocal examples in which the relationship between ancestral plasticity and that in daughter populations could be compared. Ghalambour et al. (2007) reviewed the exceptions to this generalization, cases in which ancestral patterns of plasticity can be compared to plasticity in derived populations. These examples, involving host shifts in soapberry bugs (*Jadera haematoloma*) (Carroll et al. 1997, 1998) and invasion of novel habitats by anadromous (sea run) Pacific salmon (*Oncorhynchus nerka*) and guppies (*Poecilia reticulata*), all offer evidence of initial adaptive plasticity in some traits that could enhance

persistence, followed by an evolutionary shift in the form of a reaction norm ("genetic accommodation") (West-Eberhard, 2003) or, more specifically, the constitutive expression of a previously plastic trait ("genetic assimilation") (reviewed in Crispo, 2007). A particularly elegant example demonstrated a role for genetic assimilation in the evolution of *Daphnia* populations recently exposed to novel predators (Scoville and Pfrender, 2010). Clearly, the difficulty is that we rarely have ancestral populations available for study that we can infer have changed little since giving rise to daughter populations, permitting comparison of ancestral and derived patterns of plasticity. Nevertheless, such comparisons are crucial if we are to understand the role of plasticity in evolutionary processes.

Here, we describe initial insights on the role of ancestral phenotypic plasticity in the generation of diversity in an adaptive radiation uniquely suited for the purpose: that of the threespine stickleback fish, *Gasterosteus aculeatus*. The unique feature of this radiation lies in the oceanic populations that have changed little morphologically over at least 12 million years and are unlikely to have changed in other features for the reasons described below. Thus, oceanic stickleback are likely to represent the ancestral condition relative to the postglacial freshwater radiation initiated approximately 12,000 years ago, enabling us to assess the probable pattern of ancestral plasticity in the colonists that gave rise to freshwater derivatives. The value of the radiation for this purpose is further enhanced by the existence of extensive parallelism in freshwater derivatives, such that similar, locally adapted forms (*ecotypes*) have evolved repeatedly and independently in similar habitats. We can thus ask whether patterns of plasticity have evolved similarly and repeatedly in similar contexts or whether, instead, alternative solutions to the same environmental problem are possible. Finally, the emergence of this radiation as a new model system for evolutionary developmental genetics (Foster and Baker, 2004; Gibson, 2005; Cresko et al., 2007; Kingsley and Peichel, 2007) has resulted in the development of a diversity of molecular tools that will ultimately permit us to elucidate the genomic bases of genetic accommodation and assimilation.

VOCABULARY • *The Evolution of Plasticity*

In those organisms that can respond, phenotypic plasticity occurs when the environment influences the expression of phenotype. This phenomenon is likely widespread (Gilbert and Bolker, 2003; Sultan and Stearns, 2005) and is often expressed as a significant gene × environment interaction in a quantitative genetic framework (Scheiner, 1993; Via, 1993; Windig et al., 2004). Environmental influences can initiate what become irreversible phenotypic changes, due to modification of developmental programs or relatively short-term physiological and behavioral changes that are reversible within the life span of an organism (e.g., Pigliucci, 2001). Although some emphasize this distinction, we, like others (e.g., Carroll and Corneli, 1999; Sultan and Stearns, 2005), have elected not to do so because the results we present offer evidence that the distinction is not clear—and that the evolution of plasticity can be similarly described and evaluated regardless of the reversibility of the influence of environment.

Phenotypic plasticity can be effectively expressed in the form of a *norm of reaction* (Figure 18.1A). Originally defined as the differential response of a single genotype (or individual) to different environments (Woltereck, 1909), the concept can be extended to a comparison of the influences of different environments upon families, populations, or species. In the simplest case, behavior of an individual could be tested once in two different environments, offering insight into differential effects of environment on individual behavior. To understand whether the influence of environment on individual behavior is meaningful, repeated measures on the same individual (repeatability estimates) are necessary (e.g., Boake, 1989; Brodie and Russell, 1999). To understand whether families, populations, or species differ in patterns of plasticity, single measures of individuals in each environment can be used to determine the average reaction norms for the groups of interest. Of course, in instances in which environmental induction of a phenotype is nonreversible, matched sets of individuals must be reared in alternative environments for comparison. If the evolutionary relationships among the populations or species are known, comparisons among appropriate groups can be used to assess evolutionary transformations in norms of reaction.

In an interesting and subtle paradigm shift, populations are increasingly considered to exhibit plasticity in most aspects of phenotype (Gilbert and Bolker, 2003; Sultan and Stearns, 2005). Thus, recent models exploring the evolution of phenotypic plasticity most often consider the loss of plasticity or transitions in the pattern of plasticity. Loss of plasticity, also known as "genetic assimilation" (Figure 18.1B), occurs when a population that invades a new environment is subject to strong directional selection, resulting in the constitutive expression of a favored trait value that once required environmental induction (Crispo, 2007; Ghalambor et al., 2007). Although initially discounted as an unlikely or unimportant process (e.g., Simpson, 1953; de Jong, 2005; reviewed in Pigliucci and Murren, 2003), the feasibility of the process is now widely accepted as long as the requisite variation underlying ancestral plasticity exists in the population or sufficient time elapses to permit mutation to provide the requisite variation (Pigliucci and Murren, 2003; Price et al., 2003; West-Eberhard, 2003; Crispo, 2007; Ghalambor et al., 2007).

Genetic accommodation (*sensu* West Eberhard, 2003) is the more general process in which an initial, ancestral pattern of plasticity evolves, typically in response to environmental change or invasion of a new environment. Again, this can be most simply depicted as a shift in the norm of reaction of a population resulting in a change in slope (Figure 18.2 top) or elevation of the reaction norm (Figure 18.2 bottom). In a

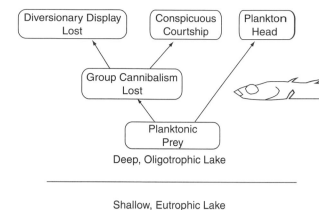

Diversionary Display Lost ← Group Cannibalism Lost → Conspicuous Courtship

Plankton Head

Group Cannibalism Lost ← Planktonic Prey

Deep, Oligotrophic Lake

Shallow, Eutrophic Lake

Benthic Prey → Group Cannibalism Persists

Group Cannibalism Persists → Diversionary Display Persists, Inconspicuous Courtship, Benthos Head

FIGURE 18.5 The benthic–limnetic differentiation of lacustrine stickleback. Persistence and loss of features are inferred with reference to oceanic fish likely to reflect phenotypes ancestral to the postglacial freshwater radiation.

regions of marine, brackish, or freshwater habitats, where they establish territories and build tubular nests of organic material that is collected from the substratum and glued together with kidney secretions. During this period males may assume the mosaic coloration usually described as typical of a courting male, although in oceanic and benthic populations drab coloration may be retained until well into the courtship phase of reproduction or even until the onset of parental care (Hulslander, 2003). Once the nest is completed, males begin courting receptive females, which are distinguishable by their distended abdomens, solitary behavior, and characteristic head-up postures when soliciting males. If the female responds positively to the male, he leads her to his nest, inserts his snout in the entrance, and turns on his side, "showing" the nest entrance. The female may enter the nest, and if she does, will usually spawn. She releases all of her eggs and returns to foraging and production of a new clutch. Males spawn multiple times, usually making the transition to parental behavior in 24–36 hours. They provide all subsequent

care of young for a 7–14 day peri○○○●
are free-swimming and leave the
male.

The courtship behavior of m ⌐
stickleback has typically been d○ ⌐
corporating primarily the prom i ▬ ▬
typical" zigzag dance, in which t ⌐▬
rapidly toward an approaching fe x ⌐
ward progress interrupted by ▬
"jumps" to the left and right (t○ ▬ ▬
Tinbergen, 1937; Rowland, 199 ▲
This very conspicuous behavior i ▬
nently displayed by males in li ▬ ▬ ▬
tions, in which cannibalistic g ⌐
form. In limnetic populations, t ⌐ ▬
nearly ubiquitous component of ○
ure 18.6), though on occasion a ⌐
stead swim directly to a nearby f○ ▬
her on the flank, then lead her to ⌐

In contrast, in nature, the r ⌐ ▬
form of courtship in benthic pop ⌐ ●
sal pricking. Typically initiated ▼
approaches a territorial male, she ⌐
self above him and presses her a ▬ ▬

Atchley, W. R., A. A. Plummer, and B. Riska. 1985. Genetics of mandible form in the mouse. *Genetics* 111:555–77.

Atchley, W. R., S. Z. Xu, and C. Vogl. 1994. Developmental quantitative genetic models of evolutionary change. *Dev Genet* 15:92–103.

Atchley, W. R., and J. Zhu. 1997. Developmental quantitative genetics, conditional epigenetic variability and growth in mice. *Genetics* 147:765–76.

Badyaev, A. V., and K. R. Foresman. 2004. Evolution of morphological integration. I. Functional units channel stress-induced variation in shrew mandibles. *Am Nat* 163:868–79.

Badyaev, A. V., K. R. Foresman, and R. L. Young. 2005. Evolution of morphological integration: Developmental accommodation of stress-induced variation. *Am Nat* 166:382–95.

Bailey, D. W. 1985. Genes that affect the shape of the murine mandible: Congenic strain analysis. *J Hered* 76:107–14.

Bailey, D. W. 1986. Genes that affect morphogenesis of the murine mandible: Recombinant-inbred strain analysis. *J Hered* 77:17–25.

Berg, R. L. 1960. The ecological significance of correlation pleiades. *Evolution* 14:171–80.

Beverdam, A., G. R. Merlo, L. Paleari, S. Mantero, F. Genova, O. Barbieri, P. Janvier, and G. Levi. 2002. Jaw transformation with gain of symmetry after *Dlx5/Dlx6* inactivation: Mirror of the past? *Genesis* 34:221–7.

Bock, W. J. 1960. A critique of morphological integration. *Evolution* 14:130–2.

Buettner-Janusch, J. 1959. Morphological integration—Olson and Miller. *Am Anthropol* 61:918–19.

Busser, B. W., M. L. Bulyk, and A. M. Michelson. 2008. Toward a systems-level understanding of developmental regulatory networks. *Curr Opin Genet Dev* 18:521–9.

Carroll, S. B., J. K. Grenier, and S. D. Weatherbee. 2001. *From DNA to Diversity.* Malden, MA: Blackwell Science.

Carter, D. R., and T. E. Orr. 1992. Skeletal development and bone functional adaptation. *J Bone Miner Res* 7:S389–95.

Caton, J., and A. S. Tucker. 2009. Current knowledge of tooth development: Patterning and mineralization of the murine dentition. *J Anat* 214:502–15.

Chai, Y., Y. Ito, and J. Han. 2003. TGF-beta signaling and its functional significance in regulating the fate of cranial neural crest cells. *Crit Rev Oral Biol Med* 14:78–88.

Cheverud, J. M. 1982. Phenotypic, genetic, and environmental morphological integration in the cranium. *Evolution* 36:499–516.

Cheverud, J. M. 1984. Quantitative genetics and developmental constraints on evolution by selection. *J Theor Biol* 110:155–71.

Cheverud, J. M. 1988. A comparison of genetic and phenotypic correlations. *Evolution* 42:958–68.

Cheverud, J. M. 1995. Morphological integration in the saddle-back tamarin (*Saguinus fuscicollis*) cranium. *Am Nat* 145:63–89.

Cheverud, J. M. 1996a. Developmental integration and the evolution of pleiotropy. *Am Zool* 36:44–50.

Cheverud, J. M. 1996b. Quantitative genetic analysis of cranial morphology in the cotton-top (*Saguinus oedipus*) and saddle-back (*S. fuscicollis*) tamarins. *J Evol Biol* 9:5–42.

Cheverud, J. M. 2004. Modular pleiotropic effects of quantitative trait loci on morphological traits. In *Modularity in Development and Evolution*, ed. G. Schlosser and G. P. Wagner, 132–53. Chicago: University of Chicago Press.

Cheverud, J. M., T. H. Ehrich, T. T. Vaughn, S. F. Koreishi, R. B. Linsey, and L. S. Pletscher. 2004. Pleiotropic effects on mandibular morphology II: Differential epistasis and genetic variation in morphological integration. *J Exp Zool B Mol Dev Evol* 302:424–35.

Cheverud, J. M., S. E. Hartman, J. T. Richtsmeier, and W. R. Atchley. 1991. A quantitative genetic analysis of localized morphology in mandibles of inbred mice using finite-element scaling analysis. *J Craniofac Genet Dev Biol* 11:122–37.

Cheverud, J. M., E. J. Routman, and D. J. Irschick. 1997. Pleiotropic effects of individual gene loci on mandibular morphology. *Evolution* 51:2006–16.

Clouthier, D. E., K. Hosoda, J. A. Richardson, S. C. Williams, H. Yanagisawa, T. Kuwaki, M. Kumada, R. E. Hammer, and M. Yanagisawa. 1998. Cranial and cardiac neural crest defects in endothelin-A receptor-deficient mice. *Development* 125:813–24.

Cobourne, M. T., and P. T. Sharpe. 2003. Tooth and jaw: Molecular mechanisms of patterning in the first branchial arch. *Arch Oral Biol* 48:1–14.

Collar, D. C., and P. C. Wainwright. 2006. Discordance between morphological and mechanical diversity in the feeding mechanism of centrarchid fishes. *Evolution* 60:2575–84.

Cowley, D. E., and W. R. Atchley. 1990. Development and quantitative genetics of correlation structure among body parts of *Drosophila melanogaster*. *Am Nat* 135:242–68.

Cowley, D. E., and W. R. Atchley. 1992. Quantitative genetic models for development, epigenetic selection, and phenotypic evolution. *Evolution* 46:495–518.

Davidson, E.H., and M.S. Levine. 2008. Properties of developmental gene regulatory networks. *Proc Natl Acad Sci USA* 105:20063–6.

Depew, M.J., and C. Compagnucci. 2008. Tweaking the hinge and caps: Testing a model of the organization of jaws. *J Exp Zool B Mol Dev Evol* 310:315–35.

Depew, M.J., J.K. Liu, J.E. Long, R. Presley, J.J. Meneses, R.A. Pedersen, and J.L. R. Rubenstein. 1999. *Dlx5* regulates regional development of the branchial arches and sensory capsules. *Development* 126:3831–46.

Depew, M.J., T. Lufkin, and J.L. R. Rubenstein. 2002. Specification of jaw subdivisions by *Dlx* genes. *Science* 298:381–5.

Depew, M.J., C.A. Simpson, M. Morasso, and J.L. R. Rubenstein. 2005. Reassessing the Dlx code: The genetic regulation of branchial arch skeletal pattern and development. *J Anat* 207:501–61.

Dobreva, G., M. Chahrour, M. Dautzenberg, L. Chirivella, B. Kanzler, I. Farinas, G. Karsenty, and R. Grosschedl. 2006. SATB2 is a multifunctional determinant of craniofacial patterning and osteoblast differentiation. *Cell* 125:971–86.

Dumont, E.R., and A. Herrel. 2003. The effects of gape angle and bite point on bite force in bats. *J Exp Biol* 206:2117–23.

Ehrich, T.H., T.T. Vaughn, S. Koreishi, R.B. Linsey, L.S. Pletscher, and J.M. Cheverud. 2003. Pleiotropic effects on mandibular morphology I. Developmental morphological integration and differential dominance. *J Exp Zool B Mol Dev Evol* 296:58–79.

Emerson, S.B., J. Travis, and M.A. R. Koehl. 1990. Functional complexes and additivity in performance—a test case with flying frogs. *Evolution* 44:2153–7.

Erickson, G.M., A.K. Lappin, and K.A. Vliet. 2003. The ontogeny of bite-force performance in American alligator (*Alligator mississippiensis*). *J Zool* 260:317–27.

Forwood, M.R., and C.H. Turner. 1995. Skeletal adaptations to mechanical usage—results from tibial loading studies in rats. *Bone* 17:S197–205.

Frazzetta, T.H. 1975. *Complex Adaptations in Evolving Populations*. Sunderland, MA: Sinauer.

Fukuhara, S., Y. Kurihara, Y. Arima, N. Yamada, and H. Kurihara. 2004. Temporal requirement of signaling cascade involving endothelin-1 endothelin receptor type A in branchial arch development. *Mech Dev* 121:1223–33.

Gerritsen, E.J. A., J.M. Vossen, I.H. G. Vanloo, J. Hermans, M.H. Helfrich, C. Griscelli, and A. Fischer. 1994. Autosomal recessive osteopetrosis—variability of findings at diagnosis and during the natural course. *Pediatrics* 93:247–53.

Gonzalez-Jose, R., S. Van der Molen, E. Gonzalez-Perez, and M. Hernandez. 2004. Patterns of phenotypic covariation and correlation in modern humans as viewed from morphological integration. *Am J Phys Anthropol* 123:69–77.

Goswami, A. 2006. Cranial modularity shifts during mammalian evolution. *Am Nat* 168:270–80.

Goswami, A. 2007. Cranial modularity and sequence heterochrony in mammals. *Evol Dev* 9:290–8.

Gould, S.J., and R.A. Garwood. 1969. Levels of integration in mammalian dentitions: An analysis of correlations in *Nesophantes micrus* (Insectivora) and *Oryzomys couesi* (Rodentia). *Evolution* 23:276–300.

Gruneberg, H. 1936. Grey-lethal, a new mutation in the house mouse. *J Hered* 27:105–9.

Habib, H., T. Hatta, J. Udagawa, L. Zhang, Y. Yoshimura, and H. Otani. 2005. Fetal jaw movement affects condylar cartilage development. *J Dent Res* 84:474–9.

Hall, B.K. 2003. Unlocking the black box between genotype and phenotype: Cell condensations as morphogenetic (modular) units. *Biol Philos* 18:219–47.

Hall, B.K., and S.W. Herring. 1990. Paralysis and growth of the musculoskeletal system in the embryonic chick. *J Morphol* 206:45–56.

Hallgrimsson, B., H. Jamniczky, N.M. Young, C. Rolian, T.E. Parsons, J.C. Boughner, and R.S. Marcucio. 2009. Deciphering the palimpsest: Studying the relationship between morphological integration and phenotypic covariation. *Evol Biol* 36:355–76.

Hallgrímsson, B., D.E. Lieberman, W. Liu, A.F. Ford-Hutchinson, and F.R. Jirik. 2007a. Epigenetic interactions and the structure of phenotypic variation in the cranium. *Evol Dev* 9:76–91.

Hallgrímsson, B., D.E. Lieberman, N.M. Young, T. Parsons, and S. Wat. 2007b. Evolution of covariance in the mammalian skull. In *Novartis Foundation Symposium*, ed. G. Bock and J. Goode, 164–85, discussion 185–90. New York: John Wiley & Sons.

Haworth, K.E., C. Healy, P. Morgan, and P.T. Sharpe. 2004. Regionalisation of early head ectoderm is regulated by endoderm and prepatterns the orofacial epithelium. *Development* 131:4797–806.

Heaney, C., H. Shalev, K. Elbedour, R. Carmi, J.B. Staack, V.C. Sheffield, and D.R. Beier. 1998. Human autosomal recessive osteopetrosis maps to 11q13, a position predicted by comparative mapping of the murine osteosclerosis (oc) mutation. *Hum Mol Genet* 7:1407–10.

Herrel, A., R. Joachim, B. Vanhooydonck, and D.J. Irschick. 2006. Ecological consequences of ontogenetic changes in head shape and bite performance in the Jamaican lizard *Anolis lineatopus*. *Biol J Linn Soc* 89:443–54.

Herrel, A., and J.C. O'Reilly. 2004. Ontogenetic scaling of bite force in lizards and turtles. Presented at the Symposium on the Ontogeny of Performance in Vertebrates, 7th International Congress of Vertebrate Morphology, Boca Raton, FL, 31–42.

Herrera, C.M. 2001. Deconstructing a floral phenotype: Do pollinators select for corolla integration in *Lavandula latifolia*? *J Evol Biol* 14:574–84.

Herrera, C.M., X. Cerda, M.B. Garcia, J. Guitian, M. Medrano, P.J. Rey, and A.M. Sanchez-Lafuente. 2002. Floral integration, phenotypic covariance structure and pollinator variation in bumblebee-pollinated *Helleborus foetidus*. *J Evol Biol* 15:108–21.

Herring, S.W. 1994. Development of functional interactions between skeletal and muscular systems. In *Bone*. Vol. 9 of *Differentiation and Morphogenesis of Bone*, ed. B.K. Hall, 165–91. Boca Raton: CRC Press.

Herring, S.W., and S.E. Herring. 1974. Superficial masseter and gape in mammals. *Am Nat* 108:561–76.

Herring, S.W., and T.C. Lakars. 1981. Craniofacial development in the absence of muscle contraction. *J Craniofac Genet Dev Biol* 1:341–57.

Herring, S.W., S.C. Pedersen, and X.F. Huang. 2005. Ontogeny of bone strain: the zygomatic arch in pigs. *J Exp Biol* 208:4509–21.

Howells, W.W. 1958. Morphological integration— Olson, E.C., Miller, R.C. *Am J Phys Anthropol* 16:371–3.

Hsu, S.C., B. Noamani, D.E. Abernethy, H. Zhu, G. Levi, and A.J. Bendall. 2006. Dlx5- and Dlx6-mediated chondrogenesis: Differential domain requirements for a conserved function. *Mech Dev* 123:819–30.

Huang, X.F., G.X. Zhang, and S.W. Herring. 1994. Age changes in mastication in the pig. *Comp Biochem Physiol A Physiol* 107:647–54.

Imbrie, J. 1958. Morphological integration. *Evolution* 12:558–9.

Jones, D.C., M.L. Zelditch, P.L. Peake, and R.Z. German. 2007. The effects of muscular dystrophy on the craniofacial shape of *Mus musculus*. *J Anat* 210:723–30.

Judex, S., T.S. Gross, R.C. Bray, and R.F. Zernicke. 1997. Adaptation of bone to physiological stimuli. *J Biomech* 30:421–9.

Kesavan, C., S. Mohan, A.K. Srivastava, S. Kapoor, J.E. Wergedal, H.R. Yu, and D.J. Baylink. 2006. Identification of genetic loci that regulate bone adaptive response to mechanical loading in C57BL/6J and C3H/HeJ mice intercross. *Bone* 39:634–43.

Kingsolver, J.G., and D.C. Wiernasz. 1991. Development, function, and the quantitative genetics of wing melanin pattern in *Pieris* butterflies. *Evolution* 45:1480–92.

Klingenberg, C.P. 2004. Integration, modules, and development: Molecules to morphology to evolution. In *Phenotypic Integration: Studying the Ecology and Evolution of Complex Phenotypes*, ed. M. Pigliucci and K. Preston, 213–30. New York: Oxford University Press.

Klingenberg, C.P. 2005. Developmental constraints, modules, and evolvability. In *Variation: A Central Concept in Biology*, ed. B. Hallgrímsson and B.K. Hall, 219–47. San Diego: Elsevier Academic Press.

Klingenberg, C.P. 2008. Morphological integration and developmental modularity. *Annu Rev Ecol Evol Syst* 39:115–32.

Klingenberg, C.P. 2009. Morphometric integration and modularity in configurations of landmarks: Tools for evaluating a priori hypotheses. *Evol Dev* 11:405–21.

Klingenberg, C.P., A.V. Badyaev, S.M. Sowry, and N.J. Beckwith. 2001a. Inferring developmental modularity from morphological integration: Analysis of individual variation and asymmetry in bumblebee wings. *Am Nat* 157:11–23.

Klingenberg, C.P., L.J. Leamy, E.J. Routman, and J.M. Cheverud. 2001b. Genetic architecture of mandible shape in mice: Effects of quantitative trait loci analyzed by geometric morphometrics. *Genetics* 157:785–802.

Klingenberg, C.P., K. Mebus, and J.C. Auffray. 2003. Developmental integration in a complex morphological structure: How distinct are the modules in the mouse mandible? *Evol Dev* 5:522–31.

Koehl, M.A.R. 1996. When does morphology matter? *Annu Rev Ecol Syst* 27:501–42.

Lakars, T.C., and S.W. Herring. 1980. Ontogeny of oral function in hamsters (*Mesocricetus auratus*). *J Morphol* 165:237–54.

LaMothe, J.M., N.H. Hamilton, and R.F. Zernicke. 2005. Strain rate influences periosteal adaptation in mature bone. *Med Eng Phys* 27:277–84.

Lande, R. 1980. The genetic covariance between characters maintained by pleiotropic mutations. *Genetics* 94:203–15.

Langenbach, G.E.J., P. Brugman, and W.A. Weijs. 1992. Preweaning feeding mechanisms in the rabbit. *J Dev Physiol* 18:253–61.

Langenbach, G.E.J., W.A. Weijs, P. Brugman, and T. van Eijden. 2001. A longitudinal electromyographic study of the postnatal maturation of mastication in the rabbit. *Arch Oral Biol* 46:811–20.

Lanyon, L.E. 1980. The influence of function on the development of bone curvature—an experimental study on the rat tibia. *J Zool* 192:457–66.

Leamy, L. 1975. Component analysis of osteometric traits in random-bred house mice. *Syst Zool* 24:176–90.

Lieberman, D. E., and B. K. Hall. 2007. The evolutionary developmental biology of tinkering: An introduction to the challenge. In *Novartis Foundation Symposium*, ed. G. Bock and J. Goode, 1–19. New York: John Wiley & Sons.

Liu, Y., P. Cserjesi, A. Nifuji, E. N. Olson, and M. Noda. 1996. Sclerotome-related helix–loop–helix type transcription factor (scleraxis) mRNA is expressed in osteoblasts and its level is enhanced by type-beta transforming growth factor. *J Endocrinol* 151:491–9.

Long, C. A., and J. Captain. 1977. Investigations on sciurid manus. 2. Analysis of functional complexes by morphological integration and coefficients of belonging. *Int J Mammal Biol* 42: 214–21.

Long, C. A., and T. Frank. 1968. Morphometric variation and function in baculum with comments on correlation of parts. *J Mammal* 49:32–43.

Long, C. A., and P. Kamensky. 1967. Osteometric variation and function of high-speed wing of free-tailed bat. *Am Midl Nat* 77:452–461.

Magwene, P. M. 2008. Using correlation proximity graphs to study phenotypic integration. *Evol Biol* 35:191–8.

Marquez, E. J. 2008. A statistical framework for testing modularity in multidimensional data. *Evolution* 62:2688–708.

Mezey, J. G., J. M. Cheverud, and G. P. Wagner. 2000. Is the genotype–phenotype map modular? A statistical approach using mouse quantitative trait loci data. *Genetics* 156:305–11.

Mina, M. 2001. Morphogenesis of the medial region of the developing mandible is regulated by multiple signaling pathways. *Cells Tissues Organs* 169:295–301.

Mina, M., Y. H. Wang, A. M. Ivanisevic, W. B. Upholt, and B. Rodgers. 2002. Region- and stage-specific effects of FGFs and BMPs in chick mandibular morphogenesis. *Dev Dyn* 223:333–52.

Mitteroecker, P., and F. Bookstein. 2007. The conceptual and statistical relationship between modularity and morphological integration. *Syst Biol* 56:818–36.

Monteiro, L. R., V. Bonato, and S. F. dos Reis. 2005. Evolutionary integration and morphological diversification in complex morphological structures: Mandible shape divergence in spiny rats (Rodentia, Echimyidae). *Evol Dev* 7:429–39.

Monteiro, L. R., and M. R. Nogueira. 2010. Adaptive radiations, ecological specialization, and the evolutionary integration of complex morphological structures. *Evolution* 64:724–44.

Moses, H. L., and R. Serra. 1996. Regulation of differentiation by TGF-beta. *Curr Opin Genet Dev* 6:581–6.

Murchison, N. D., B. A. Price, D. A. Conner, D. R. Keene, E. N. Olson, C. J. Tabin, and R. Schweitzer. 2007. Regulation of tendon differentiation by scleraxis distinguishes force-transmitting tendons from muscle-anchoring tendons. *Development* 134:2697–708.

Oka, K., S. Oka, T. Sasaki, Y. Ito, P. Bringas, K. Nonaka, and Y. Chai. 2007. The role of TGF-beta signaling in regulating chondrogenesis and osteogenesis during mandibular development. *Dev Biol* 303:391–404.

Oliveri, P., Q. Tu, and E. H. Davidson. 2008. Global regulatory logic for specification of an embryonic cell lineage. *Proc Natl Acad Sci USA* 105: 5955–62.

Olson, E. C., and R. L. Miller. 1958. *Morphological Integration*. Chicago: University of Chicago Press.

Ozeki, H., Y. Kurihara, K. Tonami, S. Watatani, and H. Kurihara. 2004. Endothelin-1 regulates the dorsoventral branchial arch patterning in mice. *Mech Dev* 121:387–95.

Reeves, G. M., B. R. McCreadie, S. Chen, A. T. Galecki, D. T. Burke, R. A. Miller, and S. A. Goldstein. 2007. Quantitative trait loci modulate vertebral morphology and mechanical properties in a population of 18-month-old genetically heterogeneous mice. *Bone* 40:433–43.

Reidl, R. 1977. A systems-analytical approach to macroevolutionary phenomena. *Q Rev Biol* 52: 351–70.

Reidl, R. 1978. *Order in Living Organisms*. New York: John Wiley & Sons.

Richtsmeier, J. T., S. R. Lele, and T. M. I. Cole. 2005. Landmark morphometrics and the analysis of variation. In *Variation: A Central Concept in Biology*, ed. B. Hallgrimsson and B. K. Hall, 49–69. Amsterdam: Elsevier.

Rivera-Perez, J. A., M. Wakamiya, and R. R. Behringer. 1999. Goosecoid acts cell autonomously in mesenchyme-derived tissues during craniofacial development. *Development* 126:3811–21.

Robling, A. G., S. J. Warden, K. L. Shultz, W. G. Beamer, and C. H. Turner. 2007. Genetic effects on bone mechanotransduction in congenic mice harboring bone size and strength quantitative trait loci. *J Bone Miner Res* 22:984–91.

Rollian, C., and K. E. Willmore. 2009. Morphological integration at 50: Patterns and processes of integration in biological anthropology. *Evol Biol* 36:1–4.

Roseman, C. C., J. P. Kenney-Hunt, and J. M. Cheverud. 2009. Phenotypic integration without modularity: Testing hypotheses about the distribution of pleiotropic quantitative trait loci in a continuous space. *Evol Biol* 36:282–91.

Rot-Nikcevic, I., K. J. Downing, B. K. Hall, and B. Kablar. 2007. Development of the mouse mandibles and clavicles in the absence of skeletal myogenesis. *Histol Histopathol* 22:51–60.

Rot-Nikcevic, I., T. Reddy, K. J. Downing, A. C. Belliveau, B. Hallgrimsson, B. K. Hall, and B. Kablar. 2006. Myf5$^{-/-}$:MyoD$^{-/-}$ amyogenic fetuses reveal the importance of early contraction and static loading by striated muscle in mouse skeletogenesis. *Dev Genes Evol* 216:1–9.

Ruest, L. B., and D. E. Clouthier. 2009. Elucidating timing and function of endothelin-A receptor signaling during craniofacial development using neural crest cell-specific gene deletion and receptor antagonism. *Dev Biol* 328:94–108.

Ruest, L. B., R. Kedzierski, M. Yanagisawa, and D. E. Clouthier. 2005. Deletion of the endothelin-A receptor gene within the developing mandible. *Cell Tissue Res* 319:447–53.

Ruest, L. B., M. L. Xiang, K. C. Lim, G. Levi, and D. E. Clouthier. 2004. Endothelin-A receptor–dependent and –independent signaling pathways in establishing mandibular identity. *Development* 131:4413–23.

Sato, T., Y. Kurihara, R. Asai, Y. Kawamura, K. Tonami, Y. Uchijima, E. Heude, M. Ekker, G. Levi, and H. Kurihara. 2008. An endothelin-1 switch specifies maxillomandibular identity. *Proc Natl Acad Sci USA* 105:18806–11.

Schmalhausen, I. I. 1949. *Factors of Evolution: The Theory of Stabilizing Selection. Philadelphia:* Blakeston.

Shefelbine, S. J., and D. R. Carter. 2004. Mechanobiological predictions of femoral anteversion in cerebral palsy. *Ann Biomed Eng* 32:297–305.

Simpson, G. G. 1958. Book Review of *Morphological Integration* by R. L. Miller and G. G Simpson. *Science* 128(3316):138.

Spencer, M. A. 1998. Force production in the primate masticatory system: Electromyographic tests of biomechanical hypotheses. *J Hum Evol* 34:25–54.

Srivastava, A. K., S. Kapur, S. Mohan, H. R. Yu, S. Kapur, J. Wergedal, and D. J. Baylink. 2005. Identification of novel genetic loci for bone size and mechanosensitivity in an ENU mutant exhibiting decreased bone size. *J Bone Miner Res* 20:1041–50.

Taylor, A. B., and C. J. Vinyard. 2009. Jaw-muscle fiber architecture in tufted capuchins favors generating relatively large muscle forces without compromising jaw gape. *J Hum Evol* 57:710–20.

Terentjev, P. 1931. Biometrical tests on the morphological characteristics of *Rana ridibunda* pall (Amphibia, Salientia). *Biometrika* 23:23–51.

Thompson, E. N., A. R. Biknevicius, and R. Z. German. 2003. Ontogeny of feeding function in the gray short-tailed opossum *Monodelphis domestica*: Empirical support for the constrained model of jaw biomechanics. *J Exp Biol* 206: 923–32.

Trumpp, A., M. J. Depew, J. L. R. Rubenstein, J. M. Bishop, and G. R. Martin. 1999. Cre-mediated gene inactivation demonstrates that FGF8 is required for cell survival and patterning of the first branchial arch. *Genes Dev* 13:3136–48.

Tucker, A., and P. Sharpe. 2004. The cutting-edge of mammalian development: How the embryo makes teeth. *Nat Rev Genet* 5:499–508.

Tucker, A. S., G. Yamada, M. Grigoriou, V. Pachnis, and P. T. Sharpe. 1999. Fgf-8 determines rostral-caudal polarity in the first branchial arch. *Development* 126:51–61.

Turner, C. H. 1998. Three rules for bone adaptation to mechanical stimuli. *Bone* 23:399–407.

van Limborgh, J. 1970. New view on control of morphogenesis of the skull. *Acta Morphol Neerland Scand* 8:143–60.

van Limborgh, J. 1972. Role of genetic and local environmental factors in the control of postnatal craniofacial morphogenesis. *Acta Morphol Neerland Scand* 10:37–47.

Van Valen, L. 1962. Growth fields in dentition of *Peromyscus*. *Evolution* 16:272–7.

Van Valen, L. 1965. The study of morphological integration. *Evolution* 19:347–9.

Wagner, G. P. 1996. Homologues, natural kinds and the evolution of modularity. *Am Zool* 36:36–43.

Wagner, G. P., and L. Altenberg. 1996. Complex adaptations and the evolution of evolvability. *Evolution* 50:967–76.

Wagner, G. P., and J. G. Mezey. 2004. The role of genetic architecture constraints in the origin of variational modularity. In *Modularity in Development and Evolution*, ed. G. Schlosser and G. P. Wagner, 338–58. Chicago: University of Chicago Press.

Wagner, G. P., J. G. Mezey, and R. Calabretta. 2005. Natural selection and the origin of modules. In *Modularity: Understanding the Development and Evolution of Natural Complex Systems*, ed. W. Callebaut and D. Rasskin-Gutman, 33–50. Cambridge, MA: MIT Press.

Wagner, G. P., M. Pavlicev, and J. M. Cheverud. 2007. The road to modularity. *Nat Rev Genet* 8:921–31.

Wainwright, P. C. 2007. Functional versus morphological diversity in macroevolution. *Annu Rev Ecol Evol Syst* 38:381–401.

Williams, S. H., E. Peiffer, and S. Ford. 2009. Gape and bite force in the rodents *Onychomys leucogaster* and *Peromyscus maniculatus*: Does jaw-muscle anatomy predict performance? *J Morphol* 270: 338–47.

Xing, W. R., D. Baylink, C. Kesavan, Y. Hu, S. Kapoor, R. B. Chadwick, and S. Mohan. 2005. Global gene expression analysis in the bones reveals involvement of several novel genes and pathways in mediating an anabolic response of mechanical loading in mice. *J Cell Biochem* 96: 1049–60.

Yamada, G., A. Mansouri, M. Torres, E. T. Stuart, M. Blum, M. Schultz, E. M. Derobertis, and P. Gruss. 1995. Targeted mutation of the murine Goosecoid gene results in craniofacial defects and neonatal death. *Development* 121:2917–22.

Young, R. L., T. S. Haselkorn, and A. V. Badyaev. 2007. Functional equivalence of morphologies enables morphological and ecological diversity. *Evolution* 61:2480–92.

Zelditch, M. L., A. R. Wood, R. M. Bonett, and D. L. Swiderski. 2008. Modularity of the rodent mandible: Integrating bones, muscles and teeth. *Evol Dev* 10:756–68.

Zelditch, M. L., A. R. Wood, and D. L. Swiderski. 2009. Building developmental integration into functional systems: Function-induced integration of mandibular shape. *Evol Biol* 36:71–87.

Zhong, N., R. A. Garman, M. E. Squire, L. R. Donahue, C. T. Rubin, M. Hadjiargyrou, and S. Judex. 2005. Gene expression patterns in bone after 4 days of hind-limb unloading in two inbred strains of mice. *Aviat Space Environ Med* 76: 530–35.

18

Epigenetic Contributions to Adaptive Radiation

INSIGHTS FROM THREESPINE STICKLEBACK

Susan A. Foster and Matthew A. Wund

CONTENTS

Phenotypic plasticity is variation in trait expression caused by influences of the environment on the expression of the phenotype. Plasticity can buffer organisms against the exigencies of environmental variation, enhancing fitness and facilitating the persistence of populations in novel environments (e.g., Baldwin, 1902; Schlichting and Pigliucci, 1998). In many con-

texts, phenotypic plasticity is unquestionably adaptive; and like other aspects of phenotype, plasticity can evolve (Scheiner, 1993; Schlichting and Pigliucci, 1998; Pigliucci, 2005, for reviews). What is less clear is how phenotypic plasticity influences evolution. On the one hand, it could shield the genome from selection, slowing genetic responses to selection (e.g., Grant, 1977; Falconer, 1981; Behera and Nanjundiah, 1995; Ancel and Fontana, 2000; Huey et al., 2003; Price et al., 2003). Alternatively, plasticity could facilitate (Hinton and Nowlan, 1987; Behera and Nanjundiah, 1995; Ancel and Fontana, 2000; Price et al., 2003) or direct evolution by determining the range of phenotypes upon which selection can act (West-Eberhard, 2003, for review). Clearly, empirical research is necessary to evaluate these possibilities if we are to understand the role of phenotypic plasticity in the real world.

The idea that patterns of phenotypic plasticity could influence subsequent evolution in novel environments is not new. The essential arguments were first articulated more than a century ago (the "Baldwin effect") (Baldwin, 1896, 1902; Morgan, 1896; Osborn, 1896) and have subsequently been refined, most recently by West-Eberhard (2003). The argument runs as follows.

When members of a population encounter novel environments, the ability to produce variant phenotypes (phenotypic plasticity) can expose previously cryptic genetic variation that can enhance or hinder persistence. In those environments in which persistence is possible, the course of subsequent evolution can be influenced by the pattern of trait expression, which can direct the course of future evolution by determining where on the new adaptive landscape a population begins its journey (e.g., Kirkpatrick, 1982; Hinton and Nowlan, 1987; Fear and Price, 1998; Pal and Miklos, 1999; Price et al., 2003). A logical extension of this argument is that the pattern of phenotypic plasticity in populations ancestral to adaptive radiations can influence the overall pattern of diversification in the radiation (West-Eberhard, 2003; Shaw et al., 2007; Wund et al., 2008).

As is so often the case, theoretical treatments outpace (and encourage) empirical research. Although a number of reviews (e.g., West-Eberhard, 1989, 1998, 2003, 2005; Pigliucci and Murren, 2003; Pigliucci, 2005, 2007) have presented examples compatible with the idea that phenotypic plasticity preceded genetic evolution, they offered few unequivocal examples in which the relationship between ancestral plasticity and that in daughter populations could be compared. Ghalambour et al. (2007) reviewed the exceptions to this generalization, cases in which ancestral patterns of plasticity can be compared to plasticity in derived populations. These examples, involving host shifts in soapberry bugs (*Jadera haematoloma*) (Carroll et al. 1997, 1998) and invasion of novel habitats by anadromous (sea run) Pacific salmon (*Oncorhynchus nerka*) and guppies (*Poecilia reticulata*), all offer evidence of initial adaptive plasticity in some traits that could enhance

persistence, followed by an evolutionary shift in the form of a reaction norm ("genetic accommodation") (West-Eberhard, 2003) or, more specifically, the constitutive expression of a previously plastic trait ("genetic assimilation") (reviewed in Crispo, 2007). A particularly elegant example demonstrated a role for genetic assimilation in the evolution of *Daphnia* populations recently exposed to novel predators (Scoville and Pfrender, 2010). Clearly, the difficulty is that we rarely have ancestral populations available for study that we can infer have changed little since giving rise to daughter populations, permitting comparison of ancestral and derived patterns of plasticity. Nevertheless, such comparisons are crucial if we are to understand the role of plasticity in evolutionary processes.

Here, we describe initial insights on the role of ancestral phenotypic plasticity in the generation of diversity in an adaptive radiation uniquely suited for the purpose: that of the threespine stickleback fish, *Gasterosteus aculeatus*. The unique feature of this radiation lies in the oceanic populations that have changed little morphologically over at least 12 million years and are unlikely to have changed in other features for the reasons described below. Thus, oceanic stickleback are likely to represent the ancestral condition relative to the postglacial freshwater radiation initiated approximately 12,000 years ago, enabling us to assess the probable pattern of ancestral plasticity in the colonists that gave rise to freshwater derivatives. The value of the radiation for this purpose is further enhanced by the existence of extensive parallelism in freshwater derivatives, such that similar, locally adapted forms (*ecotypes*) have evolved repeatedly and independently in similar habitats. We can thus ask whether patterns of plasticity have evolved similarly and repeatedly in similar contexts or whether, instead, alternative solutions to the same environmental problem are possible. Finally, the emergence of this radiation as a new model system for evolutionary developmental genetics (Foster and Baker, 2004; Gibson, 2005; Cresko et al., 2007; Kingsley and Peichel, 2007) has resulted in the development of a diversity of molecular tools that will ultimately permit us to elucidate the genomic bases of genetic accommodation and assimilation.

VOCABULARY • *The Evolution of Plasticity*

In those organisms that can respond, phenotypic plasticity occurs when the environment influences the expression of phenotype. This phenomenon is likely widespread (Gilbert and Bolker, 2003; Sultan and Stearns, 2005) and is often expressed as a significant gene × environment interaction in a quantitative genetic framework (Scheiner, 1993; Via, 1993; Windig et al., 2004). Environmental influences can initiate what become irreversible phenotypic changes, due to modification of developmental programs or relatively short-term physiological and behavioral changes that are reversible within the life span of an organism (e.g., Pigliucci, 2001). Although some emphasize this distinction, we, like others (e.g., Carroll and Corneli, 1999; Sultan and Stearns, 2005), have elected not to do so because the results we present offer evidence that the distinction is not clear—and that the evolution of plasticity can be similarly described and evaluated regardless of the reversibility of the influence of environment.

Phenotypic plasticity can be effectively expressed in the form of a *norm of reaction* (Figure 18.1A). Originally defined as the differential response of a single genotype (or individual) to different environments (Woltereck, 1909), the concept can be extended to a comparison of the influences of different environments upon families, populations, or species. In the simplest case, behavior of an individual could be tested once in two different environments, offering insight into differential effects of environment on individual behavior. To understand whether the influence of environment on individual behavior is meaningful, repeated measures on the same individual (repeatability estimates) are necessary (e.g., Boake, 1989; Brodie and Russell, 1999). To understand whether families, populations, or species differ in patterns of plasticity, single measures of individuals in each environment can be used to determine the average reaction norms for the groups of interest. Of course, in instances in which environmental induction of a phenotype is nonreversible, matched sets of individuals must be reared in alternative environments for comparison. If the evolutionary relationships among the populations or species are known, comparisons among appropriate groups can be used to assess evolutionary transformations in norms of reaction.

In an interesting and subtle paradigm shift, populations are increasingly considered to exhibit plasticity in most aspects of phenotype (Gilbert and Bolker, 2003; Sultan and Stearns, 2005). Thus, recent models exploring the evolution of phenotypic plasticity most often consider the loss of plasticity or transitions in the pattern of plasticity. Loss of plasticity, also known as "genetic assimilation" (Figure 18.1B), occurs when a population that invades a new environment is subject to strong directional selection, resulting in the constitutive expression of a favored trait value that once required environmental induction (Crispo, 2007; Ghalambor et al., 2007). Although initially discounted as an unlikely or unimportant process (e.g., Simpson, 1953; de Jong, 2005; reviewed in Pigliucci and Murren, 2003), the feasibility of the process is now widely accepted as long as the requisite variation underlying ancestral plasticity exists in the population or sufficient time elapses to permit mutation to provide the requisite variation (Pigliucci and Murren, 2003; Price et al., 2003; West-Eberhard, 2003; Crispo, 2007; Ghalambor et al., 2007).

Genetic accommodation (*sensu* West Eberhard, 2003) is the more general process in which an initial, ancestral pattern of plasticity evolves, typically in response to environmental change or invasion of a new environment. Again, this can be most simply depicted as a shift in the norm of reaction of a population resulting in a change in slope (Figure 18.2 top) or elevation of the reaction norm (Figure 18.2 bottom). In a

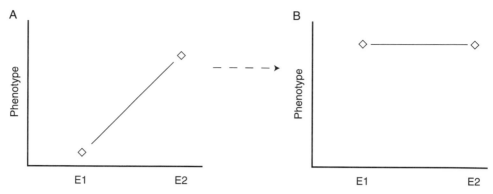

FIGURE 18.1 Genetic assimilation as represented by reaction norm diagrams. Consistent selection in environment 2 (E2) leads to a loss of ancestral plasticity when environment 1 (E1) is reimposed.

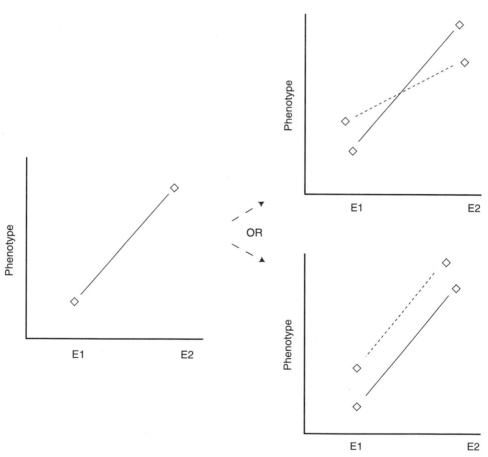

FIGURE 18.2 Genetic accommodation involves any evolutionary shift in the average reaction norm of a population. This can result in a shift in the mean phenotype overall environments (bottom, a change in reaction norm height) and/or a shift in the amount of plasticity expressed (top, a change in reaction norm slope).

broader conceptualization, genetic accommodation encompasses changes not only in the form of a plastic response but also in the timing of a plastic response, the threshold of environmental input required to elicit a plastic response, and/or the patterns of covariation among developing traits (West-Eberhard, 2003). Thus, genetic accommodation can be viewed as the adaptive evolution of Schlichting and Pigliucci's (1998) three-dimensional developmental norm of reaction, which describes the combined effects of the environment, ontogeny, and trait covariation (allometry) on phenotypic expression. Although reaction norms can be produced in many ways and can, for example, be cross-generational, iterated, and dynamic (Sultan and Stearns, 2005), for our purposes, the relatively straightforward explanation of genetic assimilation and genetic accommodation depicted in Figures 18.1 and 18.2 should suffice.

Because ancestral patterns of plasticity determine the range of phenotypes expressed in a population following environmental change or upon invasion of a novel environment, the pattern of ancestral plasticity has the potential to guide natural selection and potentially to influence the nature of the adaptations produced by selection. In revealing particular phenotypes in a novel environment, plasticity proscribes a population's starting point on the new adaptive landscape, and therefore toward which local fitness optimum the population can evolve (Figure 18.3). The relationship between the reaction norms expressed by the colonists and the shape of the landscape upon which they find themselves can lead to any of three distinct evolutionary outcomes. On the one hand, if plasticity is so highly adaptive that a population immediately finds itself at a fitness peak, then stabilizing selection ensues and no change in phenotype is expected (Price et al., 2003; Ghalambor et al., 2007). On the other hand are cases in which a population is challenged by an environment substantially different from that in which normal development occurs. Here, a trade-off can exist between the need to maintain

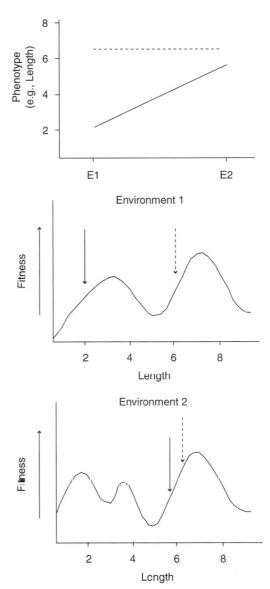

FIGURE 18.3 Plasticity can influence the outcome of evolution on novel adaptive landscapes. (Top) Reaction norms for length in two hypothetical populations. (Middle) Upon colonizing environment 1, the two populations express divergent phenotypes and, thus, would evolve toward different adaptive peaks, as indicated by solid and dashed arrows, corresponding to the solid and dashed reaction norms in the upper figure. (Bottom) In environment 2, both populations express the same phenotype and, therefore, evolve to the occupy the same adaptive peak. Thus, the environment dictates both the expressed phenotypes and the nature of selection on those phenotypes.

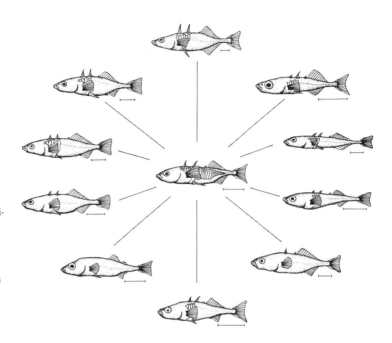

FIGURE 18.4 The adaptive radiation of the threespine stickleback. The central image represents the ancient oceanic lineage and the peripheral images, the freshwater derivatives that comprise the radiation. From Bell and Foster (1994), with permission.

homeostasis despite stress on the developmental system (i.e., canalization) and a need to respond to strong directional selection for the extreme phenotypes produced via plasticity (Ghalambor et al., 2007). The intermediate scenario, in which plasticity places a population near to (but not at) a local fitness peak, is likely the most common way for plasticity to both promote persistence under novel conditions as well as to result in genetic accommodation over time (Price et al., 2003; Ghalambor et al., 2007).

THE STICKLEBACK RADIATION

The threespine stickleback is a small fish (3–8 cm in length at breeding) that is widespread in holarctic oceanic and coastal freshwater habitats. Oceanic stickleback exhibit both anadromous and fully marine life histories (Baker, 1994). They have repeatedly given rise to resident freshwater forms in coastal areas that comprise a remarkable adaptive radiation (Figure 18.4). The most recent flush of colonization began at the onset of the last glacial recession about 12,000 years ago, giving rise to the postglacial adaptive radiation upon which we concentrate here (Bell and Foster, 1994; Malhi

et al., 2006). Oceanic stickleback are remarkably uniform morphologically over their very wide geographic range (Walker and Bell, 2000) and are unlikely to have changed in the last 12 million years (Bell, 1994). The large size of oceanic populations (Taylor and McPhail, 1999, 2000; Cresko, 2000; Hohenlohe et al., 2010) indicates that little genetic drift is likely to have occurred since the onset of postglacial colonization. Equally, the stability of the oceanic environment combined with the mobility of stickleback suggests that little adaptive change has occurred. Thus, modern oceanic stickleback are likely to closely resemble those that gave rise to the postglacial freshwater radiation, making it possible to infer, with unusual certainty, character states in an ancestor that gave rise to a remarkable and diverse adaptive radiation. By extension, the patterns of plasticity that are exhibited by oceanic stickleback exposed to novel freshwater environments offer insight into the range of phenotypic expression that would have been exhibited by oceanic stickleback immediately following invasion of novel freshwater environments.

The populations that comprise the postglacial freshwater radiation exhibit high levels of parallelism: Independently derived freshwater populations in similar habitats have evolved similar phenotypes, while those in divergent habitats exhibit consistent differences (*ecotypic variation*). Consistent phenotype–environment associations among freshwater isolates independently derived from oceanic ancestors imply that the radiation in freshwater is adaptive (Bell and Foster, 1994; Reimchen, 1994; Schluter, 2000, for reviews). The inference of adaptive cause is supported by analyses of the function and fitness effects of traits related especially to predation (Huntingford et al., 1994; Reimchen, 1994, 2000, for reviews) and foraging opportunities available in local environments (Lavin and McPhail, 1985, 1987; Schluter, 1993, 1994; Hart and Gill, 1994). Microsatellite evidence from the Cook Inlet region of Alaska (Cresko, 2000) and Canada (Taylor and McPhail, 2000) indicates that similar ecotypes within regions, but in different drainages, are typically independently derived from the oceanic ancestor, providing evolutionarily independent replication of ecotypes for comparison. These replicates can be used to evaluate whether patterns of phenotypic plasticity evolve in parallel with trait expression in populations or whether apparently similar population phenotypes can be the products of dissimilar underlying potentials for plastic responses to novel environmental conditions. Here, we focus upon the benthic and limnetic ecotypes because most of our insights into the role of phenotypic plasticity derive from research on these ecotypes. Populations that lie along other axes of ecological variation (e.g., predation intensity) will undoubtedly offer additional insights.

BENTHIC AND LIMNETIC ECOTYPES

A primary axis of ecotypic divergence in the stickleback radiation in northwestern North America is the benthic–limnetic divergence associated with foraging habit. The forms are best known in the small number of cases in which they are found as lacustrine species pairs (McPhail, 1994; Rundle et al., 2000; Gow et al., 2008), but the ecotypes are also found singly in

lakes, where phenotypes are predictably associated with lake characteristics (*allopatric ecotypes*). We will generally confine our discussion here to differences between allopatric ecotypes because it is in these that we have our best evidence concerning the influence and evolution of norms of reaction. Stickleback in small, shallow lakes have evolved a deep-bodied (*benthic*) form specialized for feeding on benthic invertebrates, while those in deep, oligotrophic lakes have evolved a slender form (*limnetic*) adapted for feeding on plankton (Figure 18.5) (Lavin and McPhail, 1985, 1987; Walker, 1997). Divergence in trophic morphology reflects differences in diet between the two ecotypes. Limnetic stickleback have long snouts with upturned mouths, which facilitate ingestion of small plankton near the water surface. Many long gill rakers within the buccal cavity strain these tiny prey before they can exit the gills. On the other hand, benthic stickleback have short snouts and wide mouths, suited to sucking large prey items from the benthos. Benthic stickleback have fewer, shorter gill rakers than do limnetic fish (Hart and Gill, 1994; Walker, 1997, for reviews).

In addition, there exist prominent behavioral differences between the ecotypes important in our discussion of the evolution of phenotypic plasticity in the radiation. Benthic populations retain ancestral group tendencies to cannibalize young in nests and court inconspicuously. In contrast, allopatric limnetics have lost cannibalistic tendencies and courtship is more conspicuous (Foster, 1994a, 1994b, 1995; Foster et al., 1998, 2003). The benthic form also exhibits the ancestral diversionary display, used to deter cannibalistic groups when they approach the nest, whereas males in limnetic populations, where groups forage exclusively on plankton, instead court females in the groups as they pass above (Figure 18.5) (Foster, 1988, 1994a, 1994b).

Most elements of the reproductive behavior of threespine stickleback are characteristic of nearly all populations, including all of those we describe here (Rowland, 1994, for review). In early spring, males move into shallow littoral

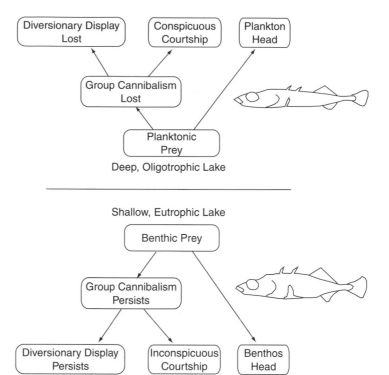

FIGURE 18.5 The benthic–limnetic differentiation of lacustrine stickleback. Persistence and loss of features are inferred with reference to oceanic fish likely to reflect phenotypes ancestral to the postglacial freshwater radiation.

regions of marine, brackish, or freshwater habitats, where they establish territories and build tubular nests of organic material that is collected from the substratum and glued together with kidney secretions. During this period males may assume the mosaic coloration usually described as typical of a courting male, although in oceanic and benthic populations drab coloration may be retained until well into the courtship phase of reproduction or even until the onset of parental care (Hulslander, 2003). Once the nest is completed, males begin courting receptive females, which are distinguishable by their distended abdomens, solitary behavior, and characteristic head-up postures when soliciting males. If the female responds positively to the male, he leads her to his nest, inserts his snout in the entrance, and turns on his side, "showing" the nest entrance. The female may enter the nest, and if she does, will usually spawn. She releases all of her eggs and returns to foraging and production of a new clutch. Males spawn multiple times, usually making the transition to parental behavior in 24–36 hours. They provide all subsequent care of young for a 7–14 day period, until the fry are free-swimming and leave the territory of the male.

The courtship behavior of male threespine stickleback has typically been described as incorporating primarily the prominent, "species-typical" zigzag dance, in which the male swims rapidly toward an approaching female, with forward progress interrupted by a pronounced "jumps" to the left and right (ter Pelwijk and Tinbergen, 1937; Rowland, 1994, for review). This very conspicuous behavior is most prominently displayed by males in limnetic populations, in which cannibalistic groups do not form. In limnetic populations, this dance is a nearly ubiquitous component of courtship (Figure 18.6), though on occasion a male will instead swim directly to a nearby female and nip her on the flank, then lead her to the nest.

In contrast, in nature, the most common form of courtship in benthic populations is dorsal pricking. Typically initiated when a female approaches a territorial male, she positions herself above him and presses her abdomen to his

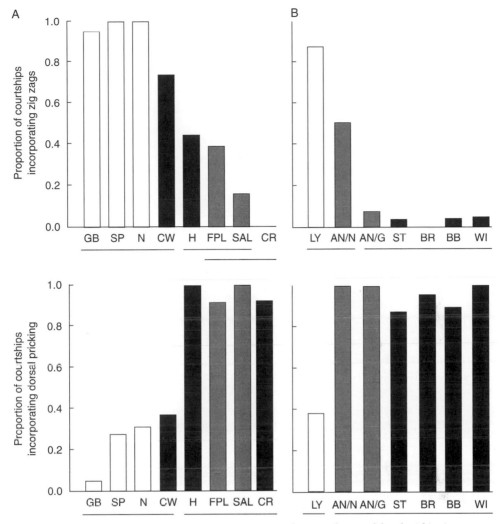

FIGURE 18.6 Proportions of courtship interactions incorporating the zigzag dance and dorsal pricking in 14 populations of threespine stickleback from (A) southern British Columbia and (B) Alaska. Open bars designate limnetic, noncannibalistic populations; shaded bars are oceanic and solid bars are benthic populations, both of which exhibit group cannibalism of young in nests guarded by males. Within histograms, bars connected by horizontal lines are not different (P > 0.05; STP contrast procedure of Sokal and Rohlf, 1995). Data include at least 16 courtships at each site. Details of populations can be found in Foster (1995) and Foster et al. (1998). Figure after Foster et al. (1998).

dorsum. Females quickly depart if the male does not raise his spines and press them back against her belly. When the male does respond, the two meander within the territory, the female above the male, and the male repeatedly presses backward into her abdomen. If a group approaches, the two break off courtship and retreat, the male defending his nest from the approaching group if necessary. Dorsal pricking has been interpreted both as enabling the pair

to survey surroundings to assess risk of intrusion (Sargent, 1982) and as discouraging the female from approaching the nest (Rowland, 1994, for review). In benthic populations, however, it may be the only form of courtship and is often broken off by the male who leads the female to his nest (Figure 18.6) (Foster, 1994a, 1995; Foster et al., 1998).

Overall, the courtship of benthic males in nature is less vigorous and noticeable than that of limnetic males, presumably reducing the risk that the nest will be attacked by visually hunting conspecifics (Foster, 1988, 1994a, 1994b, 1995; Foster et al. 1998). Males in populations at ecotypic extremes tend to perform primarily the zigzag dance (Figure 18.6: limnetics GB, LY, N, SP; benthics, BB, BR, CR, WI, ST), whereas oceanic and some freshwater populations are intermediate in expression. Here, we focus primarily on oceanic populations and extreme freshwater ecotypes.

PLASTICITY AND THE EVOLUTION OF COURTSHIP BEHAVIOR

Our first indication of ancestral plasticity in courtship behavior in the adaptive radiation came with fortuitous observations of breeding in Anchor River tide pools (anadromous, oceanic population) in 1992 and 1995. In 1995, foraging groups were abundant and routinely attacked nests, consuming any young within them (Figure 18.6: AN/G). In this year, all males performed dorsal pricking when courting, and they rarely incorporated zigzags in courtship. This is the situation typical at this and most breeding sites used by oceanic fish (Figure 18.6 and unpublished data). In contrast, in 1992 few females settled in the pools, making maximum foraging group size too small to effectively overwhelm the defenses of the males (Foster, 1985, 1988, 1995). Consequently, foraging groups were absent and male courtships included the zigzag dance significantly more often than in 1995, though not as commonly as observed in fully limnetic ecotypes (Figure 18.6: AN/N). They did, however, retain the more vigilant and less vigorous dorsal pricking in all courtships. Thus, in this oceanic population at least, courtship behavior was plastic, apparently was responsive to the presence or absence of cannibalistic foraging groups, but did not exhibit the range of plasticity needed to encompass the behavioral expression of males in the pure limnetic ecotypes.

This suggested the possibility that persistent, genetically based differences in behavior

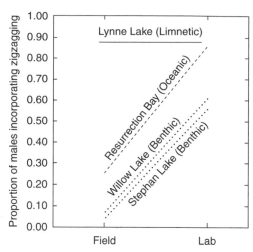

FIGURE 18.7 The proportion of males from four Alaskan threespine stickleback populations that incorporated or failed to incorporate zigzags into courtship in field and laboratory environments. Cannibalistic groups are present in the field in benthic and oceanic populations. Although historically absent in Lynne Lake, cannibalistic groups are now common there as well. Recent data support the hypothesis that the zigzag dance has undergone genetic assimilation in this population. Figure adapted from Shaw et al. (2007).

might be expressed by wild-caught stickleback from these three types of populations. Our observations on four populations offer initial insight into patterns of phenotypic plasticity in one "ancestral," one limnetic, and two benthic populations (Figure 18.7) (Shaw et al., 2007). Figure 18.7 compares observations on courtship interactions in the field, where, in all but the Lynne Lake limnetic population, foraging groups are abundant, with those in the lab, where foraging groups are absent. These observations demonstrate parallel patterns of plasticity in the oceanic and benthic populations but also that males from both benthic populations are less likely to incorporate zigzags in their courtship in either context than are oceanic males, demonstrating probable genetic accommodation of the norms of reaction.

The reaction norm for Lynne Lake males suggests genetic assimilation but could be a consequence instead of the absence of foraging groups in this lake. The field observations used in this reaction norm were made in 1992 and 1994, years in which benthic foraging and

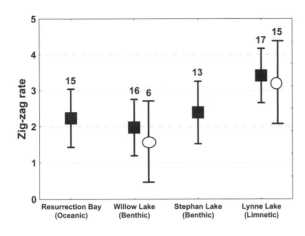

FIGURE 18.8 Mean and 95% confidence limits (square-root transformed data) for zigzag rates of wild-caught (solid squares) and laboratory-reared (open circles) stickleback. Numbers above the confidence intervals are sample sizes.

associated cannibalism of young in nests were never observed. In these years the Lynne Lake population was a pure behavioral limnetic (Foster, 1995; Foster et al., 1998). Since that time, however, mild eutrophication in the lake, apparently due to development and a blossoming of septic systems near the lake periphery, has caused an increase in individual growth rates and survival of stickleback in this population (Chock, 2008). Adult stickleback in this lake are now large enough that they have shifted to benthic foraging, as do their oceanic ancestors as adults on the breeding ground, and nest cannibalism is now common. In concert with the re-emergence of ancestral patterns of cannibalism, the diversionary display response to groups has also reappeared, although males appear to be less sensitive to the risk these groups pose than are either oceanic or benthic males. Males continue to zigzag vigorously and chronically during courtship (Chock, 2008), suggesting that genetic assimilation has occurred; indeed, in 2007, they exhibited exactly the same probability of incorporating the zigzag in courtship as in Figure 18.7. Thus, our data strongly suggest that genetic assimilation of this trait has occurred in this population.

The difference in courtship behavior between the Lynne Lake (limnetic) and Willow Lake (benthic) males does not reflect differences in experience alone. Data from a small number of families from each population reared in the laboratory demonstrate persistent differences between males from the two populations in zigzag rates within courtships. These males were reared in a uniform laboratory environment in which they had no opportunity to observe other breeding males or cannibalistic foraging groups (Figure 18.8) (Shaw et al., 2007). Here, the close match between within-courtship zigzag rates in laboratory-reared and wild-caught males under laboratory conditions suggests that the maximal zigzag rate in the absence of inhibition by foraging groups is strongly genetically determined and differs in both the direction and magnitude expected from our work with wild-caught males. Apparently, however, high encounter rates with large foraging groups in the field inhibit expression of the zigzag dance in Willow Lake (benthic), but not Lynne Lake (limnetic), males. We are now evaluating the expression of male courtship behavior before and after exposure to videotapes of foraging groups in three ancestral, three limnetic, and three benthic populations. All families are laboratory-reared and the family- and population-level norms of reaction will be contrasted with families from reciprocal crosses between benthic and limnetic population pairs, enabling us to evaluate a possible influence of maternal effects on the norms of reaction.

We have also explored the patterns of plastic response of male nuptial coloration to the presence of foraging groups in the same way and have observed similar patterns of divergence between the field and the laboratory (Hulslander, 2003; Robert, 2008). These data suggest the possibility that plasticity has been retained in some limnetic populations of stickleback but also that genetic accommodation, leading to more intense expression of red under all conditions, has occurred in limnetic populations. We are testing this possibility in the research program described above.

In summary, both sets of data suggest that genetic accommodation and/or assimilation has occurred in this adaptive radiation. They also are similar in suggesting that the pattern of plasticity in oceanic (ancestral) populations parallels that observed in the benthic–limnetic ecotypes, suggesting that the patterns of ancestral divergence could, indeed, have directed subsequent evolution in habitats favoring limnetic (noncannibalistic, plankton-feeding behavior) versus benthic (cannibalistic, bottom-feeding behavior). These and other insights from our research will be discussed after we present our results on the evolution of morphological plasticity so that we can evaluate the total evidence available to date.

PLASTICITY AND THE EVOLUTION OF ECOTYPE-SPECIFIC MORPHOLOGY

The essence of the Baldwin effect is that plasticity promotes persistence in a new environment until selection can produce traits that better match that environment. Because plasticity dictates the phenotypic variation available to selection, the resulting adaptations should bear some resemblance to those initially generated by plasticity. This process, repeated in hundreds of similar environments, should cause selection to repeatedly arrive at the same solutions to the same problems, hence producing parallel ecotypic variation. Time and again, where stickleback have colonized deep, oligotrophic lakes with a scarcity of littoral habitat, they have evolved a suite of morphological traits that are well-suited for foraging on zooplankton in the water column (Wootton, 1994). The limnetic ecotype is characterized by a long, shallow body,

many long gill rakers (which retain small food particles in the buccal cavity while water is expelled through the gills), narrow caudal peduncles, and relatively large eyes (Wootton, 1994; Walker, 1997; Spoljaric and Reimchen, 2007). The opposite traits are typically observed in fish from shallow, relatively eutrophic lakes, in which stickleback are specialized for maneuvering through a complex benthic environment where they feed on macroinvertebrates (Wootton, 1994; Walker, 1997; Spoljaric and Reimchen, 2007). In addition, the presence and type of predators in a lake can have a modifying effect on some aspects of shape (Walker, 1997).

The morphological differences between benthic and limnetic ecotypes have a significant genetic component, but phenotypic plasticity accounts for some of the variation among populations (Day et al., 1994; Spoljaric and Reimchen, 2007). Day and colleagues (Day et al., 1994; Day and McPhail, 1996) experimentally assessed the contribution of plasticity to the differences in trophic morphology between derived sympatric stickleback pairs. After rearing stickleback of each ecotype on either zooplankton or macroinvertebrates, they found that between 1% and 58% of the ecotypic differences in trophic morphology could be explained by plasticity (Day et al., 1994). Further experiments demonstrated that plasticity conferred a substantial increase in foraging efficiency and was therefore adaptive (Day and McPhail, 1996). Limnetic stickleback tended to exhibit greater plasticity than their benthic counterparts (Day et al., 1994), providing indirect evidence for genetic accommodation. Although these authors did not consider marine stickleback, differences between the two derived ecotypes indicate that one or both diverged from their common ancestral reaction norm.

If morphological plasticity promoted initial colonization of freshwater habitats and subsequently predisposed the evolution of characteristic benthic and limnetic features, plasticity in oceanic stickleback should result in phenotypes that resemble those of the derived ecotypes. In a recent experiment, we demonstrated that

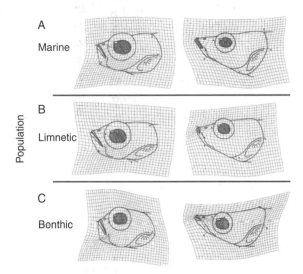

A
Marine

Population

B
Limnetic

C
Benthic

FIGURE 18.9 Deformation grids depicting the effect of diet on head shape for (A) marine fish, (B) limnetic fish, and (C) benthic fish. Dots represent the position of landmarks in the consensus configuration for each population. Vectors emanate from consensus landmarks toward their average position in the benthic (left panels) or limnetic (right panels) diet treatments. Displacements from the consensus are exaggerated four times to highlight shape differences. Using the displaced landmarks as a guide, a hand-drawn image of a stickleback was superimposed on each deformation to aid biological interpretation. Figure modified from Wund et al. (2008).

marine stickleback reared on benthic and limnetic diets, respectively, expressed variation in head shape that paralleled that of derived benthic and limnetic ecotypes (Figure 18.9) (Wund et al., 2008). Many of the plastic changes we documented were qualitatively similar to those that Day and McPhail (1996) found to be adaptive. Contrary to our expectations, gill raker length and eye size both exhibited plasticity but in a direction opposed to natural variation in these traits. Furthermore, body shape did not vary predictably (Wund et al., 2008). There are at least two interesting interpretations of these findings: (1) a single environmental cue (diet, in this case) is insufficient to produce an integrated plastic response across all traits associated with the derived ecotypes and/or (2) ancestral plasticity was not perfectly adaptive, and genetic accommodation in the alternative freshwater environments overcame this (apparently) maladaptive plasticity. This latter scenario is not strongly supported by our data, however, as derived ecotypes still expressed nonadaptive plasticity in body shape, gill rakers, and eye size. Although limited to a single environmental cue and to only a single population of each ecotype, this study has confirmed that ancestral plasticity can mirror derived, genetically based

phenotypic variation; and we advocate the need for more in-depth studies of plasticity in oceanic stickleback. The presence of a living ancestor is the most powerful aspect of this model system, yet surprisingly few studies have focused on oceanic populations.

We are currently expanding this research, rearing fish in aquaria modified to mimic not only aspects of the benthic and limnetic diets but also their structural complexity. The two main goals are to determine if a more complete set of environmental cues leads to greater adaptive integration of plastic responses and to evaluate when, during the course of ontogeny, plasticity emerges. Thus, the stickleback adaptive radiation allows us to directly examine genetic accommodation across Schlichting and Pigliucci's (1998) three axes of the developmental norm of reaction: the interacting effects of the environment, ontogeny, and allometry on the developing phenotype. Furthermore, we can address West-Eberhard's hypothesis (2003) that heterochrony has played a role in the evolution of the limnetic form. She argues that many features of the adult limnetic stickleback are essentially ancestral juvenile features, which might be expected given that adult limnetics retain the ancestrally juvenile diet of zooplankton.

STICKLEBACK-TINTED LENSES: EPIGENETIC CONTRIBUTIONS TO ADAPTIVE RADIATION

The "binary flexible stem" model of West-Eberhard (2003) proposes that adaptive radiations in which there exist high levels of parallelism, such as that of the threespine stickleback, are facilitated by the repeated emergence of alternative phenotypes as a single ancestral group repeatedly colonizes alternative environments. Our data support this model in that the oceanic (ancestral) Resurrection Bay population exhibits plastic expression of phenotypes in response to appropriate elements of benthic and limnetic habitats that mirror, but are not as extreme as, the divergence between freshwater benthic and limnetic ecotypes. Thus, the pattern of plasticity in ancestral stickleback, as reflected by this population, would have determined the expression of phenotypes in newly colonized populations and could have influenced the direction of subsequent evolutionary change and facilitated the high level of parallelism in this radiation. Although our data support most of our a priori predictions, we have tested only single populations of each ecotype in most cases. More robust conclusions clearly require evaluation of population replicates for each ecotype, research that is under way for both behavior and morphological features.

These explorations will also enable us to determine whether a common ancestor exposed to similar, novel environments will repeatedly evolve similar patterns of phenotypic plasticity. Only the benthic male courtship response to natural and laboratory environments offers insight to date, suggesting that the pair of benthic populations have evolved similarly, maintaining the ancestral pattern but with a decreased tendency to perform the zigzag dance in either environment. Ultimately, the very high level of independent replication of ecotypes in the stickleback adaptive radiation should allow us to examine not only the extent to which average population phenotypes have evolved in parallel but also the extent to which, under similar selective pressures, patterns of plasticity have evolved in parallel. The development of robust molecular technologies (Cresko et al., 2007; Kingsley and Peichel, 2007; Miller et al., 2007; Baird et al., 2008; Hohenlohe et al., 2010) should enable us to determine how underlying changes in gene expression have led to shifts in phenotypic reaction norms. Indeed, such an approach has indicated a role for the Baldwin effect and genetic assimilation of expression of a sodium-potassium ATPase as marine stickleback colonized freshwater environments (McCairns and Bernatchez, 2010). Comparative analysis of gene expression should enable us to understand the degree to which similar patterns of phenotypic plasticity are the products of similar transcriptional modifications or of different modifications that lead to similar patterns of phenotypic plasticity. This is one of the most exciting potential directions for future research involving this adaptive radiation. We can explore the extent to which similar phenotypes are the products of similar versus divergent genetic modification, and we can examine the extent to which the same genes are involved in the modification of patterns of phenotypic plasticity under opposing selective pressures.

Our data also suggest that genetic assimilation of ancestral plasticity is possible. In the Lynne Lake population, following approximately 12,000 years of relaxed selection from cannibalistic foraging groups, the conspicuous zigzag dance is consistently incorporated in male courtship, even when visually foraging cannibalistic groups are common (Shaw et al., 2007; Chock, 2008). Thus, the ancestral capacity for inhibition of conspicuous courtship by the presence of large bottom-feeding foraging groups appears to have been lost in the Lynne Lake population, presumably because the zigzag dance is noticeable to, or preferred by, females (Dzieweczynski and Rowland, 2004; Shaw et al., 2007). In contrast, the diversionary display, unlikely to have been elicited often in the 12,000 years before septic system installation, appears to have been retained. This suggests that behavioral phenotypes not expressed under novel conditions are buffered from loss, allowing long-term retention under conditions of relaxed selection.

The plasticity of several of the morphological phenotypes scored in the ancestor shifted adaptively but not sufficiently to produce derived benthic and limnetic phenotypes. This is the condition under which the rate of adaptive evolution should be maximized (e.g., Price et al., 2003; Ghalambor et al., 2007), facilitating the clear genetically based divergence in phenotypes between benthic and limnetic populations (Spoljaric and Reimchen, 2007). Surprisingly, given this expectation, genetic assimilation of male reproductive behavior appears to have occurred when the ancestral pattern of plasticity was sufficient to encompass the phenotype apparent in the limnetic, a condition that should inhibit evolutionary change (Price et al., 2003; Ghalambor et al., 2007).

The adaptive nature of plasticity in the ancestor is likely to have derived from temporal environmental variation in the oceanic environment, at least in the case of behavior. It is clear from our behavioral observations on the oceanic breeding grounds that the presence of cannibalistic foraging groups varies, a likely source of environmental heterogeneity favoring adaptive plasticity in courtship behavior. Essentially, nothing is known about the habits of oceanic stickleback outside of the breeding grounds, so at present it is impossible to evaluate whether variation in prey availability could explain the maintenance of adaptive plasticity in trophic morphology. We assume that stickleback in the open ocean are planktivores until they come into the breeding grounds. At that point, stickleback are done growing and likely not able to express any meaningful morphological plasticity before they spawn and die. Variation in prey available to fry on the breeding grounds might play an important role as well. An alternative possibility is that adaptive plasticity of trophic morphology is much more ancient, being a feature common to many vertebrates. Adaptive plasticity of trophic structures has been documented in a variety of vertebrates (e.g., fish,

Wimberger, 1992; Day et al., 1994; Wintzer and Motta, 2005; mammals, Beecher and Corruccini, 1981; Kiliaridis, 1986; reptiles, Aubret et al., 2007). Bone development responds to mechanical stress (Duncan and Turner, 1995; Turner, 1998; Nomura and Takano-Yamamoto, 2000); adaptive plasticity of craniofacial morphology in response to diet could be an ancient trait, not readily lost, even in primarily planktivorous oceanic stickleback. A growing number of examples indicate that plasticity in teleost craniofacial morphology mirrors adaptive differences among species, including the African rift lake cichlid and a number of postglacial freshwater groups (reviewed in Cooper et al., 2010). These examples offer indirect evidence in support of the flexible stem model for adaptive radiation (Cooper et al., 2010).

Predictable environmental variation favors the evolution and maintenance of plasticity (Scheiner, 1993; Schlichting and Pigliucci, 1998), but not all environments are necessarily encountered with the same frequency. When one environment is much more commonly encountered than others, one might expect plasticity to be disadvantageous, given its potential costs. However, in such circumstances, plasticity could be maintained if the costs are reduced in the common environment (Snell-Rood and Papaj, 2009). If development defaults to produce the phenotype suited to the common environment, the operating costs of producing a plastic response are incurred only in the rare environment where the benefits outweigh those costs (Snell-Rood and Papaj, 2009). This scenario might be operating in the breeding grounds of oceanic stickleback, where foraging groups are the norm but not always present. This could explain why oceanic fish always incorporate the less conspicuous dorsal pricking into their courtship, even when groups are absent. The zigzag dance is a more flexibly incorporated component of courtship, expressed only in the absence of groups. This hypothesis can be tested by examining the frequency of dorsal pricking in a range of populations along the benthic–limnetic continuum. Our prediction is that aspects of courtship that are associated with the more common environment (foraging groups present or absent) will be obligately expressed, whereas aspects of courtship behavior associated with the less common environment will be facultatively expressed. When environmental heterogeneity is completely lost, plasticity could still be retained, even if there is no cost to maintaining the *potential* to express an alternative trait. For example, in Lynne Lake, male diversionary displays were retained despite a long history without foraging groups. It might not be costly to maintain the ability to express a behavior despite a relaxation of selection on it (Coss, 1999; Lahti et al., 2009). On the other hand, assimilation might be expected when the costs of maintaining plasticity are high and the benefits are gone.

While our research has thus far involved forms of adaptive plasticity, the stickleback system certainly presents an opportunity to evaluate the role of nonadaptive plasticity in adaptive divergence. When conditions change so much that developmental homeostasis is threatened, stress can reveal cryptic genetic variation to selection (Badyaev, 2005; Ghalambor et al., 2007, for reviews). Given that past selection is unlikely to have honed this variation, it is likely to be maladaptive on average or at least random with respect to fitness (Ghalambor et al., 2007). Evaluation of the responses of stickleback to conditions not experienced by oceanic fish, such as dystrophic habitats or environments in which anadromous fish are prevented from returning to oceanic environments, could offer evidence of nonadaptive patterns of plasticity. Such research could shed important light on the influence of nonadaptive phenotypic plasticity in early stages of diversification.

These explorations will also enable us to determine whether a common ancestor exposed to similar, novel environments will repeatedly evolve similar patterns of phenotypic plasticity. Only the benthic response to natural and laboratory environments has offered insight to date, suggesting that the pair of benthic populations have evolved similarly, maintaining the ances-

tral pattern but with a decreased tendency to perform the zigzag dance in either environment. Our data suggest also that limnetic populations can lose plasticity of the response to cannibalistic foraging groups.

CONCLUSION

For more than a century, evolutionary biologists have debated the significance of phenotypic plasticity to the evolutionary process. The past 30 years have seen a dramatic increase in attention to the evolution of reaction norms, and more recently there has been a renewed interest in considering how evolution shapes developmental processes, rather than simply considering how changes in gene frequencies map directly to changes in phenotypes. The confluence of these research foci have made the time ripe to finally take serious consideration of the special ways that plasticity may be a leader, rather than a follower, in evolutionary change (West-Eberhard, 2003). Several reviews have effectively argued that understanding these processes could fundamentally enhance our understanding of the evolutionary process (Pigliucci, 2001, 2005; Schlichting 2004). We have presented the case that the threespine stickleback is especially, if not uniquely, suitable for examining the role of plasticity in evolution by virtue of the extensive phenotypic variation within the radiation, the amenability of stickleback to study, and, in particular, the presence of the living ancestor that gave rise to the adaptive radiation in freshwater. The time has come for the empirical evidence to speak to theory, and the stickleback radiation is exceptionally suited to this endeavor.

REFERENCES

Ancel, L. W., and W. Fontana. 2000. Plasticity, evolvability, and modularity in RNA. *J Exp Zoolog* 288:242–83.

Aubret, F., X. Bonnet, and R. Shine. 2007. The role of adaptive plasticity in a major evolutionary transition: Early aquatic experience affects locomotor performance of terrestrial snakes. *Funct Ecol* 21:1154–61.

Badyaev, A.V. 2005. Stress-induced variation in evolution: From behavioural plasticity to genetic assimilation. *Proc R Soc Lond B* 272:877–86.

Baird, N.A., P.D. Etter, T.S. Atwood, M.C. Currey, A.L. Shiver, Z.A. Lewis, E.U. Selker, et al. 2008. Rapid SNP discovery and genetic mapping using sequenced RAD markers. *PLoS ONE* 3:e3376.

Baker, J.A. 1994. Life history variation in female threespine stickleback. In *The Evolutionary Biology of the Threespine Stickleback*, ed. M.A. Bell and S.A. Foster, 144–87. Oxford: Oxford University Press.

Baldwin, J.M. 1896. A new factor in evolution. *Am Nat* 30:441–51.

Baldwin, J.M. 1902. *Development and Evolution*. New York: Macmillan.

Beecher, R.M., and R.S. Corruccini. 1981. Effects of diet consistency on craniofacial and occlusal development in the rat. *Angle Orthod* 51:61–9.

Behera, N., and V. Nanjundiah. 1995. An investigation into the role of phenotypic plasticity in evolution. *J Theor Biol* 172:225–34.

Bell, M.A. 1994. Paleobiology and evolution of threespine stickleback. In *The Evolutionary Biology of the Threespine Stickleback*, ed. M.A. Bell and S.A. Foster, 438–71. Oxford: Oxford University Press.

Bell, M.A., and S.A. Foster. 1994. Introduction to the evolutionary biology of the threespine stickleback. In *The Evolutionary Biology of the Threespine Stickleback*, ed. M.A. Bell and S.A. Foster, 1–27. Oxford: Oxford University Press.

Boake, C.R.B. 1989. Repeatability: Its role in evolutionary studies of mating behavior. *Evol Ecol* 3:173–82.

Brodie, E.D., and N.H. Russell. 1999. The consistency of individual differences in behaviour: Temperature effects on antipredator behaviour in garter snakes. *Anim Behav* 57:445–51.

Carroll, S.P., and P.S. Corneli. 1999. The evolution of behavioral norms of reaction as a problem in ecological genetics. In *Geographic Variation in Behavior: Perspectives on Evolutionary Mechanisms*, ed. S.A. Foster and J.A. Endler, 52–68. New York: Oxford University Press.

Carroll, S.P., H. Dingle, and S.P. Klassen. 1997. Genetic differentiation of fitness-associated traits among rapidly evolving populations of the soapberry bug. *Evolution* 51:1182–8.

Carroll, S.P., S.P. Klassen, and H. Dingle. 1998. Rapidly evolving adaptations to host ecology and nutrition in the soapberry bug. *Evol Ecol* 12:955–68.

Chock, R.L. 2008. Re-emergence of ancestral plasticity and the loss of a rare limnetic phenotype in an Alaskan population of threespine stickleback. Master's thesis, Clark University, Worcester, MA.

Cooper, W., K. Parsons, A. McIntyre, B. Kern, A. McGee-Moore, and R. Albertson. 2010. Bentho-pelagic divergence of cichlid feeding architecture was prodigious and consistent during multiple adaptive radiations within African rift-lakes. *PLoS ONE* 5:e9551.

Coss, R.G. 1999. Effects of relaxed natural selection on the evolution of behavior. In *Geographic Variation in Behavior: Perspectives on Evolutionary Mechanisms*, ed. S.A. Foster and J.A. Endler, 180–208. New York: Oxford University Press.

Cresko, W.A. 2000. The ecology and geography of speciation: A case study using an adaptive radiation of threespine stickleback in Alaska. PhD diss., Clark University, Worcester, MA.

Cresko, W., K. McGuigan, P. Phillips, and J. Postlethwait. 2007. Studies of threespine stickleback developmental evolution: Progress and promise. *Genetica* 129:105–26.

Crispo, E. 2007. The Baldwin effect and genetic assimilation: Revisiting two mechanisms of evolutionary change mediated by phenotypic plasticity. *Evolution* 61:2469–79.

Day, T., and J.D. McPhail. 1996. The effect of behavioural and morphological plasticity on foraging efficiency in the threespine stickleback (*Gasterosteus* sp.). *Oecologia* 108:380–8.

Day, T., J. Pritchard, and D. Schluter. 1994. A comparison of 2 sticklebacks. *Evolution* 48:1723–34.

de Jong, G. 2005. Evolution of phenotypic plasticity: Patterns of plasticity and the emergence of ecotypes. *New Phytol* 166:101–18.

Duncan, R.L., and C.H. Turner. 1995. Mechanotransduction and the functional response of bone to mechanical strain. *Calcif Tissue Int* 57:344–58.

Dzieweczynski, T.L., and W.J. Rowland. 2004. Behind closed doors: Use of visual cover by courting male three-spined stickleback, *Gasterosteus aculeatus*. *Anim Behav* 68:465–71.

Falconer, D.S. 1981. *Introduction to Quantitative Genetics*. New York: Longmans.

Fear, K.K., and T.D. Price. 1998. The adaptive surface in ecology. *Oikos* 82:440–8.

Foster, S.A. 1985. Group foraging in a coral reef fish: A mechanism for gaining access to defended resources. *Anim Behav* 33:782–92.

Foster, S.A. 1988. Diversionary displays of paternal stickleback: Defenses against cannibalistic groups. *Behav Ecol Sociobiol* 22:335–40.

Foster, S.A. 1994a. Evolution of the reproductive behavior of threespine stickleback. In *The Evolutionary Biology of the Threespine Stickleback*, ed. M.A. Bell and S.A. Foster, 381–98. Oxford: Oxford University Press.

Foster, S. A. 1994b. Inference of evolutionary pattern: Diversionary displays of threespine sticklebacks. *Behav Ecol* 5:114–21.

Foster, S. A. 1995. Understanding the evolution of behavior in threespine stickleback: The value of geographic variation. *Behaviour* 132:1107–29.

Foster, S. A., and J. A. Baker. 2004. Evolution in parallel: New insights from a classic system. *Trends Ecol Evol* 19:456–9.

Foster, S. A., J. A. Baker, and M. A. Bell. 2003. The case for conserving threespine stickleback populations: Protecting an adaptive radiation. *Fisheries* 28:10–18.

Foster, S. A., R. J. Scott, and W. A. Cresko. 1998. Nested biological variation and speciation. *Philos Trans R Soc Lond B Biol Sci* 353:207–18.

Ghalambor, C. K., J. K. McKay, S. P. Carrol, and D. N. Reznick. 2007. Adaptive versus non-adaptive phenotypic plasticity and the potential for contemporary adaptation in new environments. *Funct Ecol* 21:394–407.

Gibson, G. 2005. Evolution: The synthesis and evolution of a supermodel. *Science* 307:1890–1.

Gilbert, S. F., and J. A. Bolker. 2003. Ecological developmental biology: Preface to the symposium. *Evol Dev* 5:3–8.

Gow, J. L., S. M. Rogers, M. Jackson, and D. Schluter. 2008. Ecological predictions lead to the discovery of a benthic–limnetic sympatric species pair of threespine stickleback in Little Quarry Lake, British Columbia. *Can J Zool* 86:564–71.

Grant, V. 1977. *Organismic Evolution*. San Francisco: Freeman.

Hart, P. J. B., and A. B. Gill. 1994. Evolution of foraging behaviour in the threespine stickleback. In *The Evolutionary Biology of the Threespine Stickleback*, ed. M. A. Bell and S. A. Foster, 207–39. Oxford: Oxford University Press.

Hinton, G. E., and S. J. Nowlan. 1987. How learning can guide evolution. *Complex Syst* 1:497–502.

Hohenlohe, P. A., S. Bassham, P. D. Etter, N. Stiffler, E. A. Johnson, and W. A. Cresko. 2010. Population genomics of parallel adaptation in threespine stickleback using sequenced RAD tags. *PLoS Genet* 6:1–23.

Huey, R., P. Hertz, and B. Sinervo. 2003. Behavioral drive versus behavioral inertia in evolution: A null model approach. *Am Nat* 161:357–66.

Hulslander, C. L. 2003. The evolution of the male threespine stickleback colour signal. Master's thesis, Clark University, Worcester, MA.

Huntingford, F. A., P. J. Wright, and J. F. Tierney. 1994. Adaptive variation in antipredator behaviour in threespine stickleback. In *The Evolutionary Biology of the Threespine Stickleback*, ed. M. A. Bell and S. A. Foster, 277–96. Oxford: Oxford University Press.

Kiliaridis, S. 1986. The relationship between masticatory function and craniofacial morphology. III. The eruption pattern of the incisors of the growing rat fed a soft diet. *Eur J Orthod* 8:71–9.

Kingsley, D. M., and C. L. Peichel. 2007. The molecular genetics of evolutionary change in sticklebacks. In *Biology of the Three-Spined Stickleback*, ed. S. Östlund-Nilsson, I. Mayer, and F. A. Huntingford. Boca Raton, FL: CRC Press.

Kirkpatrick, M. 1982. Quantum evolution and punctuated equilibria in continuous genetic characters. *Am Nat* 119:833–48.

Lahti, D. C., N. A. Johnson, B. C. Ajie, S. P. Otto, A. P. Hendry, D. T. Blumstein, R. G. Coss, et al. 2009. Relaxed selection in the wild. *Trends Ecol Evol* 24:487–96.

Lavin, P. A., and J. D. McPhail. 1985. The evolution of freshwater diversity in the threespine stickleback (*Gasterosteus aculeatus*): Site-specific differentiation of trophic morphology. *Can J Zool* 63:2632–8.

Lavin, P. A., and J. D. McPhail. 1987. Morphological divergence and the organization of trophic characters among lacustrine populations of the threespine stickleback (*Gasterosteus aculeatus*). *Can J Fish Aquat Sci* 44:1820–9.

Malhi, R. S., G. Rhett, and A. M. Bell. 2006. Mitochondrial DNA evidence of an early Holocene population expansion of threespine sticklebacks from Scotland. *Mol Phylogenet Evol* 40:148–54.

McCairns, R. J. S., and L. Bernatchez. 2010. Adaptive divergence between freshwater and marine sticklebacks: Insights into the role of phenotypic plasticity from an integrated analysis of candidate gene expression. *Evolution* 64:1029–47.

McPhail, J. D. 1994. Speciation and the evolution of reproductive isolation in the sticklebacks (*Gasterosteus*) of south-western British Columbia. In *The Evolutionary Biology of the Threespine Stickleback*, ed. M. A. Bell and S. A. Foster, 399–437. Oxford: Oxford University Press.

Miller, M. R., J. P. Dunham, A. Amores, W. A. Cresko, and E. A. Johnson. 2007. Rapid and cost-effective polymorphism identification and genotyping using restriction site associated DNA (RAD) markers. *Genome Res* 17:240–8.

Morgan, C. L. 1896. On modification and variation. *Science* 4:733–40.

Nomura, S., and T. Takano-Yamamoto. 2000. Molecular events caused by mechanical stress in bone. *Matrix Biol* 19:91–6.

Osborn, H. F. 1896. A mode of evolution requiring neither natural selection nor the inheritance of acquired characters. *Trans N Y Acad Sci* 15:141–2.

Pal, C., and I. Miklos. 1999. Epigenetic inheritance, genetic assimilation, and speciation. *J Theor Biol* 2003:19–37.

Pigliucci, M. 2001. *Phenotypic Plasticity: Beyond Nature and Nurture: Syntheses in Ecology and Evolution*. Baltimore: Johns Hopkins University Press.

Pigliucci, M. 2005. Evolution of phenotypic plasticity: Where are we going now? *Trends Ecol Evol* 20:481–6.

Pigliucci, M. 2007. Do we need an extended evolutionary synthesis? *Evolution* 61:2743–9.

Pigliucci, M., and C. J. Murren. 2003. Genetic assimilation and a possible evolutionary paradox: Can macroevolution sometimes be so fast as to pass us by? *Evolution* 57:1455 64.

Price, T. D., A. Qvarnström, and D. E. Irwin. 2003. The role of phenotypic plasticity in driving genetic evolution. *Proc R Soc Lond B* 270:1433–40.

Reimchen, T. E. 1994. Predators and morphological evolution in threespine stickleback. In *The Evolutionary Biology of the Threespine Stickleback*, ed. M. A. Bell and S. A. Foster, 240–76. Oxford: Oxford University Press.

Reimchen, T. E. 2000. Predator handling failures of lateral plate morphs in *Gasterosteus aculeatus*: Functional implications for the ancestral plate condition. *Behaviour* 137:1081–96.

Robert, K. 2008. The role of male nuptial coloration in aggression of threespine stickleback: A two-population, context-dependent analysis. PhD diss., Clark University, Worcester, MA.

Rowland, W. J. 1994. Proximate determinants of stickleback behaviour: An evolutionary perspective. In *The Evolutionary Biology of the Threespine Stickleback*, ed. M. A. Bell and S. A. Foster, 297–344. Oxford: Oxford University Press.

Rundle, H. D., L. M. Nagel, J. W. Boughman, and D. Schluter. 2000. Natural selection and parallel speciation in sympatric sticklebacks. *Science* 287:306–8.

Sargent, R. C. 1982. Territory quality, male quality, courtship intrusions, and female nest-choice in the threespine stickleback, *Gasterosteus aculeatus*. *Anim Behav* 30:364–74.

Scheiner, S. M. 1993. Genetics and evolution of phenotypic plasticity. *Annu Rev Ecol Syst* 24:35–68.

Schlichting, C. D. 2004. The role of phenotypic plasticity in diversification. In *Phenotypic Plasticity: Functional and Conceptual Approaches*, ed. T. J. DeWitt and S. M. Scheiner, 191–200. New York: Oxford University Press.

Schlichting, C. D., and M. Pigliucci. 1998. *Phenotypic Evolution: A Reaction Norm Perspective*. Sunderland, MA: Sinauer.

Schluter, D. 1993. Adaptive radiation in sticklebacks: Size, shape and habitat use efficiency. *Ecology* 74:699–709.

Schluter, D. 1994. Experimental evidence that competition promotes divergence in adaptive radiation. *Science* 266:798–801.

Schluter, D. 2000. *The Ecology of Adaptive Radiation*. Oxford: Oxford University Press.

Scoville, A. G., and M. E. Pfrender. 2010. Phenotypic plasticity facilitates recurrent rapid adaptation to introduced predators. *Proc Natl Acad Sci USA* 107:4260–3.

Shaw, K. A., M. L. Scotti, and S. A. Foster. 2007. Ancestral plasticity and the evolutionary diversification of courtship behaviour in threespine sticklebacks. *Anim Behav* 73:415–72.

Simpson, G. G. 1953. The Baldwin effect. *Evolution* 7:110–17.

Snell-Rood, E. C., and D. R. Papaj. 2009. Patterns of phenotypic plasticity in common and rare environments: A study of host use and color learning in the cabbage white butterfly *Pieris rapae*. *Am Nat* 173:615–31.

Sokal, R. R., and F. J. Rohlf. 1995. *Biometry*. New York: W. H. Freeman and Company.

Spoljaric, M. A., and T. E. Reimchen. 2007. 10,000 years later: Evolution of body shape in Haida Gwaii three-spined stickleback. *J Fish Biol* 70:1484–1503.

Sultan, S. E., and S. C. Stearns. 2005. Environmentally contingent variation: Phenotypic plasticity and norms of reaction. In *Variation*, ed. B. Hallgrímsson and B. K. Hall, 303–32. Amsterdam: Elsevier Academic Press.

Taylor, E. B., and J. D. McPhail. 1999. Evolutionary history of an adaptive radiation in species pairs of threespine sticklebacks (*Gasterosteus*): Insights from mitochondrial DNA. *Biol J Linn Soc* 66:271–91.

Taylor, E. B., and J. D. McPhail. 2000. Historical contingency and ecological determinism interact to prime speciation in sticklebacks, *Gasterosteus*. *Proc R Soc Lond B* 267:2375–84.

ter Pelwijk, J. J., and N. Tinbergen. 1937. Eine reizbiologische analyse einiger verhaltensweisen von *Gasterosteus aculeatus* (L.). *Z Tierpsychol* 1:193–200.

Turner, C. H. 1998. Three rules for bone adaptation to mechanical stimuli. *Bone* 23:339–407.

Via, S. 1993. Adaptive phenotypic plasticity: Target or by-product of selection in a variable environment? *Am Nat* 142:352–65.

Walker, J. A. 1997. Ecological morphology of lacustrine threespine stickleback *Gasterosteus aculeatus* L. (Gasterosteidae) body shape. *Biol J Linn Soc* 61:3–50.

Walker, J. A., and M. A. Bell. 2000. Net evolutionary trajectories of body shape evolution within a microgeographic radiation of threespine stickleback (*Gasterosteus aculeatus*). *J Zool Soc Lond* 252:293–302.

West-Eberhard, M. J. 1989. Phenotypic plasticity and the origins of diversity. *Annu Rev Ecology Syst* 20:249–78.

West-Eberhard, M. J. 1998. Commentary: Evolution in light of development and cell biology, and vice versa. *Proc Natl Acad Sci USA* 95:8417–19.

West-Eberhard, M. J. 2003. *Developmental Plasticity and Evolution*. New York: Oxford University Press.

West-Eberhard, M. J. 2005. Developmental plasticity and the origin of species differences. *Proc Natl Acad Sci USA* 102:6543–9.

Wimberger, P. H. 1992. Plasticity of fish body shape: The effects of diet, development, family and age in two species of *Geophagus* (Pisces: Cichlidae). *Biol J Linn Soc* 45:197-218.

Windig, J. J., C. G. F. de Kovel, and G. De Jong. 2004. Genetics and mechanics of plasticity. In *Phenotypic Plasticity: Functional and Conceptual Approaches*, ed. T. J. DeWitt and S. M. Scheiner, 31–49. New York: Oxford University Press.

Wintzer, A. P., and P. J. Motta. 2005. Diet-induced phenotypic plasticity in the skull morphology of hatchery-reared Florida largemouth bass, *Micropterus salmoides floridanus*. *Ecol Freshwater Fish* 14:311–18.

Woltereck, R. 1909. Weitere experimentelle Untersuchungen über Artveränderung, speziell über das Wesen quantitativer Artunterschiede bei Daphniden. *Versuche Dtsch Zool Ges* 19:110–72.

Wootton, R. J. 1994. Energy allocation in the threespine stickleback. In *The Evolutionary Biology of the Threespine Stickleback*, ed. M. A. Bell and S. A. Foster, 114–43. Oxford: Oxford University Press.

Wund, M. A., J. A. Baker, B. Clancy, J. Golub, and S. A. Foster. 2008. A test of the "flexible stem" model of evolution: Ancestral plasticity, genetic accommodation, and morphological divergence in the threespine stickleback radiation. *Am Nat* 172:449–62.

19

Learning, Developmental Plasticity, and the Rate of Morphological Evolution

Christopher J. Neufeld and A. Richard Palmer

CONTENTS

Much has been written about how learning—the developmental plasticity of behavior—and morphological plasticity—the developmental plasticity of form—may individually affect the rate of morphological evolution (reviewed in Maynard Smith, 1987; Pigliucci, 2001; Weber and Depew, 2003; West-Eberhard, 2003). Surprisingly, rather little has been said explicitly about how these two kinds of plastic responses might amplify one another. Learning increases

Epigenetics: Linking Genotype and Phenotype in Development and Evolution, ed. Benedikt Hallgrímsson and Brian K. Hall.

the frequency of certain behaviors as a result of past experience. Increased frequency of a behavior may, in turn, yield developmentally plastic responses in morphology. Finally, behaviors are continually modified via learning, in response to the suitability of a given form to a particular task. This positive feedback loop should greatly accelerate the generation of morphological diversity in a novel environment, particularly where altered behaviors (via learning) or altered forms (via morphological plasticity) expose or amplify the effects of previously inconsequential heritable variation.

Although learning and morphological plasticity are both examples of developmental plasticity, one key difference between them is the time scale of response: Behaviors may be learned quickly (typically minutes to days), whereas induced changes in morphology take longer (typically weeks to months) (Pigliucci, 2001). Short-term responses to a novel environment may also involve concurrent plastic changes in both behavior and morphology, as seen in feeding behaviors and mouth form in sticklebacks (Day and McPhail, 1996).

West-Eberhard (2003, chap. 18) clearly recognized how learning might facilitate morphological evolution via what she calls "morphology-biased learning." For example "in the learned dietary specializations of some birds [genetic differences in] the form and size of the beak render some dietary choices more quickly rewarding than others" (p. 338). In other words, among an array of individuals whose bills differ slightly in size or shape due to preexisting genetic or developmental differences, an individual may quickly learn that its bill is better suited to some foods than others. These learned behaviors would amplify the fitness consequences of preexisting morphological variation and, therefore, increase the rate of morphological evolution.

We wish to draw attention to another way in which learning might influence the rate of morphological evolution: *learning-enhanced morphological plasticity*. In particular, learning may provide a mechanism whereby short-term responses to novel conditions may be "captured" in a way

that facilitates medium-term morphological responses via developmental plasticity during the life span of an individual. In addition, some plastic morphological responses—because they directly affect performance—may even accelerate the rate of learning by increasing the benefits of learned behaviors.

TWO EXAMPLES

To avoid some of the ambiguity that necessarily arises when complex concepts are discussed in general terms, we present two specific models that illustrate the interplay between learning, morphological plasticity, and rate of evolution. From these examples emerge some useful generalizations about how learning may accelerate morphological evolution by generating new phenotypic variation more rapidly and by exposing previously cryptic genetic variation.

EXAMPLE I. EVOLUTION OF INDUCED MORPHOLOGICAL DEFENSE IN RESPONSE TO A NOVEL PREDATOR

Many taxa develop predator-resistant forms when exposed to cues from potential predators. For example, gastropods develop thicker shells that are more difficult to open by shell-crushing predators, tadpoles produce a smaller body and larger tail that allow for quicker escapes, bryozoans develop spines that deter gastropod predators, and some barnacles develop highly asymmetrical shells that reduce shell-entry predation (Tollrain and Harvell, 1999). Furthermore, induced defenses can influence the strength and nature of many ecological interactions (Miner et al., 2005). However, despite many examples of induced defenses—and new examples are discovered in almost every system where they are sought (e.g., Vaughn, 2007)—we still know very little about how such defenses evolve.

The expression of an inducible morphological defense requires that two subcomponents be coupled: (1) ability to detect and correctly interpret predator-risk cues (*predator detection*) and (2) capacity to alter development to produce a predator-resistant form (*developmental flexibility*). When prey encounter a novel predator, an inducible defense will remain unexpressed (or latent) (e.g., Edgell and Neufeld, 2008) until detection of the predator can be successfully and predictably coupled to a developmental response. Such coupling has demonstrably occurred in some cases (Kiesecker and Blaustein, 1997; reviewed in Strauss et al., 2006); how these two subcomponents might become coupled, however, is largely unknown.

Given the well-documented capacity for associative learning in many taxa (Rochette et al., 1998; Gonzalo et al., 2007; Hermann et al., 2007; Epp and Gabor, 2008; Ferrari and Chivers, 2008), we suggest that learning may play an important role in correctly interpreting novel predator-risk cues and facilitating the expression of a latent inducible defense. In turn, expression of an induced defense should ensure population persistence and may simultaneously expose previously cryptic heritable variation in the inducible defense. Collectively, these effects have great potential to increase the rate of morphological evolution when prey encounter novel predators. Following a theoretical model, we review the evidence for the separate components of the model and outline the most fruitful approaches for detecting this process in nature.

A THEORETICAL MODEL

Consider how learning might influence the evolution of an inducible defense in response to a novel predator (Figure 19.1). In the absence of learning, individuals cannot correctly interpret predator-risk cues and show no change in form in the first generation despite having the capacity to do so (i.e., the induced defense is unexpressed or latent in the novel environment; Figure 19.1A,C). Therefore, presuming that high defensive trait values are favored when faced with this novel predator, selection must act on any standing variation in constitutive trait values and will favor those individuals that are most defended in the noninduced state. For an adaptive plastic response to the novel

predator to occur, some genetic change must happen that couples the novel predator cues to the existing but unexpressed developmental pathway for the induced defense. However, because all individuals will be relatively undefended against the novel predator initially, mortality would be high and populations might face local extinction before appropriate genetic modifiers arise (Figure 19.1C).

In contrast, if individuals can learn to interpret cues from the novel predator correctly, an entire population could express the induced morphological response within one generation (Figure 19.1B,D). Therefore, individuals are at least partially defended against the novel predator initially, and selection can act on any heritable variation in speed, magnitude, or nature of the induced defense or on variation in learning ability itself. In short, learned predator recognition has three evolutionarily significant consequences: (1) it brings trait distributions closer to the optimal body form in the presence of the novel predator and thus decreases the risk of local extinction, (2) it permits the adaptive plasticity necessary to cope with variation in predation risk to persist, and (3) it may expose underlying heritable variation in learning and morphological plasticity to selection.

The key requirements of this model are (1) learning enables rapid detection of novel predators; (2) learning-induced defenses are adaptive in the face of a novel predator; (3) heritable variation exists in the developmental pathways that determine the speed, magnitude, or nature of the response to predator cues (i.e., the capacity for plasticity) or in learning ability itself; and (4) expression of an induced defense exposes cryptic genetic variation. We review the evidence for each requirement below.

LEARNING ENABLES DETECTION OF NOVEL PREDATORS

Associating the scent of damaged conspecifics with the scent of a potential predator (*associative learning*) is one highly effective way for prey to detect a novel predator quickly. Furthermore, such associative learning from chemical cues

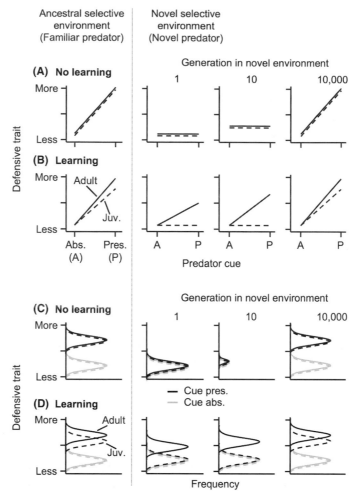

FIGURE 19.1 The impact of learning on the evolution of inducible defense in response to a novel predator. (A, B) Individual reaction norms (defensive trait values of the same genotype grown in the presence and absence of predator cues) without and with learning. Solid lines are adult forms; dashed lines are juvenile forms before any learning has taken place. Slopes of dashed lines reflect innate (unlearned) response to predatory cues; differences between dashed and solid lines reflect the component of the response triggered by learning within the lifetime of an individual. (C, D) Trait frequency distributions for populations without and with learning. Solid lines are adult forms; dashed lines are juvenile forms. Black lines show trait distributions when predator cues are present; gray lines show distributions when predator cues are absent. As in (A) and (B), differences between dashed and solid lines reflect the component of the response due to learning. (A, C) When a species not capable of learning is faced with a novel predator, individuals fail to recognize novel predatory cues and thus show no antipredator response in the novel environment. As a result, only noninduced phenotypes are expressed (generation 1) and selection favors high noninduced trait values (i.e., reaction norms with high intercepts; generation 10). Selection of individuals that are relatively poorly adapted to the novel predator may reduce the likelihood of population persistence in the novel environment. In the absence of novel predator recognition, selection must wait for mutations that act to couple the novel predator cues to the existing but unexpressed (i.e., latent) developmental pathway for the induced defense (generation 10,000). (B, D) When a species capable of learning is faced with a novel predator, individuals learn to recognize the unfamiliar predator within the first generation of exposure (generation 1). The expression of the antipredator response allows initial selection to act on standing variation in speed, magnitude, and nature of response to predator cues as well as on variation in learning ability itself. Furthermore, because learning-induced phenotypes are closer to the optimum in the novel environment (in effect, all individuals are at least partially defended from the novel predator), population size remains relatively unchanged after initial exposure to the novel predator (generation 10). Due to the learned expression of the plastic response in the novel environment, long-term selection can act on genetic variation in speed, magnitude, and nature of response to predator cues, as well as on variation in learning ability itself. If a mutation arises that yields innate recognition of novel predator cues, selection will yield an induced response that is expressed in the absence of learning (generation 10,000).

would allow prey to recognize and respond to local predation risk without having any direct predator contact. For example, the common pond snail (*Lymnaea stagnalis*) crawls above the waterline when exposed to chemical cues from a predatory fish (tench, *Tinca tinca*) (Dalesman et al., 2006). Importantly, when first exposed to tench cues paired with cues from damaged conspecifics, snails markedly increase their antipredator response to tench cues alone, suggesting that an innate antipredator behavior is augmented via associative learning. Similarly, whelks (*Buccinum undatum*) increase their antipredator behavioral response to a predatory sea star (*Leptasterias polaris*) if they are first exposed to cues from a sea star feeding on conspecific whelks (Rochette et al., 1998).

In some instances, exposure to cues released from damaged conspecifics is not necessary to detect potential predators. For example, damselfly larvae learn to recognize a predator as a risk when the predator's past diet included conspecific damselfly larvae (Chivers et al., 1996). Perhaps even more impressive, predator-naive tadpoles of the boreal chorus frog (*Pseudacris maculate*) learn to recognize predatory salamanders (*Ambystoma tigrinum*) simply by observing the behavior of predator-experienced wood frog tadpoles (*Rana sylvatica*) around them (Ferrari and Chivers, 2008).

These examples illustrate how many organisms are able to learn about the density and identity of predators in their environment without direct predator contact. Associative learning has one additional significant benefit: It permits recognition of a completely novel predator. Gonzalo et al. (2007) have shown that this capability does exist. They exposed one group of tadpoles to effluent from a novel nonpredatory fish in conjunction with effluent from damaged conspecifics and another group to the fish cue alone. When later exposed to the fish cue alone, only tadpoles conditioned with effluent of damaged conspecifics displayed antipredator behavior. Some species therefore have the capacity to learn about wholly novel cues, but more studies are clearly needed to determine how prevalent

this ability is among different taxa. Furthermore, nothing is known about whether learned predator recognition can lead to an induced *morphological* defense.

INDUCED DEFENSES ARE ADAPTIVE WHEN FACED WITH A NOVEL PREDATOR

Even if learning can facilitate expression of an induced defense in the face of a novel predator, the induced defense—the one that evolved originally to cope with *native* predators—must be at least partially adaptive against this new predator. What evidence supports this idea?

On the west coast of North America, the whelk *Nucella lamellosa* thickens its shell when exposed to effluent of a native predatory crab (*Cancer productus*) or the smell of damaged conspecifics (Appleton and Palmer, 1988) and thick-shelled snails are less vulnerable to this shell-breaking crab (Palmer, 1985). However, naive whelks exposed to the effluent of the recently introduced European green crab (*Carcinus maenas*, with which the whelks share no recent evolutionary history) exhibit no shell thickening (Edgell and Neufeld, 2008), confirming that the plastic morphological response to the introduced crab species is latent. Significantly, thick-shelled whelks are demonstrably less vulnerable to predation by this introduced crab (Edgell and Neufeld, 2008). Therefore, the existing induced defense, which evolved in response to native crabs, is sufficient to reduce vulnerability to the introduced crabs.

More generally, developmentally plastic responses that evolved to defend against one predator should also be adaptive against others that use similar modes of predation. For instance, thicker shells are a common adaptive response by many molluscs to shell-breaking predators (Vermeij, 1978). The evolution of this response in multiple independent predator–prey relationships (Vermeij, 1978) suggests that thick shells would be adaptive against any shell-breaking predator, novel or familiar. Similarly, the larger, more powerful tails developed by tadpoles exposed to dragonfly larvae (Van Buskirk and Relyea, 1998) enable faster starts and quicker turns (Wassersug and Hoff, 1985). Therefore, larger tails would likely be adaptive against any ambush predator. In the case of predator-induced chemical defenses, a substance that makes a particular organism unpalatable to one predator likely does so to others as well. Of course, the adaptive value of a particular induced defense depends on the mode of predation of the novel predator. For instance, in gastropods, an induced thicker shell would not reduce vulnerability to predators that enter through the shell aperture (Rochette et al., 2007) nor would cryptic coloration defend against a predator that hunts by olfactory cues.

HERITABLE VARIATION EXISTS IN LEARNING ABILITY AND IN CAPACITY FOR PLASTICITY

For our model to work, heritable variation must exist in the speed, magnitude, or nature of the induced defense or in learning ability itself. Selectable variation in the capacity for plasticity clearly does exist in many groups (reviewed in Scheiner, 1993). Induced morphological defenses, in particular, appear to be naturally variable and to evolve in response to different selection pressures. For example, mainland tadpoles (*Rana pirica*), which have a long history of overlap with predatory salamander larvae (*Hynobius retardatus*), develop a predator-resistant form (a bulgy body) when exposed to salamander cues. In contrast, tadpoles from islands, where salamanders are historically absent, show little response to the same predator cues (Kishida et al., 2007). Furthermore, mainland–island hybrids develop intermediate phenotypes, suggesting that these differences have a genetic basis. In a similar common garden experiment, wood frog tadpoles (*Rana sylvatica*) from different historical shade regimes exhibited different, plastic morphological responses to a predator (a larval dragonfly) (Relyea, 2002), suggesting that different historical shade regimes favor different developmental responses to the same predator. In addition, when transplanted among different ponds, tadpoles survived better in their natal pond (Relyea, 2002), confirming that

such plasticity differences are adaptive. Finally, when intertidal whelks (*Nucella lamellosa*) are exposed to cues from potential predators, individuals from relatively predator-free locations have thinner shells than those from predator-dense habitats, again suggesting that induced defenses vary among populations (Appleton and Palmer, 1988). Because thicker shells reduce vulnerability to crab predators (Palmer, 1985), such variation in plasticity among populations likely reflects past selection by predators.

The ability to learn also varies among individuals and populations, and this variation is often heritable. Numerous single-gene mutants impair learning in *Drosophila* (Davis, 1996), and artificial selection experiments can select for better learners (Mery et al., 2007b). In addition, many genes that affect memory in *Drosophila* also do so in mice (Davis, 2005), so at least some learning pathways may be conserved among taxa. Finally, variation in learning ability among natural populations of the pond snail, *Lymnaea stagnalis*, is demonstrably heritable (Orr et al., 2008), as is variation in learning within a natural population of *Drosophila*, where it has been traced to a polymorphism at a single locus (Mery et al., 2007a). Although these differences could be due either to past selection or to genetic drift, heritable variation in learning ability clearly does exist in natural populations.

DOES EXPRESSION OF AN INDUCED DEFENSE EXPOSE CRYPTIC GENETIC VARIATION?

Evidence that induced defenses expose cryptic genetic variation is limited, but one excellent example confirms that it can happen. Agrawal et al. (2002) raised 28 paternal half-sibling families of wild radish in the absence and presence of an herbivorous caterpillar. In plants exposed to caterpillars, defensive compounds (glucosinolates) varied widely among families (some families consistently produced nearly twice as much as others). In the absence of herbivory, however, not only did all families produce fewer glucosinolates but glucosinolate production did not differ among families.

Other observations hint that such cryptic variation may be widespread. Many predator-induced defenses involve the development of structures that are greatly reduced or absent during normal development. For instance, intertidal whelks (*Nucella*) develop large apertural teeth when exposed to effluent from crab predators, but these teeth are greatly reduced or absent among thin-shelled whelks grown without crabs (Appleton and Palmer, 1988; Palmer, 1990). Therefore, presuming that some of the variation in apertural tooth development is heritable, variation in the underlying norm of reaction will be visible to selection only in the presence of predators.

STUDYING THE ROLE OF LEARNING IN THE EVOLUTION OF INDUCED MORPHOLOGICAL DEFENSES

Although some evidence clearly exists for the four basic components of our model, we are not aware of studies that have directly investigated the role of learning in the evolution of induced morphological defenses. The distressing global prevalence of introduced species (Carlton and Geller, 1993) should provide many replicated, "natural" experiments to study the evolution of induced defenses (Sax et al., 2007). For example, some systems worth further study include (1) the European green crab (*Carcinus maenas*), which has invaded both North American coasts; (2) the Asian shore crab (*Hemigrapsus sanguineus*), recently introduced to the north Atlantic (McDermott, 1998); and (3) the American bullfrog (*Rana catesbeiana*), which has spread throughout much of the world (Ficetola et al., 2007). Many others undoubtedly exist.

Once systems are identified, adaptation to a novel predator may be easily studied in real time because it can happen quickly (sometimes within 20 generations) (Prentis et al., 2008). However, in other cases, where adaptation is slower, a geographic survey of populations along a continuum of exposure (ranging from completely naive populations to those with a long history of overlap) could serve as a proxy for the study of a single population through time (Trussell, 2000).

While many native species have evolved in response to recent invaders (reviewed in Strauss et al., 2006), rarely are specific aspects of the response studied in detail, making inferences about the role of learning difficult. Therefore, both learned and innate components of predator detection must be documented (for a nice example, see Epp and Gabor, 2008) along a continuum of overlap between predator and prey. Ideally, first- or second-generation lab-reared individuals should be tested along with field-caught individuals, and this should be repeated with populations from areas with and without novel predators.

In perhaps the closest study of this kind to date, Langkilde (2009a) surveyed populations of eastern fence lizards along a continuum of overlap between lizards and their invasive fire ant predators. In populations with longer exposure to fire ants, adult lizards were more likely to twitch and flee, both effective antipredator behaviors. Interestingly, when lab-reared juveniles were exposed to fire ants, all populations showed antipredator behavior, suggesting that evolution is acting on variation in the retention of the response into adulthood. However, the influence of learning in this system is still not well understood; although exposure of adult lizards to fire ants does not change the observed antipredator response (Langkilde, 2009b), the influence of repeated exposure to fire ants throughout ontogeny is still unknown and may play a role in the development of antipredator behavior.

Once similar data exist for induced defenses in multiple systems, specific questions about the role of learning in the evolution of induced defenses may be asked, including the following: (1) How often does a learned response precede an innate one? (2) Do species that learn adapt more quickly to a novel threat? and (3) Are prey populations larger where individuals can learn to recognize a novel predator?

EXAMPLE II: LEARNING, HANDEDNESS, AND THE EVOLUTION OF MORPHOLOGICAL ASYMMETRY

Despite the widespread occurrence of morphological asymmetries in many animal groups, we know surprisingly little about how asymmetrical forms evolved from symmetrical ancestors. In particular, how does novel phenotypic variation in morphological asymmetry arise that is large enough to affect performance or fitness? Learning-enhanced morphological plasticity offers an intriguing possibility.

Morphological asymmetry is an ideal trait for studying the roles of learning and developmental plasticity in the evolution of novel forms, as it is for studying the evolution of developmental pathways in general, for several reasons (Palmer, 2004; Levin and Palmer, 2007): (1) it is an easily characterized morphological state (asymmetrical vs. symmetrical); (2) morphological asymmetries have evolved from symmetrical ancestors in many taxa, so multiple independent tests are possible; and (3) much is known about the ontogeny of asymmetry at both the anatomical and molecular levels, at least in some groups. Finally, handed morphologies are often associated with handed behaviors, so the role of handed behavior in the ontogeny and evolution of morphological asymmetry may be investigated.

Asymmetrical traits (opercula in serpulid polychaetes, claws in many crustaceans, male genitalia in earwigs and caddisflies, sound-producing file-scraper systems in crickets and katydids, mouth deflection in scale-eating fishes, mate-clasping structures in male phallostethid fishes, crossed bills in some seed-eating birds) and whole-body asymmetries (gastropods, cemented bivalves, coiled polychaetes, verrucomorph barnacles, flatfishes) have evolved in many animal taxa (Ludwig, 1932; Neville, 1976; Palmer, 2005). All imply clear functional benefits to being asymmetrical in certain circumstances. That such a diverse array of morpho-

logical asymmetries has evolved is surprising, given how difficult it is to generate phenotypic variation for asymmetry that is both large enough to effect performance and heritable.

For an asymmetrical morphology to evolve, three significant hurdles must be overcome. First, subtle, random departures from bilateral symmetry in typically symmetrical species—so-called fluctuating asymmetries—are quite small, generally 1%–2% of trait size (Palmer and Strobeck, 2003). Such small phenotypic differences will not likely affect performance and, thus, will be selectively neutral. Second, even if they did affect performance, these subtle departures from symmetry *in a particular direction* are not normally heritable, at least in the very few studies that have looked (Palmer, 2004). Third, in bilaterian animals, bilateral symmetry may be a default state developmentally (Palmer, 2004) such that wholly novel signaling mechanisms or regulatory connections must be coupled to asymmetrical stimuli, either from the external environment or from the intracellular environment (Levin and Palmer, 2007), before morphological asymmetries may evolve.

A THEORETICAL MODEL

Learning, when combined with developmental plasticity, may significantly accelerate the evolution of asymmetrical morphologies from symmetrical ancestors anywhere an asymmetrical form is connected to an asymmetrical behavior (Figure 19.2). Following a model illustrating how this could occur, we will review evidence for the essential elements of the model and both ontogenetic and phylogenetic evidence for the past role of handed behavior in the evolution of morphological asymmetries. Throughout the discussion of this model, (1) "handed behavior" refers to the preferential use of the right (or left) limb *by an individual*, not to the consistent use of the right (or left) side by most or all individuals within a species, and (2) "asymmetry" refers to antisymmetry (overdevelopment of the right [or left] side in an individual but random orientation of asymmetry within a species).

First, consider the role that learning plays in the ontogeny and evolution of lateralized or handed behaviors. In the absence of learning, selection must act directly on factors that determine handed behavior constitutively in both juveniles and adults (Figure 19.2A). Although potentially possible, little opportunity for selection exists because handed behavior will vary among individuals initially only due to random binomial variation. In contrast, learning reinforces both direction (right or left) and consistency of lateralized behaviors (Biddle and Eales, 2006)—even if no preferences exist initially in juveniles—so significant variation in handed behavior may develop within one generation (Figure 19.2B, left panels). Finally, because handed behavior in this example results entirely from learning, selection for increased handed behavior actually results in selection for increased learning ability (Figure 19.2B, middle panels). In other words, degree of handed behavior in an individual can serve as a simple and valuable proxy for learning ability.

Second, consider the impact of handed behavior on the development and evolution of morphological asymmetry.[1] Preferential use of one side by an individual (i.e., handed behavior, Figure 19.2A) may induce more robust development of that side via developmental plasticity. This necessarily makes that individual morphologically asymmetrical and does so within that individual's life span, regardless of whether handed behavior is constitutive (Figure 19.2C, right panels) or entirely learned (Figure 19.2D, left panels). If more robust development of the preferentially used side enhances its performance sufficiently so that some individuals exceed a performance threshold (P in Figure 19.2C,D), then selection can begin to act directly on heritable variation affecting the development of morphological asymmetry (Figure 19.2D,

middle and right panels). Finally, heritable variation for morphological asymmetry (Figure 19.2D, lower right panel) may further facilitate the development of handed behavior (Figure 19.2B, right panels), which, in turn, will amplify morphological asymmetry (Figure 19.2D, upper right panel) via a kind of positive feedback loop.

The key elements of this model are as follows: (1) learning promotes handed behavior within the life span of an individual, (2) handed behavior can induce or amplify morphological asymmetry in an individual, (3) handed behavior or morphological asymmetry improves performance in some fashion, and (4) variation among individuals in induced morphological asymmetries may expose cryptic or previously unexpressed, heritable variation for handed behavior or morphological asymmetry to selection. What evidence supports these element?

LEARNING AND HANDED BEHAVIOR

Handed behaviors occur in many animals, including vertebrates (McManus, 2002; Rogers and Andrew, 2002; Malashichev and Wassersug, 2004) and invertebrates (Skapec and Stys, 1980; Babcock, 1993). Several lines of evidence suggest that handed behaviors are primarily learned, not innate. First, experimental attempts to select for handed behavior—*in a particular direction*—in paw use by mice (Collins, 1969) and in wing crossing or turning direction by fruit flies (Purnell and Thompson, 1973; Ehrman et al., 1978) were wholly unsuccessful, confirming that differences in the direction of handed behavior among individuals ("right-handed" vs. "left-handed") were not heritable. Although variation in the orientation of preferred side was not heritable, the strength and consistency of lateralized behavior in an individual did vary among genetic strains, so this variation was clearly heritable (Collins, 1991; Biddle et al., 1993).

Second, a fascinating recent study reveals how degree of lateralized paw use depends on learning ability in different inbred strains of mice (Ribeiro et al., 2010). Individuals of some genetically well-characterized strains (e.g.,

---

1. Throughout the discussion of this model, "asymmetry" refers to antisymmetry = random asymmetry (overdevelopment of the right [or left] side in an individual but random orientation of asymmetry within a species).

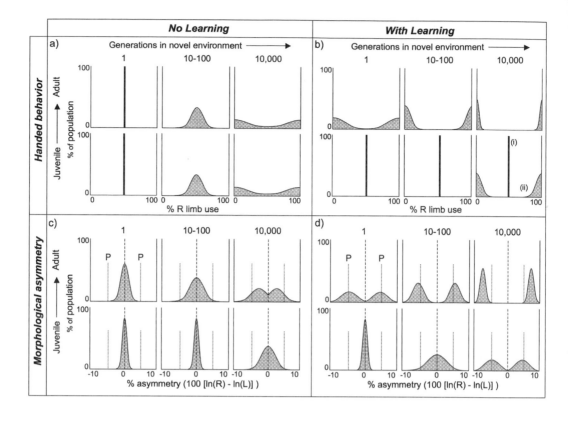

C57BL/6J) develop a clear paw preference—either right or left, at random—whereas individuals of other strains (e.g., CDS/Lay) use their paws at random (i.e., an individual mouse never develops a strong paw preference) (Figure 19.3). Both behavioral studies and modeling reveal that a key difference between strains is ability to learn from prior experience (Ribeiro et al., 2010): in strain C57BL/6J, initial use of the right paw greatly increases the likelihood of using the right paw in subsequent reaches, whereas in strain CDS/Lay, initial use of the right paw has little impact on subsequent right paw use. In addition, preferential right paw use can be induced by rearing both mice (Biddle and Eales, 1999) and rats (Wentworth, 1933) in chambers where they are constrained to use their right paw to reach when young. Furthermore, handed behaviors become more pronounced with increased experience in other ani-mals, including humans (Perelle et al., 1981) and marmosets (Hook and Rogers, 2000).

HANDED BEHAVIOR CAN INDUCE OR AMPLIFY MORPHOLOGICAL ASYMMETRY

Handed behavior induces or amplifies morphological asymmetry in many animal groups. Although the most convincing evidence comes from vertebrates, invertebrate examples exist as well.

Vertebrate bones are highly plastic structures. Load frequency or magnitude can alter mineral density, internal microstructure, and the cross-sectional size and shape of bones (Ruff et al., 2006)—even after overall body growth has ceased (Loitz and Zernicke, 1992)—and both the receptors that transduce mechanical signals and the signaling pathways that ultimately yield modified bone form are reasonably well understood (Rubin et al., 2006). Collec-

FIGURE 19.2 Effects of learning and developmental plasticity on the ontogeny and evolution of handed behavior (A, B) and morphological asymmetry (C, D). Frequency distributions for juveniles represent exclusively heritable variation in limb use (A, B) or morphological asymmetry (C, D). Differences in frequency-distribution shape between juveniles and adults arise via either learning (A, B) or morphological plasticity (C, D). R, right; L, left; P, performance threshold. (A, B) Effect of learning on the ontogeny and evolution of handed behavior in an environment where lateralized behavior is favored (*handed behavior* refers to preferential use of one limb by an individual; among individuals, either the right or left limb may be preferred, at random). (A) Before selection (generation 1) for handed behavior, juveniles exhibit no limb preference initially and, in the absence of learning, no limb preference develops during ontogeny. In the absence of learning, selection must act on handed behavior directly, yielding variation in an inborn, purely instinctive handed behavior that is weak initially (generations 10–100) and that may require many generations before strong handed behavior is evident in juveniles (generation 10,000). (B) In a species capable of learning, although juveniles exhibit no limb preference initially, a limb preference (either right or left, at random) develops during ontogeny because choice of which limb to use is reinforced by past experience (generation 1). In contrast to A, selection acts on learning ability rather than handed behavior directly, so even after selection juveniles exhibit no limb preference initially but adults develop stronger handed behavior due to more rapid or effective learning (generations 10–100). Eventually (generation 10,000), adults exhibit quite strong handed behavior, which develops either via enhanced learning ability (*i*) or via new, heritable variation for handed behavior in juveniles (*ii*) that is reinforced by learning. In the absence of learning, many generations of selection may be required to yield significant handed behavior (A), whereas organisms capable of learning develop handed behavior within the life span of an individual (B, left panels). Note that in all panels of A and B distribution widths are independent of number of observations, as the percentage of right-limb use by an individual is presumed to be measured without error or to be based on a very large number of observations. (C, D) Ontogeny and evolution of asymmetry in an environment where morphological asymmetry is favored and where developmental plasticity amplifies it in proportion to the extent of handed behavior in A and B. Vertical dashed line indicates perfect symmetry; vertical dotted line indicates an arbitrary performance threshold (P) of approximately 5% difference between sides, below which phenotypic variation has no effect on performance (i.e., is neutral with respect to selection). *Morphological asymmetry* refers to greater development of one limb in an individual; among individuals, either the right or left limb may be overdeveloped, at random. (C) Before selection, deviations from symmetry in juveniles arise solely due to developmental noise (typically 1%–2% of trait size). These may be amplified slightly by developmental plasticity due to random variation in limb use among individuals but not sufficiently to cross the performance threshold (P), so selection has limited ability to act on asymmetry directly (generation 1). After selection for handed behavior (see A), deviations from symmetry are amplified by developmental plasticity such that some individuals may cross the performance threshold as adults, though not as juveniles (generations 10–100). Eventually, selection for heritable asymmetry variation in adults exposes heritable asymmetry variation in juveniles that is further amplified by handed behavior in A during development (generation 10,000). (D) Before selection, deviations from symmetry in juveniles again arise solely due to developmental noise and learned, handed behavior (see B) amplifies these via developmental plasticity such that some individuals cross the performance threshold (P, generation 1). Selection may therefore act directly on heritable variation, affecting morphological asymmetry such that greater variation in asymmetry arises early in development and handed behavior further amplifies it via developmental plasticity (generations 10–100). Eventually, because more and more individuals exhibit asymmetries that cross the performance threshold, selection for asymmetry variation in adults exposes heritable asymmetry variation in juveniles that is further amplified by handed behavior in c during development (generation 10,000).

tively, changes in bone form alter its mechanical properties to better perform under new loading regimes (Ruff et al., 2006). For example, cross-sectional area and second moment of area (a measure of mechanical resistance to bending and shear) are higher in hand bones of technical rock climbers (Sylvester et al., 2006) and limb bones of weight lifters (Frost, 1997) compared to the normal human population.

Human limb bones are often asymmetric; however, asymmetries are most pronounced in cross-sectional areas rather than lengths. Arm bones are on average broader and heavier on the right side, while leg bones are broader and heavier (though less so) on the left (Kimura and Konishi, 1981; Auerbach and Ruff, 2006). These asymmetries appear to be induced by handed behavior. In human hand bones, both total cross-sectional area and second moment of area of the second metacarpal (the hand bone immediately proximal to the index finger) were greater in the preferred hand of a large sample of left- and right-handers (Roy et al., 1994). Remarkably, the cortical cross-sectional area of the upper arm bone (humerus) is greater in the racket arm of professional tennis players by nearly 38% (Trinkaus et al., 1994), as it is in the lower racket arm bones (by up to 12% in the radius or 10% in the ulna) in French regional-level tennis players (Ducher et al., 2004). These

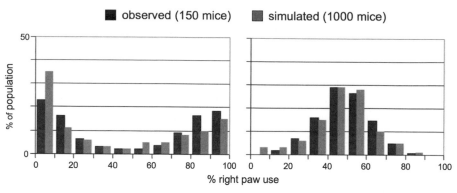

FIGURE 19.3 Frequency distributions of the consistency of right-paw use (50 reaches per mouse) among 150 individual mice from two laboratory strains (modified from Ribeiro et al., 2010) and among 1,000 simulations of mice with different learning abilities. (A) Strain C57BL/6J exhibits a strong lateralized paw use: Nearly half (49%) used the right (or left) paw at least 90% of the time (<10% and >90% pooled). (B) Strain CDS/Lay exhibits no lateralized paw use: Fewer than 5% used the right (or left) paw at least 90% of the time (<10% and >90% pooled).

differences contrast strikingly with the general human population, where such differences are generally small (1.5%–4%) (Trinkaus et al., 1994). As further evidence that developmental plasticity is responsible for these skeletal asymmetries, the difference between sides was greater the longer the individual had been playing tennis (Ducher et al., 2004).

Use-induced skeletal asymmetries occur in other vertebrates as well. Racing greyhounds run counterclockwise around the track. This yields greater stresses on the outer (right) limbs, which in turn promotes greater bone mineral density in the right central tarsal bone in the foot, a difference that is much reduced in retired racing greyhounds used for breeding (Johnson et al., 2000). Leg bone cross-sectional areas were significantly lower on the side where leg muscles had been experimentally cut when a dog was young (Howell, 1917). Handed behavior appears to induce skeletal asymmetries in crossbill finches (*Loxia*), where the upper mandible crosses to the right or the left of the lower mandible: "Because of the way the bird holds the cone, most of the strain is taken by the foot on the opposite side to that to which the lower mandible crosses [so] a left-billed bird . . . has a correspondingly longer tarsus [the right] side" (Knox, 1983, 117). Left-footed parakeets also develop slightly longer left legs, whereas the right

legs are longer on right-footed individuals (McNeil et al., 1971).

Use-induced morphological asymmetries are less widely known among invertebrates, but we suspect they are widespread in species that exhibit morphological antisymmetry (equal frequencies of right- and left-handed individuals in a species) (Palmer, 2005). An old study suggests that development of the teeth in the sound-producing file-scraper system of overlapping locust wings is induced, at least in part, by friction of one wing rubbing over another (Graber, 1872), but this hasn't been confirmed. Differential use of one of two symmetrical claws in juvenile American lobsters clearly determines the side on which the crusher claw develops (Govind and Pearce, 1986) (see details below under Ontogenetic and Phylogenetic Evidence . . .). Finally, a handed behavior induces conspicuous morphological asymmetry in male fiddler crabs and theridiid spiders. In fiddler crabs, males begin to develop two symmetrical large claws, one of which autotomizes and regenerates as a small claw and the other of which continues to develop into the hypertrophied signaling claw (Yamaguchi, 1977). In theridiid spiders, males pull off one palp immediately after the terminal molt, leaving only a single palp on the right or left side with which to mate (Knoflach and Benjamin, 2003). Whether the handed behavior is learned or simply random in these last two cases remains unknown).

FUNCTIONAL BENEFITS OF HANDED BEHAVIOR OR MORPHOLOGICAL ASYMMETRY

As with many learned behaviors, performance or skill of a handed behavior appears to increase with experience. In bumblebees (Cheverton, 1982), pigeons (Güntürkün et al., 2000), domestic chickens (Rogers, 2000), cats (Fabre-Thorpe et al., 1991), and wild chimpanzees (McGrew and Marchant, 1999), individuals with more strongly lateralized behaviors perform better than those with less strongly lateralized behavior. In other words, repeated use of the same side allows tasks to be completed more effectively because speed and dexterity are enhanced through practice. One conspicuous exception is hemipene use by lizards and snakes, which typically alternates between the right and left sides during mating, perhaps to deliver more sperm per copulation (Tokarz and Slowinski, 1990; Shine et al., 2000).

The widespread occurrence of morphological asymmetries in many animal groups (Ludwig, 1932; Neville, 1976) strongly suggests they are adaptive. Nonetheless, direct evidence of functional benefits to morphological asymmetry is sparse. In scale-eating cichlid fishes, deflection of the mouth toward one side allows them to attack prey fish more from behind (Hori, 1993). The laterally twisted bills of crossbill finches allow them to better pry apart the scales and remove seeds from closed conifer cones than if the bills are made symmetrical experimentally (Benkman, 1988). Male fiddler crabs with relatively larger master claws, which are therefore more asymmetrical, are more successful at attracting females (Oliveira and Custódio, 1998). Finally, lobsters with one large crusher claw are better able to crush hard-shelled prey than those that possess two cutter claws (Govind, 1989).

LEARNING AND MORPHOLOGICAL PLASTICITY UNMASK CRYPTIC GENETIC VARIATION

Direct evidence that handed behaviors, or induced morphological asymmetries, unmask cryptic genetic variation is limited at best. Individual mice allowed to feed in a chamber that favors use of one paw (either right or left) develop a handed behavior (Biddle and Eales, 1999). However, different domestic strains of mice develop and retain the learned handed behavior to different degrees that reflect heritable variation in learning ability (Biddle and Eales, 2006). Clearly, environments that induce handed behavior would unmask cryptic heritable variation in learning ability in these mice.

We can only speculate that use-induced morphological asymmetries unmask heritable variation for degree (though not direction) of asymmetry. Not all individuals will likely exhibit the same degree of induced morphological asymmetry in response to a lateralized behavior. If cryptic genetic variation is being unmasked, then relative variability in asymmetry should be higher among individuals where handed behaviors induce morphological asymmetries.

ONTOGENETIC AND PHYLOGENETIC EVIDENCE FOR A ROLE OF HANDED BEHAVIOR IN THE EVOLUTION OF MORPHOLOGICAL ASYMMETRY

If handed behavior facilitates the evolution of conspicuous morphological asymmetries, two kinds of evidence should emerge from studies of either ontogeny or phylogenetic history of morphological asymmetries: (1) *ontogenetic evidence*, handed behavior should precede handed morphology developmentally, where current ontogeny is presumed to retain elements of ancestral ontogeny (Gould, 1977), and (2) *phylogenetic evidence*, handed behaviors, in the absence of morphological asymmetry, should exist in sister taxa to groups in which morphological asymmetry in the structure associated with the handed behavior evolved later. Although limited, examples of both kinds of evidence exist.

The clearest evidence that handed behavior plays an essential role in the ontogeny of a morphological asymmetry comes from elegant studies of asymmetrical claw development in American lobsters (*Homarus americanus*) (Govind and Pearce, 1986; Govind, 1989). Young

TABLE 19.1

Effect of Claw Use and Handed Behavior on Direction of Claw Asymmetry in the American Lobster *Homarus Americanus*

TREATMENT DURING JUVENILE MOLT FIVE	SIDE OF CRUSHER CLAW			
	%R	%L	*n*	SOURCE
a. Soft food only	7.7	15.4	26	I
b. Soft food + oyster shell chips	42.1	50.0	18	I
c. Soft food + second juvenile lobster	39.1	47.8	23	I
d. Shell chips + autotomize left claw	95.8	4.2	24	II
e. Shell chips + denervate left claw	100	0	22	II
f. Soft food + exercise left claw	17.6	76.5	17	I

NOTE: Juvenile lobsters were held individually (except in c) in small glass bowls, were fed soft food, and experienced either (a) no additional stimuli, freedom to use either claw to manipulate hard objects (b) or to interact with another lobster (c), (d) removal of the left claw, (e) denervation of the left claw, or (f) extended artificial exercise of the left claw. R, right; L, left.
SOURCES: I, Govind and Pearce (1986); II Govind (1989).

juveniles possess symmetrical claws and pass through five postlarval intermolts before any claw asymmetry appears. Following intermolt five, one claw begins to transform into a larger, crusher-type claw and one retains the small, cutter-type form. Simple experiments (Table 19.1) confirm that whichever claw is used most during a critical window of the fifth molt transforms into the crusher claw. When allowed to use both claws freely to manipulate hard objects or to interact with another juvenile lobster, the side of the crusher claw is random (Table 19.1, b and c). Autotomy or denervation of the left claw almost exclusively induces the crusher claw to develop on the right side (Table 19.1, d and e), whereas elevated exercise of the left claw induces development of the crusher claw on the left side (Table 19.1, f). Remarkably, if neither claw is sufficiently stimulated during this critical molt, neither transforms into a crusher claw (Table 19.1, a), and that lobster will possess two small cutter-type claws for the rest of its life. Clearly, differential use of one claw—handed behavior—remains an essential inducer of handed morphology. In other words, the developmental program to produce a crusher claw still depends on handed behavior to trigger it.

Two intriguing examples suggest that handed behavior preceded handed morphology phylogenetically. First, right-handed penis use by earwigs possessing two penises appears to precede the loss of the left penis evolutionarily (Kamimura, 2006; Palmer, 2006). Males in the crown-group Eudermaptera possess only a single penis on the right side of their abdomen. Curiously, males of more basal earwig families possess two penises and, in nearly all families, mate with the left or right penis at random. Unique among paired penis–bearing earwigs, *Labidura riparia* (Labiduridae) males mate preferentially (90% of the time) with the right penis. Furthermore, 90% of males hold the right penis in the "ready" or cocked position, despite no apparent morphological differences between right and left penises and despite having a fully functional left penis. Thus, right-handed penis use in *L. riparia* appears to be exclusively behavioral (Kamimura, 2006). Significantly, the Labiduridae is the sister group to the crown-group Eudermaptera, which lack the left penis altogether. Thus, the inferred evolutionary sequence is (1) two penises, right or left used at random; (2) two penises (morphologically symmetrical), with a right-handed behavioral preference; and (3) anatomical loss of the left penis.

Second, handed morphology in the sound-producing file-scraper system on the wing covers of katydids (Tettigonoidea) also appears to have been preceded evolutionarily by symmetrical ancestors with a handed behavior (Masaki et al., 1987; Gwynne, 1995). In tettigoniid katydids, a clear morphological asymmetry exists: The left wing cover overlaps the right, and the file occurs only on the left and the scraper only on the right side. However, in haglid katydids, the sister group to the tettigoniid katydids, a "file" occurs on both the right and left wing covers of an individual but only one file can be used to sing at a time (Gwynne, 1995). Whether individual haglid katydids consistently cross one wing over the other is unknown, but singing by an individual at any one moment is necessarily a handed behavior (Gwynne, 1995). Finally, file development, at least in locusts, is initiated or enhanced, at least in part, by friction of one wing rubbing over the other (Graber, 1872).

CONCLUSIONS: LEARNING, ADAPTIVE PLASTICITY, AND RATES OF EVOLUTION

Although the two models above focus on quite different phenomena, three common themes emerge about how learning may allow morphological evolution to proceed more quickly or in a novel direction.

First, the most obvious way learning-enhanced morphological plasticity may increase the rate or direction of morphological evolution is via population persistence. Where some morphological change is necessary to permit a population to persist in a novel environment, learning may facilitate expression of inducible phenotypes that are necessary for survival. In short, a population cannot evolve if all individuals quickly die off. For example, numerous recent extinctions can be traced to the arrival of a novel predator to which prey had no effective defense (Cowie, 1992; Forys et al., 2001; Wiles et al., 2003). In many of these cases, the ability to induce an appropriate plas-

tic response might have made the difference between extinction and adaptation. Of course, the idea that plasticity may facilitate population persistence is not new. Spalding (1873) believed that survival and reproduction of the quickest learners led to the evolution of certain instinctual behaviors. Baldwin (1896) extended this idea to suggest that plastic traits in general could keep certain animals alive (something he termed "organic selection") and thus direct the course of future evolution (what he termed "orthoplasy"). More recently, learned behaviors have been shown to influence how selection acts on existing trait variation (Price et al., 2003; West-Eberhard, 2003), and this "behavioral drive" may have contributed to the greater diversification of body size in families of birds with larger brains (Sol and Price, 2008). Here, we suggest that learning-enhanced morphological plasticity provides an additional mechanism whereby learning can enhance individual survival and population persistence in a novel environment.

Second, where learning enables a population to persist in a novel environment, many individuals will likely express at least a partial morphological response (West-Eberhard, 2003). As a result, populations in a novel environment that are capable of learning should be larger than those possessing only fixed, preexisting morphological variation. Because larger populations are more likely to achieve fixation of a beneficial allele that arises either from standing genetic variation or from a new mutation (Hermisson and Pennings, 2005), populations capable of learning should evolve more quickly. Even in the absence of appropriate standing variation, larger populations have more individuals in which a new beneficial mutation might arise. Therefore, to the extent that learning-enhanced morphological plasticity allows more individuals to survive in a new environment, it will both facilitate selection on standing variation and increase the chance that a beneficial mutation arises and becomes fixed.

Third, learning may facilitate morphological evolution in novel environments by exposing otherwise hidden (or "cryptic") genetic variation to selection. Although much has been written about how developmental processes can act as capacitors of cryptic genetic variation and how environmental perturbations may expose such variation to selection (Rutherford and Lindquist, 1998; Gibson and Dworkin, 2004; Suzuki and Nijhout, 2006; Moczek, 2007), learning has received much less attention as a potential releaser of cryptic genetic variation. For example, as detailed above (see Evolution of Induced Morphological Defense in Response to a Novel Predator), in the absence of learning any underlying heritable variation in the inducible shell-thickening pathway will remain "cryptic" and may be exposed to selection only once prey learn to recognize the novel predator. A similar argument can be made for the developmental pathways that yield an asymmetrical form via developmental plasticity (see above, Learning, Handedness, and the Evolution of Morphological Asymmetry) or any other instances where learning is required for the expression of a developmental pathway.

Studies of the genes or genetic regulatory networks underlying induced responses offer another fruitful avenue for future research (e.g., Ettensohn, 2009). Such studies are essential for understanding the evolution of developmental pathways responsible for specific plastic responses. One promising system for such studies is the water flea, *Daphnia magna* (De Meester, 1993). It offers a lovely example of how phenotypic variation among individuals in degree of plastic response is greatly amplified in a changed environment such that previously cryptic genetic variation is exposed. Different *Daphnia* clones in the same lake exhibit minor differences in a phototactic behavior when grown under fish-free conditions. These differences in plastic response among clones, however, are greatly amplified when chemical cues of a fish predator are introduced, and clones that exhibit a greater response are less vulnerable to fish predation (De Meester et al., 1995). Studies of the underlying genes and developmental pathways responsible for differences among clones would provide important information about how many genes are responsible for the observed behavioral differences exposed by the presence of fish cues.

Throughout, we have argued that learning-induced expression or amplification of plastic traits should accelerate morphological evolution. However, if learning itself enables morphological adaptation to a novel environment, why should any subsequent evolution occur? The inherent limits and costs of a learned response provide a partial answer. Learning increases individual fitness by reinforcing specific behaviors with positive or negative rewards (food capture, predator avoidance, mate choice). However, the correct positive or negative behavioral response to a situation requires time to develop. Time expended before the "correct" response develops is, in effect, time spent *not* performing the most adaptive behavior. In addition, learning can be costly: It requires that time and energy be expended performing less rewarding actions, and in the most extreme cases, it may entail damage due to a failed predation attempt. Furthermore, where individuals must learn through association of predator cues with cues from damaged conspecifics, they are more likely to be eaten while still undefended. Therefore, individuals that depend on learning to evoke a response will be either more vulnerable or less effective at a task than those where the response does not depend on learning.

In summary, nearly all animals exhibit some capacity to learn, and such learning almost invariably leads to adaptive behaviors. To this uncontroversial observation we merely add that learning has great potential to accelerate morphological evolution if adaptive behaviors yield adaptive morphological differences via developmental plasticity, as illustrated in detail in the two models above. Our claim that learning-enhanced morphological plasticity can reveal previously cryptic heritable variation is largely speculative, but this remains an important topic for future studies of plasticity and learning in general.

However, one conclusion emerges loud and clear: We can learn much by studying the ontogeny of behavior *in concert with* the ontogeny of form. Such studies would help to answer key questions arising from the two models presented above. In the case of antipredator traits: In a prey species that can "recognize" novel predator cues and produce an appropriately altered defensive morphology, is this a learned response (e.g., associative learning over the life span of an individual) or a response to selection in detection or interpretation of signals (e.g., a constitutive response)? In the case of asymmetry: Does handed behavior bias the direction of asymmetry, or does morphological asymmetry bias the direction of handed behavior?

REFERENCES

Agrawal, A.A., J.K. Conner, M.T. J. Johnson, and R. Wallsgrove. 2002. Ecological genetics of an induced plant defense against herbivores: Additive genetic variance and costs of phenotypic plasticity. *Evolution* 56:2206–13.

Appleton, R.D., and A.R. Palmer. 1988. Water-borne stimuli released by predatory crabs and damaged prey induce more predator-resistant shells in a marine gastropod. *Proc Natl Acad Sci USA* 85:4387–91.

Auerbach, B., and C.B. Ruff. 2006. Limb bone bilateral asymmetry: Variability and commonality among modern humans. *J Hum Evol* 50: 203–18.

Babcock, L.E. 1993. Trilobite malformations and the fossil record of behavioral asymmetry. *J Paleontol* 67:217–29.

Baldwin, J.M. 1896. A new factor in evolution. *Am Nat* 30:354–451, 536–53.

Benkman, C.W. 1988. On the advantages of crossed mandibles: An experimental approach. *Ibis* 130:288–93.

Biddle, F.G., C.M. Coffaro, J.E. Ziehr, and B.A. Eales. 1993. Genetic variation in paw preference (handedness) in the mouse. *Genome* 36:935–43.

Biddle, F.G., and B.A. Eales. 1999. Mouse genetic model for left–right hand usage: Context, direction, norms of reaction, and memory. *Genome* 42:1150–66.

Biddle, F.G., and B.A. Eales. 2006. Hand-preference training in the mouse reveals key elements of its learning and memory process and resolves the phenotypic complexity in the behaviour. *Genome* 49:666–77.

Carlton, J.T., and J.B. Geller. 1993. Ecological roulette—the global transport of nonindigenous marine organisms. *Science* 261:78–82.

Cheverton, J. 1982. Bumblebees may use a suboptional arbitrary handedness to solve difficult foraging decisions. *Anim Behav* 30:934–5.

Chivers, D.P., B.D. Wisenden, and R.J. F. Smith. 1996. Damselfly larvae learn to recognize predators from chemical cues in the predator's diet. *Anim Behav* 52:315–20.

Collins, R.L. 1969. On the inheritance of handedness. II: Selection for sinistrality in mice. *J Hered* 60:117–19.

Collins, R.L. 1991. Reimpressed selective breeding for lateralization of handedness in mice. *Brain Res* 564:194–202.

Cowie, R.H. 1992. Evolution and extinction of Partulidae, endemic Pacific island land snails. *Philos Trans R Soc B Biol Sci* 335:167–91.

Dalesman, S., S.D. Rundle, R.A. Coleman, and P.A. Cotton. 2006. Cue association and antipredator behaviour in a pulmonate snail, *Lymnaea stagnalis*. *Anim Behav* 71:789–97.

Davis, R.L. 1996. Physiology and biochemistry of *Drosophila* learning mutants. *Physiol Rev* 76: 299–317.

Davis, R.L. 2005. Olfactory memory formation in *Drosophila*: From molecular to systems neuroscience. *Annu Rev Neurosci* 28:275–302.

Day, T., and J.D. McPhail. 1996. The effect of behavioral and morphological plasticity on foraging efficiency in the threespine stickleback (*Gasterosteus* sp.). *Oecologia* 108:380–8.

De Meester, L. 1993. Genotype, fish-mediated chemicals, and phototactic behavior in *Daphnia magna*. *Ecology* 74:1467–74.

De Meester, L., L.J. Weider, and R. Tollrian. 1995. Alternative antipredator defences and genetic polymorphism in a pelagic predator–prey system. *Nature* 378:483–5.

Ducher, G., S. Prouteau, D. Courteix, and C.-L. Benhamou. 2004. Cortical and trabecular bone at the forearm show different adaptation patterns in response to tennis playing. *J Clin Densitom* 7:399–405.

Edgell, T.C., and C.J. Neufeld. 2008. Experimental evidence for latent developmental plasticity: Intertidal whelks respond to a native but not an introduced predator. *Biol Lett* 4:385–7.

Ehrman, L., J. Thompson, I. Perelle, and B. Hisey. 1978. Some approaches to the question of *Drosophila* laterality. *Genet Res* 32:231–8.

Epp, K.J., and C.R. Gabor. 2008. Innate and learned predator recognition mediated by chemical signals in *Eurycea nana*. *Ethology* 114:607–15.

Ettensohn, C. A. 2009. Lessons from a gene regulatory network: Echinoderm skeletogenesis provides insights into evolution, plasticity and morphogenesis. *Development* 136:11–21.

Fabre-Thorpe, M., J. Fagot, and J. Vauclair. 1991. Cats paw preference in pointing towards a moving target. *C R Acad Sci Life Sci* 313:427–33.

Ferrari, M. C. O., and D. P. Chivers. 2008. Cultural learning of predator recognition in mixed-species assemblages of frogs: The effect of tutor-to-observer ratio. *Anim Behav* 75:1921–5.

Ficetola, G. F., W. Thuiller, and C. Miaud. 2007. Prediction and validation of the potential global distribution of a problematic alien invasive species—the American bullfrog. *Divers Distrib* 13:476–85.

Forys, E. A., A. Quistorff, C. R. Allen, and D. P. Wojcik. 2001. The likely cause of extinction of the tree snail *Orthalicus reses reses* (Say). *J Mollusc Stud* 67:369–76.

Frost, H. M. 1997. Why do marathon runners have less bone than weight lifters? A vital-biomechanical view and explanation. *Bone* 20:183–9.

Gibson, G., and I. Dworkin. 2004. Uncovering cryptic genetic variation. *Nat Rev Genet* 5:681–90.

Gonzalo, A., P. Lopez, and J. Martin. 2007. Iberian green frog tadpoles may learn to recognize novel predators from chemical alarm cues of conspecifics. *Anim Behav* 74:447–53.

Gould, S. J. 1977. *Ontogeny and Phylogeny*. New York: Harvard University Press.

Govind, C. K. 1989. Asymmetry in lobster claws. *Am Sci* 77:468–74.

Govind, C. K., and J. Pearce. 1986. Differential reflex activity determines claw and closer muscle asymmetry in developing lobsters. *Science* 233:354–6.

Graber, V. 1872. Uber der Tonapparat der Locustiden, ein Beitrage zum Darwinismus. *Z Wissen Zool* 22:100–19.

Güntürkün, O., B. Diekamp, M. Manns, F. Nottelmann, H. Prior, A. Schwarz, and M. Skiba. 2000. Asymmetry pays: Visual lateralization improves discrimination success in pigeons. *Curr Biol* 10:1079–81.

Gwynne, D. T. 1995. Phylogeny of the Ensifera (Orthoptera): A hypothesis supporting multiple origins of acoustical signaling, complex spermatophores and maternal care in crickets, katydids, and weta. *J Orthopteran Res* 4:203–18.

Hermann, P. M., A. Lee, S. Hulliger, M. Minvielle, B. Ma, and W. C. Wildering. 2007. Impairment of long-term associative memory in aging snails (*Lymnaea stagnalis*). *Behav Neurosci* 121:1400–14.

Hermisson, J., and P. S. Pennings. 2005. Soft sweeps: Molecular population genetics of adaptation from standing genetic variation. *Genetics* 169:2335–52.

Hook, M. A., and L. J. Rogers. 2000. Development of hand preferences in marmosets (*Callithrix jacchus*) and effects of aging. *J Comp Psychol* 114:263–71.

Hori, M. 1993. Frequency-dependent natural selection in the handedness of scale-eating cichlid fish. *Science* 260:216–19.

Howell, J. A. 1917. An experimental study of the effect of stress and strain on bone development. *Anat Rec* 13:233–52.

Johnson, K. A., P. Muir, R. G. Nicoll, and J. K. Roush. 2000. Asymmetric adaptive modeling of central tarsal bones in racing greyhounds. *Bone* 27: 257–63.

Kamimura, Y. 2006. Right-handed penises of the earwig *Labidura riparia* (Insecta, Dermaptera, Labiduridae): Evolutionary relationships between structural and behavioral asymmetries. *J Morphol* 267:1381–9.

Kiesecker, J. M., and A. R. Blaustein. 1997. Population differences in responses of red-legged frogs (*Rana aurora*) to introduced bullfrogs. *Ecology* 78:1752–60.

Kimura, K., and M. Konishi. 1981. Handedness and laterality in some measurements of the human upper limb. *Hum Ecol Race Hyg* 47:51–61.

Kishida, O., G. C. Trussell, and K. Nishimura. 2007. Geographic variation in a predator-induced defense and its genetic basis. *Ecology* 88: 1948–54.

Knoflach, B., and S. P. Benjamin. 2003. Mating without sexual cannibalism in *Tidarren sisyphoides* (Araneae, Theridiidae). *J Arachnol* 31:445–8.

Knox, A. G. 1983. Handedness in crossbills *Loxia* and the akepa *Loxops coccinea*. *Bull Br Ornithol Club* 103:114–18.

Langkilde, T. 2009a. Invasive fire ants alter behavior and morphology of native lizards. *Ecology* 90:208–17.

Langkilde, T. 2009b. Repeated exposure and handling effects on the escape response of fence lizards to encounters with invasive fire ants. *Anim Behav* 79:291–8.

Levin, M., and A. R. Palmer. 2007. Left–right patterning from the inside out: Widespread evidence for intracellular control. *Bioessays* 29:271–87.

Loitz, B. J., and R. F. Zernicke. 1992. Strenuous exercise-induced remodelling of mature bone: Relationships between in vivo strains and bone mechanics. *J Exp Biol* 170:1–18.

Ludwig, W. 1932. *Das Rechts-Links Problem im Teirreich und beim Menschen*. Berlin: Springer.

Malashichev, Y.B., and R.J. Wassersug. 2004. Left and right in the amphibian world: Which way to develop and where to turn? *Bioessays* 26:1–11.

Masaki, S., M. Kataoka, K. Shirato, and M. Nakagahara. 1987. Evolutionary differentiation of right and left tegmina in crickets. In *Evolutionary Biology of Orthopteroid Insects*, ed. B. Baccetti, 347–57. Chichester, UK: Horwood.

Maynard Smith, J. 1987. When learning guides evolution. *Nature* 329:761–2.

McDermott, J.J. 1998. The western Pacific brachyuran (*Hemigrapsus sanguineus*: Grapsidae), in its new habitat along the Atlantic coast of the United States: Geographic distribution and ecology. *Ices J Mar Sci* 55:289–8.

McGrew, W.C., and L.F. Marchant. 1999. Laterality of hand use pays off in foraging success for wild chimpanzees. *Primates* 40:509–13.

McManus, I.C. 2002. *Right Hand Left Hand. The Origins of Asymmetry in Brains, Bodies, Atoms and Cultures*. Cambridge, MA: Harvard University Press.

McNeil, R., J.R. Rodriguez, and D.M. Figuera. 1971. Handedness in the brown-throated parakeet *Aratinga pertinax* in relation with skeletal asymmetry. *Ibis* 113:494–9.

Mery, F., A.T. Belay, A.K. C. So, M.B. Sokolowski, and T.J. Kawecki. 2007a. Natural polymorphism affecting learning and memory in *Drosophila*. *Proc Natl Acad Sci USA* 104:13051–5.

Mery, F., J. Pont, T. Preat, and T.J. Kawecki. 2007b. Experimental evolution of olfactory memory in *Drosophila melanogaster*. *Physiol Biochem Zool* 80:399–405.

Miner, B.G., S.E. Sultan, S.G. Morgan, D.K. Padilla, and R.A. Relyea. 2005. Ecological consequences of phenotypic plasticity. *Trends Ecol Evol* 20:685–92.

Moczek, A.P. 2007. Developmental capacitance, genetic accommodation, and adaptive evolution. *Evol Dev* 9:299–305.

Neville, A.C. 1976. *Animal Asymmetry*. London: Edward Arnold.

Oliveira, R.F., and M.R. Custódio. 1998. Claw size, waving display and female choice in the European fiddler crab, *Uca tangeri*. *Ethol Ecol Evol* 10:241–51.

Orr, M.V., K. Hittel, and K. Lukowiak. 2008. Comparing memory-forming capabilities between laboratory-reared and wild *Lymnaea*: Learning in the wild, a heritable component of snail memory. *J Exp Biol* 211:2807–16.

Palmer, A.R. 1985. Adaptive value of shell variation in *Thais lamellosa*—effect of thick shells on vulnerability to and preference by crabs. *Veliger* 27:349–56.

Palmer, A.R. 1990. Effect of crab effluent and scent of damaged conspecifics on feeding, growth, and shell morphology of the Atlantic dogwhelk *Nucella lapillus* (L). *Hydrobiologia* 193:155–82.

Palmer, A.R. 2004. Symmetry breaking and the evolution of development. *Science* 306:828–33.

Palmer, A.R. 2005 Antisymmetry. In *Variation*, ed. B. Hallgrímsson and B.K. Hall, 359–97. New York: Elsevier.

Palmer, A.R. 2006. Caught right-handed. *Nature* 444:689–91.

Palmer, A.R., and C. Strobeck. 2003. Fluctuating asymmetry analyses revisited. In *Developmental Instability (DI): Causes and Consequences*, ed. M. Polak, 279–319. Oxford: Oxford University Press.

Perelle, I.B., L. Ehrman, and J.W. Manowitz. 1981. Human handedness: The influence of learning. *Percept Motor Skills* 53:967–77.

Pigliucci, M. 2001. *Phenotypic Plasticity. Beyond Nature and Nurture*. Baltimore: Johns Hopkins University Press.

Prentis, P.J., J.R. U. Wilson, E.E. Dormontt, D.M. Richardson, and A.J. Lowe. 2008. Adaptive evolution in invasive species. *Trends Plant Sci* 13: 288–94.

Price, T.D., A. Qvarnstrom, and D.E. Irwin. 2003. The role of phenotypic plasticity in driving genetic evolution. *Proc R Soc Lond B Biol Sci* 270: 1433–40.

Purnell, D.J., and J.N. J. Thompson. 1973. Selection for asymmetrical bias in a behavioral character of *Drosophila melanogaster*. *Heredity* 31:401–5.

Relyea, R.A. 2002. Local population differences in phenotypic plasticity: Predator-induced changes in wood frog tadpoles. *Ecol Monogr* 72:77–93.

Ribeiro, A., B.A. Eales, and F.G. Biddle. 2010. Dynamic agent-based model of hand-preference behavior patterns in the mouse. *Adaptive Behavior* 18:116–31.

Rochette, R., D.J. Arsenault, B. Justome, and J.H. Himmelman. 1998. Chemically-mediated predator-recognition learning in a marine gastropod. *Ecoscience* 5:353–60.

Rochette, R., S.P. Doyle, and T.C. Edgell. 2007. Interaction between an invasive decapod and a native gastropod: Predator foraging tactics and prey architectural defenses. *Mar Ecol Prog Ser* 330: 179–88.

Rogers, L.J. 2000. Evolution of hemispheric specialization: Advantages and disadvantages. *Brain Lang* 73:236–53.

Rogers, L.J., and R.J. Andrew, eds. 2002. *Comparative Vertebrate Lateralization*. Cambridge: Cambridge University Press.

Roy, T.A., C.B. Ruff, and C.C. Plato. 1994. Hand dominance and bilateral asymmetry in the structure of the second metacarpal. *Am J Phys Anthropol* 94:203–11.

Rubin, J., C. Rubin, and C.R. Jacobs. 2006. Molecular pathways mediating mechanical signaling in bone. *Gene* 367:1–16.

Ruff, C.B., B. Holt, and E. Trinkaus. 2006. Who's afraid of the big bad Wolff? "Wolff's law" and bone functional adaptation. *Am J Phys Anthropol* 129:484–98.

Rutherford, S.L., and S. Lindquist. 1998. Hsp90 as a capacitor for morphological evolution. *Nature* 396:336–42.

Sax, D.F., J.J. Stachowicz, J.H. Brown, J.F. Bruno, M.N. Dawson, S.D. Gaines, R.K. Grosberg, et al. 2007. Ecological and evolutionary insights from species invasions. *Trends Ecol Evol* 22:465–71.

Scheiner, S.M. 1993. Genetics and evolution of phenotypic plasticity. *Annu Rev Ecol Syst* 24:35–68.

Shine, R., M.M. Olsson, M.P. LeMaster, I.T. Moore, and R.T. Mason. 2000. Are snakes right-handed? Asymmetry in hemipenis size and usage in gartersnakes (*Thamnophis sirtalis*). *Behav Ecol* 11:411–15.

Skapec, L., and P. Stys. 1980. Asymmetry in the forewing position in Heteroptera. *Acta Entomol Bohemoslov* 77:353–74.

Sol, D., and T.D. Price. 2008. Brain size and the diversification of body size in birds. *Am Nat* 172:170–7.

Spalding, D. 1873. Instinct. With original observations on young animals. *Macmillans Magazine* 27:282–93.

Strauss, S.Y., J.A. Lau, and S.P. Carroll. 2006. Evolutionary responses of natives to introduced species: What do introductions tell us about natural communities? *Ecol Lett* 9:354–71.

Suzuki, Y., and H.F. Nijhout. 2006. Evolution of a polyphenism by genetic accommodation. *Science* 311:650–2.

Sylvester, A.D., A.M. Christensen, and P.A. Kramer. 2006. Factors influencing osteological changes in the hands and fingers of rock climbers. *J Anat* 209:597–609.

Tokarz, R.R., and J.B. Slowinski. 1990. Alternation of hemipenis use as a behavioral means of increasing sperm transfer in the lizard *Anolis sagrei*. *Anim Behav* 40:374–9.

Tollrain, R., and C.D. Harvell, eds. 1999. *The Ecology and Evolution of Inducible Defenses*. Princeton, NJ: Princeton University Press.

Trinkaus, E., S.E. Churchill, and C.B. Ruff. 1994. Postcranial robusticity in *Homo*. II: Humeral bilateral asymmetry and bone plasticity. *Am J Phys Anthropol* 93:1–34.

Trussell, G.C. 2000. Phenotypic clines, plasticity, and morphological trade-offs in an intertidal snail. *Evolution* 54:151–66.

Van Buskirk, J., and R.A. Relyea. 1998. Selection for phenotypic plasticity in *Rana sylvatica* tadpoles. *Biol J Linn Soc* 65:301–28.

Vaughn, D. 2007. Predator-induced morphological defenses in marine zooplankton: A larval case study. *Ecology* 88:1030–9.

Vermeij, G.J. 1978. *Biogeography and Adaptation. Patterns of Marine Life*. Cambridge, MA: Harvard University Press.

Wassersug, R.J., and K. Hoff. 1985. The kinematics of swimming in anural larvae. *J Exp Biol* 119:1–30.

Weber, B., and D. Depew, eds. 2003. *Evolution and Learning: The Baldwin Effect Reconsidered*. Cambridge, MA: MIT Press.

Wentworth, K.L. 1933. The effect of early reaches on handedness in the rat: A preliminary study. *J Genet Psychol* 52:429–32.

West-Eberhard, M.J. 2003. *Developmental Plasticity and Evolution*. New York: Oxford University Press.

Wiles, G.J., J. Bart, R.E. Beck, and C.F. Aguon. 2003. Impacts of the brown tree snake: Patterns of decline and species persistence in Guam's avifauna. *Conserv Biol* 17:1350–60.

Yamaguchi, T. 1977. Studies on the handedness of the fiddler crab, *Uca lactea*. *Biol Bull* 152:424–36.

20

Epigenetics: Adaptation or Contingency?

Thomas F. Hansen

CONTENTS

Living organisms are enormously complex. Through development, the complex phenotype is built by cells using the genetic information encoded in some billions of base pairs of genome. This process is amazingly accurate. Excluding extrinsic mortality, most fertilized eggs are converted into normal functional adults of the requisite type. The conversion from genotype to phenotype requires accurate orchestration of numerous events on different hierarchical scales, from molecular interactions that control gene expression through the production of intracellular structures and metabolism, generation of cell-specific morphology and behavior, organization of tissues, orchestration of coordinated growth, and induction of tissues on different distances up to the generation of whole-organism morphology, physiology, and behavior. Through interaction with the environment, this generates a life history, which ultimately determines the fitness of the organism.

In this book we discuss epigenetics, which, following Waddington's (1940) concept of the epigenetic landscape, is broadly understood as the functional pathways (sensu Houle, 2001) from genotype to phenotype. These pathways determine how molecular genetic variation is converted into structured phenotypic variation and are instrumental in understanding the variational properties and evolutionary potential of characters (e.g., Raff and Kaufman, 1983; Gerhart and Kirschner, 1997; Raff, 1996; Wilkins, 2002). The epigenetic pathways themselves are impressive examples of complex biological adaptations for producing coordinated phenotypes as quickly, efficiently, and accurately as possible. This does not mean, however, that the variational consequences of development are also adaptations. Despite this, there is a strong consensus that variational adaptations for evolvability or robustness are common in terms of gene and genome structure (Davidson, 2001; Carroll, 2005), the structure of the genotype–phenotype map (Wagner and Altenberg, 1996; Raff, 1996), gene-regulatory networks (Gerhart and Kirschner, 1997; Davidson 2001), developmental pathways (Gerhart and Kirschner, 1997;

Wilkins, 2002), and ecologically in relation to disturbed, fluctuating, or novel habitats (Lee and Gelembiuk, 2008).

In the fields of molecular and developmental evolution, it goes almost unquestioned that structural features of the organism have evolved under the influence of natural selection. In a series of papers, Michael Lynch (2005, 2007a, 2007b, 2007c) has criticized this bias for adaptive explanations and argued that many aspects of genome architecture are better explained by mutation and genetic drift. He suggested that this may extend to the variational properties of the genome, such as evolvability.

There are, however, alternatives to both Lynch's neutral model and the panselectionism of evolutionary developmental biologists. When reading the evo–devo literature, I often get the impression that the concepts of adaptation and selection are used interchangeably. Although all (genetic) adaptation is a result of selection, the converse is far from true. In particular, characters may often be under indirect selection that is not caused by interaction with the environment. There is no general reason that such indirect selection should lead to adaptation of the character. Due to the complexity and interconnectedness of the epigenetic system, every character is subject to a multitude of potential indirect selection pressures. This should make us expect indirect selection to be both common and strong. From the point of view of the individual character, these selection pressures may appear more or less stochastic, but this stochasticity is distinctly different from the kind of stochasticity that comes from the reproductive sampling we call "genetic drift." For one thing, stochastic indirect selection will not depend on population size in a similar way, if at all.

In this chapter I take a critical look at some of the theoretical arguments that form the basis of the adaptive view of variational properties in general and evolvability in particular. I ask whether the possibility of adaptation is sufficient basis for the expectation of adaptation, particularly when we take into account that adaptation in real biological systems must happen in the face of rampant indirect selection, which is usually carefully excluded from mathematical models.

THE EVOLUTION OF VARIATIONAL PROPERTIES

WHAT ARE VARIATIONAL PROPERTIES?

All evolution is based on some form of change in the functional pathways from genes to phenotype. To avoid turning the concept of epigenetic evolution into an empty generality, it needs to be restrictive. In this chapter I will restrict my attention to the evolution of what have been referred to as the "variational properties" of the genotype–phenotype map (Wagner and Altenberg, 1996). *Variational properties* are those that affect the ability of the genetic system to produce and maintain variation. Thus, a change in the genetic system that alters the rates or effects of potential future genetic or environmental changes qualifies as a change in variational properties. Changes in genotype frequencies or in means or variances of phenotypic characters do not qualify. All evolutionary change is as change of statistical distribution, and there is no deep difference between change in the mean and change in the variance of a trait. Following the distinctions made by Wagner and Altenberg (1996), our focus is on the dispositional property of variability and not on realized variation.

The interest in variability is rather recent. The modern synthesis had little interest in variability as a phenomenon and viewed it as inconsequential for evolutionary dynamics (Amundson, 2005). Even though a growing interest in variational constraints emerged during the 1970s and 1980s, variability has usually been treated as fixed biological parameters, for example, as fixed rates and effects of mutation. Serious investigations of the evolution of variational properties started only with the emergence of evolutionary developmental biology in the 1980s and 1990s. Then, aspects of variability started to enter as dynamical variables in population genetical models (e.g., Wagner et al., 1997). Today,

much effort is going into understanding the origin and maintenance of specific forms of variational coordination, as reflected in concepts such as modularity, integration, allometry, heterochrony, plasticity, continuity, and symmetry. Most of this effort is explicitly or implicitly based on the assumption that variational properties are some form of adaptations, usually for evolvability or robustness; but before evaluating these hypotheses, we need to consider more precisely what we mean by "adaptation" and "selection."

A NOTE ON SELECTION AND ADAPTATION

Sober (1984) made an important distinction between selection *for* a trait and selection *of* a trait. The redness of vertebrate blood is an example. Although redness serves no function, it is almost certainly an effect of selection; but the cause of selection in this case is the oxygen-binding capability of hemoglobin. It just happens to be that hemoglobin also reflects red light. Thus, we have direct causal selection for binding oxygen and indirect noncausal selection of redness. Oxygen binding is an adaptation, but redness is not. Direct selection leads to adaptation, but indirect selection does not.

Adaptations are traits that are adapted to perform specific functions, and when we say that trait Y is an adaptation for function X, we presuppose that Y is maintained at least in part by selection for function X. Indirect selection cannot create or maintain functionality except by circumstance and is no more a mechanism for adaptation than factors such as genetic drift or gene flow.

In most investigations of adaptation it is more or less standard to assume that the trait in question is optimal in the sense that the forces of selection acting upon it have come to a balance. The task is then to test hypotheses about what specific selective factors are involved in this balance (e.g., Mitchell and Valone, 1990). For example, the optimal color of a flower is likely to be a compromise among many selective causes stemming not only from its ability to attract the desired pollinators but also from its ability to avoid nectar thieves and herbivores,

TABLE 20.1

The Impossibility of Modifiers

	X	Y
A	a	b
B	c	d

NOTE: A "modifier" with two states, A or B, interacts with a "target" gene with two states, X and Y. The four possible phenotypic states are a, b, c, and d.

interaction with their targets, and epistasis is inherently symmetrical under very general assumptions (Hansen and Wagner, 2001a).

Even if this demonstrates that there is no clear biological separation between genes that affect variational properties of characters and genes that affect the characters themselves, this does not mean that we cannot entertain the conceptual distinction between variational evolution and character evolution. It is conceptually helpful to understand the specific direct selection pressures that act on variational properties in isolation. It is important, however, not to forget that these will be idealizations with limited biological reality and that there almost always will be strong interactions between evolution of variational properties and character evolution, even in models set up to avoid this.

CAN EPISTASIS BE THE BASIS OF VARIATIONAL ADAPTATIONS?

Changes in the phenotypic effects of gene substitution form the basis of many changes in variability. It has been assumed that systematic changes in gene effects must involve some form of gene interaction, or *epistasis* (Wagner et al., 1997; Rice, 1998; Flatt, 2005; but see Hansen, 2006, and the section on allelic effects below). If the effect of a gene depends on what other genes are present, then changes in the genetic background can lead to changes in gene effects and, thus, to changes in variational properties. An obvious, but rarely emphasized, fact is that such changes depend crucially on *how* genes interact. Epistasis is all too often treated as a single homogeneous entity, sometimes

known as "nonadditive variation"; and it is a mistake to view its evolutionary effects as unitary. My collaborators and I have studied the role of epistasis in the response to linear directional selection in some detail (Carter et al., 2005; Hansen et al., 2006). Although the role of epistasis is complex, we have found that its main short-term effects can be captured by a directionality parameter which measures whether gene substitutions on average tend to reinforce or diminish each other's effects on the trait. When epistasis is positive on average, gene substitutions will tend to increase the effects of subsequent substitutions, additive genetic variance will increase, and the response to selection will accelerate. Conversely, when epistasis is negative on average, gene substitutions will tend to decrease the effects of subsequent substitutions, additive genetic variance will decrease, and the response to selection will decelerate. These effects can be pronounced, and over more than a few generations, the response to selection may diverge greatly from the response of an additive architecture. In contrast, if epistasis is nondirectional, the response is all but indistinguishable from the response of an additive architecture over very large numbers of generations. Therefore, general measures of the amount of epistasis, such as the standard epistatic variance components, are useless for predicting evolutionary dynamics. There is no direct link between epistatic variance components and changes in additive variance under either selection or drift (Hansen and Wagner, 2001a; Barton and Turelli, 2004).

Both additive and mutational variance may increase, decrease, or remain unchanged during a response to directional selection depending on the exact genetic architecture (Hansen et al., 2006). Similarly, neither additive nor mutational variance is necessarily pushed to a minimum under stabilizing selection (Hermisson et al., 2003).

Thus, general models of gene interactions do not support a mechanism for consistently "adaptive" evolution of gene effects. Instead, there emerges a picture of genetic changes primarily driven by their effects on the trait mean and changes in genetic architecture and variational properties following along depending on the particular relationship in which they stand to genes with major effects on the trait mean.

HERITABLE ALLELIC EFFECTS AS A MECHANISM OF VARIATIONAL EVOLUTION

Epistasis is, however, not the only mechanism that can generate systematic changes in the effects of mutations and allele substitutions. Systematic changes can occur if the effects of allelic changes are heritable in the sense that a mutation changes not only the effect of the allele but also the mutational spectrum of the allele such that subsequent mutations on the same allele have altered effects. If there is a systematic relation between the direct effect on the allele and the changes in the mutational spectrum, then such "heritable allelic effects" can lead to systematic evolutionary changes in genetic architecture (Hansen, 2003, 2006). Consider, for example, a mutation that creates a new gene-regulatory module (sensu Wray et al., 2003) that drives the expression of the gene in a new spatiotemporal setting, i.e., creates an effect on a new character. Many subsequent mutations on this gene will then inherit the expression on the new character. If the new character is under directional selection, this may favor the original mutation and, thus, also indirectly select for the associated changes in the mutational spectrum. Directional selection on a character can thus act to increase its mutational target size. In this case pleiotropy will also be increased, but we can imagine similar scenarios where mutations eliminating pleiotropic constraints may be favored.

The biological basis and evolutionary consequences of heritable allelic effects remain unexplored, but it might be a more promising mechanism for coordinated changes in genetic architecture than epistasis since there is a closer link between the direct effects of the mutation and the indirect effects on variational properties. This sets up a situation where coordinated variability may evolve not only as an adaptation but also "quasi-adaptively" through indirect selection.

CRITICISM OF HYPOTHESES ABOUT VARIATIONAL ADAPTATIONS

THE CASE FOR COORDINATED VARIABILITY AS ADAPTATION

Due to the extreme complexity and high dimensionality of biological organisms, evolvability depends crucially on extensive functional coordination of the variation being presented for selection. Adaptive evolution would not be possible on a random genotype–phenotype map; without any correlation between the phenotypic values of nearby genotypes there can be no continuity or consistent path toward better adaptation (Kauffman, 1993). Genotype–phenotype maps are far from random, and the simple fact that adaptation is possible at all shows that a level of coordination must typically be present (Riedl, 1978). To explain why this is so is arguably among the most fundamental problems of theoretical biology.

As with all complex traits, the high improbability of coordinated variability shows that some form of selection must be involved and is suggestive of adaptation. Two types of adaptation have been suggested: adaptation for evolvability and adaptation for robustness. Most obviously, coordinated variability may be an adaptation to increase evolvability. If traits that need to be functionally coordinated also become variationally coordinated, then a larger fraction of new mutations are potentially adaptive. Symmetrical body plans provide an example. If bilateral symmetry, for example, is fixed in the body plan, then most mutations will have symmetric effects and will be more likely to be adaptive if functions, such as locomotion, depend on bilateral symmetry. Of course, this will also induce constraints on the evolution of asymmetric body shapes when this is advantageous, but evolvability relative to "normal" selective challenges would be enhanced. Alternatively, integrated symmetry could be seen as an adaptation to reduce the deleterious effects of new mutations or developmental perturbations, i.e., as an adaptation for robustness.

The questions we must ask of these hypotheses are how strong the adaptive selection pressures can be and how they compare to alternative evolutionary mechanisms. In the case of symmetry, the most obvious alternative hypothesis is that its variational consequence has evolved as a side effect of selection on symmetry itself caused by the direct functional advantages of symmetry for locomotion, etc. Similarly, Young and Hallgrimsson (2005) observed that the covariation between fore- and hindlimbs in vertebrates is less in groups where the limbs have different functions. Is this an adaptation for independent evolvability or a side effect of the associated morphological differences between the limbs?

MODULARITY AND PLEIOTROPY AS ADAPTATIONS FOR EVOLVABILITY

Modularity, a degree of variational independence between distinct integrated characters, has been the main focus of discussion about variational evolution (see Wagner et al., 2007, for review). The working hypothesis of evolutionary developmental biology seems to be that modularity has evolved as an adaptation to increased evolvability. This hypothesis rests on the assumptions that organisms indeed are modular and that this favors their evolvability. It is, however, not obvious to what degree, or in what sense, these assumptions are fulfilled. Lewontin (1978) pointed out that the evolvability of a biological character requires a degree of variational "quasi-independence" from the rest of the organism, and obviously, there can be no evolution of shape if characters always must change in concert; but this still leaves open the precise quantitative relation to evolvability and the question of how quasi-independence can be implemented in the genotype–phenotype map.

With my colleagues, I have attempted to develop ways of measuring quasi-independence based on simple quantitative genetics theory and the concept of "conditional evolvability"

(Hansen et al., 2003; Hansen, 2003; Hansen and Houle, 2008). *Conditional evolvability* is defined as a character's response to directional selection per strength of selection when other constraining characters, in a defined set, are not allowed to change. We showed that the conditional evolvability of a character is equal to its residual additive genetic variance when regressed on the constraining characters. Our analysis seems to confirm the intuition that increasing variational independence increases evolvability, but we have to add the caveat that this is true only if independence can be achieved without reducing the variational basis of each character. If there is a limited number of genes that must be distributed among characters, then the "optimal" solution includes pleiotropic genes (Hansen, 2003).

Following Wagner (1996; Wagner and Altenberg, 1996), modularity is often associated with a lack of pleiotropy between characters. However, quasi-independence and conditional evolvability are not confined to maps with low pleiotropy between characters. In fact, high conditional evolvability can be achieved with high levels of pleiotropy, provided the pattern of pleiotropy varies across genes so that one can find combinations of genetic changes where the effects nullify each other on all traits except the focal one (Figure 20.1). Thus, even if the evolution of the genotype–phenotype map is influenced by selection for increased evolvability, it does not follow that pleiotropic modularity will evolve. Indeed, the interest in modularity as an adaptation to evolvability is poorly focused, and our interest should be shifted toward the evolution of pleiotropy and to the relationship between patterns of pleiotropy and evolvability.

There are many arguments in the literature pertaining to modularity or the pattern of pleiotropy evolving as an adaptation to increased evolvability. This is for the most part based on verbal intuition or complex simulation studies where the causal basis of selection, direct or indirect, is confounded. This is nicely illustrated by Gardner and Zuidema (2003), who show that the evolution of modularity in a fluctuating

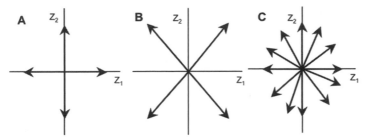

FIGURE 20.1 Pleiotropy does not preclude variational modularity. The figure shows different patterns of pleiotropy on two characters, z_1 and z_2. The arrows represent the effects of genetic changes. In A there is no pleiotropy; only genetic change affecting one character at a time is possible. In B all changes are pleiotropic, but they have either an equal or an opposite effect on the characters. In C many different types of pleiotropic change are possible. Although A corresponds to complete modularity and all mutations in B are pleiotropic, the variational potential of these two cases is similar. If there are equal amounts of variation along the two arrows in B, the two traits will be uncorrelated like the traits in A and can evolve independently of each other. In C any evolutionary change could be made in a number of ways based on changes in different combinations of genes.

environment in the simulations of Lipson et al. (2002) is not due to an effect on evolvability but to an implicit direct correlation between modularity and fitness.

In an interesting study, Draghi and Wagner (2008) compared the evolution of pleiotropy under stabilizing selection and fluctuating selection in a simple two-trait, two-locus model. They found that random fluctuations of the trait optimum favored changes in pleiotropy toward increasing angle between the vector effects of the two loci and that this enhances average evolvability in random directions. In contrast, stabilizing selection favors the angle staying constant at whatever value it has. Draghi and Wagner showed that genetic architectures with larger angle are more likely to generate beneficial mutations and that the more evolvable architectures are less likely to be invaded by less evolvable architectures. They argued that both of these mechanisms are likely to operate in their simulations and, thus, that the evolution of pleiotropy could be an adaptation to evolvability. I note, however, that mutations changing the pleiotropic angle will also have large direct effects on the trait and be under selection to bring the trait closer to the optimum. This induces indirect selection on pleiotropy, and although it is hard to

disentangle the contributions of direct and indirect selection on evolvability, it is not obvious that direct selection is the dominant factor.

The potential for indirect selection is also underscored in another recent theoretical study of the evolution of pleiotropy. Jones et al. (2007) studied the evolution of mutational correlations between two traits in balance between mutation and stabilizing selection. They showed that if the mutational variances of the two traits are kept constant, then there is disruptive selection on the mutational correlation between the traits, making it evolve toward +1 or −1. In either case, increasing integration is predicted under stabilizing selection. They also showed, however, that selection on the mutational correlation is weak and vulnerable to stochastic fluctuations. For this reason, they concluded that genetic drift is important and that systematic evolution of pleiotropy may be most likely in large populations. Note that this also makes the evolution of pleiotropy extremely sensitive to any direct effects of the genes affecting the mutational correlation. The selection pressures identified in this model are simply too weak to allow robust biological predictions. Thus, although this model is an important contribution to our understanding of the evolution of pleiotropy, I see it as an argument *against* variational adaptation.

The results of these two studies are also not consistent with each other, and they depend on specific model assumptions. For example, the disruptive selection in the model of Jones et al. (2007) stems from the fact that trait-specific mutational variances are kept constant so that the mutational variance in fitness is minimized when the two traits become integrated. This suggests that evolution toward integration may not generalize to other assumptions about the mutational variance, such as keeping the magnitude of the mutation vector constant. Jones et al. also show that stronger selection on mutational correlation may appear if traits are displaced from their optima, but this selection is complex and depends on the exact directions of the displacement.

Cheverud and coworkers have proposed a more general model for the evolution of pleiotropy in terms of differential epistasis (Cheverud, 2001; Cheverud et al., 1997, 2004; Wolf et al., 2005; Wagner et al., 2007; Pavlicev et al., 2008). The idea is that the specific pleiotropic effects of allele substitutions may be different in different genetic backgrounds so that evolutionary changes in pleiotropy can happen when the genetic background evolves. As noted in my comments on epistasis above, this mechanism will require specific systematic patterns of epistasis to generate favorable changes in pleiotropy, and there remains to be shown that such patterns are common. As far as we currently know, selection on epistatic genetic architectures may be as likely to generate unfavorable as favorable changes of pleiotropy.

MODULAR GENE REGULATION AS AN ADAPTATION FOR EVOLVABILITY

The modularity of gene regulation is one of the core paradigmatic assumptions of evolutionary developmental biology as a field (e.g., Raff, 1996; Von Dassow and Munro, 1999; Stern, 2000; Davidson, 2001; Carroll, 2005). Gene expression is thought to be regulated by the binding of specific combinations of transcription factors to regulatory modules that, in conjunction with a basal promoter, are independently capable of driving the expression of the gene (Wray et al., 2003). This sets up a situation where modules controlling different functions of the gene, such as expression in different tissue types or developmental stages, can adapt quasi-independently of each other. The evolvability this confers has led to the expectation that most or all adaptive evolution of morphology is based on the modification of *cis*-regulatory modules (e.g., Carroll, 2005). This position has been criticized by pointing to evidence for structural or *trans*-regulatory evolution (Hoekstra and Coyne, 2007; Lynch and Wagner, 2008; Lynch et al., 2008), but *cis*-regulatory modularity still seems to be a prime example of an architectural feature of the genotype–phenotype map well suited for adaptive evolution.

The origins of modular *cis*-regulatory gene control go back at least to the origin of eukaryotes, and it is hard to formulate testable hypotheses for the mechanisms involved. It seems, however, possible to investigate what forces maintain the system and to ask whether group or individual selection for evolvability may be involved. There has been little theoretical work on this problem, and it has yet to be formally shown whether or how a *cis*-regulatory system enhances evolvability. The arguments that exist are based on claiming (or showing) that it is easy for new regulatory modules to appear (e.g., Stone and Wray, 2001), but this does not demonstrate that adaptive changes in gene regulation could not happen by other means such as by modifying protein–protein interaction among transcription factors (Lynch and Wagner, 2008). Also, even if *cis*-regulatory modules are well suited for evolutionary change, it does not follow that direct natural selection could be strong enough to maintain the system. It is difficult to see how selection could act independently on every locus to maintain *cis*-regulatory modularity, and no genomewide mechanism for selection to act upon has been proposed. Thus, the idea that *cis*-regulatory modularity is an adaptation for evolvability is both without supporting evidence and implausible on first principles. Given that Lynch (2005, 2007b) has outlined plausible nonadaptive scenarios for modularization of gene-regulatory regions based on either near-neutral accretion of specific modules with subsequent degeneration of general modules or duplication and subfunctionalization, it is clear that adaptation for evolvability should not be our default hypothesis.

COORDINATED VARIABILITY AS AN ADAPTATION FOR ROBUSTNESS

Another general benefit of coordinated variability is that it may help to reduce the deleterious effects of mutations or developmental disturbances. Andreas Wagner (2005) has presented a forceful case for robustness as a fundamental property of biological structures of many different types and, in this, presumably confirmed the intuition of most biologists. The evolutionary basis of this robustness has been a topic of extensive empirical and theoretical investigations (for reviews, see de Visser et al., 2003; Flatt, 2005; Wagner, 2005). Robustness has been considered both as an adaptation (e.g., Waddington, 1942; Wagner et al., 1997; Rutherford and Lindquist, 1998) and as an emergent property of living systems (e.g., Kacser and Burns, 1981; von Dassow et al., 2000; Lynch, 2007a). Hypotheses about an adaptive basis for robustness have a history going back at least to the ideas of Waddington (1942) on canalization (reduced variability) as an outcome of stabilizing selection.

Wagner et al. (1997) brought this into a population-genetic framework and studied the evolution of modifiers of gene effect on a polygenic trait in mutation–selection balance. They found that some degree of canalization could evolve under stabilizing selection of intermediate strength. This is due to a reduction of the deleterious effects of new mutations, and since the strength of selection is then proportional to the mutation rate, it is necessary with modifiers acting on many loci to produce a reasonable direct selection strength (Wagner, 1999; Proulx and Phillips, 2005). It is also necessary with a high rate of recombination between the modifiers and the target loci to decouple the good effects of canalization from the nullifying long-term effects of increasing the frequency of segregating deleterious alleles (Gardner and Kalinka, 2006).

The reason that genetic canalization gets weaker under strong stabilizing selection in the Wagner et al. (1997) model may be due to the inevitable direct trait effects of the postulated modifiers. This illustrates the main issue with adaptive explanations for canalization and robustness: There are correlations between the direct and canalizing effects of the mutations, and it needs to be established that the canalizing selection pressures constitute a significant force in balance with other selective forces. Formal decompositions of these selection pressures have been presented by Rice (1998, 2002) and Hermisson et al. (2003), and these indeed confirm that direct "canalizing" selection arises under stabilizing selection on epistatic genetic architectures and can be strong enough to make adaptation for robustness a viable hypothesis. Hermisson et al. (2003) also found that this force pulls in a direction to reduce segregating additive genetic variance, and this may not be the direction that minimizes mutational variability. Hence, although some adaptive canalization is expected, mutational robustness will not be maximized.

Robustness can be adaptive in two senses, as an adaptation either against genetic disturbances, such as deleterious mutations or recombination with unfavorable alleles, or against developmental or environmental disturbances. The latter case corresponds to the so-called congruence hypothesis, where genetic robustness evolves as a side effect of adaptation to environmental disturbances (de Visser et al., 2003). There is obviously a strong selective premium for robustness to environmental variation, and some adaptation in this direction is to be expected. If there is a correlation between genetic and environmental robustness, this also favors the congruence hypothesis for the evolution of genetic robustness (Wagner et al., 1997; de

Visser et al., 2003; Wagner, 2005). Such correlations are almost inevitable since any change that improves the stability of a developmental pathway is likely to buffer against both genetic and environmental perturbations. Direct evidence for a correlation between mutational and environmental variation in gene expression has been found by Rifkin et al. (2005).

There are also links between robustness and evolvability. One consequence of genetic robustness is that "hidden" variation will accumulate (Hermisson and Wagner, 2004; Wagner, 2005). If this variation could be revealed, it could fuel rapid evolutionary changes (Waddington, 1953; see Le Rouzic and Carlborg, 2008, for recent review). The idea that this could happen in an adaptive manner is the essence of the so-called capacitance hypothesis (Rutherford and Lindquist, 1998). In particular, the discovery that perturbations of certain heat-shock proteins could act to release hidden variation in other genes suggested the possibility of adaptive decanalization during stress (Rutherford and Lindqvist, 1998). Masel (2005) has shown that direct selection for a "revealing" mechanism can exist when the population is rendered maladapted by periodic stressors. Masel has shown that this selection can overcome genetic drift for realistic parameter values, but in my view it seems less likely to overcome the effects of indirect selection. The capacitance hypothesis is also challenged by the finding of Hermisson and Wagner (2004) that a release of hidden variation under stress is a generic property of epistatic systems, and they thus provide a nonadaptive alternative hypothesis.

PLASTICITY AS AN ADAPTATION FOR ENVIRONMENTAL UNCERTAINTY

Organisms are born into unpredictable environments, and a degree of robustness is necessary to produce the required phenotype in the face of the environmental variation that the organism may encounter. It is, however, also the case that different environmental conditions will favor different phenotypes, and many organisms display *adaptive phenotypic plasticity*, where the same genotype may produce alternative phenotypes in relation to environmental cues. This includes such phenomena as sun and shade leaves in plants, muscle and bone development in relation to use in vertebrates, and the development of a variety of antipredator or antiherbivore responses in a range of organisms. Such plasticity is a variational property, and the ability to produce alternative phenotypes that fit the environment is clearly adaptive. The rather subtle argument we have to enter is whether the plasticity itself is the target of selection or if the plasticity is better considered a side effect of selection acting differently on trait expression in different environments. For example, Via and Lande (1985) developed a model for the evolution of phenotypic plasticity based on treating trait expression in two different environments as two traits under different selection pressures. Note that this presupposes a degree of modularity between trait expression in the two environments. If the "traits" are fully integrated, plasticity in the sense of differential trait expression cannot evolve. From the variational point of view, the question then becomes whether there is direct selection for modularity or whether the modularity arises as a side effect of selection on the "traits." While it is easy to define a plasticity parameter, such as the slope of the reaction norm, and describe selection acting on this parameter, there cannot be selection on plasticity without selection on some of the traits. In many cases, then, the causal basis of selection is to be found in the interaction between particular expressions of the trait and the state of the environment, and in such cases plasticity is better seen as a side effect of trait adaptation than as an adaptation on its own. True adaptive plasticity occurs in situations where there are many and varied environmental situations and the adaptation is achieved through a clearly individualized plastic character. For example, the plastic growth of muscle in response to use is an individualized trait and not a separately evolved ability of each muscle in the body.

ALTERNATIVE HYPOTHESES FOR VARIATIONAL EVOLUTION

COORDINATED VARIABILITY AS A GROUP ADAPTATION

Given that direct individual selection on variational properties is often a weak second-order effect, we must also consider the possibility of variational adaptations as group adaptations. Here, I understand group adaptation in the sense of Sober (1984; Sober and Wilson, 1998) as an adaptation based on selection that is *caused* by differential fitness of permanent or temporal groups of individuals. We now understand that the traditional objections to group selection based on group traits being less heritable than genes were in fact objections to groups as units of evolution and not to groups as units of selection. There is no reason that there cannot be strong selection among even ephemeral "trait groups," and this selection may influence the course of evolution even if the groups themselves dissolve (as indeed do individuals). Group adaptations in the sense of Sober and Wilson (1998) may in fact be widespread in biology (Wilson and Wilson, 2007).

Most obviously, variational properties could be group adaptations for evolvability. It is easy to imagine that local populations with higher evolvabilities may be more successful in adapting to shifting local conditions and that this may affect their persistence and/or the rate at which they deliver migrants to the general metapopulation. Gerhart and Kirschner (1997, chap. 11) indeed followed their argument for adaptive coordinated variability with the hypothesis that this is maintained by clade selection for evolvability.

On some level, group selection must act to maintain evolvability as species or populations that lose their evolvability are prone to disappear. Sexual recombination is a variational property related to evolvability that is likely to be maintained by some form of group or clade selection. There are many well-established population-level advantages to sex and recombination. These include reducing the fixation load of deleterious mutations that result from Muller's ratchet and mutational meltdown (Gabriel et al., 1993; Lynch et al., 1993, 1994), reducing the mutation load in the presence of synergistic epistasis (Kondrashov 1988; Hansen and Wagner 2001b) and increasing the rate at which rare advantageous mutations are combined (Maynard Smith, 1976). The maintenance of sexual recombination as a group adaptation is strongly supported by the near absence of long-lived asexual species (Bell, 1982; Gladyshev et al., 2008).

Most of these advantages of sex and recombination would take a long time to exert themselves in a population and may thus be selected only if groups are isolated over a relatively long time. Sex and recombination are nevertheless normally kept at high rates even in systems where lower rates could easily evolve (Williams, 1975). Since sexual reproduction often has high costs at the individual or gene level, this suggests that there must also be short-term advantages. These short-term advantages could be on either the group or the individual level, but regardless of which, they imply that evolvability is not the cause of selection. For example, the tangled-bank hypothesis (Ghiselin, 1974; Bell, 1982) postulates that the advantage stems from making variable offspring, thus increasing the probability that some offspring are well adapted to a complex, uncertain environment. This mechanism could also be a trait-group advantage by reducing competition among group members, but in either case, evolvability is not the direct cause of selection but a side effect.

In summary, evolvability as a group adaptation is a serious hypothesis with some support. For example, the Red Queen hypothesis for maintenance of sex implies direct group selection for evolvability. On the other hand, other hypotheses for the maintenance of sex imply indirect selection on evolvability; and outside the evolution of sex and recombination, group-adaptation hypotheses for evolvability and other variational properties have hardly been investigated.

THE NEUTRAL HYPOTHESIS FOR EVOLUTION OF VARIATIONAL PROPERTIES

Michael Lynch (2005, 2007a, 2007b, 2007c) has proposed a neutral alternative to the adaptationist theories of variational evolution. More generally, he has argued that many features of genome organization and the genotype–phenotype map make little sense as adaptations and are better explained by the neutral theory, i.e., as an outcome of imperfect stabilizing selection, where near-neutral changes slip by in finite-sized populations. The core of Lynch's argument is that alleles will behave as if they are neutral when their fitness differences are less than $1/4N_l$, where N_l is the long-term effective population size taking account of selective sweeps. As large multicellular organisms have lower population size, Lynch explains their more complex genomes filled with nontranscribed and nontranslated regions, as well as all kinds of redundant or nonfunctional duplicates, as a consequence of ineffective selection. This increase in the genome complexity of small populations also has variational consequences. In particular, selection is often too weak to eliminate redundant regulatory sites and gene duplicates, and this can lead to increasing robustness and evolvability.

For example, Lynch (2007b) argues that the "distributed robustness" (Wagner, 2005) of gene-regulatory networks, i.e., the tendency for such networks to remain functional after single-gene knockouts as a consequence of excess connections, can be understood as a consequence of accumulating near-neutral redundancies. This is both a direct and an indirect consequence of small population size as the accumulation of nonfunctional DNA in and around genes increases the probability of random occurrence of new regulatory sites. Similarly, the near-neutral accumulation of gene duplicates can provide a substrate for increased evolvability by allowing one initially redundant duplicate to take on a new function or

to specialize on one subfunction (Lynch, 2007a).

Lynch's model is built on the low power of direct selection on the aspects of genetic architecture that he considers. His model is therefore convincing only insofar as there is no indirect selection on the aspects he considers. This is plausible in many cases, as with the evolution of untranslated regions of genes, but perhaps open to debate in the case of many gene duplications and novel regulatory sites. In any case, it seems plausible that neutral evolution of the type envisioned by Lynch is an important component of the evolution of variational properties, but its applicability is limited to cases where there are no strong direct effects of the genetic changes on the trait. The evolution of a new pleiotropic link, for example, has an unavoidable effect on the new character and will experience a selection strength on the order of this effect times the strength of selection on the character (Lynch, 1984). A neutral model for changes in pleiotropy must be based on the accumulation of very small changes.

Note that Lynch's model is not invalidated by the observation that selection on molecular variants is very common (Hahn, 2008), as Lynch's concept of long-term effective population size accounts for indirect selection resulting from linkage to selective sweeps (Lynch, 2007a, chap. 4). Random selective sweeps have effects that are similar to those of reduced effective population size.

Regardless of its ultimate causes, the robustness of gene networks and developmental systems can itself provide a substrate for quasi-neutral evolution as many different genotypic states can code for the same phenotype. If sets of genotypes with equivalent phenotypic effects form connected subsets of genotype space, then we expect neutral divergence, or "developmental systems drift" (True and Haag, 2001), on these "neutral spaces" (Wagner, 2005). This process has consequences for the evolution of reproductive isolation (Gavrilets and Gravner, 1997; Gavrilets, 2004) and may in itself provide a basis for the adaptive evolution of variational properties as second-order selection pressures may be able to influence evolution on a subspace without direct trait effects. Van Nimwegen et al. (1999) show how adaptive robustness can evolve on a neutral network due to this mechanism, but robustness may also evolve indirectly through simple mutational bias for robust regions of the neutral space (Wagner, 2005).

VARIATIONAL PROPERTIES AS SIDE EFFECTS OF TRAIT EVOLUTION

Even if direct selection on variability is weak, variational properties are likely subject to a variety of indirect selection pressures. Allelic substitutions that affect variability will inevitably have other effects in the organism. In particular, it is hard to imagine genetic changes in the variability of character without effects on the character itself. To include indirect selection from the character itself in our theory of variational evolution, we need to know two things: (1) What is the pattern of selection on the trait(s) in question? and (2) What is the link between the trait and the trait's variational properties?

In general, trait selection is idiosyncratic and variable; but as argued above, any complex trait must experience long-term stabilizing selection, and short-term fluctuating directional selection is probably common. Wagner (1996) presented an interesting hypothesis for the evolution of modularity under such a selection regime. After arguing that modularity was unlikely to evolve under stabilizing selection, which, if anything, would simply favor general canalization, he argued that a combination of fluctuating directional selection on one character with stabilizing selection on another would lead to selection against alleles with pleiotropic effects on both characters. This hypothesis remains to be tested in formal models but seems plausible if there are heritable allelic effects on the traits. Let us say we have two integrated characters, A and B. There is stabilizing selection on A and fluctuating directional selection on B. Any mutation with a favorable effect on B

but with little or no effect on A will be favored. For this to lead to modularity, the mutation must also change the mutational spectrum such that subsequent mutations are more likely to have independent effects. In principle, this can happen either through epistasis modifying other loci in this direction or through heritable mutational effects on the locus itself. As argued above, we have established no reason to expect epistasis to generally behave in this way, but heritable allelic effects of this sort are plausible; for example, if the mutation consists of acquiring a new regulatory module that governs expression solely on B, further mutations to this module are then likely to affect B but not A. Similarly, subfunctionalization of duplicate genes (Force et al., 1999) may be viewed as a heritable modularization of the gene that will be favored by divergent selection on the subfunctionalized gene duplicate.

Wagner's (1996) model may provide the beginnings of what a general theory for the evolution of coordinated variability might look like. The missing point is to understand the correlation between character value and character variability. One possibility is that there are no systematic patterns in this relation and, consequently, that the evolution of variability is erratic and idiosyncratic. The other possibility is that there exist general mechanisms that generate systematic correlations between character values and character variability. The development and testing of hypotheses of this sort may be the path forward for understanding the evolution of evolvability.

An example of an indirect-selection hypothesis for robustness is the Kacser and Burns (1981) hypothesis for the evolution of dominance as a side effect of selection for increased flux in metabolic pathways. As selection increases flux in a linear pathway, the effects of individual enzyme activities on flux will reach a plateau and, consequently, become more robust toward genetic changes that reduce activity. This robustness will manifest itself as dominance of wild-type (high-activity) alleles over mutant (low-activity) alleles. While more detailed analyses have shown that the evolution of dominance and robustness is not inevitable in nonlinear pathways (Bagheri et al., 2003; Bagheri and Wagner, 2004; Bagheri, 2006) and is dependent on patterns of epistasis in a similar way as is the evolution of individual gene effects, this much-discussed example shows that we may expect systematic correlations between the state of traits and their variational properties.

VARIATIONAL PROPERTIES AS SIDE EFFECTS OF DEVELOPMENTAL-PATHWAY EVOLUTION

Above, I focused on indirect selection on the variability of a character caused by specific direct selection on the character itself. An alternative indirect-selection hypothesis is that the variational properties of specific characters are formed by selection on generic features of the developmental architecture of the whole organism. There is, of course, no sharp distinction between phenotypic characters and developmental characters, but we can imagine a continuum from adult shapes and sizes on the one end to developmental mechanisms such as pattern formation and tissue interactions on the other. It seems plausible that developmental mechanisms are adapted for general properties such as efficiency, speed, accuracy, and plasticity. This generic selection of developmental pathways will induce indirect selection on the variability of the characters they produce. This indirect selection may be of a different and more general form than the indirect selection induced from adaptation of specific characters. For example, one can imagine that selection for general developmental accuracy would indirectly reduce the variability of many characters and, thus, be a general explanation for both environmental and genetic robustness.

A fundamental question is to what degree the variational properties of conventional phenotypic characters such as sizes, forms, and behaviors are constrained by the developmental process. The adaptive view assumes that the variational properties of a character can be shaped not only quasi-independently of the character itself but also quasi-independently from the variational properties of other characters. If variational properties are largely determined by fundamental developmental processes, then adaptations for evolvability, robustness, etc. of individual characters will be less feasible.

CONCLUSIONS

The variational properties of phenotypes are almost certainly products of selection. The genotype–phenotype map is highly nonrandom and appears to produce highly coordinated variation (Riedl, 1978; Raff, 1996; Dennett, 1995; Gerhart and Kirschner 1997; Dawkins, 1996; Wagner, 2001; Pigliucci and Preston, 2004; Hallgrimsson and Hall, 2005). Any trait of such complexity would degenerate in the absence of stabilizing selection. I have argued, however, that this does not necessarily imply that the genotype–phenotype map is adapted to increase variability, evolvability, robustness, or plasticity. My argument rests on three ideas: (1) direct individual selection for variability, evolvability, and robustness is often weak; (2) variational properties are correlated with trait values, and direct selection for variational adaptations is unlikely to match strong selection on the traits themselves; (3) epistasis, thought to be the main basis for variational evolution, may be as likely to generate negative as positive responses to selection.

In this light, variational evolution through indirect selection from the traits themselves emerges as the leading hypothesis. Accepting this position suggests a rather dramatic shift in both theoretical and empirical work on variational evolution. First of all, theoretical work must abandon the near exclusive focus on showing that adaptive evolution of evolvability is possible when everything else is held equal or in comparison with pure genetic drift. Instead, it is necessary to evaluate how selection for evolvability fares in a balance with indirect selection from the trait. Indeed, many simulation studies could have been more illuminating if the causes of selection had been properly identified. To study indirect selection and to understand how indirect selection may cause coordinated variability, we need to focus on the relation between character values (the target of selection) and the variability of the character. We need to know how allelic and mutational effects correlate with character states. Finally, the study of epistasis needs to focus on identifying systematic patterns of interactions. It is not enough to show that genes interact; we need to know how they interact.

ACKNOWLEDGMENTS

I thank Brian Hall, Benedikt Hallgrímsson, David Houle, and Mihalea Pavlicev for comments on the manuscript and Antonieta Labra for making the figure.

REFERENCES

Amundson, R. 2005. *The Changing Role of the Embryo in Evolutionary Thought: Roots of Evo–Devo.* Cambridge Studies in Philosophy and Biology. Cambridge: Cambridge University Press.

Armbruster, W. S., V. S. Di Stilio, J. D. Tuxill, T. C. Flores, and J. I. Velásquez Runke. 1999. Covariance and decoupling of floral and vegetative traits in nine neotropical plants: A re-evaluation of Berg's correlation-pleiades concept. *Am J Bot* 86:39–55.

Bagheri, H. C. 2006. Unresolved boundaries of evolutionary theory and the question of how inheritance systems evolve: 75 years of debate on the evolution of dominance. *J Exp Zoolog B Mol Dev Evol* 306:329–59.

Bagheri, H. C., J. Hermisson, J. R. Vaisnys, and G. P. Wagner. 2003. Effects of epistasis on phenotypic robustness in metabolic pathways. *Math Biosci* 184:27–51.

Bagheri, H. C., and G. P. Wagner. 2004. The evolution of dominance in metabolic pathways. *Genetics* 168:1713–35.

Barton, N. H., and M. Turelli. 2004. Effects of genetic drift on variance components under a general model of epistasis. *Evolution* 58:2111–32.

Bell, G. 1982. *The Masterpiece of Nature.* Berkeley: University of California Press.

Berg, R. L. 1959. A general evolutionary principle underlying the origin of developmental homeostasis. *Am Nat* 93:103–5.

Carroll, S. B. 2005. *Endless Forms Most Beautiful: The New Science of Evo Devo.* New York: Norton.

Carter, A. J. R., J. Hermisson, and T. F. Hansen. 2005. The role of epistatic gene interactions in the response to selection and the evolution of evolvability. *Theor Popul Biol* 68:179–96.

Cheverud, J. M. 2001. The genetic architecture of pleiotropic relations and differential epistasis. In *The Character Concept in Evolutionary Biology,* ed. G. P. Wagner, 411–33. San Diego: Academic Press.

Cheverud, J. M., T. H. Ehrich, T. T. Vaughn, S. F. Koreishi, R. B. Linsey, and S. Pletscher. 2004. Pleiotropic effects on mandibular morphology II: Differential epistasis and genetic variation in morphological integration. *J Exp Zoolog B Mol Dev Evol* 302:424–35.

Cheverud, J. M., E. J. Routman, and D. J. Irschick. 1997. Pleiotropic effects of individual gene loci on mandibular morphology. *Evolution* 51:2006–16.

Davidson, E. H. 2001. *Genomic Regulatory Systems: Development and Evolution.* San Diego: Academic Press.

Dawkins, R. 1996. *Climbing Mount Improbable.* London: Viking.

Dennett, D. C. 1995. *Darwin's Dangerous Idea: Evolution and the Meanings of Life.* New York: Simon and Schuster.

de Visser, J. A. G. M., J. Hermisson, G. P. Wagner, L. Ancel Meyers, H. Bagheri-Chaichian, J. Blanchard, L. Chao, et al. 2003. Evolution and detection of genetic robustness. *Evolution* 57:1959–72.

Draghi, J., and G. P. Wagner. 2008. Evolution of evolvability in a developmental model. *Evolution* 62:301–15.

Flatt, T. 2005. The evolutionary genetics of canalization. *Q Rev Biol* 80:287–316.

Force, A., M. Lynch, F. B. Pickett, A. Amores, Y.-L. Yan, and J. Postlethwait, 1999. Preservation of duplicate genes by complementary degenerative mutations. *Genetics* 151:1531–45.

Gabriel, W., M. Lynch, and R. Bürger. 1993. Muller's ratchet and mutational meltdowns. *Evolution* 47:1744–57.

Gardner, A., and A. T. Kalinka. 2006. Recombination and the evolution of mutational robustness. *J Theor Biol* 241:707–15.

Gardner, A., and W. Zuidema. 2003. Is evolvability involved in the origin of modular variation? *Evolution* 57:1448–50.

Gavrilets, S. 2004. *Fitness Landscapes and the Origin of Species.* Monographs in Population Biology. ed. S. A. Levin and H. Horn. Princeton, NJ: Princeton University Press.

Gavrilets, S., and J. Gravner. 1997. Percolation on the fitness hypercube and the evolution of reproductive isolation. *J Theor Biol* 184:51–64.

Gerhart, J., and M. Kirschner. 1997. *Cells, Embryos and Evolution: Towards a Cellular and Developmen-*

tal Understanding of Phenotypic Variation and Evolutionary Adaptability. Oxford: Blackwell.

Ghiselin, M.T. 1974. *The Economy of Nature and the Evolution of Sex.* Berkeley: University of California Press.

Gladyshev, E.A., M. Meselson, and I.R. Arkipova. 2008. Massive horizontal gene transfer in bdelloid rotifers. *Science* 320:1210–13.

Hahn, M.W. 2008. Toward a selection theory of molecular evolution. *Evolution* 62:255–65.

Hallgrimsson, B., and B.K. Hall. 2005. *Variation: A Central Concept in Biology.* New York: Academic Press.

Hansen, T.F. 2003. Is modularity necessary for evolvability? Remarks on the relationship between pleiotropy and evolvability. *Biosystems* 69:83–94.

Hansen, T.F. 2006. The evolution of genetic architecture. *Annu Rev Ecol Evol Syst* 37:123–57.

Hansen, T.F., J.M. Alvarez-Castro, A.J. R. Carter, J. Hermisson, and G.P. Wagner. 2006. Evolution of genetic architecture under directional selection. *Evolution* 60:1523–36.

Hansen, T.F., W.S. Armbruster, M.L. Carlson, and C. Pélabon. 2003. Evolvability and genetic constraint in *Dalechampia* blossoms: Genetic correlations and conditional evolvability. *J Exp Zoolog B Mol Dev Evol* 296:23–39.

Hansen, T.F., and D. Houle. 2008. Measuring and comparing evolvability and constraint in multivariate characters. *J Evol Biol* 21:1201–19.

Hansen, T.F., C. Pelabon, and W.S. Armbruster. 2007. Comparing variational properties of homologous floral and vegetative characters in *Dalechampia scandens*: Testing the Berg hypothesis. *Evol Biol* 34:86–98.

Hansen, T.F., and G.P. Wagner. 2001a. Modeling genetic architecture: A multilinear model of gene interaction. *Theor Popul Biol* 59:61–86.

Hansen, T.F., and G.P. Wagner. 2001b. Epistasis and the mutation load: A measurement-theoretical approach. *Genetics* 158:477-85.

Hereford, J., T.F. Hansen, and D. Houle. 2004. Comparing strengths of directional selection: How strong is strong? *Evolution* 58:2133–43.

Hermisson, J., T.F. Hansen, and G.P. Wagner. 2003. Epistasis in polygenic traits and the evolution of genetic architecture under stabilizing selection. *Am Nat* 161:708–34.

Hermisson, J., and G.P. Wagner. 2004. The population genetic theory of hidden variation and genetic robustness. *Genetics* 168:2271–84.

Hoekstra, H., and J.A. Coyne. 2007. The locus of evolution: Evo devo and the genetics of adaptation. *Evolution* 61:995–1016.

Houle, D. 1998. How should we explain variation in the genetic variance of traits? *Genetica* 102/103:241–53.

Houle, D. 2001. Characters as the units of evolutionary change. In *The Character Concept in Evolutionary Biology*, ed. G.P. Wagner, 109–40. San Diego: Academic Press

Houle, D., B. Morikawa, and M. Lynch. 1996. Comparing mutational variabilities. *Genetics* 143:1467–83.

Jones, A.G., S.J. Arnold, and R. Burger. 2007. The mutation matrix and the evolution of evolvability. *Evolution* 61:727–45.

Kacser, H., and J.A. Burns. 1981. The molecular basis of dominance. *Genetics* 97:639–66.

Kauffman, S.A. 1993. *The Origins of Order: Self-Organization and Selection in Evolution.* New York: Oxford University Press.

Kondrashov, A.S. 1988. Deleterious mutations and the evolution of sexual reproduction. *Nature* 336:435–40.

Lee, C.E., and G.W. Gelembiuk. 2008. Evolutionary origins of invasive populations. *Evol Appl* 1:427–48.

Le Rouzic, A., and O. Carlborg. 2008. Evolutionary potential of hidden genetic variation. *TREE* 23:33–7.

Lewontin, R.C. 1978. Adaptation. *Sci Am* 239:212–31.

Lipson, H., J.B. Pollack, and N.P. Suh. 2002. On the origin of modular variation. *Evolution* 56:1549–56.

Lynch, M. 1984. The selective value of alleles underlying polygenic traits. *Genetics* 108:1021–33.

Lynch, M. 1988. The rate of polygenic mutation. *Genet Res* 51:137–48.

Lynch, M. 1990. The rate of morphological evolution in mammals from the standpoint of the neutral expectation. *Am Nat* 136(6):727–41.

Lynch, M. 2005. The origins of eukaryotic gene structure. *Mol Biol Evol* 23:450–68.

Lynch, M. 2007a. *The Origins of Genome Architecture.* Sunderland, MA: Sinauer.

Lynch, M. 2007b. The evolution of genetic networks by non-adaptive processes. *Nat Rev Genet* 8:803–13.

Lynch, M. 2007c. The frailty of adaptive hypotheses for the origins of organismal complexity. *Proc Natl Acad Sci USA* 104:8597–604.

Lynch, M., R. Bürger, D. Butcher, and W. Gabriel. 1993. Mutational meltdowns in asexual populations. *J Hered* 84:339–44.

Lynch, M., J. Conery, and R. Bürger. 1994. Mutational meltdowns in sexual populations. *Evolution* 49:1067–80.

Lynch, V.J., A. Tanzer, Y. Wang, F.C. Leung, B. Gellersen, D. Emera, and G.P. Wagner. 2008.

Adaptive changes in the transcription factor HoxA-11 are essential for the evolution of pregnancy in mammals. *Proc Natl Acad Sci USA* 105:14928–33.

Lynch, V. J., and G. P. Wagner. 2008. Perspective: Resurrecting the role of transcription factor change in developmental evolution. *Evolution* 62:2131–54.

Masel, J. 2005. Evolutionary capacitance may be favored by natural selection. *Genetics* 170:1359–71.

Maynard Smith, J. 1976. *The Evolution of Sex.* Cambridge: Cambridge University Press.

Mitchell, W. A., and T. J. Valone. 1990. The optimization research program: Studying adaptations by their function. *Q Rev Biol* 65:43–52.

Pavlicev, M., J. P. Kenney-Hunt, E. A. Norgard, C. C. Roseman, J. B. Wolf, and J. M. Cheverud. 2008. Genetic variation in pleiotropy: Differential epistasis as a source of variation in the allometric relationship between long bone lengths and body weight. *Evolution* 62:199–213.

Perez-Barrales, R., J. Arroyo, and W. S. Armbruster. 2007. Differences in pollinator faunas may generate geographic differences in floral morphology and integration in *Narcissus papyraceus* (Alarcissiopapyraceris). *Oikos* 116:1904–18.

Pigliucci, M., and K. Preston, eds. 2004. *Phenotypic Integration: Studying the Ecology and Evolution of Complex Phenotypes.* New York: Oxford University Press.

Proulx, S. R., and P. C. Phillips. 2005. The opportunity for canalization and the evolution of genetic networks. *Am Nat* 165:147–62.

Raff, R. A. 1996. *The Shape of Life: Genes, Development, and the Evolution of Animal Form.* Chicago: University of Chicago Press.

Raff, R. A., and T. C. Kaufman. 1983. *Embryos, Genes, and Evolution: The Developmental-Genetic Basis of Evolutionary Change.* Bloomington: Indiana University Press.

Rice, S. H. 1998. The evolution of canalization and the breaking of von Baer's laws: Modeling the evolution of development with epistasis. *Evolution* 52(3):647–56.

Rice, S. H. 2002. A general population genetic theory for the evolution of developmental interactions. *Proc Natl Acad Sci USA* 99:15518–23.

Riedl, R. 1978. *Order in Living Organisms: A Systems Analysis of Evolution.* New York: Wiley.

Rifkin, S. A., D. Houle, J. Kim, and K. P. White. 2005. A mutation accumulation assay reveals a broad capacity for rapid evolution of gene expression. *Nature* 438:220–3.

Rutherford, S., and S. Lindquist. 1998. Hsp90 as a capacitor for morphological evolution. *Nature* 396:336–42.

Sober, E. 1984. *The Nature of Selection: Evolutionary Theory in Philosophical Focus.* Cambridge, MA: Bradford Books.

Sober, E., and D. S. Wilson. 1998. *Unto Others: The Evolution and Psychology of Unselfish Behavior.* Cambridge, MA: Harvard University Press.

Stern, D. 2000. Perspective: Evolutionary developmental biology and the problem of variation. *Evolution* 54:1079–91.

Stone, J., and G. A. Wray. 2001. Rapid evolution of cis-regulatory sequences via local point mutations. *Mol Biol Evol* 18:1764–70.

True, J. R., and E. S. Haag. 2001. Developmental systems drift and flexibility in evolutionary trajectories. *Evol Dev* 3:109–19.

van Nimwegen, E., W. Y. Crutchfield, and M. Huynen. 1999. Neutral evolution of mutational robustness. *Proc Natl Acad Sci USA* 96:9716–20.

Via, S., and R. Lande. 1985. Genotype–environment interaction and the evolution of phenotypic plasticity. *Evolution* 39:505–22.

Von Dassow, G., E. Meir, E. M. Munro, and G. M. Odell. 2000. The segment polarity network is a robust developmental module. *Nature* 406:188–92.

Von Dassow, G., and E. Munro. 1999. Modularity in animal development and evolution: Elements of a conceptual framework for evodevo. *J Exp Zoolog B Mol Dev Evol* 285:307–25.

Waddington, C. H. 1940. *Organizers and Genes.* London: Cambridge University Press.

Waddington, C. H. 1942. Canalization of development and the inheritance of acquired characters. *Nature* 150:563–5.

Waddington, C. H. 1953. Genetic assimilation of an acquired character. *Evolution* 7:118–26.

Wagner, A. 1999. Redundant gene functions and natural selection. *J Evol Biol* 12:1–16.

Wagner, A. 2005. *Robustness and Evolvability in Living Systems.* Princeton, NJ: Princeton University Press.

Wagner, G. P. 1989. Multivariate mutation-selection balance with constrained pleiotropic effects. *Genetics* 122:223–34.

Wagner, G. P. 1996. Homologues, natural kinds and the evolution of modularity. *Am Zool* 36:36–43.

Wagner, G. P. 2001. *The Character Concept in Evolutionary Biology.* San Diego: Academic Press.

Wagner, G. P., and L. Altenberg. 1996. Complex adaptations and evolution of evolvability. *Evolution* 50:967–76.

Wagner, G. P., G. Booth, and H. Bagheri-Chaichian. 1997. A population genetic theory of canalization. *Evolution* 51:329–47.

Wagner, G. P., and R. Bürger. 1985. On the evolution of dominance modifiers II. A non-equilibrium approach to the evolution of genetic systems. *J Theor Biol* 113:475–500.

Wagner, G. P., M. Pavlicev, and J. M. Cheverud. 2007. The road to modularity. *Nat Rev Genet* 8:921–31.

Wilkins, A. 2002. *The Evolution of Developmental Pathways*. Sunderland, MA: Sinauer.

Williams, G. C. 1975. *Sex and Evolution*. Princeton, NJ: Princeton University Press.

Wilson, D. S., and E. O. Wilson. 2007. Rethinking the theoretical foundation of sociobiology. *Q Rev Biol* 82:327–48.

Wolf, J., L. J. Leamy, E. Routman, and J. M. Cheverud. 2005. Epistatic pleiotropy and the genetic architecture of covariation within early- and late-developing skull trait complexes in mice. *Genetics* 171:683–94.

Wray, G. A., M. W. Hahn, E. Abouheif, J. P. Balhoff, M. Pizer, M. V. Rockman, and L. A. Romano. 2003. The evolution of transcriptional regulation in eukaryotes. *Mol Biol Evol* 20:1377–1419.

Young, N. M., and B. Hallgrimsson. 2005. Serial homology and the evolution of mammalian limb covariation structure. *Evolution* 59: 2691–704.

21

The Epigenetics of Dysmorphology

CRANIOSYNOSTOSIS AS AN EXAMPLE

Christopher J. Percival and Joan T. Richtsmeier

CONTENTS

Important discoveries in the field of evolutionary developmental biology have added significantly to our understanding of the evolution of developmental processes and the production of novel phenotypes, thereby advancing our knowledge of the molecular bases of genetic disease. This is especially true in the field of craniofacial biology, where anomalies of the head and neck account for 75% of all congenital birth defects (Chai and Maxson, 2006). Advances in the understanding of craniofacial dysmorphogenesis have come from varying approaches. Perhaps the most successful approach over the past 20 years has been the genetic mapping of diseases using human samples followed by a focused investigation of the function of the identified gene(s) and its association with craniofacial growth and development. These approaches have resulted in the identification of disease-causing mutations, but only in a few cases have these identifications brought us closer to an understanding of the actual production of dysmorphology.

The overall absence of satisfactory explanations of the genotype–phenotype continuum in craniofacial anomalies reveals a general lack of understanding of *phenogenetics*, the connection between biological phenotypes and their underlying genetic bases (Weiss, 2005; Weiss and Buchanan, 2004). Phenogenetic phenomena, like hierarchies of regulatory cascades or nested epigenetic networks, provide a structural connection between genes and phenotypes (Carroll

Epigenetics: Linking Genotype and Phenotype in Development and Evolution, ed. Benedikt Hallgrímsson and Brian K. Hall.
Copyright © by The Regents of the University of California. All rights of reproduction in any form reserved.

377

et al., 2001; Davidson, 2001; Wilkins, 2002; Weiss and Buchanan, 2004; Weiss, 2005). At a more coarse scale, phenogenetic phenomena include processes such as regional differentiation by dynamic inductive signaling and repetitive patterning by quantitative interactions, which affect interactions among populations of cells (Weiss, 2005). The complexity of these phenomena and the processes that underlie them explains why the proximate function of a single mutation is rarely sufficient to account for more than a subset of associated syndromic phenotypes even in diseases that are plainly genetic. True understanding of a genetic disease requires the design of a process-based strategy for a solution to the genotype–phenotype conundrum (Weiss, 2005; Buchanan et al., 2009). Our current state of knowledge and technical capacity requires that we discover genotype–phenotype relationships in a piece-by-piece fashion (but see Chen et al., 2008). Epigenetics provides a theoretical framework from which to study the intricate networks of interactions that make up development, to design experiments to uncover these relationships, and ultimately to join them together.

Epigenetics, in its strictest sense, refers to the interactions between genes, with other transcribed portions of the genome, and between the products of both (for further discussion, see Chapter 3, this volume). The emergent properties of the complex networks of molecular interactions occurring through a variety of regulatory mechanisms may be the starting point for an explanation of the phenotypic variation that exists among individuals with the same "genes." However, interactions at higher levels, between cells, tissues, and the external environment, contribute in a major way to the production of the phenotype. By extending the definition of *epigenetics* to include interactions at levels above the gene, the complexity of the interactions involved in the production of phenotypic variation is revealed. Physical interactions between cells can lead to changes in gene expression and the movement of cell populations (Radlanski and Renz, 2006). Mechanical forces of strain and compression have been shown to regulate the speed of transcription in certain cells and the form of associated skeletal elements (Mao et al., 2003; Sun et al., 2004; Opperman and Rawlins, 2005). Environmental perturbations can lead to shifts in gene expression (Lopez-Maury et al., 2008; Li et al., 2008), influencing the phenotypes of individuals. These environmentally based phenotypic variants are often discussed in terms of norms of reaction and phenotypic plasticity (Sultan and Stearns, 2005; Grether, 2005). While the engine for development may be based on the molecular machinery of transcription, regulation, and expression of genes and gene products, the production of gross phenotypic form is driven by interactions at intercellular and higher levels.

THE EPIGENETICS OF CRANIOFACIAL DYSMORPHOLOGY

This chapter focuses on the epigenetic basis of craniofacial dysmorphology. With the understanding that certain processes are yet to be discovered, we argue that an understanding of normative processes already in the developmental biologist's catalogue will provide the best information about the epigenetics of dysmorphology. Consideration of the epigenetic processes that could produce a localized dysmorphology requires a structurally broad view that considers interactions of many types and scales. This view includes interactions within tissues local to and at a distance from the dysmorphology in question and interactions before and during the physical manifestation of that dysmorphology.

Normal phenotypic variation is produced by genetic and environmental input through the lens of development. From an epigenetic perspective, all dysmorphologies stem from modifications of particular contexts within the vast network of developmental interactions. The degree of epigenetic disruption and the developmental context of that disruption are major factors in determining the nature of the subsequent dysmorphology. There are two broad mechanisms of the epigenetics of dysmorphology. First, dysmorphology can occur due to changes in normal epigenetic processes of development. These changes might be related to modifications of timing, rate, or quality of some factor involved in an epigenetic interaction. Examples of changes in the regulation of existing factors include mutations within regulatory elements and changes in environmental conditions that manifest as movement along a norm of reaction. Changes in the regulation of existing processes can have a major phenotypic effect, as shown by the recent discovery of an inhibitory cascade that controls molar morphology as well as the presence or absence of the third molar in mammals (Kavanagh et al., 2007). Second, dysmorphology occurs due to the introduction of new or the removal of resident interactions. These more drastic changes introduce new dynamics to the organism's epigenetic system and could be based, for example, on the insertion, deletion, or modification of protein-coding genes or the introduction of a major environmental factor, such as a teratogen early in development.

In the context of broader developmental networks, the distinction between changes in the regulation of existing epigenetic interactions and the creation of novel interactions is not always clear. A new interaction at one scale might be a mere change in the regulation of interactions at a coarser scale. For instance, the appearance of a gene that codes for micro-RNA would provide a completely new interaction within the cell (Eberhart et al., 2008) but might change the regulation of the production of certain gene products only at the tissue level. In addition, we expect that novel interactions utilize mechanisms of interaction already in place within developmental contexts at the scale of development in question. In the simplest case, we expect that the expression of novel mutations will still be modified by the expression of a regulatory sequence and will still produce gene products. The introduction of novel mechanical forces is expected to influence development as mechanical forces normally do. That is, a new source of mechanical force may act on a certain cell population, but the reaction of that cell population to the force will be the same type of reaction that the cell population would have to any mechanical force of the same type. Therefore, an understanding of the complexities of normal development is necessary to understand epigenetic change leading to dysmorphology, regardless of the nature or scale of that change.

The major attributes that determine the nature and degree of dysmorphology caused by a developmental modification are the context of the developmental modification and the strength of that modification. A *developmental context* (DC) is defined as a set of epigenetic interactions between genetic, epigenetic, and/or environmental inputs that produce epigenetic and/or phenotypic outputs at a specific time and place (Figure 21.1). Epigenetic input is the epigenetic output of other DCs and can come in the form of molecular products moving spatially between contexts, physical forces generated between contexts, or merely the changing conditions in the same space between different points of time. Phenotypic output is the measurable modification of a morphological trait of interest. Phenotypic and epigenetic outputs broadly overlap. Each DC exists at a specific scale of interaction, at a specific time during development, and in a particular spatial context within the developing organism. In reality, all DCs are transient and alterable with continuous interactions. Developmental processes are therefore less dissectible and compartmentalized than usually described, but some simplifications must be made in order to describe developmental interactions.

The magnitude of the perturbation of a normal developmental interaction within a DC will influence the output of that DC. The strength and context of a developmental perturbation will determine its effect on the phenotype. The strength of a particular modification on the phenotype can be measured by the magnitude of the particular phenotypic change produced in an individual compared to equivalent individuals unaltered by the modification.

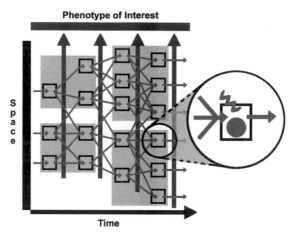

Phenotype of Interest

S p a c e

Time

FIGURE 21.1 Developmental contexts (DCs) exist within and are limited by space and time. The DCs (hollow boxes) are connected by epigenetic input/output (small arrows) and add to the combined output of all DCs of a specific time (large arrows). DCs are defined according to their location in developmental time and anatomical space. The phenotype of interest is the measurable phenotype of a specific trait. The magnification of a DC on the right side of this figure displays the factors that epigenetically interact within a DC to produce an epigenetic output (exiting small arrow). These factors are epigenetic input from other DCs (entering small arrow), local gene expression (circle), and external environmental input (wave). DCs can be examined at different spatial scales (e.g., cell population, tissue, system) or temporal scales of varying frequency. DCs at a finer scale (hollow boxes) are components of DCs at a coarser scale (light gray boxes).

The DC of a perturbation has an impact on the nature and degree of dysmorphology produced because DCs exist upstream of and in parallel with a variable number of other DCs, which in turn are related to a variable number of output phenotypes. When a DC with many downstream output phenotypes is perturbed, this single modification has the potential to create dysmorphology in all of the associated downstream output phenotypes (Figure 21.1), though this is not a necessary outcome. Simply stated, the farther upstream a change is, the more widespread its potential—but not necessarily its realized phenotypic influence. The context of a DC is obviously more complex than this. First, a particular epigenetic output may have a minor effect on the output of downstream DCs. This might be because the interactions of downstream DCs are not significantly influenced by that output. It is also possible that

a particular phenotypic trait is simply more canalized than others. Some phenotypes are better canalized because the network of DCs upstream of phenotypic output has greater redundancy, where some DCs compensate for a change in other DCs. For instance, parallel or downstream DCs might produce outputs, potentially as a reaction to abnormal upstream epigenetic outputs, which mask the effects of the initial abnormal output of the upstream DC (Figure 21.2).

A BROAD VIEW OF CRANIOFACIAL DEVELOPMENT

The head is an intricate composite of soft tissue and hard tissue elements, each with its own pattern of cellular organization, tissue generation, regulation, and growth. During head development, suites of genes interact under strict

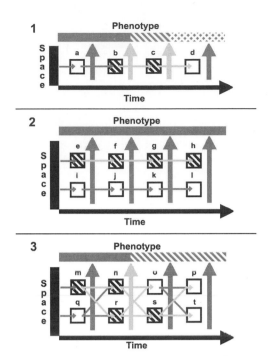

FIGURE 21.2 Some potential ways in which the influence of a developmental perturbation (DP) might affect a phenotype or disappear within a developmental system. Symbols and axes are based on those provided for Figure 21.1. The time and space axes represent the context of a DC. The phenotype (horizontal line at top of each panel) is a measurable phenotype of interest, within normal limits when solid. Hollow boxes are DCs operating within normal limits; hatched boxes are DCs with abnormal interactions. Small dark arrows are normal epigenetic input/output, large dark arrows are normal combinations of phenotypic output, and both are significantly modified when lighter. (1) Potential downstream influence of a DP at *b*. Abnormal epigenetic interactions of *b* produce abnormal epigenetic output to *c*, causing abnormal epigenetic interactions of *c* and abnormal phenotypic output. Further abnormal epigenetic output from *c* to *d* does not significantly modify the epigenetic interactions of *d* because of other developmental factors. Therefore, the influence of the DP within the DC network ceases. (2) Potential canalizing interactions of parallel DCs. A DP at *e* leads to abnormal epigenetic output of *e* and all downstream DCs (*f, g, h*), but unaffected parallel DCs (*i, j, k, l*) continue to produce normal epigenetic outputs. Potentially, the normal phenotypic output of some DCs can mask or repair the abnormal phenotypic output of other DCs, leading to overall normal phenotypic output. In this case, the parallel DCs are redundant, thereby enabling normal phenotypic output. (3) Potential canalizing interactions due to a simple feedback interaction. A DP on *m* leads to abnormal interactions within *n* and *r* through abnormal epigenetic output. As in network 2, both parallel DCs need to be modified before the combined phenotypic output leads to an abnormal phenotype. A potential feedback mechanism: The subsequent abnormal epigenetic output of

spatiotemporal control. Many of these genes are pleiotropic and are also involved in the organization, growth, and development of other noncranial systems and tissues (Francis-West et al., 2003). In a perfect world, we would be able to trace the cause of a particular dysmorphology through the network of DCs to a particular developmental perturbation. Although our current methods do not enable a phenotype–genotype map of this accuracy, it is crucial to keep the broader epigenetic networks and scales of development in mind as we pursue the basis of dysmorphology and phenotypic form in general. A broad view of craniofacial development provides a context in which to discuss the role of epigenetic interactions in craniofacial dysmorphogenesis.

The mammalian head can be modeled as a series of concentrically arranged tissue layers with an attached facial component (Figure 21.3). At the center, a developing hollow neural tube gains complexity as it folds in on itself, becoming internal complexes of neural cells and organs of cerebrospinal fluid production and circulation. These are enveloped by an external layer called the "leptomeninx" (pia and arachnoid mater) that closely adheres to the outer surface of the brain and a layer of neural crest–derived tissue called the "pachymeninx" (dura mater). A protective skeletal layer forms ectocranial to the dura mater. The cranial base skeleton is established inferior to the neural mass, forming around large neurovascular bundles (including the spinal cord) that connect the central nervous system with special sense organs in the head and to the peripheral nervous system. The cranial vault bones surround the more superior surfaces of the brain. The facial skeleton forms anteroinferior to the cranial base and vault, within the pharyngeal arch complex and

r to *o* counteracts the abnormal epigenetic output from *n* to *o* and serves to normalize the epigenetic interactions of the top row of DCs. As in the interaction between *c* and *d* of network 1, the abnormal epigenetic output of *s* does not lead to abnormal epigenetic interactions of *p* or *t*.

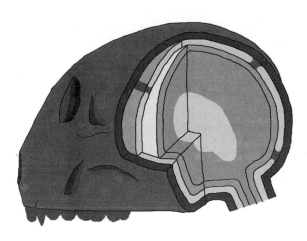

FIGURE 21.3 The head is a volume of integrated tissue layers. This organization imparts a logical conduit for the exchange of information not only within but also between various tissue layers, with the potential for linked processes across tissues. From the center of the head are the neural tissues, dura mater, bones, sutures, and more superficial tissues like muscle, tendon, and skin. The face is hafted onto the cranium, and this volume includes its own suite of tissue layers.

around additional important sensory organs (e.g., globe of the eye, naso- and oropharynx). The three skeletal portions of the head (cranial base, cranial vault, and facial skeleton) eventually fuse, forming a complete bony skull (Figure 21.4). Ectocranial to the skull is a layer of non-osteogenic mesenchyme. Certain muscles develop within this layer, inserting directly into the bones. However, it has been shown that the connective tissues of some muscles of the avian vault differentiate within the same group of cells that differentiate to become the bone into which they insert (Kontges and Lumsden, 1996). The ectodermally derived skin is the most exterior tissue layer of the head.

The importance of this highly simplified model of the head is that it forces us to broaden our definition of local (and nonlocal) factors that influence cranial development. These layers are often thought of as isolated surfaces across which processes and interactions take place. Importantly, the basic quasi-concentric arrangement of head tissues (Figure 21.3) provides a unique design for developmental relationships to become established. This organization imparts a logical conduit for the exchange of information between, across, and through various tissue layers with the potential for linked processes among tissues. For example, the developmental reaction of one tissue layer might be converted into the stimulus for a process in adjacent tissues. As seen for molecular

gene products, mechanical strain, and environmental factors, the properties of each layer may limit the types of epigenetic input and output that can be transmitted across them.

THE EPIGENETICS OF CRANIOSYNOSTOSIS

Craniosynostosis (CS), the premature closure of cranial sutures, results in characteristic craniofacial phenotypes and can profitably be considered from an epigenetics perspective. In humans, CS usually commences prenatally, but it can occur postnatally, including incidences of iatrogenic CS due to changes in cerebrospinal fluid volume and the associated changes in pressure (Carson and Dufresne, 1992). CS can occur as an isolated anomaly or as a feature of a syndrome. At least 100 syndromes with CS are known (Rice, 2008), and a significant proportion of CS cases are syndromic (Cohen and Maclean, 2000; Rice, 2008). The incidence of CS has been associated with genetic, environmental, and mechanical factors (Rasmussen et al., 2008). Certainly, the most familiar factors are genetic (Table 21.1). There has been great success in mapping mutations of genes responsible for specific CS syndromes in humans and in discovering the effect of these mutations on specific aspects of molecular signaling by studying mouse models carrying orthologues of these mutations. Although considerable progress is

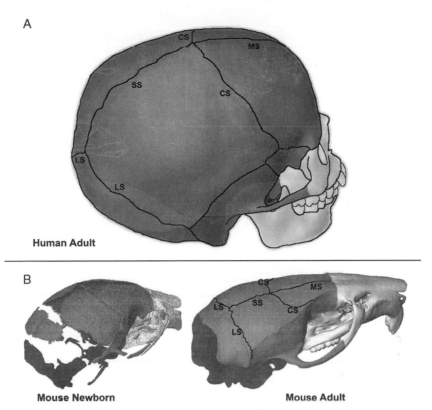

FIGURE 21.4 The mammalian skull is made up of the intramembranous splanchnocranium or facial skeleton (light gray) and the neurocranium, which is divided into the primarily intramembranous cranial vault (medium gray) and the endochondral cranial base (dark gray), though the boundaries between the skull divisions are imprecise. Vault sutures include the metopic suture (MS) that fuses early in humans, the paired coronal sutures (CS), the sagittal suture (SS), and the paired lambdoid sutures (LS). SS is homologous to the interparietal suture, and MS is homologous to the interfrontal sutures of other mammals. Images are not to scale. (A) Developmental origin of skull bones in humans. (B) Developmental origin of skull bones in newborn and adult mice.

being made at the molecular level, current CS research indicates few clear causal relationships between discovered mutations, precise developmental processes, and cranial vault shape variation in CS. In short, we know little of the developmental mechanisms involved in the production of CS phenotypes.

The incidence of isolated CS varies by suture (Figure 21.4). Isolated sagittal synostosis occurs more than twice as often as any other type (estimated as 1.9–2.3 per 10,000 live births). Isolated coronal synostosis (0.8–1.0 per 10,000) occurs significantly more often than either metopic (1 per 10,000-15,000) or lambdoid (quite rare) synostosis (Boyadjiev, 2007). Lambdoid synostosis occurs infrequently and is often mis-diagnosed, leading to an artificially inflated prevalence in some studies. The difference in incidence of CS of the various sutures suggests that the mechanisms involved in suture creation and/or persistence are more difficult to modify for some sutures than others or that they are at least influenced by different factors. When more than one cranial suture is involved, the affected sutures are always contiguous, suggesting that mechanisms of normal suture development and closure may covary spatially. In addition, there is significant sex bias in the expression of premature closure of some calvarial sutures, suggesting that developmental interactions essential to suture patency are different between males and females (Cohen and Maclean, 2000).

TABLE 21.1

A Partial List of Genes and Phenotypes Associated with Craniosynostosis

GENE	GENE SYMBOL	PHENOTYPES
Fibroblast growth factor receptor 1	*FGFR1*	Pfeiffer syndrome, osteoglophonic dysplasia
Fibroblast growth factor receptor 2	*FGFR2*	Crouzon syndrome Crouzon syndrome with scaphocephaly Jackson-Weiss syndrome Pfeiffer syndrome Apert syndrome Saethre-Chotzen syndrome-like cutis gyrata syndrome of Beare-Stevenson syndrome Antley-Bixley syndrome, nonclassifiable syndromes with craniosynostosis
Fibroblast growth factor receptor 3	*FGFR3*	Muenke syndrome Crouzon with acanthosis nigricans Saethre-Chotzen syndrome-like thanatophoric dysplasia types I and II
Twist homolog *Drosophila* 1	*TWIST1*	Saethre-Chotzen syndrome, nonsyndromic craniosynostosis
Ephrin-B1	*EFNB1*	Craniofrontonasal syndrome
Ras-associated protein RAB23	*RAB23*	Carpenter syndrome
Muscle segment homeobox homolog *Drosophila* 2	*MSX2*	Craniosynostosis Boston type
Transforming growth factor-β receptor type I	*TGFBRI*	Loeys-Dietz syndrome
Transforming growth factor-β receptor type II	*TGFBRII*	Loeys-Dietz syndrome
Cytochrome P-450 reductase gene	*POR*	Antley-Bixley syndrome
Fibrillin	*FBN1*	Shprintzen-Goldberg craniosynostosis syndrome

SOURCE: Adapted from Passos-Bueno et al. (2008).

From an epigenetics perspective, classifications of CS should distinguish between incidences of CS on the basis of developmental perturbations that cause the dysmorphology. The distinction between syndromic and isolated CS seems to serve this purpose and might be further refined. Syndromic individuals exhibit CS and dysmorphology of a number of other complex traits in diverse bodily systems, while individuals with isolated CS simply display CS. The list of dysmorphologies associated with any CS syndrome suggests that a causal developmental modification occurs in a DC upstream of many other DCs and phenotypic outputs, allowing that developmental modification to influence a large number of cell populations, which in turn causes additional downstream modifications. If more interactions are modified, we would expect to see more serious phenotypic effects. In fact, individuals with syndromic CS show higher incidences of resynostosis and higher rates of reoperation than individuals with simple CS (McCarthy et al., 1995; Williams et al., 1997; Foster et al., 2008). This suggests that more serious modifications to the cranial growth process occur. In cases of syndromic CS, it seems that developmental perturbations affect cell populations more broadly or that more than one developmental perturbation is involved. One or more mutations have been associated with many CS syndromes, while few cases of isolated CS have been shown to have a heritable basis (Cohen and Maclean, 2000). However, it is generally believed that genetic perturbations play a role in at least a few familial cases of isolated CS.

For CS, local gene-expression levels and particular cell types or cell processes at the suture are routinely analyzed, and the changes in certain local factors are usually implicated as a cause of the observed dysmorphology. These studies provide information about potential epigenetic mechanisms in DCs adjacent to the area under study but are limited in both time and space. Certainly, changes occur locally to a closed or closing suture; but what other processes, local and nonlocal, current and preceding, might contribute to the processes that eventually lead to premature suture closure (Martínez-Abadías et al., 2010)? The study of any dysmorphology should be expanded temporally and spatially (anatomically) to include the full complement of epigenetic interactions.

Normal development of the embryonic skull is dependent on proper coordination between bony, neural, muscle, and other tissue precursors; but the relationship among these growing tissues is poorly understood. The underlying factors influencing the relationship include pleiotropic effects of specific genes, local and global biomechanical influences, tissue interactions, and most assuredly an ontogenetically sensitive combination of these and other factors. We consider examples of genetic, mechanical, and external environmental perturbations as they might affect development of the skull and contribute to CS as a window into the more general area of the epigenetic basis of craniofacial dysmorphogenesis.

GENES AND GENE NETWORKS

Perturbations of genetic interactions come in the form of mutant alleles or unusual combinations of alleles. We define genes broadly for this discussion as any stretch of DNA that codes for protein or RNA or is a regulatory element because all play roles in epigenetic interactions. Populations contain allelic variants, some of which we refer to as "mutations" because they are rare and associated with significant dysmorphology or disease. Perturbations of normal epigenetic interactions occur in all DCs in which the gene with the mutation is expressed. DCs that receive downstream epigenetic inputs from processes affected by the mutation have the potential to be secondarily affected. Epigenetic output and inputs connect DCs on the basis of chemical and physical interactions so that spatial proximity can be important, but not necessary, for an effect to occur. This means that mutated genes expressed at a distance from the location of future dysmorphology are potential triggers of that dysmorphology. Until the processes behind the connection of DCs through epigenetic output and input are understood, these types of interactions will escape the researcher's scrutiny and intuition. On the other hand, the expression of an unusual allele that is associated with dysmorphology is not necessarily the "cause" of dysmorphology. The effect of an inborn mutation is not isolated in space or time: It occurs in every cell in the individual from conception onward. The effects of the mutation are marked on the phenotype through the lens of the interactions of all DCs upstream of the phenotypic output under investigation—in our example, CS.

Some individuals with a particular mutation may exhibit a less serious dysmorphology or none at all because of other alleles or nongenetic factors in the DCs of gene expression. The other alleles within an individual's genotype are critical for determining the nature of expression of a given complex trait in the presence of a particular mutation. Recognizing that mutations are nothing but alternate alleles, an unusual combination of alleles that individually might not be implicated as causal of dysmorphology could lead to dysmorphology if their interactions produced an abnormal epigenetic output.

Given known pleiotropy and assuming that mutations associated with CS syndromes are found in every cell after conception, most syndromic CS cases are likely to have phenotypes that arise from perturbations of multiple upstream DCs. Mutations of fibroblast growth factor receptors *FGFR1*, *FGFR2*, and *FGFR3* as well as *TWIST1* and *MSX2* are associated with CS and are all pleiotropic (Cohen and Maclean, 2000; McIntosh et al., 2000; Morriss-Kay and Wilkie, 2005; Passos-Bueno et al., 2008). These genes are known to be involved in many developmental processes and are expressed in multiple tissues throughout most bodily systems (Figure 21.5). For instance, *FGFR2* expression is important for the proper development of the midbrain (Saarimäki-Vire et al., 2007) and the lungs (White et al., 2006) and for the regulation of osteoblasts across the developing individual (Yu et al., 2003). Two neighboring mutations in *FGFR2* are known to be responsible for >99% of all cases of Apert syndrome, and it is believed that these mutations are direct causative agents of the CS phenotype (Wang et al., 2005, 2010). While *FGFR2* is expressed at all vault suture margins, only the coronal suture generally exhibits CS in Apert syndrome. Moreover, coronal CS in Apert syndrome can occur on the right side, the left side, or bilaterally. Thus, if *FGFR2* is expressed at all vault suture margins but the coronal suture is specifically affected, it follows that the expression of a particular mutant allele in some sutures is not enough to produce CS. This suggests that although the known mutation causes a genetic perturbation within a large number of DCs, the epigenetic outputs of DCs upstream of the phenotypic output of some su-

tures do not tip the balance toward premature suture closure, while others do. Any disease-causing mutation has the potential to impact phenotypic output of a large number of DCs but does not necessarily do so.

Though specific combinations of phenotypes enable classification and therefore diagnosis, there is significant phenotypic variation within CS syndromes, even among individuals who share the same major mutation (Doherty et al., 2007). Clinically different phenotypes can be caused by the same mutation, and a particular clinical phenotype can be expressed in individuals carrying different mutations (Passos-Bueno et al., 2008). All of these observations support a role for additional genetic or nongenetic factors contributing significantly to determining phenotypes.

Despite many years of determined research efforts, the genetic basis of isolated (nonsyndromic) CS remains unknown (Boyd et al., 2002; Boyadjiev, 2007). It is more difficult to find genetic bases of isolated CS because instead of a combination of dysmorphologies to associate with a mutation, there is simply synostosis. Also, the potentially numerous factors that contribute to suture fusion lead to the same general phenotype. Abnormal osteocyte migration to the suture, increased osteocyte proliferation, increased osteoblast differentiation, and decreased osteoblast apoptosis at a suture have all been proposed as processes responsible for both normal and premature suture fusion (Opperman, 2000; Marie et al., 2008). The resulting closed suture phenotype will be similar regardless of the mechanism underlying increased osteogenesis within the previously fibrous sutures. Many studies have demonstrated associations between certain gene mutations and different mechanisms leading to CS. For instance, mutations of *FGFR2* have been associated with perturbations in cell proliferation, cell differentiation, and apoptosis (Marie et al., 2008). A review of the literature demonstrates that different mutations of the same gene can lead to the same dysmorphic phenotype through the modification of different

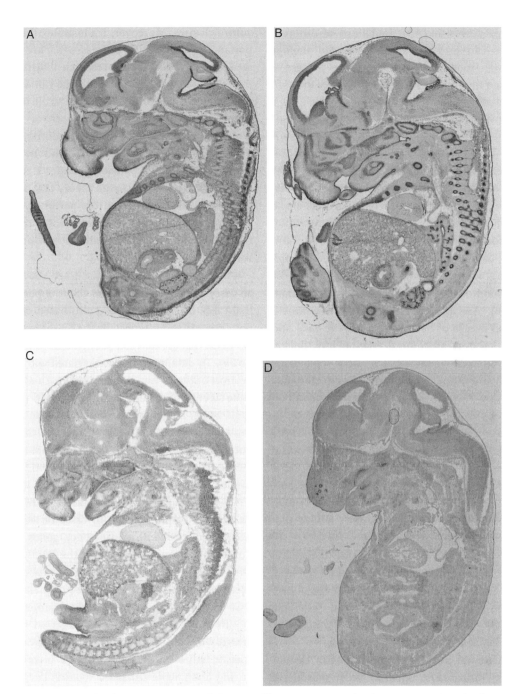

FIGURE 21.5 The pleiotropic expression (dark areas) of four genes associated with craniosynostosis at developmental stage E14.5. *Fgfr1* (A), *Fgfr2* (B), and *Twist1* (C) exhibit strong expression across the body. The exception, *Msx2* (D), is expressed in only a few places at this developmental stage. Each image is an equivalent parasagittal slice of a developing mouse from Genepaint (http://www.genepaint.org/). Color images can be found there (set IDs: FG34, FG35, EB1518, MH3086).

McCarthy, J. G., S. B. Glasberg, C. B. Cutting, F. J. Epstein, B. H. Grayson, G. Ruff, C. H. Thorne, J. Wisoff, and B. M. Zide. 1995. Twenty-year experience with early surgery for craniosynostosis: II. The craniofacial synostosis syndromes and pansynostosis—results and unsolved problems. *Plast Reconstr Surg* 96:284–95.

McCarthy, J. G., and C. A. Reid. 1980. Craniofacial synostosis in association with vitamin D–resistant rickets. *Ann Plast Surg* 4:149–53.

McIntosh, I., G. A. Bellus, and E. W. Jabs. 2000. The pleiotropic effects of fibroblast growth factor receptors in mammalian development. *Cell Struct Funct* 25:85–96.

Meikle, M. C., J. K. Heath, and J. J. Reynolds. 1984. The use of in vitro models for investigating the response of fibrous joints to tensile mechanical stress. *Am J Orthod* 85:141–53.

Menking, M., J. Wiebel, W. U. Schmid, W. T. Schmidt, K. D. Ebel, and R. Ritter. 1972. Premature craniosynostosis associated with hyperthyroidism in 4 children with reference to 5 further cases in the literature. *Monatsschr Kinder* 120:106–10.

Milunsky, A., J. W. Graef, and M. F. Gaynor. Jr. 1968. Methotrexate-induced congenital malformations. *J Pediatr* 72:790–5.

Mooney, M. P., M. I. Siegel, A. M. Burrows, T. D. Smith, H. W. Losken, J. Dechant, G. Cooper, and M. R. Kapucu. 1998. A rabbit model of human familial, nonsyndromic unicoronal suture synostosis. I. Synostotic onset, pathology, and sutural growth patterns. *Child Nerv Syst* 14:236–46.

Morimoto, R. I. 1993. Cells in stress: Transcriptional activation of heat shock genes. *Science* 259:1409–10.

Morriss-Kay, G. M., and A. O. Wilkie. 2005. Growth of the normal skull vault and its alteration in craniosynostosis: Insights from human genetics and experimental studies. *J Anat* 207:637–53.

Moss, M. L. 1975. Functional anatomy of cranial synostosis. *Child Brain* 1:22–33.

Moss, M. L., and R. W. Young. 1960. A functional approach to craniology. *Am J Phys Anthropol* 18:281–92.

Opperman, L. A. 2000. Cranial sutures as intramembranous bone growth sites. *Dev Dyn* 219:472–85.

Opperman, L. A., and J. T. Rawlins. 2005. The extracellular matrix environment in suture morphogenesis and growth. *Cells Tissues Organs* 181:127–35.

Passos-Bueno, M. R., A. L. Sertié, F. S. Jehee, R. Fanganiello, and E. Yeh. 2008. Genetics of craniosynostosis: Genes, syndromes, mutations, and genotype–phenotype correlations. In *Craniofacial Sutures*, ed. D. P. Rice, 107–43. Basel: Karger.

Radlanski, R. J., and H. Renz. 2006. Genes, forces, and forms: Mechanical aspects of prenatal craniofacial development. *Dev Dyn* 235:1219–29.

Rafferty, K. L., and S. W. Herring. 1999. Craniofacial sutures: Morphology, growth, and in vivo masticatory strains. *J Morphol* 242:167–79.

Rasmussen, S. A., M. M. Yazdy, J. L. Frías, and M. A. Honein. 2008. Priorities for public health research on craniosynostosis: Summary and recommendations from a Centers for Disease Control and Prevention–sponsored meeting. *Am J Med Genet A* 146:149–58.

Reilly, B. J., J. M. Leeming, and D. Fraser. 1964. Craniosynostosis in the rachitic spectrum. *J Pediatr* 64:396–405.

Rice, D. P. 2008. Clinical features of syndromic craniosynostosis. In: *Craniofacial Sutures*, ed. D. P. Rice, 91–106. Basel: Karger.

Richtsmeier, J. T., K. Aldridge, V. B. DeLeon, J. Panchal, A. A. Kane, J. L. Marsh, P. Yan, and T. M. Cole III. 2006. Phenotypic integration of neurocranium and brain. *J Exp Zoolog B Mol Dev Evol* 306:360–78.

Riggs, W., Jr., R. S. Wilroy, Jr., and J. N. Etteldorf. 1972. Neonatal hyperthyroidism with accelerated skeletal maturation, craniosynostosis, and brachydactyly. *Radiology* 105:621–5.

Saarimäki-Vire, J., P. Peltopuro, L. Lahti, T. Naserke, A. A. Blak, D. M. Vogt Weisenhorn, K. Yu, D. M. Ornitz, W. Wurst, and J. Partanen. 2007. Fibroblast growth factor receptors cooperate to regulate neural progenitor properties in the developing midbrain and hindbrain. *J Neurosci* 27:8581–92.

Schowing, P. J. 1968. Mise en evidence du role inducteur de l'encephale dans l'osteogenese du crane embryonnaire du poulet. *J Embryol Exp Morphol* 19:83–93.

Shashi, V., and T. C. Hart. 2002. Environmental etiologies of orofacial clefting and craniosynostosis. In *Understanding Craniofacial Anomalies: The Etiopathogenesis of Craniosynostosis and Facial Clefting*, ed. M. P. Mooney and M. I. Siegel, 163–206. New York: John Wiley & Sons.

Sidor, C. A. 2001. Simplification as a trend in synapsid cranial evolution. *Evolution* 55:1419–42.

Smith, J. M., R. Burian, S. Kauffman, P. Alberch, J. Campbell, B. Goodwin, R. Lande, D. Raup, and L. Wolpert. 1985. Developmental constraints and evolution: A perspective from the Mountain Lake Conference on Development and Evolution. *Q Rev Biol* 60:265–87.

Sultan, S. E., and S. C. Stearns. 2005. Environmentally contingent variation: Phenotypic plasticity and norms of reaction. In *Variation*, ed. B. Hallgrímsson and B. K. Hall, 303–32. Burlington, MA: Elsevier.

Sun, Z., E. Lee, and S. W. Herring. 2004. Cranial sutures and bones: Growth and fusion in relation to masticatory strain. *Anat Rec A Discov Mol Cell Evol Biol* 276:150–61.

Sun, Z., E. Lee, and S. W. Herring. 2007. Cell proliferation and osteogenic differentiation of growing pig cranial sutures. *J Anat* 211:280–9.

Ting, M.-C., L.-Y. Wu, P. G. Roybal, J. Sun, L. Liu, Y. Yan, and R. E. Maxson, Jr. 2009. EphA4 as an effector of Twist1 during neural crest–mesoderm boundary formation and the guidance of migratory osteogenic precursor cells in calvarial bone growth and craniosynostosis. *Development* 136: 855–64.

Wang, Y., R. Xiao, F. Yang, B. O. Karim, A. J. Iacovelli, J. Cai, C. P. Lerner, J. T. Richtsmeier, J. M. Leszl, and C. A. Hill. 2005. Abnormalities in cartilage and bone development in the Apert syndrome FGFR2+/S252W mouse. *Development* 132:3537–48.

Wang, Y., M. Sun, V. L. Uhlhorn, X. Zhou, I. Peter, N. Martínez-Abadías, C. A. Hill, et al. 2010. Activation of p38 MAPK pathway in the skull abnormalitics of Apert syndrome Fgfr2+/P253R mice. *BMC Dev Biol* 10:22.

Weiss, K. M. 2005 The phenogenetic logic of life. *Nat Rev Genet* 6:36–45.

Weiss, K. M., and A. V. Buchanan. 2004. *Genetics and the Logic of Evolution.* New York: Wiley-Liss.

White, A. C., J. Xu, Y. Yin, C. Smith, G. Schmid, and D. M. Ornitz. 2006. FGF9 and SHH signaling coordinate lung growth and development through regulation of distinct mesenchymal domains. *Development* 133:1507–17.

Wilkins, A. 2002. *The Evolution of Developmental Pathways.* Sunderland, MA: Sinauer Associates.

Williams, J. K., S. R. Cohen, F. D. Burstein, R. Hudgins, W. Boydston, and C. Simms. 1997. A longitudinal, statistical study of reoperation rates in craniosynostosis. *Plast Reconstr Surg* 100:305–10.

Yu, J. C., J. H. Lucas, K. Fryberg, and J. L. Borke. 2001. Extrinsic tension results in FGF-2 release, membrane permeability change, and intracellular Ca++ increase in immature cranial sutures. *J Craniofac Surg* 12:391–8.

Yu, K., J. Xu, Z. Liu, D. Sosic, J. Shao, E. N. Olson, D. A. Towler, and D. M. Ornitz. 2003. Conditional inactivation of FGF receptor 2 reveals an essential role for FGF signaling in the regulation of osteoblast function and bone growth. *Development* 130:3063–74.

Zeiger, J. S., T. H. Beaty, J. B. Hetmanski, H. Wang, A. F. Scott, L. Kasch, G. Raymond, E. W. Jabs, and C. VanderKolk. 2002. Genetic and environmental risk factors for sagittal craniosynostosis. *J Craniofac Surg* 13:602–6.

22

Epigenetics of Human Disease

Peter D. Gluckman, Mark A. Hanson, Alan S. Beedle,
Tatjana Buklijas, and Felicia M. Low

CONTENTS

EPIGENETIC DISORDERS ARE FUNDAMENTALLY DISEASES OF DEVELOPMENT

Research into the ways in which vulnerability to chronic noncommunicable disease in humans is governed by patterns of gene expression determined by molecular epigenetic marks established during development in large part follows from, and is closely linked to, the growing interest in the "developmental origins of health and disease" (DOHaD). Historically, development played a limited role in medical thought on the causation of human disease. This has dramatically changed in the last two decades, with the accumulation of epidemiological and animal experimental data pointing to early life as a critical period when much of future susceptibility to disease is established. It is now clear that a molecular epigenetic perspective is essential to understanding development: Such stable modifications of gene expression that do not alter the genome are important mechanisms of cell differentiation (Reik, 2007). Across species, organisms respond to environmental input by using mechanisms termed together "developmental plasticity" to produce a range of different phenotypes. There is increasing evidence that this plasticity is also dependent, at least in part, on molecular epigenetic processes (Kucharski et al., 2008).

The developing organism has a number of potential responses to environmental cues, and these can be both adaptive and nonadaptive in origin. The former induce teratogenic and the latter plastic responses; but both, in particular contexts, may lead to pathological outcomes. The main focus of this chapter will be on the latter class of responses, although the division between the two groups is not always clear (Bateson et al., 2004; Gluckman et al., 2005). A key focus of research into the origin of human disease in the postgenomic era is the extent to which epigenetic processes play a role. Here, we restrict our definition of *epigenetics* to those processes that alter gene expression at the transcriptional level through changes in promoter state or chromatin packaging or posttranscriptionally through effects on messenger RNA.

In this chapter, we review the current state of knowledge about the epigenetics of human disease. A brief review of the historical relationship between development and disease is followed by a discussion of basic concepts of developmental plasticity and disruption and the role of epigenetic mechanisms. This is followed by a discussion of the DOHaD paradigm. These introductory sections are succeeded by consideration of the epigenetics of specific disease groups, including the heritability and timing of the induced marks. We focus primarily on diseases that may be interpreted as normal responses to environmental variations during development (metabolic, cardiovascular, mental illness) and follow this with a discussion of other diseases arising from disruption of molecular epigenetic processes and from epimutations.

HISTORY OF DEVELOPMENT AS A CONTRIBUTOR TO HUMAN DISEASE

Development in the modern sense dates back to the decades around 1800 (Hopwood, 2009). Human embryos had earlier been depicted as miniature children increasing in size, but now they were pictured as progressively changing their shape as well (Hopwood, 2007). These representational changes reflected the contemporary victory of *epigenesis*, which argued that the developed form arose from unformed matter, over the *preformationist* view that the complete form was present in the egg or sperm cell, only much smaller (Pinto-Correia, 1997; Roe, 1981). Malformed children and animals, once seen as portents of divine messages or wonders of nature, were now recast as evidence of misguided development and objects of a new science of teratology (Daston and Park, 1998; Leroi, 2003). While traditional explanations of monsters had invoked the concept of "maternal imagination" to emphasize the impact of maternal environment, teratology, relying on easily available animal models such as the chick embryo to produce malformations experimentally, reduced the natural complexity to mechanical factors such as heat and shaking (Appel, 1987).

Embryology was a key science of life in the nineteenth century and was taught in medical schools but overall remained excluded from medicine's then contemporary drive to define causes of disease. This trend was reinforced in the early twentieth century with the acceptance of August Weismann's doctrine, according to which the environment did not affect hereditary material in the germ cells, and the subsequent lack of interest in development within the rapidly growing field of genetics. For much of the twentieth century human embryology focused on collecting human embryos to define the standards of normal human development (Hopwood, 2007).

Teratology, now a well-established discipline examining disruptive influences that result in structural malformations, received a boost from the political and social shifts of the mid-twentieth century, especially war, the growth of the pharmaceutical industry, and the medicalization of pregnancy and childbirth. The increased number of birth defects in children born after the nuclear bombing of Hiroshima and Nagasaki drew attention to the use of X-rays in pregnancy, leading to Alice Stewart's famous 1956 study showing a higher incidence of childhood cancer in exposed children (Oakley, 1984). In the 1950s and 1960s, the widespread use of two drugs marketed to pregnant women, thalidomide (to relieve nausea) and the synthetic hormone diethylstilbestrol (to prevent miscarriage), resulted in high incidences in the offspring of phocomelia (limb shortening) and vaginal adenocarcinoma, respectively. In the early 1970s, the controversial East German endocrinologist Günther Dörner, studying how maternal androgen and estrogen levels influence the development and later function of the "hypothalamo–hypophyseal–gonadal" system of the child, proposed to expand the remit of teratology beyond structural malformations (Dörner, 1973; Dörner et al., 1973; Dörner and Mohnike, 1973). His suggestions that the cause(s) of adult obesity, diabetes mellitus, and atherosclerosis lay

in early development were mostly ignored, at least in the West. There, however, at about the same time, the American diabetologist Norbert Freinkel developed the concept of metabolic teratogenesis, by which maternal hyperglycemia induced permanent changes in fetal metabolic function (Freinkel, 1980), while in Belgium Frans Van Assche and colleagues reported that in rats fetal growth retardation led to offspring with metabolic compromise (Aerts and Van Assche, 1979). From the 1980s, the DOHaD paradigm, which will be discussed in the following section, resurrected interest in this earlier work. The recent engagement with development is supported and inspired by the contemporary reintegration of genetics and development into developmental biology, as well as by the marked social interest in reproduction. The focus is now on how processes acting during development mediate gene–environmental influences and how their long-term consequences mediated by epigenetic changes affect organ structure or function to produce disease.

FETAL RESPONSES TO DEVELOPMENTAL CUES

Across species, a single genotype may produce a range of phenotypes in different circumstances. The fetus has a number of responses to environmental cues imposed during development (Bateson et al., 2004; Gluckman et al., 2005). Some early-life environmental cues—such as thalidomide and diethylstilbestrol, discussed above—may be disruptive and cause irreparable damage. Others induce potentially adaptive responses, the phenotypic effects of which may become manifest either immediately or later in life. While natural selection requires long time spans, these types of "developmentally plastic" responses are mechanisms that operate over a single lifetime and can assist in protecting or promoting fitness (Gluckman et al., 2007a).

The evolutionary origin of developmentally plastic processes lies in the potential fitness advantage which they provide through ensuring that the organism survives long enough to reproduce (Horton, 2005). A commonly occurring adverse environment is poor nutrition, caused variously by inadequacies of maternal food intake, maternal metabolism, cardiovascular function, and, especially, placental function (Gluckman and Hanson, 2004a). The fetal response to insufficient supply of nutrients may be a reduction of its anabolic demand, partly by lowering plasma concentrations of insulin and insulin-like growth factor I (IGF-I) (Gluckman, 1995). Growth may be reduced or disproportionate if fetal blood flow is redistributed, to preserve supply to vital organs. Another mode of response, observed in unfavorable circumstances such as infection and poor nutrition, may be premature birth. However, while growth retardation and prematurity increase the chance of survival to birth, both are associated with a higher risk of postnatal morbidity and mortality. Short-term survival is thus traded off against long-term health.

Adaptive responses need not be evident immediately during pregnancy or at birth: In some cases, the adaptive advantage of a phenotypic response that occurred during development becomes evident only after a delay. We term such mechanisms "predictive adaptive responses" (PARs), without intending to suggest a conscious foresight on the part of the embryo. The onset of puberty is a valuable example, for it has a clear evolutionary significance and is highly sensitive to environmental influences such as climate and nutrition. It has been reported that girls adopted from a developing into a developed country experience menarche earlier than their peers (Parent et al., 2003). Following clinical studies that showed that girls born small but who became large at 8 years reached menarche earlier (Sloboda et al., 2007), animal studies confirmed that caloric restriction antenatally or during lactation accelerates reproductive maturation (Sloboda et al., 2009). The largest effect was manifest in the group undernourished in pregnancy and lactation but fed a high-fat diet afterward. The life-history explanation suggests that an organism which spends its early development and childhood in adverse nutritional circumstances may advance sexual maturation in order to ensure reproduction (Gluckman and Hanson, 2006). Importantly, immediate and predictive adaptive responses may overlap, with the same cue producing two kinds of phenotypic changes. Thus, early puberty may be associated with small size at birth.

Predictive responses are adaptive as long as their forecast is overall correct, but where a mismatch between the predicted and the eventual environment occurs, disease may develop in later life. After the following review of epigenetic mechanisms, we shall discuss how the environment during the "plastic period" shapes future susceptibility to chronic metabolic and cardiovascular disease and examine the current knowledge of epigenetic markers that stably direct the genome to produce such "risky" phenotypes.

EPIGENETIC MECHANISMS

Epigenetics in the context of this chapter is used to describe the molecular mechanisms that establish and maintain mitotically stable patterns of gene expression, including DNA methylation, chromatin remodeling, and activity of small noncoding RNAs such as micro-RNAs. These mechanisms are often interlinked and able to act in concert to modulate the expression of relevant genes.

DNA METHYLATION

DNA methylation involves the covalent addition of a methyl group to carbon-5 of the cytosine ring, yielding 5-methylcytosine. This is catalyzed by various DNA methyltransferase isozymes. DNMT3a and DNMT3b appear to be responsible for de novo methylation, facilitated by the regulatory factor DNMT3L. Conversely, DNMT1 has greater affinity for hemimethylated DNA as a substrate and, thus, serves to maintain preestablished methylation patterns. In mammals, methylation occurs only on cytosine that is 5′ adjacent to guanine. These CpG dinucleotides (where "p" represents the intervening phosphate group) are underrepresented in the mammalian genome; 5-methylcytosine is a mutational hotspot due to its propensity to undergo spontaneous deamination to yield thymine. However, they are found at the predicted frequency in the promoter regions of many genes, and these so-called CpG islands can extend up to hundreds of base pairs. In humans, CpG islands are present in ~60% of promoter regions, but most are hypomethylated (Antequera, 2003).

Due to the double-stranded nature of DNA and the ability of cytosine to form base pairs with guanine, CpGs are always found at the same site on both DNA strands. Thus, after one round of DNA replication, both daughter strands possess CpG sites that can be methylated in the same pattern as the parent strand. In this way, the set of epigenetic modifications throughout the genome, known as the *epigenome*, is preserved following cell division, assuming that DNMT1 is functional and that there is an adequate supply of methyl groups. This stable inheritance through mitotic division imposes a form of cellular memory that is maintained over successive cell divisions.

DNA methylation is typically associated with decreased expression of relevant genes, achieved via decreased binding affinity of transcription factors and other transcriptional components. The first maps of cytosine methylation in humans have recently been reported for embryonic stem cells and fibroblasts (Lister et al., 2009; Laurent et al., 2010). Unexpectedly, close to one-quarter of methylation in stem cells occurred in a non-CpG context, but this pattern was not seen upon differentiation, raising the possibility of different methylation mechanisms for different cell types. DNA methylation plays a fundamental role in many epigenetic phenomena such as cell differentiation, genomic imprinting, X-chromosome inactivation, and retrotransposon repression.

CHROMATIN REMODELING

Chromatin is the DNA–histone protein assembly which, when condensed, forms chromosomes in eukaryotic cells. Chromatin remodeling involves alteration of its higher-order structure and constitutes another major mechanism of epigenetic control. The basis for this is that the more compact the chromatin, the less accessible the DNA to transcriptional machinery and regulatory factors, favoring silencing of the gene. Conversely, any factors that open up the chromatin structure generally promote transcriptional activity.

Specific amino acid residues on histone tails are susceptible to covalent modifications such as methylation, acetylation, phosphorylation, and poly(ADP-ribosyl)ation. These different enzyme-mediated modifications have differing effects on chromatin structure. For example, acetylation introduces a negative charge that interferes with the histone–DNA interaction, generally resulting in reduced compactness and thus gene activation; however, methylation serves to recruit other effector proteins that either unwind or compact chromatin, leading to gene activation or repression. Furthermore, DNA methylation can trigger recruitment of methyl-CpG-binding domain proteins which, together with other repressor proteins, modify the structure of chromatin. The myriad forms and combinations of histone modifications can function in a cumulative fashion to remodel chromatin, possibly reflecting what has been dubbed a "histone code" (Jenuwein and Allis, 2001). It should be noted that, in contrast to DNA methylation, strong support for the mitotic transmission of histone modifications is not yet available.

A well-studied mechanism of chromatin modeling is the polycomb/trithorax group of proteins, which have opposing abilities to modulate the degree of chromatin fiber compaction (Mohd-Sarip and Verrijzer, 2004). Polycomb group proteins (PcGs) function as complex and specifically condensed nucleosomes, while the trithorax proteins use energy provided by ATP hydrolysis to unravel chromatin. In mice, humans, and *Drosophila*, PcGs bind preferentially to genes encoding transcription factors known to target homeotic genes in *Drosophila* (Schuettengruber et al., 2007). Interestingly, trimethylation of a histone lysine residue is required to serve as a binding site for a component of the complex (Cao and Zhang, 2004), and the colocalization of several small RNAs facilitates appropriate nuclear organization of PcG chromatin targets. This highlights the intricacy and interdependence of these mechanisms.

MICRO-RNAs

Micro-RNAs (miRNAs) provide another mechanism of epigenetic regulation, at the posttranscriptional level rather than during gene expression. These endogenous, noncoding RNA molecules, 20–22 nucleotides long, bind to target messenger RNA (mRNA) via complementary base pairing. Depending on the degree of complementarity, this results in either inhibition of translation or total mRNA degradation. To date, descriptions for more than 720 human miRNA genes have been deposited in databases (MiRBase, available online at http://mirbase.org), and it has been estimated that 30% of human genes are regulated by miRNAs (Li et al., 2008). The expression of miRNAs can in turn be modulated by DNA methylation and chromatin modifications (Barski et al., 2009), thus providing an additional layer of regulation of gene expression which nevertheless is still poorly understood. Interestingly, miRNAs may have a modulatory role in the development of cancer (see Epigenetics in Cancer section, p. 413).

Currently, the mechanisms involved in the reversal of epigenetic marks are not well characterized. Indeed, histone methylation was thought to be irreversible until the recent identification of histone deacetylases and demethylases (Cloos et al., 2008). The mechanisms by which active DNA demethylation occurs in mammals have also been controversial due to conflicting biochemical and knockout mouse model data (Ooi and Bestor, 2008), although recent data suggest that DNMT3a and DNMT3b may in fact possess dual methylation/demethylation properties (Métivier et al., 2008).

The processes of genomic imprinting, differentiation of pluripotent cells, and developmental plasticity may utilize different sets of epigenetic mechanisms and marks, depending on whether flexibility or stability is required. For example, DNA methylation generally imposes long-term, stable gene silencing, while chromatin remodeling is associated with shorter-term, flexible gene repression (Reik, 2007). It should be borne in mind that not all mechanisms of developmental plasticity necessarily have a basis in these molecular epigenetic mechanisms—endocrine and paracrine extracellular signals could also play a role (Waterland and Michels, 2007).

EPIGENETICS DURING MAMMALIAN DEVELOPMENT

TIMELINE OF MARKS

From studies on mice, it is known that epigenetic processes operating during zygote fertilization, early embryogenesis, and gametogenesis display a high level of dynamism. Erasure or reestablishment of epigenetic marks is required for conferring a toti- or pluripotent or a differentiated state, as necessary for the specific stage of development (Farthing et al., 2008; Morgan et al., 2005).

After fertilization, global erasure of both parental epigenomes occurs. Erasure of each genome proceeds differently—the paternal genome is rapidly demethylated by unknown mechanisms even before the first cell division, while the maternal genome undergoes gradual, passive demethylation, probably due to the exclusion of DNMT1 from the nucleus, precluding maintenance methylation activity. Notably, imprinted genes are protected from demethylation during preimplantation development by the somatic isoform of DNMT1 (Hirasawa et al., 2008). Total levels of methylation continue to decrease until the morula stage is reached and pluripotency is restored (Reik, 2007). Development into the blastocyst, which involves differentiation into two distinct cell lineages, is paralleled by de novo methylation by DNMT3a and DNMT3b. A new pattern of epigenetic marks is thus established. Methylation patterns of the blastocyst are cell lineage–specific: The inner cell mass, which is destined to develop into embryonic and adult tissues, shows hypermethylation; in contrast, the trophectoderm, which develops into placental tissue, is hypomethylated. This distinction is preserved in the tissues after complete development. Once cells are differentiated, genes associated with pluripotency become epigenetically repressed (Reik, 2007).

Epigenetic reprogramming reoccurs at the onset of gametogenesis. The primordial germ cell (PGC), which is derived from the blastocyst inner cell mass, rapidly loses most of its methylation marks in a cytidine deaminase–dependent process (Popp et al., 2010). As with early embryogenesis reprogramming, pluripotency is reconferred; but a significant difference is that imprinted methylation marks at imprinted control regions are erased (Farthing et al., 2008). The PGCs migrate to the genital ridges, and new epigenetic marks including sex-specific imprints are laid down during spermatogenesis and oogenesis. This process is complete prior to birth in the male germ line but commences only after birth in the female germ line.

Silencing of transposable elements is crucial for gamete integrity, and recent research has shed light on the mechanism by which specific retrotransposons in the germ line are recognized and targeted for methylation. A class of small RNAs, known as piRNAs, appear to be involved in guiding de novo DNA methylation of retrotransposons in mouse fetal male germ cells (Kuramochi-Miyagawa et al., 2008). piRNAs are so named due to their interaction with the germ line–specific piwi proteins (it should be noted that they are distinct from the miRNAs described earlier). Loss of two piwi homologues, MILI and MIWI2, in null mice decreases de novo DNA methylation at the regulatory regions of their target retrotransposons, thus increasing expression of those genes (Kuramochi-Miyagawa et al., 2008). piRNAs complementary to transposable element sites are abundant in the germ cells, and this is thought to help recruit chromatin modifiers to effect DNA methylation (Aravin and Bourc'his, 2008). It is notable that the repertoire of fetal piRNAs is distinct from

that expressed in neonatal and adult germ cells (Kuramochi-Miyagawa et al., 2008).

ENVIRONMENTAL LABILITY

Generally, epigenetic marks that are established after differentiation are stably maintained through the organism's life. However, the period during erasure and reestablishment of DNA methylation in the early embryo is susceptible to environmental influences. Numerous studies of metastable epialleles in rodents have demonstrated the effects of the availability or lack of nutritional methyl donors on epigenetic marks. A widely used model is the agouti viable yellow mouse as hypomethylation of the promoter of the *agouti* gene enhances gene expression, and the resulting phenotype includes easily monitored characteristics such as a yellow coat color and obesity. Conversely hypermethylation suppresses gene expression, resulting in a brown coat color and normal weight. Agouti offspring that have been subjected in utero to maternal deficiency of methyl donors and cofactors show promoter hypomethylation and develop a yellow coat color and obesity (Waterland and Jirtle, 2003). Addition of an endocrine disruptor in the form of the soy isoflavone genistein to the maternal diet induces an opposite effect of *agouti* promoter hypermethylation, brown coat color, and decreased rates of obesity (Dolinoy et al., 2006). Similarly in the axin-fused mouse strain, ample maternal consumption of methyl donors during the periconceptional and prenatal periods increases DNA methylation of the axin-fused gene promoter. As a result, offspring show decreased incidence of the tail kinking that is characteristic of the strain (Waterland et al., 2006a). Hypermethylation occurs at midgestation as it is observed only in the tail, and this indicates that epigenome plasticity extends past the early stages of embryonic development.

Indeed, there is experimental evidence that newly established epigenetic patterns in the embryo continue to exhibit environment-mediated plasticity into the postnatal period and throughout the organism's lifetime. For example, postweaning mice fed a diet deficient in methyl donors showed loss of imprinting of IGF2 (Waterland et al., 2006b). Human monozygotic twins, who by definition possess identical genotypes, often are not phenotypically identical and show discordance in frequency or onset of diseases. In a study on 15 female and 25 male pairs of monozygotic twins aged 3–74 years, epigenetic differences between the members of each pair increased with age. For example, a representative pair of 3-year-old twins had highly similar patterns of global and locus-specific DNA methylation, whereas a representative pair of 50-year-old twins showed substantial variation in methylation (Fraga et al., 2005). Significantly, older twin pairs showed a fourfold increase in differentially expressed genes. The degree of discordance was correlated with lifestyle differences between twins, suggesting an effect of the environment. It has been postulated that this age-dependent "epigenetic drift" is a result of occasional errors in transmitting and maintaining epigenetic marks over many cell divisions.

While it remains unclear if observations of epigenetic drift reflect normative processes or pathology, there are data demonstrating a relationship between age and dysregulation of methylation levels. Global DNA hypomethylation has been found in various tissues in aging humans and is associated with many age-related disease states. This is possibly due to loss of DNMT1 efficacy or fidelity. Even so, concurrent age-dependent hypermethylation of a wide array of gene promoters has been documented (reviewed in Fraga and Esteller, 2007).

CHRONIC NONCOMMUNICABLE DISEASE

METABOLIC AND CARDIOVASCULAR DISEASES: EPIDEMIOLOGICAL AND CLINICAL STUDIES

It has recently been emphasized that while the Human Genome Project has resulted in the discovery of single-nucleotide polymorphisms (SNPs) and haplotypes associated statistically with major chronic noncommunicable diseases, these genetic factors together account for a very small fraction of the heritable attributable risk of such disease (Maher, 2008). This reinforces, by the process of exclusion, the idea that other components of human life must contribute substantially to the heritable risk of chronic disease. Development ranks high on that list.

The current interest in developmental origins of chronic metabolic and cardiovascular diseases may be traced back to the epidemiological studies of the 1970s showing that cohorts that had experienced high infant mortality exhibited higher cholesterol levels and cardiovascular disease rates in adulthood (Forsdahl, 1977; Forsdahl, 1978). These observations were further refined in studies that revealed an inverse relationship between birth weight and susceptibility to hypertension, cardiovascular morbidity, insulin resistance, type 2 diabetes mellitus, hyperlipidemia, and obesity (Barker et al., 1989). A hypothesis was proposed that fetal metabolic adjustments in nutritionally adverse circumstances, aimed at restricting growth to safeguard the brain, may result in a higher risk of chronic disorders in later life (Hales and Barker, 1992). However, the relationship between birth weight and later cardiovascular disease shows continuity rather than a threshold effect (Godfrey, 2006). Furthermore, data from the survivors of the Dutch "hunger winter" (an intense, short-term famine in 1944–1945) indicate that individuals exposed to adverse conditions in utero need not have low birth weight to exhibit later effects (Painter et al., 2005). Altered birth proportions—small head circumference combined with a short body—have been shown to predict later pathological outcomes better than birth weight (Godfrey, 2006).

Prospective studies have monitored specific anthropometric and laboratory parameters, rather than relying on proxy measures such as birth size. A study in Cebu, Philippines, found that maternal third-trimester energy level (as estimated by upper arm fat) was inversely related to low-density lipoprotein and total cholesterol levels and positively related to high-density lipoprotein levels in adolescent sons. The strong effect was, however, not found in female offspring (Kuzawa and Adair, 2003). Another study, in Pune, India, found that newborns, who are otherwise lighter and shorter than their British counterparts, nonetheless have higher umbilical leptin concentrations and a higher percentage of visceral adipose tissue at birth (Yajnik, 2004). Visceral fat is thought to influence metabolic pathways involved in insulin action as well as the synthesis of lipids and inflammatory substances, all underlying the appearance of type 2 diabetes and cardiovascular disease later. Babies born small for gestational age, while insulin-sensitive at birth, develop insulin resistance by 3 years of age (Mericq et al., 2005); this has been linked to fast neonatal growth and is associated with an increased risk of later obesity (Singhal, 2006). Increased fat deposition in the conditions of starvation is not the only metabolic mechanism that lays the foundations for future disease. Children exposed to hyperglycemia in utero or whose mothers are obese are at increased risk of developing metabolic disorders, especially type 2 diabetes mellitus (Boney et al., 2005).

Recent preliminary work has matched the methylation status of specific CpG sites in umbilical cord DNA from healthy neonates with phenotype during childhood, and it was shown that methylation levels of endothelial nitric oxide synthase (*eNOS*) and retinoid X receptor-alpha (*RXRα*) at birth can predict greater than 25% of the variance in fat mass at age 9 years (Godfrey et al., 2009). eNOS regulates vasodilation and has been found to be downregulated in animal models of hypertension, while RXRα is a transcription factor that has been implicated in insulin sensitivity and adipogenesis. The potential therefore exists of using epigenetic status at birth as a prognostic tool for susceptibility to later metabolic disease.

METABOLIC AND CARDIOVASCULAR DISEASES: EXPERIMENTAL STUDIES

Findings in humans have been further elucidated using animal models of the developmental origins of disease. These have been constructed to simulate human pathology by altering maternal nutrition (protein or energy intake) during pregnancy, by impairing placental perfusion, or by administering glucocorticoids to the pregnant mother. Researchers have measured the activity of relevant metabolic pathways as well as changes in gene expression and epigenetic marks.

In a primate model, maternal high-fat diet led to impaired fetal lipid metabolism in association with increased histone H3 acetylation and decreased histone deacetylase activity (both indicative of increased transcription) together with increased expression of several relevant genes (Aagaard-Tillery et al., 2008).

Feeding a low-protein diet to pregnant rats resulted in offspring that had hypertension and endothelial dysfunction, accompanied by metabolic and gene-expression changes, including overexpression of the hepatic glucocorticoid receptor (*GR*) and peroxisome proliferator–activated receptor α (*PPARα*) genes. The accompanying epigenetic changes that facilitate transcription of the receptors included, for the *GR*, histone modifications, hypomethylation of the $GR1_{10}$ promoter, and reduced expression of DNMT1 (Lillycrop et al., 2007). Promoter hypomethylation of *PPARα* was associated with reduced methylation of individual CpG dinucleotides (Lillycrop et al., 2008). In a similar low maternal protein model, increased expression of adrenal AT_{1b} angiotensin receptor from the first week of life was accompanied by hypomethylation of the proximal promoter of the receptor gene (Bogdarina et al., 2007).

The model of uteroplacental insufficiency was used to address the etiology of hypertension. Bilateral uterine artery ligation in pregnant rats caused increased kidney *p53* expression in association with reduced methylation of its promoter and reduced DNMT1 activity. Increased *p53* gene expression was predicted to increase renal apoptosis and reduce the number of glomeruli (Pham et al., 2003). Uteroplacental insufficiency was also used to generate a rodent model of beta-cell dysfunction which develops diabetes in adulthood. The molecular lesion underlying this pathology is underexpression of the pancreas-specific transcription factor *Pdx1*, and the study traced the ontogeny of the underlying epigenetic changes. Neonatal (and reversible) histone modifications reducing *Pdx1* expression were followed in adulthood, after development of diabetes, by methylation of the CpG island in the *Pdx1* promoter and permanent gene silencing (Park et al., 2008). Finally, neonatal changes in hepatic gene expression following uteroplacental insufficiency in the rat have also been shown to correlate with changes in the binding of acetylated histone H3 to the respective promoters (Fu et al., 2004).

Epigenetic changes and the associated metabolic phenotype in the offspring of maternally undernourished rats can be prevented by the administration of leptin in the neonatal period (Gluckman et al., 2007c; Vickers et al., 2005). While the proximate explanation may be a change in hypothalamic development, the suggested ultimate explanation is a modification in the developmental prediction of the future nutritional environment as leptin is a hormone made by fat. Interestingly, all aspects of the phenotype are reversed by leptin, suggesting that the change in phenotype induced by the early environment is integrated, rather than affecting isolated traits. The metabolic syndrome may thus represent the mismatch between a developmentally induced phenotype appropriate for a low-nutrient environment and living in a high-energy environment (Gluckman et al., 2007a; Gluckman and Hanson, 2004b).

STRESS AND MENTAL ILLNESS

Epidemiological studies have long linked the quality of family life with later susceptibility to illness in childhood and adulthood. Parental neglect was found to greatly increase the offspring's risk of obesity at around 20 years of age, with an odds ratio of 7.1 in comparison with harmonious support (Lissau and Sorensen, 1994). A public-health study revealed a strong graded relationship between the breadth of exposure to childhood abuse or household dysfunction and risk behaviors as well as the prevalence of chronic diseases, such as ischemic heart disease, cancer, and stroke (Felitti et al., 1998). The persuasive explanation for these findings was that early life experiences modulate neural and hormonal responses to stressful events, through the hypothalamic–pituitary–adrenal (HPA) axis. While stress is a physiological mechanism, evolved to provide the organism with a quick response to an immediate threat, high or erratic levels of catecholamines and glucocorticoids, with the increased cardiovascular stress and catabolic metabolism, may give rise to disease (Meaney, 2001).

Animal and human studies have shown that early life stress affects the HPA axis by modulating the central corticotropin-releasing factor (CRF) system. For instance, while a preceding traumatic event is a necessary criterion for the diagnosis of posttraumatic stress disorder (PTSD), only one-quarter of those traumatized develop this illness, characterized, among other things, by a hyperregulated and hyperdynamic HPA axis, low cortisol levels, and high GR levels. In babies of mothers who had experienced the 2001 terrorist attacks on the United States during the third trimester of their pregnancies, there was a significant association between the severity of maternal symptoms and infants' cortisol levels at awakening (Yehuda et al., 2005). A study measuring cortisol levels every 30 minutes over a 24-hour period found the offspring of Holocaust survivors with PTSD to have lower mean values (and changed associated chronobiological parameters) than the children of Holocaust survivors without PTSD or comparison subjects (Yehuda et al., 2007). The association was especially strong for maternal PTSD, thus reinforcing the hypothesis of perinatal effects on the stress response.

Animal studies have modeled maternal stress and variations in offspring care. Macaques reared by mothers under either consistently low (easily accessible food) or consistently high (unpredictable access to food) foraging demands showed, respectively, low and high CRF levels in young adulthood (Heim et al., 1997). Rats that were handled—briefly separated from the mother—neonatally were shown to have decreased stress reactivity in later life in comparison to the nonhandled group (Meaney et al., 1989). A comparison of the offspring of two groups of rat dams, one characterized by frequent licking and grooming behavior and arched-back nursing (high LG-ABN) and the other showing opposite characteristics and, thus, a lower level of maternal care (low LG-ABN), showed that the difference in hippocampal *GR* expression served as a mechanism for the effect of early experience on the development of individual differences in HPA responses to stress (Meaney, 2001).

Epigenetic studies in the last few years have shown that the adult offspring of low-LG-ABN mothers have a hypermethylated hippocampal exon I$_7$ *GR* promoter sequence, associated with hypoacetylated histone H3-lysine and reduced binding to nerve growth factor–inducible protein A; in the offspring of high-LG-ABN mothers, the findings were reversed (Weaver et al., 2004). These differences emerged over the first week of life, were reversed with cross-fostering, and persisted into adulthood. Further studies have begun to link variations in maternal behavior with other differences in gene expression and epigenetic marking, most importantly with differences in estrogen receptor α (ER$_α$) expression in the medial preoptic area. Maternal care was found to be associated with cytosine methylation of the *ER$_α$* promoter, which can be seen as a potential mechanism for the programming of individual differences in ER$_α$ expression and maternal behavior in the female offspring (Champagne et al., 2006).

Epigenetic mechanisms thus link early life experiences with later behavioral responses. Research in this field currently follows several major directions. One concerns transgenerational transmission of behavioral patterns—maternal care, stress response, cognitive ability, and response to reward—through epigenetic alterations to steroid receptors (Champagne, 2008). Another focuses on the stability of epigenetic marks that determine patterns of behavior throughout a lifetime (Weaver, 2007). Both epigenetic stability and transgenerational

transmission will be discussed in following sections. Finally, knowledge acquired in animal studies has begun to be applied to the understanding of human mental illness. A recent study compared suicide victims who had previously suffered from depression with control subjects who died suddenly of causes other than suicide (Poulter et al., 2008). Strikingly, suicide victims showed upregulation of *DN-MT3b* in the frontopolar cortex and hypermethylation of the γ-aminobutyric acid (GABA)$_A$ receptor α1 subunit gene promoter region, the product of which is known to be downregulated in suicide victims. This may indicate that long-term reprogramming of gene expression constitutes the basis of chronic mental illness. The effects of poor maternal care in humans have been hinted at in a similar study investigating the epigenetic characteristics in brain tissue of suicide victims who have had a history of childhood abuse (McGowan et al., 2009). Interestingly, higher levels of the neuron-specific GR gene *NR3C1* methylation, coincident with decreased hippocampal expression, were observed. In alignment with this, prenatal exposure to maternal depression during the third trimester of pregnancy was associated with neonatal methylation of *NR3C1*; furthermore, increased HPA stress reactivity, as measured by salivary cortisol stress response, was also seen at age 3 months (Oberlander et al., 2008).

IMMUNE SYSTEM AND INFLAMMATORY DISEASES

A research area that has emerged over the past few years focuses on the ways in which epigenetic changes effected by various maternal exposures shape offspring susceptibility to diseases of the immune system. In mice, maternal intake of the methyl donor folate resulted in hypermethylation of *Runx3*, a gene previously associated with downregulation of allergic airway inflammation, in offspring lung tissue (Hollingsworth et al., 2008). This result is consistent with clinical findings that folate supplements in pregnancy are associated with increased risk of wheeze and respiratory tract infections up to 18 months of age (Håberg et al., 2009). Also, in humans, prenatal exposure to airborne polycyclic aromatic hydrocarbons has been associated with increased risk of childhood asthma, for which the methylation of *ACSL3* has been implicated (Perera et al., 2009). Other studies have uncovered the potential role of histone acetylation: Inhibition of histone deacetylases has been shown to lead to expression of genes involved in inflammation, as well as upregulation of Th2 cytokine responses and specific T-cell responses (reviewed in Prescott and Clifton, 2009). Reduced histone deacetylase activity has also been found in various tissues of patients with chronic obstructive pulmonary disease (Ito et al., 2005).

DEVELOPMENTAL WINDOWS FOR ESTABLISHING LATER PATTERNS OF DISEASE

As discussed earlier, epigenetic marks are established by and large in the period around conception. The importance of the periconceptional period in relation to future metabolic and cardiovascular disease was first suggested by the clinical studies of humans prenatally exposed to famine in the Dutch "hunger winter." The short, sharply delineated duration of famine allowed for division of the affected into those who had been exposed in the first, second, or third trimester of pregnancy. Those exposed in late pregnancy had impaired glucose tolerance, while those affected in the second trimester had higher prevalence of obstructive airways disease and microalbuminuria. However, the worst affected were those whose mothers had starved in early pregnancy. Although at birth these children were not lighter (or significantly heavier) than unexposed controls, as adults they experienced a threefold increase in coronary heart disease as well as more obesity and atherogenic lipid profiles, among other effects (Painter et al., 2005). Individuals exposed in the periconceptional period also showed decreased methylation of specific CpG dinucleotides in the imprinted *IGF2* gene when measured nearly 60 years later (Heijmans et al., 2008). Methylation of other genes associated with growth and metabolic disease was bidirectionally different depending on gender and the gestational timing of exposure (Tobi et al., 2009). Taken together, these studies indicate that early-life nutrition can cause persistent changes in the human epigenome.

Recent animal and human studies have pursued this research avenue by asking which nutrients, if lacking or available in excess, may have permanent effects on the epigenome and on disease-related metabolic pathways. In sheep, periconceptional restriction of maternal folate, vitamin B_{12}, and methionine intake led to widespread changes in the fetal epigenome and to adult offspring that started with similar birth weights as controls yet grew heavier, fatter, more insulin-resistant, and relatively hypertensive and had a changed immune response to antigenic challenges (Sinclair et al., 2007). Of particular interest is the recent finding that in humans periconceptional folate supplementation increases *IGF2* methylation in children aged 17 months (Steegers-Theunissen et al., 2009). While the implications for later-life disease risk have yet to be determined, another study has found that maternal folate levels during pregnancy positively correlate with offspring adiposity and insulin resistance at 6 years of age, whereas vitamin B_{12} levels negatively correlate with insulin resistance (Yajnik et al., 2008). The greatest degree of insulin resistance was observed in children whose mothers were both folate-replete and vitamin B_{12}-deficient during pregnancy. The association was stronger for the levels of folate and B_{12} measured at 18 weeks than at 28 weeks of pregnancy. These results underline the importance of the timing of the insult and of epigenetic considerations in informing public-health recommendations on maternal diet during pregnancy.

Earlier we stated that the epigenome, for all its overall stability throughout life, remains malleable by environmental influences. Studies of the effects of maternal behavior on the HPA axis and GR expression showed that methionine infusion reversed the hypomethylated status of the exon I_7 *GR* in the adult offspring of high-LG-ABN mothers (Weaver et al., 2005). Some of the changes may lead to disease. Studies on monozygotic twins have reported an association between epigenetic differences and discordance in diseases such as Beckwith-Wiedemann syndrome (Weksberg et al., 2002) and bipolar disorder (Rosa et al., 2008). The influence of age has also been demonstrated in another monozygotic twin study, which examined COX7A1, a component of the mitochondrial oxidative phosphorylation cascade shown to be downregulated in skeletal muscle of patients with type 2 diabetes mellitus. Older twins had much higher methylation of the promoter compared with younger twins, and the investigators related the correspondingly lower gene expression to metabolic dysfunction in older age (Rönn et al., 2008).

More generally, it appears that many of the epigenetic changes described in aging—such as the above-described gradual global DNA hypomethylation, subsequent overexpression of *DNMT3b*, and related hypermethylation of promoter CpG islands that are usually unmethylated in normal cells (Fraga and Esteller, 2007)—are remarkably similar to those evident in cancer, which will be discussed in detail later (see Epigenetics in Cancer).

TRANSGENERATIONAL TRANSMISSION OF EPIGENETIC MARKS

There is some evidence that changes occurring in the epigenome during development may be inherited by subsequent generations. For example, in rodent studies, nutritional or endocrinological interventions in pregnant F_0 animals can cause phenotypic and/or epigenetic changes that persist to at least the F_2 generation (Burdge et al., 2007; Drake et al., 2005; Gluckman et al., 2007b). It is not clear whether this represents true transgenerational transmission or simply effects on the germ cells of the F_1 generation that are exposed in utero. Transgenerational inheritance of environmentally malleable epigenetic marks in the mouse metastable epiallele models

(*agouti* and *axin-fused*) also remains controversial (Rakyan et al., 2003; Waterland et al., 2007).

There is inconclusive epidemiological evidence of nongenomic inheritance of disease risk in humans, for instance, in the children of the Dutch "hunger winter" cohort (Painter et al., 2008). A study using historical harvest and food price records for an isolated Swedish community found association between paternal grandparents' food supply and the relative mortality risk of their grandchildren. Paternal grandmother's good or poor food supply in early childhood seems to have, respectively, halved or doubled the granddaughter's relative mortality risk (Pembrey et al., 2006). Molecular evidence for transgenerational epigenetic inheritance in humans is very limited, although there is a report of a germ-line epimutation in the promoter of the DNA mismatch-repair gene *MLH1* that led to maternal-line inheritance of tumor susceptibility (Hitchins et al., 2007).

The specific mechanisms by which epigenetic information, whether environmentally imposed methylation patterns or genomic imprinting, persists through the reprogramming that occurs during embryogenesis and gametogenesis remain unresolved. However, small RNAs in the sperm have been increasingly indicated as the mechanism underlying non-Mendelian inheritance of tail coloration in the mouse (Rassoulzadegan et al., 2006). Injection of an miRNA that targets a key regulator of cardiac growth into embryos resulted in cardiac hypertrophy, and this effect was efficiently transmitted to progeny over at least three generations (Wagner et al., 2008). The role of sperm in transmitting the pathology to progeny was supported by observations of miRNA accumulation in the sperm of paramutated males. An alternate mechanism, as demonstrated in the studies of maternal behavioral induction of epigenetic changes in the brains of infant mice, may not involve survival of the epigenetic mark beyond gametogenesis. Rather, in each generation reestablishment of the behavior or environment that induces the mark—known as *niche recreation*—occurs (Champagne and Curley,

2008; Weaver et al., 2004). Similarly, girls born small for age grow to have a proportionately smaller uterus and in turn have smaller offspring themselves (Ibáñez et al., 2003). Thus, both generations might have similar epigenetic changes present without germ-line transmission having occurred.

IMPRINTING DISORDERS

The diseases discussed above for the most part arise from maladaptive developmental responses to an imperfect environment. Imprinting disorders, however, fall into a different category. In diploid organisms, most autosomal genes are expressed from both the paternal and maternal alleles; but in mammals, a small proportion of genes are expressed in a haploid fashion from only one of the parental alleles. The monoallelic expression of these so-called imprinted genes results from the establishment in the germ line of parent-of-origin-specific epigenetic marks on regions of the gametic chromosomes, resulting in (generally) the silencing of the imprinted gene.

The majority of imprinted genes occur in clusters of a few protein-coding genes associated with a differentially DNA-methylated region (DMR) or imprinting centre (IC) that carries the parent-specific imprinting mark established during gametogenesis (Reik and Walter, 2001). The cluster also typically contains one or more genes for noncoding (nc) RNAs. Although the function of the ncRNAs remains unclear, it is a general pattern that their expression is reciprocal to that of the mRNA genes in the cluster, suggesting that the ncRNAs may be involved in silencing the expression of the imprinted protein-coding genes. An alternative mechanism for gene silencing by imprinting is that the DMR may exhibit methylation state–specific binding of regulatory proteins, thereby acting as an insulator between the enhancer and promoter regions of the protein-coding genes in the cluster.

Probably 80–150 genes are imprinted in mammals (Morison et al., 2005). Importantly, a high proportion of imprinted genes are involved in the regulation of fetal or neonatal growth and development (Smith et al., 2006). This, together with the evolutionary appearance of imprinting only in the placental mammals (Edwards et al., 2007) among animal taxa (and flowering plants), suggests that imprinting evolved in response to the problems of a reproductive strategy in which the fetus grows within the mother's body and directly extracts resources from her. In this context, two specific hypotheses for the evolution of imprinting in mammals have been proposed. First, the observation that imprinted genes expressed from the paternal allele tend to promote fetal growth whereas imprinted genes expressed from the maternal allele tend to constrain fetal growth prompted David Haig (1992; Haig and Westoby, 2006), following the work of Trivers (1974), to propose that "parental conflict" underlies imprinting. Since a male can father offspring in several different females, it is in his evolutionary interest that fetuses carrying his genome extract as much resource as possible from their mothers, thereby enhancing their (and therefore his) fitness. Conversely, the mother's fitness is dependent on the average fitness of all her offspring (which may have different fathers), and therefore, it is in her interest to limit her investment in any particular offspring. Although more recent work emphasizes coadaptation (Curley et al., 2004; Wolf and Hager, 2006) rather than conflict, the fundamental idea that imprinting serves in some way to align parental resources with offspring demands remains valid. A second hypothesis for the development of imprinting notes the invasive nature of the trophoblast and that several genes promoting development of the trophoblast and placenta are silenced by imprinting of the maternal genome and expressed only from the paternal genome. Imprinting therefore serves as a "safety catch," ensuring that trophoblast tissue can develop only after introduction of the paternal genome at fertilization, protecting the female from the consequences (such as ovarian trophoblastic disease) of parthenogenetic activation of an oocyte (Varmuza and Mann, 1994).

Whatever the evolutionary origins of imprinting (Wilkins and Haig, 2003), the benefits must outweigh the disadvantages, which include the complexity of the mechanisms involved and, in particular, the generation of a functionally haploid state for a number of developmentally important genes. Nevertheless, occasional defects in the mechanisms of establishment or erasure of imprinting marks, sometimes in combination with genetic mutations that acquire functional dominance in the context of defective imprinting (resulting in nonexpression of both alleles of a gene), result in human disease (Ubeda and Wilkins, 2008). The following sections describe examples of such disorders.

PRADER-WILLI SYNDROME AND ANGELMAN SYNDROME

Prader-Willi syndrome (PWS) and Angelman syndrome (AS) arise from genetic or epigenetic disruption of a cluster of imprinted genes on chromosome 15q (Buiting et al., 2003; Horsthemke and Buiting, 2008). PWS, caused by loss of the paternal alleles, is characterized by developmental delay, short stature, obesity, hyperphagia, and mild or moderate cognitive impairment. The molecular defect in PWS appears to be loss of expression of several paternally expressed ncRNAs, which presumably in turn affects expression of protein-coding genes in the cluster. AS, caused by loss of the maternal alleles, is characterized by microcephaly, severe developmental delay, severe cognitive impairment, and impaired coordination. The molecular defect in AS is loss of expression of the ubiquitin E3 ligase (*UBE3A*) gene, which targets proteins for degradation; the gene shows tissue-specific imprinting, being expressed only from the maternal allele in the brain, which may account for the major defects in AS being related to cognition and coordination (Horsthemke and Buiting, 2008).

The majority of cases of PWS and AS result from large (several million base pairs) deletions in the cluster, but a significant proportion arise from imprinting defects. Of these, most are de novo epimutations that appear to arise from failure to erase grandparental imprints during formation of the parental germ cells or from failure to maintain imprints during the widespread demethylation that occurs after fertilization. A small proportion of imprinting defects arise from small deletions in the IC that result in faulty imprint maintenance across the whole imprinted gene cluster.

BECKWITH-WIEDEMANN SYNDROME AND SILVER-RUSSELL SYNDROME

Beckwith-Wiedemann syndrome (BWS) and a proportion of cases of Silver-Russell syndrome (SRS) arise from imprinting defects in a cluster of imprinted genes on chromosome 11p15 that result in overgrowth or growth failure, respectively (Eggermann et al., 2008). BWS is characterized by asymmetrical somatic overgrowth and increased susceptibility to developmental tumors such as Wilms tumor. Conversely, SRS is characterized by growth retardation and craniofacial anomalies. The imprinted gene cluster on 11p contains several growth-related genes, of which the most likely to be significant for BWS and SRS are the paternally expressed gene for IGF-II (*IGF2*), a major regulator of fetal growth, and the maternally expressed gene for cyclin-dependent kinase inhibitor 1C (*CDKN1C*), a negative regulator of cell proliferation and a putative tumor-suppressor gene (Horsthemke and Buiting, 2008).

About two-thirds of cases of BWS arise from imprinting defects in one of the two imprinting control regions in this cluster, causing biallelic expression of *IGF2* or silencing of *CDKN1C*; the remainder arise from genetic defects, particularly uniparental disomy, that lead to similar patterns of gene expression. Cases of SRS arising from lesions in 11p15 predominantly show epimutations in ICR1, resulting in silencing of *IGF2* expression; interestingly, these epimutations show a mosaic distribution, suggesting a postfertilization defect in establishment of the imprinting mark. This is supported by the high rate of discordance of SRS between monozygotic twins (Bailey et al., 1995).

PSEUDOHYPOPARATHYROIDISM

The stimulatory G protein $G_s\alpha$ is involved in intracellular signal transduction of many hormones. The *GNAS1* gene on chromosome 20q13 displays alternative splicing, leading to several different protein products; and the expression of these transcripts is controlled by tissue-specific imprinting (Bastepe, 2008). Dysregulation of this complex system leads to a spectrum of clinical variants, typified by resistance to parathyroid hormone in the proximal renal tubule, which causes hypocalcemia and hyperphosphatemia in spite of otherwise normal renal function. The maternal allele is preferentially expressed in renal tubules, and maternal-line mutations or imprinting defects cause haploinsufficiency for $G_s\alpha$ and defective signaling. In other tissues, where biallelic expression of $G_s\alpha$ occurs, expression from the paternal allele is presumably adequate to maintain normal signal transduction.

ENVIRONMENTAL MODULATION OF THE PREVALENCE OF IMPRINTING DISORDERS

DNA methylation is vulnerable to environmental changes in the availability of methylation substrates and cofactors, such as folates discussed above; and there is some evidence that environmental factors can modulate the incidence of epimutations. 5,10-Methylenetetrahydrofolate reductase (MTHFR) is a key enzyme in folate metabolism. The 677C→T polymorphism of *MTHFR* reduces its activity and decreases the supply of substrate for DNA methylation. Maternal homozygosity for the 677T allele of *MTHFR* increases the risk of AS in the offspring (Zogel et al., 2006), and the phenotypic effects of severe *MTHFR* deficiency overlap with those of AS (Arn et al., 1998).

Although the incidence of imprinting disorders is low, the incidence of BWS and AS is increased severalfold in children conceived by assisted reproductive technology. Consistent with the information on *MTHFR* polymorphism, this increased incidence appears to be restricted to syndromes where the imprinting change takes the form of hypomethylation on the maternal allele, although in vitro studies of isolated gametes suggest that the paternal epigenome may also be subjected to imprinting defects and that hypermethylation can also occur (Amor and Halliday, 2008). These observations, together with the epigenetic lability of the early conceptus (Morgan et al., 2005) and its susceptibility to nutritional perturbations (Watkins et al., 2007, 2008), underline the need for careful control of the environmental conditions for oocyte maturation and embryo culture.

The so-called endocrine disruptors appear to have transgenerational effects that may involve perturbation of imprinting. For example, exposure of pregnant rats to the antiandrogenic fungicide vinclozolin during the period of fetal sex determination and germ-cell development (embryonic days 12–15), but not later in pregnancy, was reported to cause defects in sperm formation and fertility in their male offspring, which were transferred through the male line to at least the fourth generation (Anway et al., 2005). These defects correlated with epigenetic (DNA methylation) alterations in the testis. A similar transgenerational effect has been reported in humans after exposure of the mother in utero to the synthetic estrogen diethylstilbestrol (Brouwers et al., 2006). If confirmed, these observations would provide examples of transient exposure to a hormonally active substance establishing an impaired pattern of development that can be transmitted to subsequent generations.

EPIGENETICS IN CANCER

Cancer can be considered a disorder of development in which inappropriate control of gene expression leads to dysregulation of cellular proliferation, adhesion, and motility. Typically, this involves increased expression of genes that promote these processes (oncogenes) or decreased expression of genes that inhibit them (tumor-suppressor genes).

The role of epigenetic processes in the regulation of gene expression has focused attention on the roles of epigenetics in cancer, and numerous epigenetic alterations have now been discovered in cancer cells. Nevertheless, there appears to be an overriding pattern of change in the epigenome during tumorigenesis, which can be summarized as a global hypomethylation in conjunction with specific hypermethylation in the promoters of tumor-suppressor genes resulting in gene silencing (Esteller, 2008). These changes in the pattern of DNA methylation are accompanied by associated changes in histone modification and miRNA expression. The following sections consider these changes in more detail.

GLOBAL HYPOMETHYLATION

In the 1980s, hypomethylation of the bulk of the genome was the first epigenetic alteration observed in human cancer (Feinberg and Vogelstein, 1983), and more recent studies have demonstrated that the extent of hypomethylation increases during tumor progression (Fraga et al., 2004). Hypomethylation occurs both in repetitive DNA sequences and in coding regions. Since DNA methylation contributes to chromosomal stability and silencing of retrotransposons, the consequences of hypomethylation may include chromosomal aberrations and reactivation of parasitic elements, further contributing to loss of control of cell proliferation. Such changes can be demonstrated in cancer tissue and in cultured cells in which expression of DNA methyltransferases has been knocked down (Karpf and Matsui, 2005). Hypomethylation around coding regions may also result in loss of imprinting, leading to biallelic expression of growth-related genes, as occurs with the embryonic tumors associated with BWS (Horsthemke and Buiting, 2008) and in activation of other proliferation-related genes such as *Pax2* (Wu et al., 2005).

SPECIFIC HYPERMETHYLATION

Gene silencing by specific hypermethylation of CpG islands in their promoter regions is a significant contributor to tumorigenesis. Among the genes silenced are classic tumor-suppres-

sor genes such as the retinoblastoma tumor-suppressor (*Rb*) gene, the adenomatous polyposis coli (*APC*) gene in colon cancer, and the *BRCA1* gene in breast cancer, as well as DNA repair genes such as *MLH1* and *MGMT* and adhesion molecules such as E-cadherin (Esteller, 2008). Many such genes are found to be silenced during the early phases of tumor development, suggesting a fundamental contribution to neoplastic transformation. For example, silencing of the tumor-suppressor gene $p16^{INK4a}$ by promoter hypermethylation is frequently found in the early stages of several human cancers (Belinsky et al., 1998), and in cancer cell culture forced expression of this gene by demethylating agents abrogates the cancer cell phenotype (Bender et al., 1998). Indeed, gene silencing by hypermethylation can be the "second hit" in Knudson's two-hit model for carcinogenesis (Grady et al., 2000; Knudson, 1996). Recent mapping studies have revealed that hundreds of promoter CpG islands may become methylated in a given tumor type (Esteller, 2007); the mechanisms underlying this tumor-specific promoter hypermethylation remain unknown. However, some subgroups of a specific tumor type appear to display a CpG island methylator phenotype (CIMP) with high levels of tumor-specific methylation that correlate with poorer prognosis (Teodoridis et al., 2008).

HISTONE MODIFICATIONS AND miRNA EXPRESSION

Cancer-related histone modifications are in general consistent with those expected from the changes in DNA methylation, with histone marks typical of a repressive chromatin state (H3 and H4 deacetylation and H3 lysine-9 methylation) being found in association with hypermethylated CpG islands and increased levels of histone-modifying enzymes such as deacetylases and methyltransferases (Esteller, 2007, 2008).

As described above, miRNAs are short ncRNAs that negatively regulate gene expression by binding to mRNA and appear to play roles in the control of cell proliferation and differentiation. Changes in miRNA expression in cancer tissues have been reported, often in a direction that implies that the relevant miRNAs act as tumor-suppressors by targeting expression of oncogenes (Ma and Weinberg, 2008). In turn, the reduced expression of miRNAs appears to result from aberrant hypermethylation of their respective promoters (Guil and Esteller, 2009).

EPIGENETIC THERAPY OF CANCER

Modulation of gene expression by epigenetic processes is, at least in principle, reversible, suggesting that alterations in the cancer epigenome may be viable targets for therapeutic intervention.

Demethylating agents such as 5-azacytidine can restore DNA methylation–silenced gene expression in cultured cells and have antitumor activity in animal models (Mund et al., 2006). A number of such drugs are currently under investigation, and two, azacitidine and decitabine, have now been approved for myelodysplastic syndromes (Ellis et al., 2009). Histone deacetylation is a hallmark of epigenetically silenced regions of chromatin, and deacetylase inhibitors, which also target nonhistone proteins that may be involved in modulating gene expression, have been observed to reactivate gene expression in cultured cells and to induce cell cycle arrest, differentiation, and cell death. One nonspecific deacetylase inhibitor, valproic acid, has a long history of use as an anticonvulsant drug and is now undergoing clinical trials as an adjunctive therapy in various tumors (Duenas-Gonzalez et al., 2008). Another more specific histone deacetylase inhibitor, vorinostat, has been approved for the treatment of cutaneous T-cell lymphoma (Duvic and Vu, 2007). Trials are also under way to study the synergistic effects of DNA methylation inhibitors and histone deacetylase inhibitors (Bishton et al., 2007). Further research is aimed at testing histone methylation inhibitors and combining different classes of epigenetic therapies as well as epigenetic agents with other types of anticancer treatments. There is currently a wide array of epigenetic-based therapeutics in preclinical and clinical development (Ellis et al., 2009).

One major concern of current epigenetic therapies for cancer is their lack of specificity. In principle, these therapies promote reactivation of epigenetically silenced gene expression and reestablishment of normal patterns of growth control in cancer cells. However, theoretically at least, such approaches may lead to adverse effects arising from unintended gene activation, such as of oncogenes or other growth-controlling pathways. Current research is aimed at finding more directed therapies that can effect epigenetic modification of specific genes (Smith et al., 2008).

DISORDERS ARISING FROM DEFECTS IN EPIGENETIC PATHWAYS

Some rare disorders stem from genetic defects in the proteins controlling epigenetic pathways. In this section, we focus on two so-called chromatin disorders because of their effect on chromatin structure and gene expression: Rett syndrome and immunodeficiency, centromeric instability, and facial anomalies (ICF) syndrome. These diseases are caused by mutations in genes coding for proteins that establish or transduce DNA methylation marks; given the apparently central role of these processes in cellular regulation across the metazoa, the compatibility of such defects with life, with only rather specific phenotypic deficits, is evidence for considerable redundancy in this highly conserved system of molecular machinery.

RETT SYNDROME

Rett syndrome is an X-linked disorder with predominantly neurological manifestations. Rett syndrome occurs in 1 in 10,000 live births, almost exclusively in females because of lethality in male embryos. Girls with Rett syndrome appear normal at birth, but cognitive and motor functions regress in early childhood to leave a severe intellectual disability. Brain stem control of breathing is particularly affected, and respiratory failure is a major cause of death in these individuals (Chahrour and Zoghbi, 2007; Matarazzo et al., 2009).

The molecular defect in Rett syndrome is in the gene for methyl-CpG binding protein 2 (*MeCP2*) located at Xq28. This protein binds to methylated DNA sequences and recruits corepressors of transcription such as histone deacetylases. The majority of cases of Rett syndrome show protein-coding mutations in this gene (Chahrour and Zoghbi, 2007).

Gene dosage of *MeCP2* is regulated by X-chromosome inactivation, and Rett syndrome patients are mosaic for *MeCP2* expression, with skewed X inactivation accounting for variations in disease severity and for occasional reports of asymptomatic female carriers (Ishii et al., 2001). The importance of correct *MeCP2* gene dosage is shown by animal models of Rett syndrome in which either double expression in all cells (Collins et al., 2004) or half-expression in all cells (Samaco et al., 2008) leads to neurological disease.

Given the apparent central role of *MeCP2* in transducing DNA methylation into chromatin structure, it was initially surprising that no general dysregulation of gene expression arises from abrogation of MeCP2 activity in Rett syndrome (Tudor et al., 2002). Rather, it seems that the function of MeCP2 can be assumed by other proteins with a methyl-binding domain and that the critical defect in gene expression caused by MeCP2 deficiency is limited to a small subset of cell types in the brain. One putative target gene is brain-derived neurotrophic factor (*BDNF*), which encodes a growth factor involved in neural development. Interaction between *MeCP2* and the *BDNF* promoter is dependent on neural activity (Zhou et al., 2006). Normal mice made *BDNF* null in the brain display symptoms similar to those of Rett syndrome, whereas forced expression of *BDNF* in the brain of *MeCP2* mutant mice relieves their symptoms (Chang et al., 2006). Dysregulation of *BDNF* expression is also a feature of Huntington disease (Zuccato and Cattaneo, 2007), offering the possibility of synergies in the approach to research and treatment of these disorders.

ICF SYNDROME

ICF syndrome is characterized clinically by severe immunodeficiency together with facial anomalies, delayed development, and some degree of cognitive impairment. Patients show DNA hypomethylation, particularly in pericentromeric regions of specific chromosomes, that leads to chromosomal instability in these regions and cytological abnormalities, particularly in cells of the lymphoid lineage. This in turn causes defective immunoglobulin secretion and immunodeficiency (Ehrlich et al., 2008; Matarazzo et al., 2009).

The primary molecular defect in ICF syndrome is aberrant DNA methylation caused by mutations in the gene for DNMT3b on chromosome 20p. DNMT3b is one of the de novo methyltransferases that establish methylation patterns on repetitive sequences and imprinted genes. The disorder is inherited in autosomal recessive fashion, with most patients being homozygous or compound heterozygous for mutations in the region coding for the catalytic domain of the protein. Since homozygous deletion of the catalytic domain of DNMT3b in mice is embryonically lethal, it is likely that the mutations in humans leave some residual enzyme activity (Ehrlich et al., 2008).

As well as chromosomal abnormalities, the DNA hypomethylation caused by deficiency in DNMT3b activity in ICF syndrome is associated with aberrant expression of genes involved in development and immune function. Upregulated genes lose DNA methylation and repressive chromatin marks and gain active chromatin marks (Jin et al., 2008). In general, however, the changes in gene expression are subtle and not consistently associated with promoter hypomethylation (Ehrlich et al., 2008); and it is possible that the cellular defect in DNMT3b deficiency is a result of generalized alterations in chromatin architecture rather than aberrant methylation of particular target genes. Although

DNA hypomethylation is a consistent feature of carcinogenesis (see above), patients with ICF syndrome do not appear to be at particular risk of developing tumors (Ehrlich et al., 2008).

CONCLUSION

The relevance of molecular epigenetics to human disease is only now emerging. The role of somatic epimutations in the etiology of many cancers is now relatively well established, and parental imprinting disorders, while rare, are well recognized. However, it has been the emergence of the study of developmental pathways to human disease that has led to a major focus on epigenetic processes as a contributor to the etiology of noncommunicable chronic disease. While the evidence is still largely based on animal studies and clinical correlates, it seems probable that epigenetic mechanisms play a significant role in influencing disease risk. Based on epidemiological associations, we anticipate that epigenetic mechanisms will be shown to play a key role in a range of disorders including cardiovascular disease, adult-onset diabetes mellitus, obesity, osteoporosis, mood disorders, atopic disease, and possibly aging itself. A possible scenario is that epigenetic biomarkers in early life may be useful prognostic indicators of individuals or population groups at particular risk. Moreover, understanding these processes may lead to interventions based on manipulation of the epigenetic state.

REFERENCES

Aagaard-Tillery, K., K. Grove, J. Bishop, X. Ke, Q. Fu, R. McKnight, and R. Lane. 2008. Developmental origins of disease and determinants of chromatin structure: Maternal diet modifies the primate fetal epigenome. *J Mol Endocrinol* 41:91–102.

Aerts, L., and F.A. Van Assche. 1979. Is gestational diabetes an acquired condition? *J Dev Physiol* 1:219–25.

Amor, D.J., and J. Halliday. 2008. A review of known imprinting syndromes and their association with assisted reproduction technologies. *Hum Reprod* 23:2826–34.

Antequera, F. 2003. Structure, function and evolution of CpG island promoters. *Cell Mol Life Sci* 60:1647–58.

Anway, M.D., A.S. Cupp, M. Uzumcu, and M.K. Skinner. 2005. Epigenetic transgenerational actions of endocrine disruptors and male fertility. *Science* 308:1466–9.

Appel, T.A. 1987. *The Cuvier-Geoffroy Debate: French Biology in the Decades Before Darwin*. New York: Oxford University Press.

Aravin, A.A., and D. Bourc'his. 2008. Small RNA guides for de novo DNA methylation in mammalian germ cells. *Genes Dev* 22:970–5.

Arn, P.H., C.A. Williams, R.T. Zori, D.J. Driscoll, and D.S. Rosenblatt. 1998. Methylenetetrahydrofolate reductase deficiency in a patient with phenotypic findings of Angelman syndrome. *Am J Med Genet* 77:198–200.

Bailey, W., B. Popovich, and K.L. Jones. 1995. Monozygotic twins discordant for the Russell-Silver syndrome. *Am J Med Genet* 58:101–5.

Barker, D.J.P., P.D. Winter, C. Osmond, B. Margetts, and S.J. Simmonds. 1989. Weight in infancy and death from ischaemic heart disease. *Lancet* 2:577–80.

Barski, A., R. Jothi, S. Cuddapah, K. Cui, T.-Y. Roh, D.E. Schones, and K. Zhao. 2009. Chromatin poises miRNA- and protein-coding genes for expression. *Genome Res* 19:1742–51.

Bastepe, M. 2008. The GNAS locus and pseudohypoparathyroidism. *Adv Exp Med Biol* 626: 27–40.

Bateson, P., D. Barker, T. Clutton-Brock, D. Deb, B. D'Udine, R.A. Foley, P. Gluckman, et al. 2004. Developmental plasticity and human health. *Nature* 430:419–21.

Belinsky, S.A., K.J. Nikula, W.A. Palmisano, R. Michels, G. Saccomanno, E. Gabrielson, S.B. Baylin, and J.G. Herman. 1998. Aberrant methylation of p16INK4a is an early event in lung cancer and a potential biomarker for early diagnosis. *Proc Natl Acad Sci USA* 95:11891–6.

Bender, C.M., M.M. Pao, and P.A. Jones. 1998. Inhibition of DNA methylation by 5-aza-2´-deoxycytidine suppresses the growth of human tumor cell lines. *Cancer Res* 58:95–101.

Bishton, M., M. Kenealy, R. Johnstone, W. Rasheed, and H.M. Prince. 2007. Epigenetic targets in hematological malignancies: Combination therapies with HDACis and demethylating agents. *Expert Rev Anticancer Ther* 7:1439–49.

Bogdarina, I., S. Welham, P.J. King, S.P. Burns, and A.J. Clark. 2007. Epigenetic modification of the renin–angiotensin system in the fetal programming of hypertension. *Circ Res* 100: 520–6.

Boney, C. M., A. Verma, R. Tucker, and B. R. Vohr. 2005. Metabolic syndrome in childhood: Association with birth weight, maternal obesity, and gestational diabetes mellitus. *Pediatrics* 115: 290–6.

Brouwers, M. M., W. F. J. Feitz, L. A. J. Roelofs, L. A. L. M. Kiemeney, R. P. E. de Gier, and N. Roeleveld. 2006. Hypospadias: A transgenerational effect of diethylstilbestrol? *Hum Reprod* 21:666–9.

Buiting, K., S. Grob, C. Lich, G. Gillessen-Kaesbach, O. El-Maarri, and B. Horsthemke. 2003. Epimutations in Prader-Willi and Angelman syndromes: A molecular study of 136 patients with an imprinting defect. *Am J Hum Genet* 72:571–7.

Burdge, G. C., J. L. Slater-Jefferies, C. Torrens, E. S. Phillips, M. A. Hanson, and K. A. Lillycrop. 2007. Dietary protein restriction of pregnant rats in the F_0 generation induces altered methylation of hepatic gene promoters in the adult male offspring in the F_1 and F_2 generations. *Br J Nutr* 97:435–9.

Cao, R., and Y. Zhang. 2004. The functions of E(Z)/ EZH2-mediated methylation of lysine 27 in histone H3. *Curr Opin Genet Dev* 14:155–64.

Chahrour, M., and H. Y. Zoghbi. 2007. The story of Rett syndrome: From clinic to neurobiology. *Neuron* 56:422–37.

Champagne, F. A. 2008. Epigenetic mechanisms and the transgenerational effects of maternal care. *Front Neuroendocrinol* 29:386–97.

Champagne, F. A., and J. P. Curley. 2008. Maternal regulation of estrogen receptor α methylation. *Curr Opin Pharmacol* 8:1–5.

Champagne, F. A., I. C. G. Weaver, J. Diorio, S. Dymov, M. Szyf, and M. J. Meaney. 2006. Maternal care associated with methylation of the estrogen receptor-α1b promoter and estrogen receptor-α expression in the medial preoptic area of female offspring. *Endocrinology* 147:2909–15.

Chang, Q., G. Khare, V. Dani, S. Nelson, and R. Jaenisch. 2006. The disease progression of Mecp2 mutant mice is affected by the level of BDNF expression. *Neuron* 49:341–8.

Cloos, P., J. Christensen, K. Agger, and K. Helin. 2008. Erasing the methyl mark: Histone demethylases at the center of cellular differentiation and disease. *Genes Dev* 22:1115–40.

Collins, A. L., J. M. Levenson, A. P. Vilaythong, R. Richman, D. L. Armstrong, J. L. Noebels, J. D. Sweatt, and H. Y. Zoghbi. 2004. Mild overexpression of MeCP2 causes a progressive neurological disorder in mice. *Hum Mol Genet* 13: 2679–89.

Curley, J. P., S. Barton, A. Surani, and E. B. Keverne. 2004. Coadaptation in mother and infant regulated by a paternally expressed imprinted gene. *Proc R Soc Lond B* 271:1303–9.

Daston, L., and K. Park. 1998. *Wonders and the Order of Nature, 1150–1750.* New York: Zone Books.

Dolinoy, D. C., J. R. Weidman, R. A. Waterland, and R. L. Jirtle. 2006. Maternal genistein alters coat color and protects A^{vy} mouse offspring from obesity by modifying the fetal epigenome. *Environ Health Perspect* 114:567–72.

Dörner, G. 1973. Die mögliche Bedeutung der prä- und/oder perinatalen Ernährung für die Pathogenese der Obesitas. *Acta Biol Med Ger* 30: K19–22.

Dörner, G., H. Haller, and M. Leonhardt. 1973. Zur möglichen Bedeutung der prä- und/ oder frühpostnatalen Ernährung für die Pathogenese der Arteriosklerose. *Acta Biol Med Ger* 30:K31–5.

Dörner, G., and A. Mohnike. 1973. Zur möglichen Bedeutung der prä- und/oder frühpostnatalen Ernährung für die Pathogenese des Diabetes mellitus. *Acta Biol Med Ger* 30:K7–10.

Drake, A. J., B. R. Walker, and J. R. Seckl. 2005. Intergenerational consequences of fetal programming by in utero exposure to glucocorticoids in rats. *Am J Physiol Regul Integr Comp Physiol* 288: R34–8.

Duenas-Gonzalez, A., M. Candelaria, C. Perez-Plascencia, E. Perez-Cardenas, E. de la Cruz-Hernandez, and L. A. Herrera. 2008. Valproic acid as epigenetic cancer drug: Preclinical, clinical and transcriptional effects on solid tumors. *Cancer Treat Rev* 34:206–22.

Duvic, M., and J. Vu. 2007. Vorinostat: A new oral histone deacetylase inhibitor approved for cutaneous T-cell lymphoma. *Expert Opin Investig Drugs* 16:1111–20.

Edwards, C. A., W. Rens, O. Clarke, A. J. Mungall, T. Hore, J. A. M. Graves, I. Dunham, A. C. Ferguson-Smith, and M. A. Ferguson-Smith. 2007. The evolution of imprinting: Chromosomal mapping of orthologues of mammalian imprinted domains in monotreme and marsupial mammals. *BMC Evol Biol* 7:157.

Eggermann, T., K. Eggermann, and N. Schönherr. 2008. Growth retardation versus overgrowth: Silver-Russell syndrome is genetically opposite to Beckwith-Wiedemann syndrome. *Trends Genet* 24:195–204.

Ehrlich, M., C. Sanchez, C. Shao, R. Nishiyama, J. Kehrl, R. Kuick, T. Kubota, and S. M. Hanash. 2008. ICF, an immunodeficiency syndrome: DNA methyltransferase 3B involvement, chromosome anomalies, and gene dysregulation. *Autoimmunity* 41:253–71.

Ellis, L., P.W. Atadja, and R.W. Johnstone. 2009. Epigenetics in cancer: Targeting chromatin modifications. *Mol Cancer Ther* 8:1409–20.

Esteller, M. 2007. Cancer epigenomics: DNA methylomes and histone-modification maps. *Nat Rev Genet* 8:286–98.

Esteller, M. 2008. Epigenetics in cancer. *N Engl J Med* 358:1148–59.

Farthing, C.R., G. Ficz, R.K. Ng, C.-F. Chan, S. Andrews, W. Dean, M. Hemberger, and W. Reik. 2008. Global mapping of DNA methylation in mouse promoters reveals epigenetic reprogramming of pluripotency genes. *PLoS Genet* 4:e1000116.

Feinberg, A.P., and B. Vogelstein. 1983. Hypomethylation distinguishes genes of some human cancers from their normal counterparts. *Nature* 301:89–92.

Felitti, V.J., R.F. Anda, D. Nordenberg, D.F. Williamson, A.M. Spitz, V. Edwards, M.P. Koss, and J.S. Marks. 1998. Relationship of childhood abuse and household dysfunction to many of the leading causes of death in adults—the Adverse Childhood Experiences (ACE) study. *Am J Prev Med* 14:245–58.

Forsdahl, A. 1977. Are poor living conditions in childhood and adolescence an important risk factor for arteriosclerotic heart disease? *Br J Prevent Soc Med* 31:91–5.

Forsdahl, A. 1978. Living conditions in childhood and subsequent development of risk factors for arteriosclerotic heart disease. The cardiovascular survey in Finnmark 1974–75. *J Epidemiol Community Health* 32:34–7.

Fraga, M.F., E. Ballestar, M.F. Paz, S. Ropero, F. Setien, M.L. Ballestar, D. Heine-Suñer, et al. 2005. Epigenetic differences arise during the lifetime of monozygotic twins *Proc Natl Acad Sci USA* 102:10604–9.

Fraga, M.F., and M. Esteller. 2007. Epigenetics and aging: The targets and the marks. *Trends Genet* 23:413–18.

Fraga, M.F., M. Herranz, J. Espada, E. Ballestar, M.F. Paz, S. Ropero, E. Erkek, et al. 2004. A mouse skin multistage carcinogenesis model reflects the aberrant DNA methylation patterns of human tumors. *Cancer Res* 64:5527–34.

Freinkel, N. 1980. Banting lecture 1980: Of pregnancy and progeny. *Diabetes* 29:1023–35.

Fu, Q., R.A. McKnight, X. Yu, L. Wang, C.W. Callaway, and R.H. Lane. 2004. Uteroplacental insufficiency induces site-specific changes in histone H3 covalent modification and affects DNA-histone H3 positioning in day 0 IUGR rat liver. *Physiol Genomics* 20:108–16.

Gluckman, P.D. 1995. The endocrine regulation of fetal growth in late gestation: The role of insulin-like growth factors. *J Clin Endocrinol Metab* 80:1047–50.

Gluckman, P.D., and M.A. Hanson. 2004a. Maternal constraint of fetal growth and its consequences. *Semin Fetal Neonatal Med* 9: 419–25.

Gluckman, P.D., and M.A. Hanson. 2004b. Living with the past: Evolution, development, and patterns of disease. *Science* 305:1733–6.

Gluckman, P.D., and M.A. Hanson. 2006. Evolution, development and timing of puberty. *Trends Endocrinol Metab* 17:7–12.

Gluckman, P.D., M.A. Hanson, and A.S. Beedle. 2007a. Early life events and their consequences for later disease: A life history and evolutionary perspective. *Am J Hum Biol* 19:1–19.

Gluckman, P.D., M.A. Hanson, and A.S. Beedle. 2007b. Non-genomic transgenerational inheritance of disease risk. *Bioessays* 29:149–54.

Gluckman, P.D., M.A. Hanson, H.G. Spencer, and P. Bateson. 2005. Environmental influences during development and their later consequences for health and disease: Implications for the interpretation of empirical studies. *Proc R Soc Lond B* 272:671–7.

Gluckman, P.D., K.A. Lillycrop, M.H. Vickers, A.B. Pleasants, E.S. Phillips, A.S. Beedle, G.C. Burdge, and M.A. Hanson. 2007c. Metabolic plasticity during mammalian development is directionally dependent on early nutritional status. *Proc Natl Acad Sci USA* 104:12796–800.

Godfrey, K. 2006. The "developmental origins" hypothesis: Epidemiology. In *Developmental Origins of Health and Disease*, ed. P.D. Gluckman and M.A. Hanson, eds. 6–32. Cambridge: Cambridge University Press.

Godfrey, K.M., P.D. Gluckman, K.A. Lillycrop, G.C. Burdge, J. Rodford, J.L. Slater-Jefferies, C. McLean, et al. 2009. Epigenetic marks at birth predict childhood body composition at age 9 years. *J Dev Orig Health Dis* 1:S44.

Grady, W.M., J. Willis, P.J. Guilford, A.K. Dunbier, T.T. Toro, H. Lynch, G. Wiesner, et al. 2000. Methylation of the CDH1 promoter as the second genetic hit in hereditary diffuse gastric cancer. *Nat Genet* 26:16–17.

Guil, S., and M. Esteller. 2009. DNA methylomes, histone codes and miRNAs: Tying it all together. *Int J Biochem Cell Biol* 41:87–95.

Håberg, S., S. London, H. Stigum, P. Nafstad, and W. Nystad. 2009. Folic acid supplements in pregnancy and early childhood respiratory health. *Arch Dis Child* 94:180–4.

Haig, D. 1992. Genomic imprinting and the theory of parent–offspring conflict. *Semin Dev Biol* 3:153–60.

Haig, D., and M. Westoby. 2006. An earlier formulation of the genetic conflict hypothesis of genomic imprinting. *Nat Genet* 38:271.

Hales, C. N., and D. J. Barker. 1992. Type 2 (non-insulin-dependent) diabetes mellitus: The thrifty phenotype hypothesis. *Diabetologia* 35:595–601.

Heijmans, B. T., E. W. Tobi, A. D. Stein, H. Putter, G. J. Blauw, E. S. Susser, P. E. Slagboom, and L. H. Lumey. 2008. Persistent epigenetic differences associated with prenatal exposure to famine in humans. *Proc Natl Acad Sci USA* 105:17046–9.

Heim, C., M. J. Owens, P. M. Plotsky, and C. B. Nemeroff. 1997. The role of early adverse life events in the etiology of depression and posttraumatic stress disorder—focus on corticotropin-releasing factor. *Ann N Y Acad Sci* 821:194–207.

Hirasawa, R., H. Chiba, M. Kaneda, S. Tajima, E. Li, R. Jaenisch, and H. Sasaki. 2008. Maternal and zygotic Dnmt1 are necessary and sufficient for the maintenance of DNA methylation imprints during preimplantation development. *Genes Dev* 22:1607–16.

Hitchins, M., J. Wong, G. Suthers, C. Suter, D. Martin, N. Hawkins, and R. Ward. 2007. Inheritance of a cancer-associated MLH1 germ-line epimutation. *N Engl J Med* 356:697–705.

Hollingsworth, J. W., S. Maruoka, K. Boon, S. Garantziotis, Z. Li, J. Tomfohr, N Bailey, et al. 2008. In utero supplementation with methyl donors enhances allergic airway disease in mice. *J Clin Invest* 118:3462–9.

Hopwood, N. 2007. A history of normal plates, tables and stages in vertebrate embryology. *Int J Dev Biol* 51:1–26.

Hopwood, N. 2009. Embryology. In *The Modern Biological and Earth Sciences*. Vol. 6 of *The Cambridge History of Science*, ed. P. J. Bowler and J. V. Pickstone, 285–315. Cambridge: Cambridge University Press.

Horsthemke, B., and K. Buiting. 2008. Genomic imprinting and imprinting defects in humans. *Adv Genet* 61:225–46.

Horton, T. H. 2005. Fetal origins of developmental plasticity: Animal models of induced life history variation. *Am J Hum Biol* 17:34–43.

Ibáñez, L., N. Potau, G. Enriquez, M. V. Marcos, and F. DeZegher. 2003. Hypergonadotrophinaemia with reduced uterine and ovarian size in women born small-for-gestational-age. *Hum Reprod* 18:1565–9.

Ishii, T., Y. Makita, A. Ogawa, S. Amamiya, M. Yamamoto, A. Miyamoto, and J. Oki. 2001. The role of

different X-inactivation pattern on the variable clinical phenotype with Rett syndrome. *Brain Dev* 23:S161–4.

Ito, K., M. Ito, W. M. Elliott, B. Cosio, G. Caramori, O. M. Kon, A. Barczyk, et al. 2005. Decreased histone deacetylase activity in chronic obstructive pulmonary disease. *N Engl J Med* 352:1967–76.

Jenuwein, T., and C. D. Allis. 2001. Translating the histone code. *Science* 293:1074–80.

Jin, B., Q. Tao, J. Peng, H. M. Soo, W. Wu, J. Ying, C. R. Fields, et al. 2008. DNA methyltransferase 3B (DNMT3B) mutations in ICF syndrome lead to altered epigenetic modifications and aberrant expression of genes regulating development, neurogenesis and immune function. *Hum Mol Genet* 17:690–709.

Karpf, A. R., and S. Matsui. 2005. Genetic disruption of cytosine DNA methyltransferase enzymes induces chromosomal instability in human cancer cells. *Cancer Res* 65:8635–9.

Knudson, A. G. 1996. Hereditary cancer: Two hits revisited. *J Cancer Res Clin Oncol* 122:135–40.

Kucharski, R., J. Maleszka, S. Foret, and R. Maleszka. 2008. Nutritional control of reproductive status in honeybees via DNA methylation. *Science* 319:1827–30.

Kuramochi-Miyagawa, S., T. Watanabe, K. Gotoh, Y. Totoki, A. Toyoda, M. Ikawa, N. Asada, et al. 2008. DNA methylation of retrotransposon genes is regulated by Piwi family members MILI and MIWI2 in murine fetal testes. *Genes Dev* 22:908–17.

Kuzawa, C. W., and L. S. Adair. 2003. Lipid profiles in adolescent Filipinos: Relation to birth weight and maternal energy status during pregnancy. *Am J Clin Nutr* 77:960–6.

Laurent, L., E. Wong, G. Li, T. Huynh, A. Tsirigos, C. T. Ong, H. M. Low, et al. 2010. Dynamic changes in the human methylome during differentiation. *Genome Res* 20:320–31.

Leroi, A. M. 2003. Mutants: On the Form, Varieties and Errors of the Human Body. London: HarperCollins.

Li, M., C. Marin-Muller, U. Bharadwaj, K.-H. Chow, Q. Yao, and C. Chen. 2008. MicroRNAs: Control and loss of control in human physiology and disease. *World J Surg* 33:667–84.

Lillycrop, K. A., E. S. Phillips, C. Torrens, M. A. Hanson, A. A. Jackson, and G. C. Burdge. 2008. Feeding pregnant rats a protein-restricted diet persistently alters the methylation of specific cytosines in the hepatic PPARα promoter of the offspring. *Br J Nutr* 100:278–82.

Lillycrop, K. A., J. L. Slater-Jefferies, M. A. Hanson, K. M. Godfrey, A. A. Jackson, and G. C. Burdge.

2007. Induction of altered epigenetic regulation of the hepatic glucocorticoid receptor in the offspring of rats fed a protein-restricted diet during pregnancy suggests that reduced DNA methyltransferase-1 expression is involved in impaired DNA methylation and changes in histone modifications. *Br J Nutr* 97:1064–73.

Lissau, I., and T. I. A. Sorensen. 1994. Parental neglect during childhood and increased risk of obesity in young adulthood. *Lancet* 343:324–7.

Lister, R., M. Pelizzola, R. H. Dowen, R. D. Hawkins, G. Hon, J. Tonti-Filippini, J. R. Nery, et al. 2009. Human DNA methylomes at base resolution show widespread epigenomic differences. *Nature* 462:315–22.

Ma, L., and R. A. Weinberg. 2008. Micromanagers of malignancy: Role of microRNAs in regulating metastasis. *Trends Genet* 24:448–56.

Maher, B. 2008. Personal genomes: The case of the missing heritability. *Nature* 456:18–21.

Matarazzo, M. R., M. L. De Bonis, M. Vacca, F. Della Ragione, and M. D'Esposito. 2009. Lessons from two human chromatin diseases, ICF syndrome and Rett syndrome. *Int J Biochem Cell Biol* 41:117–26.

McGowan, P. O., A. Sasaki, A. C. D'Alessio, S. Dymov, B. Labonté, M. Szyf, G. Turecki, and M. J. Meaney. 2009. Epigenetic regulation of the glucocorticoid receptor in human brain associates with childhood abuse. *Nat Neurosci* 12:342–8.

Meaney, M. J. 2001. Maternal care, gene expression, and the transmission of individual differences in stress reactivity across generations. *Annu Rev Neurosci* 24:1161–92.

Meaney, M. J., D. H. Aitken, V. Viau, S. Sharma, and A. Sarrieau. 1989. Neonatal handling alters adrenocortical negative feedback sensitivity and hippocampal type II glucocorticoid receptor binding in the rat. *Neuroendocrinology* 50:597–604.

Mericq, V., K. K. Ong, R. Bazaes, V. Peña, A. Avila, T. Salazar, N. Soto, G. Iñiguez, and D. B. Dunger. 2005. Longitudinal changes in insulin sensitivity and secretion from birth to age three years in small- and appropriate-for-gestational-age children. *Diabetologia* 48:2609–14.

Métivier, R., R. Gallais, C. Tiffoche, C. Le Peron, R. Z. Jurkowska, R. P. Carmouche, D. Ibberson, et al. 2008. Cyclical DNA methylation of a transcriptionally active promoter. *Nature* 452:45–50.

Mohd-Sarip, A., and C. P. Verrijzer. 2004. A higher order of silence. *Science* 306:1484–5.

Morgan, H. D., F. Santos, K. Green, W. Dean, and W. Reik. 2005. Epigenetic reprogramming in mammals. *Hum Mol Genet* 14:R47–58.

Morison, I. M., J. P. Ramsay, and H. G. Spencer. 2005. A census of mammalian imprinting. *Trends Genet* 21:457–65.

Mund, C., B. Brueckner, and F. Lyko. 2006. Reactivation of epigenetically silenced genes by DNA methyltransferase inhibitors: Basic concepts and clinical applications. *Epigenetics* 1:7–13.

Oakley, A. 1984. The Captured Womb: A History of the Medical Care of Pregnant Women. Oxford: Basil Blackwell.

Oberlander, T. F., J. Weinberg, M. Papsdorf, R. Grunau, S. Misri, and A. M. Devlin. 2008. Prenatal exposure to maternal depression, neonatal methylation of human glucocorticoid receptor gene (*NR3C1*) and infant cortisol stress responses. *Epigenetics* 3:97–106.

Ooi, S. K., and T. H. Bestor. 2008. The colorful history of active DNA demethylation. *Cell* 133:1145–8.

Painter, R. C., C. Osmond, P. Gluckman, M. Hanson, D. I. W. Phillips, and T. J. Roseboom. 2008. Transgenerational effects of prenatal exposure to the Dutch famine on neonatal adiposity and health in later life. *BJOG* 115:1243–9.

Painter, R. C., T. J. Roseboom, and O. P. Bleker. 2005. Prenatal exposure to the Dutch famine and disease in later life: An overview. *Reprod Toxicol* 20:345–52.

Parent, A.-S., G. Teilmann, A. Juul, N. Skakkebaer, J. Toppari, and J.-P. Bourguignon. 2003. The timing of normal puberty and the age limits of sexual precocity: Variations around the world, secular trends, and changes after migration. *Endocr Rev* 244:668–93.

Park, J. H., D. A. Stoffers, R. D. Nicholls, and R. A. Simmons. 2008. Development of type 2 diabetes following intrauterine growth retardation in rats is associated with progressive epigenetic silencing of Pdx1. *J Clin Invest* 118:2316–24.

Pembrey, M. E., L. O. Bygren, G. Kaati, S. Edvinsson, K. Northstone, M. Sjöström, J. Golding, and the ALSPAC Study Team. 2006. Sex-specific, male-line transgenerational responses in humans. *Eur J Hum Genet* 14:159–66.

Perera, F., W.-Y. Tang, J. Herbstman, D. Tang, L. Levin, R. Miller, and S.-M. Ho. 2009. Relation of DNA methylation of 5'-CpG island of *ACSL3* to transplacental exposure to airborne polycyclic aromatic hydrocarbons and childhood asthma. *PLoS One* 4:e4488.

Pham, T. D., N. K. MacLennan, C. T. Chiu, G. S. Laksana, J. L. Hsu, and R. H. Lane. 2003. Uteroplacental insufficiency increases apoptosis and alters *p53* gene methylation in the full-term IUGR rat kidney. *Am J Physiol Regul Integr Comp Physiol* 285:R962–70.

Pinto-Correia, C. 1997. *The Ovary of Eve: Egg and Sperm and Preformation*. Chicago: Chicago University Press.

Popp, C., W. Dean, S. Feng, S. J. Cokus, S. Andrews, M. Pellegrini, S. E. Jacobsen, and W. Reik. 2010. Genome-wide erasure of DNA methylation in mouse primordial germ cells is affected by AID deficiency. *Nature* 463:1101–5.

Poulter, M. O., L. S. Du, I. C. G. Weaver, M. Palkovits, G. Faludi, Z. Merali, M. Szyf, and H. Anisman. 2008 GABA$_A$ receptor promoter hypermethylation in suicide brain: Implications for the involvement of epigenetic processes. *Biol Psychiatry* 64:645–52.

Prescott, S. L., and V. Clifton. 2009. Asthma and pregnancy: Emerging evidence of epigenetic interaction in utero. *Curr Opin Allergy Clin Immunol* 9:417–26.

Rakyan, V. K., S. Chong, M. E. Champ, P. C. Cuthbert, H. D. Morgan, K. V. K. Luu, and E. Whitelaw. 2003. Transgenerational inheritance of epigenetic states at the murine *Axin^{Fu}* allele occurs after maternal and paternal transmission. *Proc Natl Acad Sci USA* 100:2538–43.

Rassoulzadegan, M., V. Grandjean, P. Gounon, S. Vincent, I. Gillot, and F. Cuzin. 2006. RNA-mediated non-Mendelian inheritance of an epigenetic change in the mouse. *Nature* 441:469–74.

Reik, W. 2007. Stability and flexibility of epigenetic gene regulation in mammalian development. *Nature* 447:425–32.

Reik, W., and J. Walter. 2001. Genomic imprinting: Parental influence on the genome. *Nat Rev Genet* 2:21–32.

Roe, S. 1981. *Matter, Life and Generation: Eighteenth-Century Embryology and the Haller-Wolff Debate*. Cambridge: Cambridge University Press.

Rönn, T., P. Poulsen, O. Hansson, J. Holmkvist, P. Almgren, P. Nilsson, T. Tuomi, et al. 2008. Age influences DNA methylation and gene expression of *COX7A1* in human skeletal muscle. *Diabetologia* 51:1159–68.

Rosa, A., M. Picchioni, S. Kalidindi, C. Loat, J. Knight, T. Toulopoulou, R. Vonk, et al. 2008. Differential methylation of the X-chromosome is a possible source of discordance for bipolar disorder female monozygotic twins. *Am J Med Genet B Neuropsychiatr Genet* 147:459–62.

Samaco, R. C., J. D. Fryer, J. Ren, S. Fyffe, H. T. Chao, Y. Sun, J. J. Greer, H. Y. Zoghbi, and J. L. Neul. 2008. A partial loss of function allele of methyl-CpG-binding protein 2 predicts a human neurodevelopmental syndrome. *Hum Mol Genet* 17:1718–27.

Schuettengruber, B., D. Chourrout, M. Vervoort, B. Leblanc, and G. Cavalli. 2007. Genome regulation by polycomb and trithorax proteins. *Cell* 128:735–45.

Sinclair, K. D., C. Allegrucci, R. Singh, D. S. Gardner, S. Sebastian, J. Bispham, A. Thurston, et al. 2007. DNA methylation, insulin resistance, and blood pressure in offspring determined by maternal periconceptional B vitamin and methionine status. *Proc Natl Acad Sci USA* 104:19351–6.

Singhal, A. 2006. Early nutrition and long-term cardiovascular health. *Nutr Rev* 64:S44–9.

Sloboda, D. M., R. Hart, D. A. Doherty, C. E. Pennell, and M. Hickey. 2007. Age at menarche: Influences of prenatal and postnatal growth. *J Clin Endocrinol Metab* 92:46–50.

Sloboda, D. M., G. J. Howie, A. Pleasants, P. D. Gluckman, and M. H. Vickers. 2009. Pre- and postnatal nutritional histories influence reproductive maturation and ovarian function in the rat. *PLoS One* 4:e6744.

Smith, A. E., P. J. Hurd, A. J. Bannister, T. Kouzarides, and K. G. Ford. 2008. Heritable gene repression through the action of a directed DNA methyltransferase at a chromosomal locus. *J Biol Chem* 283:9878–85.

Smith, F. M., A. S. Garfield, and A. Ward. 2006. Regulation of growth and metabolism by imprinted genes. *Cytogenet Genome Res* 113:279–91.

Steegers-Theunissen, R. P., S. A. Obermann-Borst, D. Kremer, J. Lindemans, C. Siebel, E. A. Steegers, P. E. Slagboom, and B. T. Heijmans. 2009. Periconceptional maternal folic acid use of 400 μg per day is related to increased methylation of the *IGF2* gene in the very young child. *PLoS One* 4:e7845.

Teodoridis, J. M., C. Hardie, and R. Brown. 2008. CpG island methylator phenotype (CIMP) in cancer: Causes and implications. *Cancer Lett* 268:177–86.

Tobi, E. W., L. H. Lumey, R. P. Talens, D. Kremer, H. Putter, A. D. Stein, P. E. Slagboom, and B. T. Heijmans. 2009. DNA methylation differences after exposure to prenatal famine are common and timing- and sex-specific. *Hum Mol Genet* 18:4046–53.

Trivers, R. L. 1974. Parent–offspring conflict. *Am Zool* 14:249–64.

Tudor, M., S. Akbarian, R. Z. Chen, and R. Jaenisch. 2002. Transcriptional profiling of a mouse model for Rett syndrome reveals subtle transcriptional changes in the brain. *Proc Natl Acad Sci USA* 99:15536–41.

Ubeda, F., and J. F. Wilkins. 2008. Imprinted genes and human disease. *Adv Exp Med Biol* 626:101–15.

Varmuza, S., and M. Mann. 1994. Genomic imprinting—defusing the ovarian time bomb. *Trends Genet* 10:118–23.

Vickers, M. H., P. D. Gluckman, A. H. Coveny, P. L. Hofman, W. S. Cutfield, A. Gertler, B. H. Breier, and M. Harris. 2005. Neonatal leptin treatment reverses developmental programming. *Endocrinology* 146:4211–16.

Wagner, K. D., N. Wagner, H. Ghanbarian, V. Grandjean, P. Gounon, F. Cuzin, and M. Rassoulzadegan. 2008. RNA induction and inheritance of epigenetic cardiac hypertrophy in the mouse. *Dev Cell* 14:962–9.

Waterland, R. A., D. C. Dolinoy, J.-R. Lin, C. A. Smith, X. Shi, and K. G. Tahiliani. 2006a. Maternal methyl supplements increase offspring DNA methylation at *Axin Fused*. *Genesis* 44:401–6.

Waterland, R. A., and R. L. Jirtle. 2003. Transposable elements: Targets for early nutritional effects on epigenetic gene regulation. *Mol Cell Biol* 23:5293–300.

Waterland, R. A., J. R. Lin, C. A. Smith, and R. L. Jirtle. 2006b. Post-weaning diet affects genomic imprinting at the insulin-like growth factor 2 (Igf2) locus. *Hum Mol Genet* 15:705–16.

Waterland, R. A., and K. B. Michels. 2007. Epigenetic epidemiology of the developmental origins hypothesis. *Annu Rev Nutr* 27:363–88.

Waterland, R. A., M. Travisano, and K. G. Tahiliani. 2007. Diet-induced hypermethylation at *agouti viable yellow* is not inherited transgenerationally through the female. *FASEB J* 21:3380–5.

Watkins, A. J., D. Platt, T. Papenbrock, A. Wilkins, J. J. Eckert, W. Y. Kwong, C. Osmond, M. Hanson, and T. P. Fleming. 2007. Mouse embryo culture induces changes in postnatal phenotype including raised systolic blood pressure. *Proc Natl Acad Sci USA* 104:5449–54.

Watkins, A. J., A. Wilkins, C. Cunningham, V. H. Perry, M. J. Seet, C. Osmond, J. J. Eckert, et al. 2008. Low protein diet fed exclusively during mouse oocyte maturation leads to behavioural and cardiovascular abnormalities in offspring. *J Physiol* 586:2231–44.

Weaver, I. C. G. 2007. Epigenetic programming by maternal behavior and pharmacological intervention—nature versus nurture: Let's call the whole thing off. *Epigenetics* 2:22–8.

Weaver, I. C. G., N. Cervoni, F. A. Champagne, A. C. D'Alessio, S. Sharma, J. R. Seckl, S. Dymov, M. Szyf, and M. J. Meaney. 2004. Epigenetic programming by maternal behavior. *Nat Neurosci* 7:847–54.

Weaver, I. C. G., F. A. Champagne, S. E. Brown, S. Dymov, S. Sharma, M. J. Meaney, and M. Szyf. 2005. Reversal of maternal programming of stress responses in adult offspring through methyl supplementation: Altering epigenetic marking later in life. *J Neurosci* 25:11045–54.

Weksberg, R., C. Shuman, O. Caluseriu, A. C. Smith, Y. L. Fei, J. Nishikawa, T. Stockley, et al. 2002. Discordant KCNQ1OT1 imprinting in sets of monozygotic twins discordant for Beckwith-Wiedemann syndrome. *Hum Mol Genet* 11:1317–25.

Wilkins, J. F., and D. Haig. 2003. What good is genomic imprinting: The function of parent-specific gene expression. *Nat Rev Genet* 4:1–10.

Wolf, J. B., and R. Hager. 2006. A maternal–offspring coadaptation theory for the evolution of genomic imprinting. *PLoS Biol* 4:2238–43.

Wu, H., Y. Chen, J. Liang, B. Shi, G. Wu, Y. Zhang, D. Wang, et al. 2005. Hypomethylation-linked activation of PAX2 mediates tamoxifen-stimulated endometrial carcinogenesis. *Nature* 438: 981–7.

Yajnik, C. S. 2004. Early life origins of insulin resistance and type 2 diabetes in India and other Asian countries. *J Nutr* 134:205–10.

Yajnik, C. S., S. S. Deshpande, A. A. Jackson, H. Refsum, S. Rao, D. J. Fisher, D. S. Bhat, et al. 2008. Vitamin B_{12} and folate concentrations during pregnancy and insulin resistance in the offspring: The Pune Maternal Nutrition Study. *Diabetologia* 51:29–38.

Yehuda, R., S. M. Engel, S. R. Brand, J. Seckl, S. M. Marcus, and G. S. Berkowitz. 2005. Transgenerational effects of posttraumatic stress disorder in babies of mothers exposed to the world trade center attacks during pregnancy. *J Clin Endocrinol Metab* 90:4115–8.

Yehuda, R., M. H. Teicher, J. R. Seckl, R. A. Grossman, A. Morris, and L. M. Bierer. 2007. Parental posttraumatic stress disorder as a vulnerability factor for low cortisol trait in offspring of Holocaust survivors. *Arch Gen Psychiatry* 64: 1040–8.

Zhou, Z., E. J. Hong, S. Cohen, W. Zhao, H. y. H. Ho, L. Schmidt, W. G. Chen, et al. (2006) Brain-specific phosphorylation of MeCP2 regulates activity-dependent Bdnf transcription, dendritic growth, and spine maturation. *Neuron* 52:255–69.

Zogel, C., S. Bohringer, S. Grosz, R. Varon, K. Buiting, and B. Horsthemke. 2006. Identification of cis- and trans-acting factors possibly modifying the risk of epimutations on chromosome 15. *Eur J Hum Genet* 14:752–8.

Zuccato, C., and E. Cattaneo. 2007. Role of brain-derived neurotrophic factor in Huntington's disease. *Prog Neurobiol* 81:294–330.

23

Epigenetics: The Context of Development

Benedikt Hallgrímsson and Brian K. Hall

CONTENTS

The term *epigenetics* was coined by Conrad Hal Waddington in 1957 as a merger of *epigenesis* with *genetics*. Waddington (1957) defined *epigenetics* as the causal control of development or the causal control of gene action, without reference to specific mechanisms. As implied by the prefix *epi*, his intent was to convey mechanisms acting above the gene level. His metaphor of the epigenetic landscape provided a framework for conceptualizing how such mechanisms operate to bridge genotype and phenotype. At the time, the molecular basis for genetics and develop-ment was very poorly understood. For this rea-son, Waddington's original concept of epi-genetics corresponds to what is now a wide and heterogeneous range of mechanisms that oper-ate at various levels of development. In the ma-jority of current contexts, *epigenetics* has a much narrower meaning than Waddington intended, referring to mechanisms that regulate (often by suppressing) transcription. At the same time, the term is used to refer to other kinds of mech-anisms such as physical interactions during de-velopment (Newman and Muller, 2000; New-man et al., 2006) and the morphological or spatiotemporal context of development (see Chapter 7).

The different senses in which the epigenetics concept is applied in modern biology seem on the surface to have little in common, yet they are largely consistent with Waddington's origi-nal intent. What they have in common, beyond a general sense of invoking mechanisms that are above the level of the gene, is an emphasis on the developmental or molecular context within which genes function. Thus, this empha-sis on context is a common thread, whether the context in which DNA is modified (*epigenetic modification*) or the context in which cells, tissues, or organs develop, function, and interact in space and time (*epigenetic interaction*). The central thesis implicit in this edited volume is that these two different conceptualizations of epigenetics spring from common theoretical ground and that, taken together, both enrich and broaden our view of development, evolution, and disease.

The study of epigenetic modification and epigenetic interactions reminds us that in multicellular organisms there is never a one-to-one coupling of genotype and phenotype. The origination and modification of morphology are not solely the products of heritable genetic mutations but a dynamic interplay between many players. Changes during development are fundamental to evolutionary change and novelty. Many changes are based on epigenetic mechanisms. In this chapter, we enumerate what is meant by epigenetics at various levels of the biological hierarchy to show the threads that link the various approaches and topics discussed in this volume.

EPIGENETIC MODIFICATION · *Epigenetics at the Gene Level*

TYPES OF CHROMATIN MODIFICATION: METHYLATION AND IMPRINTING

Gene transcription can be, and often is, controlled epigenetically. One important type of epigenetic transcriptional regulation is the *methylation* (addition of a methyl group) of cytosine (C) residues of DNA, which usually restricts gene activity in a tissue-specific manner (Ohlsson et al., 1995; Ohlsson, 1999). Methylation of regulatory regions of genes, for example, of the promoter, prevents regulatory factors from binding to the gene. In general, the greater the methylation, the greater the gene repression or the lower the transcriptional activity. Methylation has been shown to be related to polymorphisms for flower structures in land plants (Cubas et al., 1999). The first methylation map was published by Zhang et al. (2006). It is for the watercress, *Arabidopsis thaliana*, in which widespread methylation was mapped.

Because methylated C residues usually lie adjacent to a G residue, methylation is maintained when cells divide; enzymatic recognition of methylated C adjacent to G on one strand of DNA allows methylation of the C residue on the second strand. Methylation also regulates imprinted genes (see below). Hypoacetylation of histone and silencing of RNA are two other means by which genomes are imprinted, the term *imprinting* referring to those classes of epigenetic control by which genomes are differentially regulated in different cells, tissues, or organs.

Imprinting (genomic or parental imprinting) refers to the situation in which the activation of genes received from the male or female parent differs in the offspring. One allele (e.g., that from the female) is active; the other (e.g., from the male) is silenced permanently. Active alleles are often hypomethylated (see above); silenced alleles are usually hypermethylated. Mouse zygotes constructed so as to contain either two male or two female nuclei do not develop normally because normal embryonic development requires differentially imprinted genes.

Methylation and imprinting are epigenetic because both mechanisms involve modification beyond the level of the primary DNA sequences that comprise the genes. As forms of epigenetic modification, they stand in contrast to epigenetic interactions by which cells, tissues, and organs regulate the activity of other cells, tissues, or organs (see Epigenetic Interactions at the Cellular Level and Epigenetic Interactions at the Tissue Level, below). Epigenetic modification (the modern use of the term *epigenetics* for changes that modify the function of DNA) and epigenetic interactions (the older Waddingtonian usage of *epigenetics* for the causal control of development) are both epigenetic because they modify function in ways that cannot be predicted from the nucleotide sequence of the gene (methylation, imprinting) or from spatial or temporal relationships between cells, tissues, or organs during development (e.g., the juxtaposition of lens and overlying epidermis during eye development in vertebrates is insufficient to predict that the lens will induce the epidermis to transform into the transparent cornea).

EPIGENETIC INTERACTIONS AT THE CELLULAR LEVEL

A cell's environment, location, and surroundings also provide epigenetic factors that influence the cell's identity and activities. Osteoblasts remain at the osteoblast stage of differentiation until the bony matrix they secrete surrounds them and they become entombed, after which they become osteocytes (Franz-Odendaal et al., 2006).

Often, the role of the extracellular matrix can be resolved into individually important components, as is true for collagen in bone matrix. Collagen, the most abundant protein of bone, self-assembles into woven plywood–like assemblages that provide resistance to mechanical stresses. Although cells secrete the collagen, individual molecules self-assemble because of intrinsic physical properties and because of external factors such as pH or mechanical stress and/or interactions with forming collagen that are epigenetic. Once assembled, collagen becomes an epigenetic factor influencing osteogenic cell activity and function. This was shown, for example, through the work of Harris and others on the alignment of fibroblasts along collagen fibers in culture and their reciprocal effect of fibroblasts on collagen (Harris et al., 1980, 1981, 1984; Stopak and Harris, 1982).

The extracellular matrix of cartilage similarly exerts influences on the activity of chondrocytes and/or on the overall growth of cartilage via physical influences. The *Brachymorph* mouse mutant, for instance, has an autosomal recessive mutation in the phosphoadenosine-phosphosulfate synthetase 2 gene that results in undersulfation of glycosaminoglycans in the cartilage matrix, producing instability of the cartilaginous matrix, which in turn produces disorganized growth plates that exhibit reduced growth. In some cases, the resulting degenera-tion of the matrix results in premature onset of ossification and premature fusion of the growth plate or synchondrosis, producing a dramatic change in bone growth. Reduced bone length and altered craniofacial shape are among the major phenotypic effects of the brachymorph mutation (Ford-Hutchinson et al., 2005a, 2005b; Hallgrímsson et al., 2006). However, this effect is epigenetic in the sense that it emerges from the physical interaction of the extracellular matrix and the proliferating chondrocytes rather than from a direct influence on cartilage growth. The effect can be understood only in the context of the physical interaction between chondrocytes and the molecular constituents of their extracellular matrix.

Epigenetic interactions can also take place *within* cells. Heterogeneity in gene protein expression among genetically identical cells, for example, is well known and has been proposed to be a source of stochastic variation in developmental systems (Kaern et al., 2005). Much of this variation is due to short-term temporal fluctuations that manifest as cell-level variation in a cell population frozen at some particular sampling time. However, Huang and others have shown that some of this variation is not due to such fluctuations, is often bi- or multimodal, and appears to be stable (Brock et al., 2009; Huang, 2009a). They argue that at least some of this variation reflects the presence of multiple stable states that arise from the dynamics of complex cell-regulatory networks with measures of phenotypic variation in cells (such as patterns of gene or protein expression) forming peaks with variation around these multiple stable states (Brock et al., 2009; Huang, 2009a, 2009b). In this line of thinking, Huang places one of Waddington's original descriptions of epigenetics in a modern systems biology context. In *The Strategy of the Genes*, Waddington follows a short introduction to catastrophe theory with the argument that cell differentiation in development can be seen as a process in which cells are steered along one of multiple canalized paths to alternative end states (Waddington, 1957, 13–29).

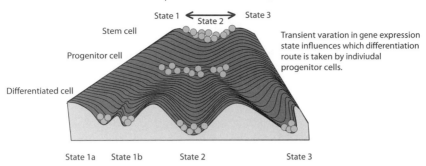

Transient variation in gene expression state represents noise within attractor basin

State 1 ←→ State 3
State 2

Stem cell

Transient varation in gene expression state influences which differentiation route is taken by indiviudal progenitor cells.

Progenitor cell

Differentiated cell

State 1a   State 1b   State 2   State 3

FIGURE 23.1 Simplified version of Figure 3 from Huang (2003, 8), illustrating how epigenetic noise in gene-expression states in cells might impact cell-differentiation pathways. Here, cell differentiation is depicted as a series of probabilistic steps superimposed on the epigenetic landscape. Gene-expression state is driven by some underlying gene-regulatory network which exhibits oscillatory activity between multiple metastable states. A network is not shown here, but this is illustrated in Huang (2003). In the case illustrated here, there are three states, but that is an arbitrary choice for the purpose of illustration. At each time point, there is transient variation or noise in the gene-expression state of the cells. At the stem-cell stage, this noise can influence which progenitor cell groove (or attractor) the cell moves toward on the landscape. The same scenario is played out at the second differentiation step. The landscape itself represents the degree of instability (or stochastic variation) in expression states through the steepness of the sides of the grooves or the sharpness of the cusps that separate them. The epigenetic landscape is thus a probabilistic metaphor for thinking about cell differentiation.

Huang (2009b) has argued recently that cell types evolved stable gene-expression patterns. The stability of these patterns and the instability of the patterns in between these stable states are emergent in the sense that they make sense only in light of interactions among elements of the gene-regulatory network. In this sense, these states are epigenetic (Figure 23.1). This controversial idea turns the explanatory role of chromatin modification in cell differentiation upside down. Instead of being the primary cause of differentiation, chromatin modification occurs as a secondary consequence of the overall "network state" and functions to stabilize a gene-expression pattern that is already in place (Huang, 2009b).

EPIGENETIC INTERACTIONS AT THE TISSUE LEVEL

Embryonic inductions, usually between epithelial and mesenchymal cells, are a common form of epigenetic interaction in developing vertebrate embryos. Induction requires both a group of cells or a tissue to produce a signal and the ability (competence) of responding cells or tissues to receive and respond to the inductive signal. Embryonic cells may be competent to respond to multiple signals, their spatial and temporal context determining the pathway of differentiation. Optic cups growing out from the developing forebrain induce overlying ectoderm to form a lens. However, placing an optic cup elsewhere in the embryo can elicit lens formation in that ectopic location but only during a defined window of developmental time (Hall and Miyake, 1997). The concept of a developmental window of timing applies equally to the ability of the inductive cells to produce their signal and to the competent cells to respond; induction and competence have start and end points. Consequently, embryos develop in an integrated and predictable way, with variation around the norm essentially set by the duration and strength of the inductive cell or tissue interactions, i.e., set epigenetically.

Self-organization, or the emergence of order within developing tissues, is a topic that is attracting increasing interest. Epithelial cells, for instance, take on characteristic hexagonal shapes as and form tissue-level morphology simply as a consequence of cell proliferation (Gibson et al., 2006). Gibson et al. (2006) have argued that this tendency for self-organization is subsumed in developmental processes related to pattern formation and epithelial histogenesis in diverse contexts. This mechanical influence of cell division is an emergent property in that it is explicable only through understanding how individual cell behavior influences histogenesis through interaction with other cells doing the same thing. The discussion of pigment patterns by Olsson in this volume (Chapter 10) provides other examples of related phenomena.

As cells and tissues develop, interactions between adjacent tissues are initiated. Among the better-studied examples are the interactions that occur between the developing muscular and skeletal systems. This topic is discussed more fully by Herring (Chapter 13) and by Kablar (Chapter 15) in this volume. These interactions can be investigated experimentally in chicken embryos, which can be paralyzed pharmacologically without stopping development (Hall and Herring, 1990; Hall, 2005). Similarly, individual components (modules) of the mammalian dentary bone that constitutes the bony skeleton of the lower jaw respond to epigenetic interactions from different tissues. Areas of the dentary associated with the teeth respond to movement associated with the developing teeth; the posterior process of the dentary (condylar, coronoids, and angular), onto which muscles insert, responds to muscle action in ways that are specific to sets of muscles and the processes with which they interact (Atchley and Hall, 1991; Atchley, 1993).

Mechanical effects on development within tissues are another group of epigenetic mechanisms. Examples of such processes are discussed by Larsen and Atallah (Chapter 7) in this

volume. An interesting, albeit poorly established, example of this is the potential role of mechanical factors in determining organ size. Hufnagel et al. (2007), for example, have proposed that wing size in flies may be determined by an interaction between morphogen gradients and compressive stress within the imaginal disk. The stress factor, in this case, is emergent in that it arises from the anatomical context of cell proliferation within the imaginal disk.

During embryonic development, growing tissues and organs can also alter the conditions of adjacent structures. As the neural tube grows in the vertebrate embryo, for instance, adjacent tissues are displaced and altered in shape, affecting the relative positioning of the neural crest, for instance. Alteration of adjacent tissues in shape and position, in turn, can influence mechanisms such as morphogen gradients that are dependent on specific three-dimensional morphology (Jaeger et al., 2008). Larsen, in this volume and elsewhere (Larsen et al., 1996), has argued, for instance, that morphogenetic fields must be understood within their morphogenetic spatiotemporal context. In this way, the morphogenetic context exerts an emergent effect on developmental–genetic processes.

EPIGENETIC INTERACTIONS AT THE ORGAN LEVEL

Epigenetic interactions also occur at higher levels in organismal development. In vertebrates, for example, the growing brain has long been proposed to influence the development of the skull, presumably through mechanical interactions. This and related topics are discussed in Chapter 16 by Lieberman. De Beer (1937) argued, for example, that the relative size of the brain was a major determinant of overall cranial morphology and, in particular, of the basicranial angle. Extending this idea to human evolution, Biegert (1963) argued that as the hominid brain expanded relative to the length of the cranial base, the basicranial angle became more acute as the expanded neurocranium was forced to basically wrap around the relatively smaller cranial base. There is good comparative evidence in primates to support this claim (Ross and Ravosa, 1993; Ross and Henneberg, 1995).

For human evolution, this argument is profoundly important because it bears on the need to generate an independent explanation for the unusual position of the human face, which results from the highly flexed cranial base. With Dan Lieberman, one of us (Hallgrímsson) has tested this idea in mice using mutations which alter brain size or cranial base size independently (Hallgrímsson and Lieberman, 2008; Lieberman et al., 2008). This work shows that changes in brain size in mice produce the expected alteration of the cranial base angle. Equally interesting is the result that the size of the brain relative to the size of the cranial base explains a very large proportion (31%) of the total shape variation in a large sample of mice from several inbred strains with various and unrelated mutations (Hallgrímsson and Lieberman, 2008; Lieberman et al., 2008) (Figure 23.2).

A related example is the interaction between the chondrocranium and dermatocranium across the vertebrates. Pointing out that achondroplastic dwarves in different mammalian species exhibit qualitatively similar phenotypes, De Beer (1937) argued that this interaction was a major determinant of overall craniofacial shape in mammals. Analyses of mouse mutants support this idea (Hallgrímsson and Lieberman, 2008). As Trish Parsons (unpublished data) has shown in her dissertation work, mice which exhibit reduced chondrocranial growth via very different developmental mechanisms, for instance, share a similar phenotype, which appears to result from the mechanical interaction between the growth of the chondrocranium and the rest of the skull (Figure 23.3). One of these, the brachymorph phenotype, is discussed by Hallgrímsson et al. (2006), while the other, the SectRNA mouse, is discussed in general terms by Downey et al. (2009) and a further analysis

of these two strains is presented in Hallgrímsson et al. (2009).

EPIGENETIC INTERACTIONS AT THE INDIVIDUAL LEVEL

Epigenetic interactions link ecology to development and to evolution and, therefore, provide mechanisms underpinning phenotypic plasticity and life-history strategies. Indeed, the same mechanisms that operate between cells or tissues *within* embryos operate *between* individuals, often between individuals of different species (Hall, 1992, 1999, 2004), as exemplified in the following three categories of epigenetic interactions:

Epigenetic interactions within individuals: embryonic inductions, interactions between tissues and organs leading to the origination of cell types, tissues, and organs. Included within this category are all the classic embryonic inductions through which vertebrate embryos generate the majority of their cell types and all tissues and organs. The series of interactions set in motion by contact of the optic cup with the overlying ectoderm initiating the development of the lens was one of the first described and is one of the best analyzed, both in normal development (Henry and Grainger, 1990) and in the context of eye reduction in blind cavefish. Many tissues and organs once established interact with one another, a well-studied example being the one-way interactions between the muscular and skeletal systems by which onset of differentiation, morphogenesis, and/or growth may be regulated (Hall and Herring, 1990; Atchley and Hall, 1991; Hall, 2005).

Epigenetic interactions between individuals of the same species: interactions that are fundamentally similar to embryonic inductions but in which the signal is released by one individual and acts on another individual of the same species. Density-dependent signaling among

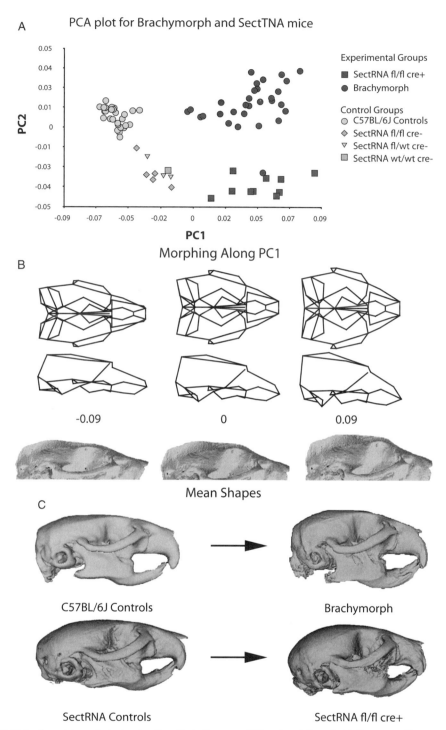

FIGURE 23.2 Morphometric analysis showing the effect of two very different mutations that influence cartilage growth in the skull. The brachymorph mutation influences the sulfation of glycosaminoglycans in cartilage, while the SectRNA mouse has a cartilage-specific knockout of the Selenocysteine tRNA Gene (*Trsp*), which is required for selenoprotein function. This produces chondronecrosis and, hence, reduction in cartilage growth during development. (A) Principal component analysis (PCA) plot of these two mutants and their controls. This plot shows that PC1 corresponds to the morphological differences caused by these mutations. (B) Morphing along PC1, both as wireframe deformations and as deformations of a 3D object map. (C) Mean shapes of the mutants and their respective controls, obtained using the rigid superimposition technique of Kristensen et al. (2008).

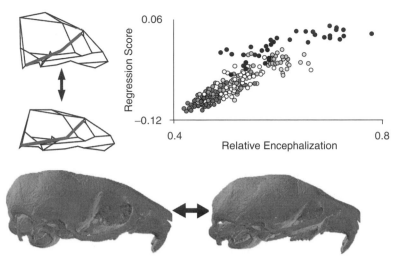

FIGURE 23.3 Multivariate regression of total craniofacial shape against brain size relative to cranial base length (relative encephalization) for 344 mice from 15 different genotypes and strains. In this sample, relative encephalization is calculated as the cube root of endocranial volume divided by cranial base length (basion to the sphenothmoidal synchondrosis to the foramen cecum). This regression explains 31% of the total variation in cranial shape in this sample.

larvae leading to polymorphism and reflecting phenotypic plasticity among amphibians has been well studied. Increase in numbers of tadpoles or decrease in food supply results in tadpoles of some species developing rapidly and producing enlarged bodies, hypertrophied jaw muscles, and aggressive behavior that transforms a plant-eating tadpole into a cannibal that devours its siblings (Collins and Cheek, 1983; Hall, 2004). Production of soldiers in colonies of the ant *Pheidole bicarinata* is another well-studied example (Nijhout and Wheeler, 1996; Nijhout, 1999; Abouheif and Wray, 2002).

Epigenetic interactions between individuals of different species: predator–prey interactions in which the predator releases a chemical signal to which eggs, embryos, or larvae of the prey respond, resulting in polymorphism, either because of enhancement of an existing structure (as in the enlargement and altered shape of the helmet of the water midge Daphnia) or because of the production of a novel structure(s). Formation of additional spines in rotifers

in response to the presence of predators is a well-studied example (Gilbert, 1966, 1980; Stemberger and Gilbert, 1987). In both cases the enlarged helmet or additional spines protect the individuals from being eaten by the predator. In both examples only some individuals respond to the signal emitted by the predator, ensuring that neither prey nor predator species is eliminated (as could occur if no prey responded or if all prey responded, respectively). Longer-term adaptive value of this phenotypic plasticity is evident in increased fecundity, shorter generation times, and/or increased life span.

These three levels of epigenetic interactions represent a single class of mechanisms linking embryonic inductions, life-history morphs and ecological adaptation, and evolutionary plasticity. All share the property of activating canalized developmental pathways to produce new embryonic, larval, or adult morphologies. All reveal the central role played by epigenetic interactions in linking genotypes and phenotypes in development and in evolution.

THE ROLE OF EPIGENETICS IN DEVELOPMENT, EVOLUTION, AND DISEASE

Epigenetics, as defined by Waddington, is about aspects in developmental systems that can be understood only by recourse to processes acting above the gene level. Emergent properties, morphological and spatiotemporal contexts, and the physics of development are all epigenetic phenomena in this sense. At some level, these phenomena apply to almost any aspect of development. All developmental processes take place in context. As Larsen and Atallah point out in this volume (Chapter 7), a fertilized egg, once cracked open and stirred in a bowl, can no longer produce an embryo, even though its entire genome is intact. Without the anatomical context of a normal intact egg, the genome is insufficient to allow development to occur.

This broad view of epigenetics has fallen out of favor in modern biology and has been replaced by the much narrower one which deals only with chromatin modification. One reason for this may be that the mechanisms that are included within the broader epigenetics concept are so diverse that they appear to capture all of development while the mechanisms of chromatin modification are clearly delineated. Another reason, however, is undoubtedly the short memory of modern biology and the unfortunate and increasing tendency to overlook older literature. This is unfortunate because, diffuse as it may be, much is lost both in our understanding of development and in its implications for evolution by throwing out Waddington's epigenetic view of development (Jamniczky et al., 2010).

To understand what is lost, it is useful to remind ourselves what explanations of development are intended to accomplish. Developmental biology seeks to explain how organisms develop, while evolutionary biology aims to explain the origin and diversity of life. While much work in developmental biology is focused on fleshing out the roles of specific genes or components of gene-regulatory networks, the ultimate goal of this work is to understand how organisms develop. As postgenomic biology gives rise to an increasingly systems-oriented view of development, the concepts of emergence, context, and physical interactions find a place in developmental explanations. In most of these cases, the term *epigenetics* is not used. However, the concepts presaged by Waddington's epigenetic view of development are finding new currency in a modern guise (Jamniczky et al., 2010).

In evolutionary biology, epigenetic mechanisms may have more direct relevance. In our construction, evolutionary developmental biology deals with the ways in which development impinges on evolutionary explanation (Hendrikse et al., 2007). One of us (Hallgrímsson), along with others (Hendrikse et al., 2007), has argued that identifying the developmental determinants of evolvability is the central question in evolutionary developmental biology. Development is relevant to evolutionary explanation because it structures the phenotypic variation on which natural selection acts. It does so in three fundamental ways: the first is that it structures variation into modules, which exhibit internal covariation and partial independence from the rest of the system; the second is that it suppresses or enhances the expression of genetic variation; and the third is the tendency to express or suppress noise (undesirable variation?) that is not of genetic or environmental origin. These three concepts—modularity and integration, canalization, and developmental stability—do not refer to mechanisms or patterns. Rather, they are dispositional concepts in that they refer to the *tendency* for developmental systems to exhibit covariation or suppress variation (Wagner et al., 1997; Hallgrímsson et al., 2009).

These tendencies to structure or suppress variation arise from developmental mechanisms, many of which are likely epigenetic in origin. Of these concepts, Waddington devoted by far the most thought to *canalization*, which refers to the tendency to suppress variation of either genetic or environmental origin. In Waddington's view, canalization was clearly an emergent property of complex systems, and he thought of its mechanistic basis in terms of attractor states or in terms of gene interactions (Waddington, 1942, 1953, 1957). Canalization is relevant to evolution in three ways. One is that evolved insensitivity to genetic or environmental perturbation is likely a prerequisite to complex organismal life. Complex developmental systems which involve many genes and large genomes have a great potential to vary. Selection tends to produce canalization in such systems because of the high potential cost of this variation, given that most mutations are deleterious if expressed (Wagner et al., 1997). The second is that canalization suppresses the expression of genetic variation, thus hiding it from the view of selection until it is expressed through interaction with a mutation or genetic insult (Waddington, 1957). The third is genetic assimilation. As Waddington showed experimentally, phenotypic states that are initially environmentally induced can become inherited when subjected to stabilizing selection. Waddington's experiments with canalization are well known and spectacular. It is not known, though, how commonly this actually occurs in nature. Calluses on newborn human hands and feet or the ventral callosities of ostriches may well be examples of traits that have evolved via this mechanism (Waddington, 1957).

Integration is relevant to evolutionary explanation in much more direct and obvious ways. In a general sense, integration and modularity are key determinants of evolvability (Hallgrímsson et al., 2009). Integration, or the tendency for covariation, results from selection acting on functionally related traits (Cheverud, 1996). This favors developmental processes that affect the functionally related traits in common, leading to a developmental tendency for covariation or developmental integration. Once established by selection, this tendency to covary allows functionally related traits to coevolve along optimized paths. When variation in the length of the forelimb produces correlated changes in the hindlimb in a cursorial mammal, for example, evolutionary changes in limb length effectively

involve selection on one trait as opposed to two. Selection for divergent functions in related traits produces the opposite outcome. Selection for functional divergence favors developmental factors that act individually on the traits as opposed to those that influence them in common, thus reducing the tendency for the traits to exhibit covariation. This reduction in integration, in turn, facilitates further divergence of the traits by allowing selection to act more independently on the traits. The interplay of integration and modularity in hominoid hands and feet and mammalian fore- and hindlimbs demonstrates well how these phenomena influence evolutionary change (Young and Hallgrímsson, 2005; Rolian et al., 2010; Young et al., 2010b).

In a more specific sense, integration and modularity influence evolutionary explanation because understanding their interplay helps to determine what natural selection acted on to produce the evolutionary change in question. If, as Rolian et al. (2010) argue, hominid thumbs evolved initially toward human proportions as the result of selection acting on the hallux, then selection for hominid bipedalism produced changes in the hand as a secondary consequence that just happened to be preadaptive for manipulative ability and tool use. The assumption that the hands and feet vary independently would require a very different evolutionary explanation.

Not all instances of integration involve an epigenetic mechanism, but many do. The examples involving the vertebrate skull, discussed above, clearly involve epigenetic interactions between the developmental components of the skull. These epigenetic interactions may be important determinants of craniofacial variation patterns that have profound implications for how natural selection acting on aspects of craniofacial function produce changes in craniofacial form (Hallgrímsson et al., 2007, 2009; Lieberman et al., 2008). If the position of the face relative to the braincase is altered by changes in the size of the brain, then one does not need to posit independent evolutionary explanations for evolutionary changes in the cranial base angle.

Similarly, reduction in limb length via overall cartilage growth producing a reduction in chondrocranial growth relative to the rest of the skull will produce a correlated set of cranial shape side effects that do not require additional evolutionary explanation.

In those cases where integration is epigenetically based, epigenetic interactions contribute directly to the evolvability of complex systems. Lieberman makes this argument in this volume (Chapter 16) as well as elsewhere (Lieberman, 2010), mainly in the context of the mammalian skull. In the skull, he argues, epigenetic interactions allow the accommodation of variation in individual components because adjacent or functionally related structures respond epigenetically to that variation in adaptive ways. This situation is very common and is seen in a diversity of phylogenetic and anatomical contexts. The bones of the orbit, for example, are influenced by the growth of the eye (Cepela et al., 1992), so the epigenetic interaction between eye growth and its bony housing helps to ensure the appropriate size relations of these very different organs in most vertebrate species. The same argument can be made about the brain and the skull or about many of the individual bony elements of the skull. In such cases, integration is produced because natural selection has favored the ability to respond appropriately to variation in neighboring or functionally related elements. This greatly enhances the evolvability of such systems by reducing the need to select for functionally concordant variation in multiple structures.

An alternative point of view, based on quantitative genetic arguments, is presented by Hansen in this volume (Chapter 20), who argues that integration may not necessarily be adaptive even if it arises indirectly through natural selection. This argument and its implications for the relationship between integration and evolvability are also discussed in more detail by Zelditch and Swiderski in this volume (Chapter 17).

Understanding the epigenetic basis for development also has a more general implication for developmental and evolutionary explanation. The epigenetic view in which developmental systems are hierarchically structured into processes that exist and act at multiple levels begs the need to choose the relevant explanatory level for different phenomena. As Minelli (2003) and Newman and Muller (Newman and Muller, 2000; Muller and Newman, 2005; Newman et al., 2006) have argued, if physical interactions among cells, due to processes such as differential adhesion, explain the morphogenetic events that underlie early metazoan forms, then it makes little sense to skip that level and reduce explanations to gene function or even gene-regulatory networks. In the case of the role of cell adhesion and early metazoan form, the genes that code for the proteins that constitute the junctional complexes or produce the cell membrane properties that underlie adhesion are certainly interesting. To understand why a group of dividing cells forms a sheet or a hollow sphere, however, it is sufficient to understand how the mechanical properties of cell adhesion determine morphogenesis. Multiple genetic and molecular mechanisms, moreover, might well correspond to the same mechanical properties. The genetic level is thus only one of many levels at which explanations can be posited. It is a privileged level because it is responsible for inheritance. However, it cannot generate complete or even always relevant explanations of developmental or evolutionary developmental phenomena. The epigenetic view of development is thus a healthy counterbalance to the gene-centric view that has become so dominant in modern biology.

Epigenetic explanations of development may also offer a level at which explanations can be generalized in a comparative context in which the genetic or even the developmental–genetic particulars may be different. The use of model organisms in evolutionary developmental biology hinges on identifying levels of relevant generalization (Hallgrímsson and Lieberman, 2008). To use mice, for example, to study the impact of increasing brain size on cranial shape, it is not necessary or probably even possible to recreate the exact genetic and developmental steps involved in evolutionary increases in human brain size. Indeed, changes in brain size via very different developmental mechanisms are still informative here. The relevant level of generalization in this case is not the underlying developmental genetics but, rather, the epigenetic interaction between the growth of the brain and the surrounding skull.

A similar argument about the relevant level of generalization can also be made for the use of model organisms to study the developmental basis for malformation or disease. Ralph Marcucio and colleagues have used a chick model to investigate how signals emanating from the forebrain influence the outgrowth of the midface (Marcucio et al., 2005; Foppiano et al., 2007; Hu and Marcucio, 2009; Young et al., 2010a). This work is relevant to understanding the etiology of craniofacial malformations such as holoprosencephaly and cleft lip and palate. This work has shown how signals emanating from the forebrain regulate mesenchymal proliferation in the midface and set up an epithelial signaling center in the upper lip, the frontonasal ectodermal zone (FEZ), which further regulates the outgrowth of the face. They have shown how perturbing this system can generate a range of phenotypic variation in chick embryos that mimics the human holoprosencephaly spectrum (Young et al.,). Holoprosencephaly does not need to occur in chicks via the same mutations and developmental–genetic mechanisms for this work to be relevant to understanding the malformation in humans. The key here is to consciously choose relevant levels of generalization that are sufficiently general to apply both in humans and in this model. The interaction of many of the molecular players, and the role of sonic hedgehog in particular, likely generalizes. Importantly, so do processes such as the relationship between the growth of the brain and the size of the *Shh* signaling domain that sets these events in motion, the relationship between mesenchymal proliferation in the midface and brain growth that drives the formation of the face, and the size of the field within the midfacial mesenchyme over which the signals emanating from the FEZ and the brain must act. In the sense that Waddington originally intended, these are all phenomena that exist above the level of the gene. In that sense, the use of animal models to understand human malformations often involves choices about the appropriate level of epigenetic generalization.

In all three of these domains, development, evolution, and disease, the epigenetic perspective described in this volume is about the explanatory value of mechanisms acting above the gene level for both evolution and development. It deals with the emergence of phenomena that can be understood only through epigenetic interactions and the context within which genes act and function. As such, it does not deal with a readily definable list of processes but, rather, with the architecture of the systems that translate the genotype–phenotype map. Chromatin modification is certainly an important part of this, but it is only one level at which the genotype–phenotype map is complicated (enriched, enhanced, effected, realized) by epigenetic processes. Whether or not the term *epigenetics* retains its original meaning and the one argued for here or becomes restricted to the narrower set of processes that relate to chromatin modification remains to be seen. Either way, however, the rich systems-level view of development that was implied by Waddington's original use of the term will return to favor as systems biology approaches gain traction and generate mechanistic explanations of phenomena, such as canalization, that exist only above the level of the gene.

REFERENCES

Abouheif, E., and G. A. Wray. 2002. Evolution of the gene network underlying wing polyphenism in ants. *Science* 297(5579):249–52.

Atchley, W. R. 1993. Genetic and developmental aspects of variability in the mammalian mandible. In *The Skull: Development*, ed. J. Hanken and B. K. Hall, 207–47. Chicago: University of Chicago Press.

Atchley, W. R., and B. K. Hall. 1991. A model for development and evolution of complex morphological structures. *Biol Rev* 66:101–57.

Biegert, J. 1963. The evaluation of characteristics of the skull, hands and feet for primate taxonomy. In *Classification and Human Evolution*, ed. S. L. Washburn, 116–45. Chicago: Aldine.

Brock, A., H. Chang, and S. Huang. 2009. Non-genetic heterogeneity—a mutation-independent driving force for the somatic evolution of tumours. *Nat Rev Genet* 10(5):336–42.

Cepela, M. A., W. R. Nunery, and R. T. Martin. 1992. Stimulation of orbital growth by the use of expandable implants in the anophthalmic cat orbit. *Ophthal Plast Reconstr Surg* 8(3):157–67, discussion 168–9.

Cheverud, J. M. 1996. Developmental integration and the evolution of pleiotropy. *Am Zool* 36: 44–50.

Collins, J. P., and J. E. Cheek. 1983. Effect of food and density of development of typical and cannibalistic salamander larvae in *Ambystoma tigrinum nebulosum*. *Am Zool* 23:77–84.

Cubas, P., C. Vincent, and E. Coen. 1999. An epigenetic mutation responsible for natural variation in floral symmetry. *Nature* 401(6749):157–61.

DeBeer, G. 1937. *The Development of the Vertebrate Skull*. London: Oxford University Press.

Downey, C. M., C. R. Horton, B. A. Carlson, T. E. Parsons, D. L. Hatfield, B. Hallgrímsson, and F. R. Jirik. 2009. Osteo-chondroprogenitor-specific deletion of the selenocysteine tRNA gene, *Trsp*, leads to chondronecrosis and abnormal skeletal development: A putative model for Kashin-Beck disease. *PLoS Genet* 5(8):e1000616.

Foppiano, S., D. Hu, and R. S. Marcucio. 2007. Signaling by bone morphogenetic proteins directs formation of an ectodermal signaling center that regulates craniofacial development. *Dev Biol* 312(1):103–14.

Ford-Hutchinson, A. F., Z. Ali, A. Seerattan, D. M. L. Cooper, B. Hallgrímsson, P. T. Salo, and F. R. Jirik. 2005a. Premature degenerative joint disease of the knee in mice lacking phosphoadenosine-phosphosulfate synthetase 2 activity (Papss2): A model of human PAPSS2 deficiency. *J Bone Miner Res* 20(9):S200.

Ford-Hutchinson, A. F., Z. Ali, R. A. Seerattan, D. M. Cooper, B. Hallgrímsson, P. T. Salo, and F. R. Jirik. 2005b. Degenerative knee joint disease in mice lacking 3′-phosphoadenosine 5′-phosphosulfate synthetase 2 (Papss2) activity: A putative model of human PAPSS2 deficiency-associated arthrosis. *Osteoarthritis Cartilage* 13(5): 418–25.

Franz-Odendaal, T. A., B. K. Hall, and P. E. Witten. 2006. Buried alive: How osteoblasts become osteocytes. *Dev Dyn* 235(1):176–90.

Gibson, M. G., A. K. Patel, R. Nagpal, and N. Perrimon. 2006. The emergence of geometric order in proliferating metazoan epithelia. *Nature* 443:1038–41.

Gilbert, J. J. 1966. Rotifer ecology and embryological induction. *Science* 151(715):1234–7.

Gilbert, J. J. 1980. Female polymorphism and sexual reproduction in the rotifer *Asplanchna*. Evolution of their relationship and control by dietary tocopherol. *Am Nat* 116:409–31.

Hall, B. K. 1992. *Evolutionary Developmental Biology*. London: Chapman & Hall.

Hall, B. K. 1999. *Evolutionary Developmental Biology*. Dordrecht: Kluwer.

Hall, B. K. 2004. Evolution as the control of development by ecology. In *Environment, Evolution and Development: Towards a Synthesis*, ed. B. K. Hall, R. Pearson, and G. B. Müller, ix–xxiii. Cambridge, MA: MIT Press.

Hall, B. K. 2005. *Bones and Cartilage: Developmental and Evolutionary Skeletal Biology*. New York: Elsevier Academic Press.

Hall, B. K., and S. W. Herring. 1990. Paralysis and growth of the musculoskeletal system in the embryonic chick. *J Morphol* 206:45–56.

Hall, B. K., and T. Miyake. 1997. How do embryos tell time. In *Evolution Through Heterochrony*, ed. K. J. McNamara, 1–20. Chichester, UK: John Wiley & Sons.

Hallgrímsson, B., J. J. Y. Brown, A. F. Ford-Hutchinson, H. D. Sheets, M. L. Zelditch, and F. R. Jirik. 2006. The brachymorph mouse and the developmental-genetic basis for canalization and morphological integration. *Evol Dev* 8(1):61–73.

Hallgrímsson, B., H. Jamniczky, N. Young, C. Rolian, T. Parsons, J. Boughner, and R. Marcucio. 2009. Deciphering the palimpsest: studying the relationship between morphological integration and phenotypic covariation. *Evol Biol* 36(4): 355–76.

Hallgrímsson, B., and D. E. Lieberman. 2008. Mouse models and the evolutionary developmental biology of the skull. *Integr Comp Biol* 48(3):373–84.

Hallgrímsson, B., D. E. Lieberman, W. Liu, A. F. Ford-Hutchinson, and F. R. Jirik. 2007. Epigenetic interactions and the structure of phenotypic variation in the cranium. *Evol Dev* 9(1): 76–91.

Harris, A. K., D. Stopak, and P. Warner. 1984. Generation of spatially periodic patterns by a mechanical instability: A mechanical alternative to the Turing model. *J Embryol Exp Morphol* 80:1–20.

Harris, A. K., D. Stopak, and P. Wild. 1981. Fibroblast traction as a mechanism for collagen morphogenesis. *Nature* 290(5803):249–51.

Harris, A. K., P. Wild, and D. Stopak. 1980. Silicone rubber substrata: A new wrinkle in the study of cell locomotion. *Science* 208(4440):177–9.

Hendrikse, J. L., T. E. Parsons, B. Hallgrímsson. 2007. Evolvability as the proper focus of evolutionary developmental biology. *Evol Dev* 9(4):393–401.

Henry, J. J., and R. M. Grainger. 1990. Early tissue interactions leading to embryonic lens formation in *Xenopus laevis*. *Dev Biol* 141(1):149–63.

Hu, D., and R. S. Marcucio. 2009. A SHH-responsive signaling center in the forebrain regulates craniofacial morphogenesis via the facial ectoderm. *Development* 136(1):107–16.

Huang, S. 2009a. Non-genetic heterogeneity of cells in development: More than just noise. *Development* 136(23):3853–62.

Huang, S. 2009b. Reprogramming cell fates: Reconciling rarity with robustness. *Bioessays* 31:1–15.

Hufnagel, L., A. A. Teleman, H. Rouault, S. M. Cohen, and B. I. Shraiman. 2007. On the mechanism of wing size determination in fly development. *Proc Natl Acad Sci USA* 104(10):3835–40.

Jaeger, J., D. Irons, and N. Monk. 2008. Regulative feedback in pattern formation, towards a general relativistic theory of positional information. *Development* 135:3175–83.

Jamniczky, H. A., J. C. Boughner, C. Rolian, P. N. Gonzalez, C. D. Powell, E. J. Schmidt, T. E. Parsons, F. L. Bookstein, and B. Hallgrímsson. 2010. Rediscovering Waddington in the post-genomic age. *Bioessays* 32:553–8.

Kaern, M., T. C. Elston, W. J. Blake, and J. J. Collins. 2005. Stochasticity in gene expression: From theories to phenotypes. *Nat Rev Genet* 6:451–64.

Kristensen, E., T. E. Parsons, B. Hallgrímsson, and S. K. Boyd. 2008. A novel 3-D image–based morphological method for phenotypic analysis. *IEEE Trans Biomed Eng* 55(12):2826–31.

Larsen, E., T. Lee, and N. Glickman. 1996. Antenna to leg transformation, dynamics of developmental competence. *Dev Genet* 19:333–9.

Lieberman, D. E. 2010. *The Evolution of the Human Head*. Cambridge, MA: Harvard University Press.

Lieberman, D. E., B. Hallgrímsson, W. Liu, T. E. Parsons, and H. A. Jamniczky. 2008. Spatial packing, cranial base angulation, and craniofacial shape variation in the mammalian skull: Testing a new model using mice. *J Anat* 212(6):720–35.

Marcucio, R. S., D. R. Cordero, D. Hu, and J. A. Helms. 2005. Molecular interactions coordinating the development of the forebrain and face. *Dev Biol* 284(1):48–61.

Minelli, A. 2003. *The Development of Animal Form: Ontogeny, Morphology, and Evolution*. Cambridge: Cambridge University Press.

Muller, G. B., and S. A. Newman. 2005. The innovation triad: An evodevo agenda. *J Exp Zoolog B Mol Dev Evol* 304(6):487–503.

Newman, S. A., G. Forgacs, and G. B. Muller. 2006. Before programs: The physical origination of multicellular forms. *Int J Dev Biol* 50(2–3):289–99.

Newman, S. A., and G. B. Muller. 2000. Epigenetic mechanisms of character origination. *J Exp Zool* 288(4):304–17.

Nijhout, H. F. 1999. Hormonal control in larval development and evolution—insects. In *The Origin and Evolution of Larval Form*, ed. B. K. Hall and D. B. Wake, 217–54. San Diego: Academic Press.

Nijhout, H. F., and D. E. Wheeler. 1996. Growth models of complex allometries and evolution of form: An algorithmic approach. *Syst Zool* 35:445–57.

Ohlsson, R. 1999. *Genomic Imprinting: An Interdisciplinary Approach*. New York: Springer.

Ohlsson, R., K. Hall, and M. Ritzén. 1995. *Genomic Imprinting: Causes and Consequences*. Cambridge: Cambridge University Press.

Rolian, C., D. E. Lieberman, and B. Hallgrímsson. 2010. The co-evolution of human hands and feet. *Evolution* 64:1558–68.

Ross, C., and M. Henneberg. 1995. Basicranial flexion, relative brain size, and facial kyphosis in *Homo sapiens* and some fossil hominids. *Am J Phys Anthropol* 98(4):575–93.

Ross, C. F., and M. J. Ravosa. 1993. Basicranial flexion, relative brain size, and facial kyphosis in nonhuman primates. *Am J Phys Anthropol* 91(3):305–24.

Stemberger, R. S., and J. J. Gilbert. 1987. Multiple-species induction of morphological defences in the rotifer *Keratella testudo*. *Ecology* 68:370–8.

Stopak, D., and A. K. Harris. 1982. Connective tissue morphogenesis by fibroblast traction. I. Tissue culture observations. *Dev Biol* 90(2):383–98.

Waddington, C. H. 1942. The canalisation of development and the inheritance of acquired characters. *Nature* 150:563.

Waddington, C. H. 1953. The genetic assimilation of an acquired character. *Evolution* 7:118–26.

Waddington, C. H. 1957. *The Strategy of the Genes*. New York: MacMillan.

Wagner, G. P., G. Booth, and H. Bagheri-Chaichian. 1997. A population genetic theory of canalization. *Evolution* 51(2):329–47.

Young, N. M., H. J. Chong, H. Hu, B. Hallgrímsson, and R. S. Marcucio. Activation of Shh-signaling in the brain predicts variation in the shape of the vertebrate midface.

Young, N.M., and B. Hallgrímsson. 2005. Serial homology and the evolution of mammalian limb covariation structure. *Evolution* 59(12): 2691–704.

Young, N.M., H.J. Chong, B. Hallgrímsson, and R.S. Marcucio. 2010a. *Development* 137(20): 3405–09.

Young, N.M., G.P. Wagner, and B. Hallgrímsson. 2010b. Development and the evolvability of human limbs. *Proc Natl Acad Sci USA* 107: 3400–5.

Zhang, X., J. Yazaki, A. Sundaresan, S. Cokus, S.W. Chan, H. Chen, I.R. Henderson, et al. 2006. Genome-wide high-resolution mapping and functional analysis of DNA methylation in *Arabidopsis*. *Cell* 126(6):1189–1201.

INDEX

accommodation of variation, 434
acetylation, 44, 50, 53–54, 140, 145, 147, 150, 153, 402
achaete-scute cluster (AS-C), 145
AciI, 73
Acoela, 124
activator-inhibitor process/model, 166–68
activators in activator-inhibitor models, 166–68
adaptation, 209, 321, 358
  definition of, 359
  direction of relative to the origin of variation, 17
  and epigenetic inheritance, 34
  through epigenetic interaction, 306
  epigenetic phenomena as, 15
  evolutionary causes of, 26
  of gene effects, 362
  and integration, 291
  and morphological integration, 291
  to novel predators, 343
  and phenotypic plasticity, 321, 328, 351
  relation to selection, 358
  and selection, 26
adaptive decanalization, 368
adaptive landscape, 318, 321
adaptive phenotypic plasticity, 368
adaptive plasticity, 306, 310, 318, 331, 332
adaptive radiation, 318, 322, 330
  of threespine stickleback, 322–23, 326–28
  post-glacial radiation, 322
adaptive robustness, 371
adaptive topography, 295–97
adenomatous polyposis coli (APC), 414
adipocytes, 257
adrenal angiotensin receptor, 406
age and bone remodeling, 224
aging, 224, 405, 410

agouti mutation, 90, 92, 98, 404
airborne hydrocarbons, 409
algae, 126
Alk5, 187
alkaline phosphatase, 202
allergy, 408
Allium tuberosum, 97
allometry, 280, 321, 329
alpine newts, 167
alternation of haploid and diploid generations, 94
alternation of ploidy, 94
Alx 4, 211
Ambystoma mexicanum, 167–72
Ambystoma tigrinum, 171, 341
aminopterin, epigenetic effects of maternal
  exposure, 391
amniotes, apical ectodermal ridge, 250
amphibians, 107, 202, 239
  limb development in, 249–50
  pigment patterns, 167
Amphimixis, 89, 90
amphipod crustaceans, 120
amyotrophic lateral sclerosis (ALS), 257, 264–65
amystomatids, pigment patterns in, 167–75
androgens, 223
anencephaly, 208
aneuploidy, 98
angelfish, 168
Angelman Syndrome, 412
angiosperms, 57–58, 94
anisogamy, 94
Annelida, 124
annelids, 119
antennae, 129
antennipedia, 106

439

cell–cell interactions, 305
   in cardiac development, 181–91
   as epigenetics, 181, 216, 276
   in development of pigment patterns, 174–75, 177
cell density, 206
cell differentiation, 138, 150, 196, 199, 203, 204, 208, 258
cell division (proliferation), 105
cell–extracellular matrix interactions, in development of pigment patterns, 174–75
cell migration, 107, 206, 276
   and cardiac development, 182–91
   and cranial sutures, 388
   and epigenetics of pigmentation patterns, 165–71, 175
   of neural crest, 170
   of osteogenic cells in craniosynostosis, 386
   and skeletogenic condensations, 196, 200
cell motility, 413
   and twist and snail, 128
cell movement, 209
cell proliferation, 108, 120, 183, 199, 232, 276, 412–14, 428
   and cancer, 413
   and epigenetics of lung development, 257
   and FGF function, 241–42, 248, 386
   as inertial condition, 119
   and mechanical loading of bone, 221, 230, 389
   multiple focal points of, 122
   of osteocytes, 386
   and self-organization, 428
   and skeletogenic condensations, 202–3
   of stem cells, 143, 154
   and suture closure, 393
   and Tgf-β function and chondrocyte proliferation, 302
   and twist and snail, 128
cell signaling, 209
cell survival, and FGF8 240
cellular automata, 166, 174–75, 177
cellular competition, 127
cellular dynamics, within condensations, 206
cellular parasitism, 127
cellular responses to mechanical forces, in cranial sutures, 389
cellular transition, 138
centipedes, 128
central nervous system, 223
centrally conserved domain (A6-A4), 49
centromeres, 82
centromeric instability, 415
Ceratodontiformes, limb development in, 244
cerebral palsy, 297
cerebrospinal fluid volume, as epigenetic factor, 382
Cetartiodactyla, limb development in, 245

Chelonia, limb development in, 244
Cheverud, Jim, 294, 366
chewing performance, 300
chick, 182, 186, 188, 201, 207, 211, 212, 215, 225, 239, 242, 250, 260, 435
childhood abuse, 407
chimeric mouse, 143
chimpanzee, 272, 280, 283, 349
Chiroptera, limb development in, 245, 247–48
choanocytes, 130
choanoflagellates, 124
cholesterol level, 405
chondrocranium, 429, 433, 431
chondrocyte differentiation, 205
chondrocytes, 203, 213, 302
   hypertrophic, 199
chondrogenesis, 231
chondrogenic condensations, 204
chondroitin sulfate proteoglycans, 204
chromatin
   architecture, 155
   marks, 87, 93
   modification, 95, 139, 425, 432
   remodeling, 142, 148, 152, 402
   remodeling factors, 153
   tertiary structure of, 142
chromosome looping, 49
chronic obstructive pulmonary disease, 409
cichlid fish, and asymmetry, 349
cichlids, 331
cilia, 119
ciliary neurotrophic factor (CNTF), 149, 150
circulatory system, 128
circumpharyngeal region, 188
cis-regulatory modularity, 366
citodieresis, 127
cladistics, 118, 180
clavicle dependence on mechanical environment, 225, 260
cleavage, 126
cleft lip, 213, 274, 279
cleft palate, role of mechanical factors, 262
clitellates, 123
clonal lineage, 88
Cnidaria, 124, 126
coadaptation, 411
coat color, 404
Coccoidea, 54–56
cochlea, 260–61
coevolution, 222
cognitive impairment, 416
collagen, 202, 205, 426
combined bisulfite restriction assay (COBRA), 74
comparative approach, 180
compartments, 128

competence of cells, 427
complex adaptations, 292, 358
complex phenotype, 358
complex systems, 432
complexity
of cranial morphology, 272
in development, 298
computer simulation of development for reaction-
diffusion models, 166
conceptual mismatch, 16
conceptus, 96
condensations, 196, 199, 210, 216, 309
boundaries, 205–6
digital condensations, 248
interaction with blood vessels, 208
partitioning, 207
size, 205
in skeletal development, 202, 203, 275
condylar elongation, 228–32
condylar process, 302
connective tissue, 183
connective tissue growth factor (*Ctgf*), 258
conservative research, 29
consilience, 25
convergence in evolution, 250
coordinated variability
as adaptation, 363
due to indirect selection, 371
as a group adaptation, 369–70
and robustness, 367
Copidosoma, 122–23
coral, 94
coregulation, 222
coronoid process (of mandible), 228
correlations (and integration), 293
cortical cytoplasm, 11
cortical development, 153
cortical inheritance, 11
cortical (neural) stem cells, 145
corticotropin-releasing factor (CRF), 407
courtship behavior in threespine stickleback, 323–28
covariation, 432
in the skull, 276, 279, 285
covariation and integration, 293
CpG dinucleotides, 44
CPG island clones, 73
CpG islands (CGIs), 71, 142, 402, 407, 414
crabs 341, 343
cranial base, 196, 274, 282, 283, 429
angle, 282, 428–29, 433
flexion, 282
cranial fossae, 278–79
cranial growth and craniosynostosis, 385
cranial morphology (of vertebrates), 222, 226, 271–85,
296, 331, 428, 433

cranial vault, 206, 208
growth, 388
craniofacial development, review of, 380–82
craniofacial dysmorphogenesis, 377
craniofacial growth, 226, 276–77, 280
craniofacial morphogenesis, 274
craniofacial tissues, 301
craniosynostosis, 278, 381, 382–85
genetic basis, 386
cribriform plate, 278
Crick, Francis, 26
Crick's central dogma, 27
crip2, 184
crista ampullaris, 260–61
Crocodilia limb, development in, 245
crossbill finches, and handed behavior, 348
crusher claws
and handed behavior, 348, 350
in lobsters, 348
crustaceans, 119, 123
cryptic variation, 318, 332, 338, 339, 342–44, 349
and learning, 351–52
CTCF protein, 47–50
Ctenophora, 124
cubozoans, 126
Cunina, 122
Cunochtantha, 122
CXXC domain, 73
cyclin-dependent kinase inhibitors (CDKIs), 154
cyclin-dependent kinase inhibitor 1C (CDKN1C), 412
Cycliophora, 124
Cynolebias, 122
Cypriniformes, limb development in, 244
Cyprinodontiformes, limb development in, 244
cytokine responses, 409
cytokine signaling, 149, 150
cytosene methylation, 87, 88, 93, 95, 97, 98
cytosines, 44, 71, 91
unmethylated, 71

Danio albolineatus, 176
Danio nigrofasciatus, 176
Danio rerio, 175–76, 246
Daphnia, 318, 352, 431
Darwin, Charles, 24, 26–27, 106, 129
Darwinian evolution, 26
Darwinian theory, 25, 35
Dasypus, 122
De Beer, Gavin, 428
deacetylation, 142
deaminated 5-methylcytosine, 90
deamination rate, 78
decapod crustaceans, 119
dedifferentiation, 152

epithelial–mesenchymal interaction, 201, 213, 290, 292, 298, 301
epithelium, 239, 292
Escherichia coli, 74
estrogen, 222–23
  receptor, 408
Euprymna scolopes, 111
European Bioinformatics Institute, 73
European Conditional Mouse Mutagenesis (EUCOMM) Consortium, 258–61
evolution
  role of phenotypic plasticity, 317–18
  of sex, 90
evolutionary developmental biology, 359
  central question of, 432
  of the vertebrate limb, 238–52
evolutionary diversification, 318
evolutionary explanation, 433
evolutionary innovation, 167
evolutionary modules, 300
evolutionary reversals in pigment patterns, 173
evolutionary theory, 90
evolvability, 4, 117, 272, 273, 276, 279, 280, 285, 360, 363, 366, 432, 433, 431
  conditional evolvability, 364
  and developmental pathway evolution, 372
  as a group adaptation, 369
  and modularity, 364
  and pleiotropy, 364
expanded evolutionary synthesis, 18
explanatory strategies, 16
external environmental interactions as epigenetic factors, 391
extinction and phenotypic plasticity, 351
extracellular matrix, 200
  epigenetic role, 203, 216, 426
extraocular muscles, 258
eye, 213, 277, 278, 431
  development, 426
  growth as epigenetic factor, 278

F statistic (Wright), 93
face, 274, 429, 433
facial anomalies (ICF) syndromes, 415
facial growth, 276
facial musculature, 260
facial prognathicism, 280
facial size, 283
famine, 405, 409
feeding, 292
feet, 433
felids, 165, 180
femoral anteversion angle, 297
femoral condyles, 297
Fertilization Independent Seed2 (FIS2), 58

Fertilization-Independent Endosperm (FIE), 58
fetal akinesia, 258
fetal growth, 411
fetal growth retardation, 400
fetal lipid metabolism, 406
fetal ocular movements, as epigenetic factor, 258
fetal sex determination, 413
Fgf/Gremlin1 inhibitory feedback loop, 240
Fibrillin, 384
fibroblast growth factors (FGFs), 201, 206, 209, 239, 241–42, 247–48, 251, 292
  FGF 17 (*Fgf17*), 240
  FGF 2 (*Fgf2*), 150, 208, 276, 280
  FGF 4 (*Fgf4*), 241
  FGF 8 (*Fgf8*), 186, 187, 189, 241–42, 243, 309
fibroblast growth factor receptors
  FGFr1, 384, 386, 393
  FGFr2, 384, 386, 393
  FGFr3, 384, 386, 393
fibroblasts , 402, 426
fibronectin, 206
fibula, 207
fiddler crabs, and asymmetry, 348–49
FIE1 histone modifications, 61
fins, 129
  development, 246–47, 250
  fin-to-limb transition, 250
  mutation in zebrafish, 176
first branchial arch, 212
FIS complex, 58–60
fish, 107, 200
  cardiac anatomy, 184
  pigment patterns, 165
fitness, 128, 292, 294–95, 298, 299, 305, 317, 322, 361, 369, 400
  maternal versus offspring, 411
  optimum, 321, 361
  of variational properties, 370
flatworm, 119, 122, 124, 126
  rhabditophoran, 122
floral symmetry, 92
flower traits, 360
fluctuating asymmetry, and evolution of asymmetry, 344
fluctuating selection, 371
flukes, 122
flying vertebrates, loading of bone, 223
focal cell-labeling studies, 182
folate, 408, 409, 413
food consistency, 310
foramen cecum, 282
foramina (in bone), 306
force reception, in the skeleton, 223
forebrain, 154
forelimb, 246
fossil record, 233
FoxN1, 168

molecular responses to biomechanical stimulus, 389

mollusks, 119, 126

monozygotic twins, 91, 97, 405, 412

morphogen gradients, 428

morphogenesis of the head, 274, 275–76

morphogenesis of pigmentation, 164–81

morphogenetic consequence, effects of function on, 297

morphogenetic fields, 110–12, 428
  in cardiac development, 190

morphogenetic gradients, 111

morphogenetic mutations, 361

morphogens, origin of concept, 166

morphological adaptation and learning, 352

morphological complexity, and epigenetic interactions, 272

morphological evolution, 105

morphological integration, 1, 4, 273, 276, 278, 290–310, 432, 433, 431

morphological plasticity, 338. *See also* phenotypic plasticity

morphology and function, 299

morphology-biased learning, 338

morphostasis, 130

mosaicism of epimutations, 412

Moss, Melvin, 273, 277, 390

motor neuron, 263

motor neuron diseases (MNDs), 263–65

mouse
  striped coat patterns, 168
  models for craniosynostosis, 382
  See also mice

mouthparts, 129

mRNA degredation, 403

Msx, 206

Msx1, 187, 213

Msx2, 208, 213

MSX2, 386

Muller, Gerd, 431

Multicopy Suppressor of Ira 1 (MSI1), 58

multipotency, 152

muscle
  attachments, 226
  contraction, 228
  contraction patterns, 222
  development, influence on bone, 307
  effect of complete absence of, 261–63
  and evolutionary inference, 232–33
  growth, 223
  mass, 296, 307
  muscle action as epigenetic factor, 256–65
  muscle–bone interactions, 208, 221–33, 261–65, 279, 290–310, 428
  muscle–bone interface, 226

muscle-derived loading, 230
  role on shaping skeletal morphoplogy, 261–63
  tissue, 125, 221

muscular dystrophy, 230

mutation(s), 90
  and genome architecture, 358
  canalizing, 367
  novel mutations, 379

mutation rate, 90

mutational correlations, 365

mutational effects, 295

mutational spectrum, 363

mutational target, 363

mutational variance, 362

Mutela bourguignati, 126

Mutelidae, 126

myelin, 258

myeloblastosis oncogene (*Myb*), 258

myelodysplastic syndromes, 414

Myf5, 256

Myf5:MyoD compound null mice, 257, 261–65

myoblasts, 149, 223

myocardial cells, 181, 186, 189

myocardium, 184, 186, 188

MyoD, 121, 256, 263–65

myogenesis, 149, 261

myogenic differentiation, 222

myogenic regulatory factors, 256

myostatin, 223, 227. *See also* GDF-8

myostatin deficient mice, 229

myotome, 211

myotubes, 223

nasal capsule, 278

nasal cavity, 277

nasopharynx, 277, 275

natural selection, 305, 306, 240–41, 432, 433

NcoR, 146

ncRNA, 411, 412, 414

nematodes, 121

Nemertea, 124

neoblasts, 125

Neoceratodus, 247

neocortex, 152

neo-Darwinian, 117

nerve cord, 125

nerve–bone interactions, 307

network of epigenetic interaction, 272–73

neural cell adhesion molecule (N-CAM), 203, 204, 210

neural crest, 167, 176, 181–91, 196, 200, 206, 212, 221, 227, 273
  cranial streams, 182
  migration, 181, 182, 187, 188, 189, 190, 200, 201

sialoprotein, 205
signal-receptor interactions, 292
signaling pathways, 186, 190, 191
Silver-Russell Syndrome, 412
Simon, Herbert, 111
simplification, evolutionary, 129
Sipuncula, 124
skeletal calcium channels, 223
skeletal condensations, 202, 209, 275
skeletal defects, 292
skeletal loading effects on bone mass, 223
skeletal morphogenesis, 225
skeletal morphology, 222
    influence of muscle loading, 228, 232
    phenotypic variation in, 209
skeletal mutants, 210–11
skeletogenesis, 195–216
skeleton, mammalian, 195–216
skull, 196, 202, 222, 226, 271–86
    embryonic development, 385
small interfering RNA, 96
smoking, 92
smooth muscle, 183–84, 188
SMRT, 146
snail chirality, 120
snakes, 209, 242
soapberry bugs, 318
Sober, Elliot, 359, 369
solider larvae, 123
somatomedin, 223, 401
somatotropin, 223
somites, 182, 211
sonic hedgehog (*Shh*), 148, 188, 209, 215, 242, 435
Southern blot analysis, 81
Sox transcription factors, 152
    Sox9, 200, 203–5, 210, 213
spatial packing hypothesis, 282
special AT-rich sequence binding protein 1 (*Satb1*), 258
species-specific phenotypes, 107
specific hypermethylation, 414
sperm, 76, 94, 107
spermatocytes, 76
spermatogenesis, 77, 79, 404
spermatogonia, 76
spiny legs, 121
splanchnic mesoderm, 184, 188
Squamata, limb development in, 245, 247
squamosal, 211
squid, 111
squirrels, 209
stabilizing selection, 307, 361, 362, 371
    developmental determinants of, 392–93
    for pigment patterns, 173, 295
    of variational properties, 370
starvation, 406

stem cells, 127, 138
    embryonic, 143
    proliferation, 143
Stenella attenuate, 239
sternum, 225, 262
steroid receptors, 408
stochastic processes, 112, 426
stomatopods, 123
strain and stress as epigenetic factor, 200
stress, 407–8
    and cryptic variation, 368
striped patterns
    emerging from self-organization, 165
    in fish, 168
stroke, 407
subtelomeric region, 82
suicide, 408
sulfation, 426
sumoylation, 140
Sus scrofa, 226, 230–32, 239
suture
    closure of, 388
    cranial, 276, 281, 381, 382–85
swarm behavior, 165
symmetry, 120, 121
    breaking of, 127
    of body plans, 363
synapsid's skull, 393
syncitia, 119–20, 127
syndecan, 203, 204, 206
syngamy, 94, 96, 97
synovial joints, 262
syntaxin binding protein 1 (*Stixbp1*), 265
systems biology, 426

T-cell receptor β, variable 13 (*Tcrb-V13*), 258
tadpole, 431
Tail, 73
Talpid, 211
Tandem Repeat Finder, 81
tapeworms, 122
teeth, 428
telencephalon, 149, 150
teleosts
    aortic arches in, 185
    fin development in, 246–47
    mechanical adaptation, 224
telomere, 88
temperature-dependent sex determination, 107
temporalis muscle, 300
temporomandibular joint, 263, 277, 285
tenascin C, 203, 204, 206, 210
teratogens, epigenetic effects of maternal exposure, 391
teratology, 399–400

COMPOSITION: Bytheway Publishing Services
TEXT: 9.5/13 Scala
DISPLAY: Scala Sans
PRINTER AND BINDER: Sheridan Books